Essentials of
WORLD REGIONAL
GEOGRAPHY

Michael Bradshaw
College of St. Mark and St. John, Plymouth, U.K.

George W. White
Frostburg State University

Joseph P. Dymond
The George Washington University

Elizabeth Chacko
The George Washington University

 **McGraw-Hill**
Higher Education

Boston   Burr Ridge, IL   Dubuque, IA   New York   San Francisco   St. Louis
Bangkok   Bogotá   Caracas   Kuala Lumpur   Lisbon   London   Madrid   Mexico City
Milan   Montreal   New Delhi   Santiago   Seoul   Singapore   Sydney   Taipei   Toronto

The McGraw·Hill Companies

McGraw-Hill
Higher Education

ESSENTIALS OF WORLD REGIONAL GEOGRAPHY

2 3 4 5 6 7 8 9 0 DOW/DOW 0 9 8

ISBN 978–0–07–352281–4
MHID 0–07–352281–3

Publisher: *Thomas D. Timp*
Executive Editor: *Margaret J. Kemp*
Senior Developmental Editor: *Joan M. Weber*
Senior Marketing Manager: *Lisa Nicks*
Project Manager: *April R. Southwood*
Lead Production Supervisor: *Sandy Ludovissy*
Lead Media Project Manager: *Tammy Juran*
Media Producer: *Daniel M. Wallace*
Designer: *John Joran*
Cover/Interior Designer: *Rokusek Design, Inc.*
Cover image: © *Gavin Hellier/The Image Bank, Getty Images*
Back cover image: © *Larry Brownstein/Getty Images*
Lead Photo Research Coordinator: *Carrie K. Burger*
Compositor: *Electronic Publishing Services Inc., NY*
Typeface: *10/12 Times Roman*
Printer: *R. R. Donnelley Willard, OH*

The credits section for this book begins on page 347 and is considered an extension of the copyright page.

Library of Congress Cataloging-in-Publication Data

Essentials of world regional geography / Michael Bradshaw ... [et al.]. -- 1st ed.
    p. cm.
Includes index.
  ISBN 978-0-07-352281-4 --- ISBN 0-07-352281-3 (hard copy : alk. paper) 1. Geography--Textbooks. I. Bradshaw, Michael J. (Michael John), 1935-
G116.E87 2008
910--dc22
                        2007024213
www.mhhe.com

To Valerie, Paul, and John,
Emily and Geo,
Maureen and Madison,
Thomas, Rebecca, and Abraham

BRIEF CONTENTS

TABLE OF CONTENTS

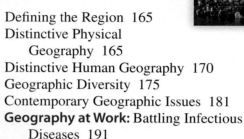
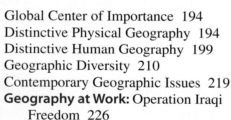
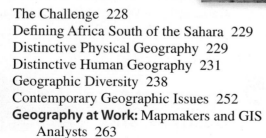

Michael Bradshaw

Michael Bradshaw and his wife live in Canterbury, England, and have two sons and three grand-children. Michael taught for 25 years at the College of St. Mark and St. John, Plymouth, as Geography Department chair and dean of the humanities course. He has written texts for British high schools and colleges since the 1960s. In 1985, he was awarded a Ph.D. from Leicester University for his study on the impacts of federal grant-aid in Appalachia. His book, *The Appalachian Regional Commission: Twenty-Five Years of Government Policy*, was published in 1992. Since 1991, he has written for U.S. students and has been responsible for two physical geography texts and the successful world regional geography text, *The New Global Order*. Michael believes that we should all be better equipped to live in the modern, increasingly global world. Understandings of geographic differences should make us more able to assess crucial issues and value other people who bring varied resources and who face pressures that we find difficult to imagine.

Contemporary World Regional Geography: Global Connections, Local Voices was awarded McGraw-Hill's First Edition of the Year in 2004 and is now in its second edition. Michael collaborated with the three coauthors who now join him in presenting *Essentials of World Regional Geography*. As well as retaining responsibility for overall coordination of the project, he is the lead author for the first chapter and those on Southeast Asia, Australia, and Oceania, and Africa South of the Sahara. *Essentials of World Regional Geography* draws on previous work, but is a totally new approach to the subject.

George W. White

George W. White grew up in Oakland, California. He pursued graduate work in Eugene, Oregon, completing a Ph.D. at the University of Oregon. He then moved to Frostburg, Maryland, where he met his wife. George is currently an associate professor and chair of the Department of Geography at Frostburg State University. Political geography and Europe are two of his primary interests. He authored a book entitled *Nationalism and Territory: Constructing Group Identity in Southeastern Europe* (2000) and another one entitled *Nation, State, and Territory. Vol. 1. Origins, Evolutions, and Developments* (2004).

After meeting Michael Bradshaw, George was impressed by Michael's long and distinguished career of teaching, research, and publication. He accepted the opportunity to join Michael in his plans to write a new world regional geography text, taking lead authorship for the chapters on Europe, Russia and Neighboring Countries, and Northern Africa and Southwestern Asia, as well as contributing to other areas of the text.

George became a geographer because he believes that the field of geography is alive and dynamic, attuned to our ever-changing world and its great diversity. The world regional approach represents the breadth of the field of geography, and world regional geography texts are the epitome of the geographer's art. George White chose to collaborate with Michael Bradshaw on this project because the text combines local practices with global processes, and explains interactions between the two as they shape each other.

Joe Dymond

Joe Dymond earned a master of science degree from Pennsylvania State University in 1994, and a master of natural sciences degree from Louisiana State University in 1999. He taught world regional geography courses for the Louisiana State University Department of Geography and Anthropology from 1995 through 2000. During Joe's six years at LSU, he instructed thousands of students and was recognized in the spring of 1997, fall of 1999, and fall of 2000 for superior instruction to freshman students by the Louisiana State University Freshman Honor Society, Alpha Lamba Delta. Joe currently lives in suburban Washington, D.C., with his wife and daughter, and is an assistant adjunct professor in the Department of Geography at The George Washington University (GWU). In the fall of 2006, Joe was honored by GWU as a recipient of a 2006 Morton A. Bender Teaching Award.

Joe is the lead author for the final two chapters on Latin America and North America. Joe is interested in providing students with the geographic tools that will help them to better understand the human and environmental patterns and relationships present in their world. His greatest concern for geography students is that they obtain a comprehensive and fair perspective when learning about the people and places comprising the regions of the world.

Elizabeth Chacko

Elizabeth Chacko was born and raised in Calcutta (Kolkata), India. She received her undergraduate degree in geography (with Honors) from the University of Calcutta. Moving to the United States for further study, she got a master's degree in geography from Miami University, Ohio. She also obtained a graduate degree in public health and a Ph.D. in geography from the University of California, Los Angeles (UCLA). Elizabeth taught geography at the college level at various institutions including Loreto College, Kolkata; UCLA; and The George Washington University, where she is associate professor of geography and international affairs.

Elizabeth was selected as Professor of the Year from the District of Columbia in 2006 by the Carnegie Foundation for the Advancement of Teaching and the Council for the Advancement and Support of Education (CASE). She teaches courses on South Asia, globalization, medical and population geography, and development. Elizabeth's research interests include women's health and the role of culture in health and health care. She is currently engaged in research on transnationalism, the African immigrant community in the United States, and the return migration of Asian Indian professionals to India. Elizabeth is on the editorial board of the *Journal of Cultural Geography*.

In this edition, Elizabeth is the lead author for the chapters on South Asia and East Asia and also contributed to other chapters. She is delighted to be part of an author team of committed geographers. She enjoys helping students understand the dynamic interactions between humans and the earth's surface and comprehend the interplay of economic, sociocultural, and political forces that impact globalization and the spatial variations that result at local, regional, and global scales. She hopes that this book will raise students' appreciation of the relevance and significance of geography in their lives.

As the authors of *Contemporary World Regional Geography,* we are pleased to bring you *Essentials of World Regional Geography.* We created this new, rewritten text because many instructors teach under circumstances that make a shorter, streamlined text more desirable than the standard world regional text. It has its own unique features. In preparing this text, we adopted a fresh approach that combines fundamental geographical elements, internal regional diversity, and contemporary issues. These allow serious discussion of cultural and environmental issues, together with those of the political and economic changes, for example, in Russia and China. The shorter length and distinctive approach were received enthusiastically by nearly all our reviewers.

The main innovations are in the ordering of the text. Each of the nine regional chapters opens with a page-size (or larger) map of the region, short accounts of people or events to provide a personal flavor of the region, an outline of the chapter contents, and a short section placing the region in its wider global context. Each chapter has three further sections. The first summarizes the distinctive physical and human geographies of the region; the second explores the internal diversity of the region at subregional, selected country, and local scales. These take forward our commitment to the comparative nature of world regional geography. The third section focuses on a selection of contemporary issues that are important to the people of each region and frequently have implications for the rest of the world. Many of these issues are highly contested with opposing factions having dramatically differing viewpoints. We have outlined these views in debate formats to help students understand them. Reviewing instructors were enthusiastic about the teaching value of this overall approach.

The opening chapter contains a discussion of the basis and value of world regional geography and overviews of the main relevant aspects of physical and human geography. Students are introduced to the maps and diagrams that are features of each chapter to encourage the comparative studies and familiarization with the set of illustrations chosen. Maps take a prominent role, and photos throughout the text provide windows on the elements of the regional geographies. We know that students are interested in how geography can be used in the workplace, and so each chapter ends with a personal example of work in progress.

We have retained the normal dimensions of a textbook page size, although the length is 384 pages rather than 650, and the price is considerably less! We have adopted a totally new design for the book. All these features make *Essentials of World Regional Geography* particularly suitable for semester-length classes.

Chapter Highlights

Chapter 1 defines geography and regional geography and introduces the concept of globalization-localization tensions. This is followed by an overview of physical geography with a focus on earth's interior forces, climate, ecosystems, and human impact on the physical environment. The human geography overview is led by a consideration of how culture influences regional character, followed by major aspects of population geography, political geography, and economic geography. These aspects are connected in a summary of approaches to human development and human rights. At the end of the chapter, the world regions are defined with a summary of their main distinctive characteristics.

Chapter 2, Europe, begins a world region tour where many modern global processes and innovations began, often building upon previous African, Asian, and Arab achievements. The contemporary issues include the development and future of the European Union, and an exploration of the growing multicultural nature of European society.

We next move eastward in **Chapter 3, Russia and Neighboring Countries**. This is a study of the geographical impacts of European-origin communist principles adopted by governments for most of the 1900s, followed by massive changes as the Soviet Union broke up in the early 1990s. The Russian "Empire" remains a political and economic reality in the region. Contemporary issues include questions of human rights and environmental problems, together with the region's wealth derived from oil and natural gas.

In **Chapter 4, East Asia**, we enter a region of cultural contrast to Europe, but one that contains the world's most significant emerging countries: Japan, China, South Korea, and Taiwan. These contrast with North Korea and Mongolia. The contemporary issues include the emergence of China as a world power and its distinctive population policy; human rights; local multinational corporations; and the globally connected cities.

Chapter 5, Southeast Asia, Australia, and Oceania, is a diverse region, in which elements of European and Asian human geography are increasingly brought together. The contemporary issues reflect this process by focusing on regional cooperation through ASEAN (Association of Southeast Asian Nations) and APEC (Asia-Pacific Economic Cooperation forum) and conflicts over ocean space and piracy. Australian identity is considered as the country changes its outlook on its European heritage and Asian neighbors. Singapore is examined as a small country with a key role in global exchanges.

Moving westward, we reach **Chapter 6, South Asia** with its distinctive cultural background, including the origins of Hinduism and Buddhism, and colonial experiences. On gaining independence from the British Raj in 1947, the new countries attempted self-sufficiency, avoiding close relations with other regions and leading a group of nonaligned countries. This policy was partly successful, but switched in the 1990s to a more global outlook. The contemporary issues include ethnic conflicts and environmental problems, alongside considerations of population and urban growth.

Further westward we enter **Chapter 7, Northern Africa and Southwestern Asia** at the junction of Asia, Europe, and Africa. Although a mainly arid region, its people initiated, influenced, and passed on many cultural and technical innovations to the surrounding regions. Today it is the world's center of the Islamic religion and has the world's largest oil resources. However, its fragmented and conflicting peoples tend toward political instability. The contemporary issues include the Israeli-Palestinian conflict, the Iraq situation, and aspects of human rights.

Southward is the subject of **Chapter 8, Africa South of the Sahara**, the world's poorest region despite its leading role at the outset of human history. After experiencing major migrations of African peoples, Muslim and European influences took control of much of the region. Most countries gained independence mainly in the 1960s, but struggled through internal political conflict and poverty. The contemporary issues include the role of HIV/AIDS, the culture shocks of global elements, exploding city populations, and the question of this century's challenge to Africans.

Crossing the South Atlantic Ocean to **Chapter 9, Latin America**, we find a world region where many indigenous peoples remain, but enjoy little political or economic power in contrast to the descendants of European colonists and those of mixed ethnic groups. Contemporary issues include the deforestation of Amazon rain forest, the international drug trade based on the northern Andes, and the growth of huge cities in Mexico and Brazil.

Finally, we reach **Chapter 10, North America**, with the world's most affluent societies in the United States and Canada. The United States in particular sets the conditions of globalization, although not all the impacts feed back better livelihoods to all Americans. The contemporary issues include the impacts of immigration, the North American Free Trade Agreement, and the role of French-speaking Québec in Canada.

A Text for Students

Students are encouraged to think about what it means to be part of a global community and to develop their geographical understandings of world events. This text features:

- **Accessibility.** Reviewers commented on the clarity of writing, clear definition of terms, and up-to-date illustrations.
- **Consistent structure.** The clear and consistent structure within each chapter encourages readers to compare world regions.
- **Superior illustrations.** Straightforward maps and diagrams with styles that are repeated in each chapter, allow students to easily compare regions.
- **An efficient and economic option.** A book of fewer pages encourages student participation.

Acknowledgments

The authors wish to express special thanks to McGraw-Hill for editorial support through Marge Kemp and Joan Weber; the marketing expertise of Lisa Nicks; and the production team led by April Southwood, John Joran, Carrie Burger, Sandy Ludovissy, Dan Wallace, Tammy Juran, and Judi David.

We greatly appreciate the help of all of the reviewers listed here—committed teachers who also have similar aims to the authors and editors:

John All
Western Kentucky University

Jeff Arnold
Southwestern Illinois College

Christopher Badurek
Appalachian State University

Bradley H. Baltensperger
Michigan Technological University

Richard W. Benfield
Central Connecticut State University

Sarah A. Blue
Northern Illinois University

Patricia Boudinot
George Mason University

Jeremy J. Brigham
Kirkwood Community College

Stanley D. Brunn
University of Kentucky

Craig S. Campbell
Youngstown State University

Keith Kyongyup Chu
Bergen Community College

Deborah Corcoran
Missouri State University

James Craine
California State University—Northridge

Robert Adam Dastrup
Salt Lake Community College

Bruce E. Davis
Eastern Kentucky University

L. Scott Deaner
Kansas State University

Jason Dittmer
Georgia Southern University

Cheryl Morse Dunkley
University of Vermont

Dennis Ehrhardt
University of Louisiana—Lafayette

Kenneth W. Engelbrecht
Metropolitan State College of Denver

Femi Ferreira
Hutchinson Community College

John H. Fohn II
Missouri State University—West Plains

Francis A. Galgano, Jr.
United States Military Academy

Matthew J. Gerike
Kansas State University

Joshua Hagen
Marshall University

Carol L. Hanchette
University of Louisville

Douglas Heffington
Middle Tennessee State University

David A. Iyegha
Alabama State University

Edward L. Jackiewicz
*California State University—
Northridge*

Cub Kahn
Oregon State University

Robert M. Kerr
University of Central Oklahoma

Margo Kleinfeld
University of Wisconsin—Whitewater

Peter Konovnitzine
Chaffey College

Chris Laingen
Kansas State University

James M. Leonard
Marshall University

Elizabeth J. Leppman
Eastern Kentucky University

Joshua Long
University of Kansas

Ju Luo
Missouri State University

Donald I. Lyons
University of North Texas

Kent Mathewson
Louisiana State University

Linda McCarthy
University of Wisconsin—Milwaukee

Brent McCusker
West Virginia University

Dianne E. Meredith
*California State University—
East Bay*

Dalton W. Miller, Jr.
Mississippi State University

Steven Oluic
United States Military Academy

Kefa M. Otiso
Bowling Green State University

Lynn M. Patterson
Kennesaw State University

Jim Penn
Grand Valley State University

Paul E. Phillips
Fort Hays State University

Gabriel Popescu
Indiana University—South Bend

Erik Prout
Texas A&M University

Rhonda E. Reagan
Blinn College

Paul Rollinson
Missouri State University

Yda Schreuder
University of Delaware

Kathleen Schroeder
Appalachian State University

Sinclair A. Sheers
George Mason University

Susan Siemens
Ozarks Technical Community College

Dean Sinclair
Northwestern State University

Peter P. Siska
Austin Peay State University

Cynthia L. Sorrensen
Texas Tech University

Jacob Sowers
Kansas State University

Dean B. Stone
Clinton Community College

Ray Sumner
Long Beach City College

Harold R. (Harry) Trendell
Kennesaw State University

Angie E. Wood
Blinn College

Laura A. Zeeman
Red Rocks Community College

Teaching and Learning Supplements

McGraw-Hill offers various tools and technology products to support *Essentials of World Regional Geography*. Students can order supplemental study materials by contacting their local bookstore or by calling 800-262-4729. Instructors can obtain teaching aids by calling the Customer Service Department at 800-338-3987, visiting the McGraw-Hill website at www.mhhe.com, or by contacting their local McGraw-Hill sales representative.

Teaching Supplements for Instructors

McGraw-Hill's ARIS—Assessment, Review, and Instruction System

(http://www.mhhe.com/bradshawessentials1e) for *Essentials of World Regional Geography* is a complete online tutorial, electronic homework, and course management system, designed for greater ease of use than any other system available. Instructors can create and share course materials and assignments with colleagues with a few clicks of the mouse. All PowerPoint

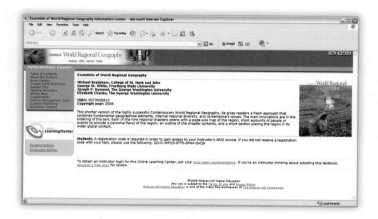

erPoint lectures, assignments, quizzes, tutorials, and interactives are directly tied to text-specific materials in *Essentials of World Regional Geography*; but instructors can also edit questions, import their own content, and create announcements and due dates for assignments. ARIS features automatic grading and reporting of easy-to-assign homework, quizzing, and testing. All student activity within McGraw-Hill's ARIS website is automatically recorded and available to the instructor through a fully integrated grade book that can be downloaded to Excel. *Instructors:* To access ARIS, request registration information from your McGraw-Hill sales representative.

ARIS Presentation Center (found at www.mhhe.com/bradshawessentials1e)

Build instructional materials where-ever, when-ever, and how-ever you want!

ARIS Presentation Center is an online digital library containing assets such as photos, artwork, animations, PowerPoint presentations, and other media types that can be used to create customized lectures, visually enhanced tests and quizzes, compelling course websites, or attractive printed support materials.

Access to your book, access to all books!

The **Presentation Center library** includes thousands of assets from many McGraw-Hill titles. This ever-growing resource gives instructors the power to utilize assets specific to an adopted textbook as well as content from all other books in the library.

Nothing could be easier!

Accessed from the instructor side of your textbook's ARIS website, the Presentation Center's dynamic search engine allows you to explore by discipline, course, textbook chapter, asset type, or keyword. Simply browse, select, and download the files you need to build engaging course materials. All assets are copyrighted by McGraw-Hill Higher Education but can be used by instructors for classroom purposes.

Instructors will find the following digital assets for *Essentials of World Regional Geography* at ARIS Presentation Center:

- **Color Art.** Full-color digital files of all illustrations in the text can be readily incorporated into lecture presentations, exams, or custom-made classroom materials.
- **Photos.** Digital files of photographs from the text can be reproduced for multiple classroom uses.
- **Additional Photos.** 386 full-color bonus photographs are available in a separate file. These photos are searchable by content and will add interest and contextual support to your lectures.
- **Tables.** Every table that appears in the text is provided in electronic format.
- **Global Base Maps.** 18 base maps of the major world regions are offered in full color and black-and-white versions. These choices allow instructors the flexibility to plan class activities, quizzing opportunities, study tools, and PowerPoint enhancements.
- **Population Density Maps.** 10 population density maps of the major world regions are offered in full color and provide an accurate illustration of the population distribution in the world. Instructors have the flexibility in planning class activities, quizzing opportunities, study tools, and PowerPoint enhancements.
- **PowerPoint Lecture Outlines.** Ready-made presentations that combine art and photos and lecture notes are provided for each of the 10 chapters of the text. These outlines can be used as they are or tailored to reflect your preferred lecture topics and sequences.
- **PowerPoint Slides.** For instructors who prefer to create their lectures from scratch, all illustrations, photos, and tables are preinserted by chapter into blank PowerPoint slides for convenience.

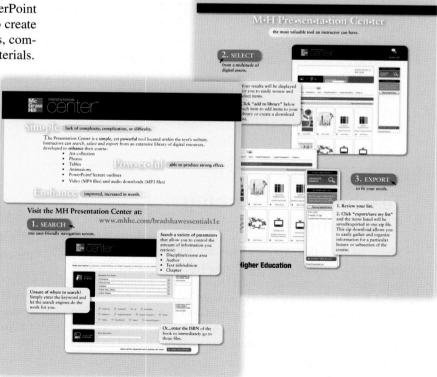

NEW! Klett-Perthes World Atlas
(ISBN 978-0-07-329073-7; MHID 0-07-329073-4)

This high-quality world atlas by well-known mapmaker Klett-Perthes features over 300 pages of detailed climate, physical, political, population, regional, and thematic maps as well as statistical information for all countries. Also included is a glossary of important terms, an extensive index, and instructions on how to read a map—a must for all introductory geography students! Ask your McGraw-Hill sales representative for further details on how to package this outstanding atlas with *Essentials of World Regional Geography*. Explore the many products and resources available from Klett-Perthes by visiting their website at www.klettmaps.com. Examples of these maps are included in the foldout section at the end of this text.

Earth and Environmental Science DVD by Discovery Channel Education
(ISBN 978-0-07-352541-9; MHID 0-07-352541-3)

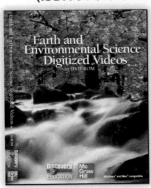

Begin your class with a quick peek at science in action. The exciting NEW DVD by Discovery Channel Education offers 50 short (3–5 minute) videos on topics ranging from conservation to volcanoes. Search by topic and download into your PowerPoint lecture.

Sights & Sounds CD-ROM by David Zurick, Eastern Kentucky University
(ISBN 978-0-07-312210-6; MHID 0-07-312210-6)

This new CD-ROM offers a unique opportunity in "seeing and hearing" the music and cultural perspectives of 10 regions:

- North America: Appalachia
- Central America: Oaxaca, Mexico
- South America: Ecuador
- Europe: British Isles
- Africa South of the Sahara: Tanzania
- South Asia: Nepal
- Middle East
- Insular Southeast Asia: Bali, Indonesia
- East Asia: China
- South Pacific: Samoa

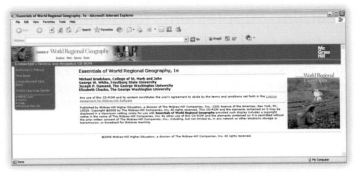

Instructor's Testing and Resource CD-ROM

This CD-ROM contains a wealth of cross-platform (Windows and Macintosh) resources for the instructor. Supplements featured on this CD-ROM include a computerized test bank, which utilizes EZ Test software to quickly create customized exams. This flexible and user-friendly program allows instructors to search for questions by topic, format, or difficulty level, and edit existing questions or add new ones. Multiple versions of the test can be created, and any test can be exported for use with course management systems such as WebCT and Blackboard. Word files of the test bank are included for those instructors who prefer to work outside of the test-generator software. Other assets on the Instructor's Testing and Resource CD-ROM are grouped within easy-to-use folders.

Course Delivery Systems

With help from WebCT, Blackboard, and other course management systems, professors can take complete control of their course content. Course cartridges containing website content, online testing, and powerful student tracking features are readily available for use within these platforms.

The Power of Place: Geography for the 21st Century STUDENT STUDY GUIDE and FACULTY GUIDE
Student Study Guide (ISBN 978-0-07-281855-0; MHID 0-07-281855-7) Faculty Guide (ISBN 978-0-07-281856-7; MHID 0-07-281856-5)

This revised and updated version of the World Regional Geography telecourse funded by Annenberg/CPB Projects, offers a revised Faculty Guide and Student Study Guide, providing insights into a wide range of geographic issues around the globe.

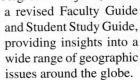

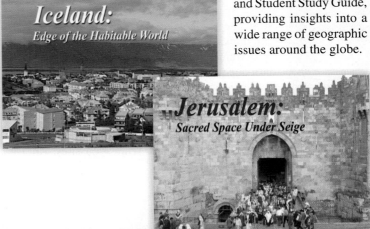

Learning Supplements for Students

ARIS (http://www.mhhe.com/bradshawessentials1e)

This site includes quizzes for each chapter, interactive base maps, and much more. Learn more about the exciting features provided for students through the *Essentials of World Regional Geography* ARIS site.

Additional Teaching/Learning Tools

Students of geography and other disciplines, as well as the general reader, will find these unique guides invaluable to their understanding of current world countries and events.

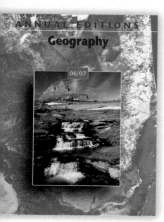

The **Annual Editions** series is designed to provide students with convenient, inexpensive access to current, carefully selected articles from the public press. They are updated regularly through continuous monitoring of over 300 periodicals. Each volume presents over 40 articles written for a general audience by experts and authorities in their fields. Organizational features include an annotated listing of selected World Wide Web sites, an annotated table of contents, a topic guide, a general introduction, and brief overviews for each section. Each title offers an instructor's resource guide containing test questions and a helpful user's guide called "Using Annual Editions in the Classroom."

Annual Editions: Developing World 07/08 by Griffiths (ISBN 978-0-07-351624-0; MHID 0-07-351624-4)
Annual Editions: Geography 06/07 by Pitzl (ISBN 978-0-07-354567-7; MHID 0-07-354567-8)
Annual Editions: Global Issues 07/08 by Jackson (ISBN 978-0-07-339728-3; MHID 0-07-339728-8)
Annual Editions: Violence and Terrorism 07/08 by Badey (ISBN 978-0-07-351619-6; MHID 0-07-351619-8)
Annual Editions: World Politics 07/08 by Purkitt (ISBN 978-0-07-339744-3; MHID 0-07-339744-X)

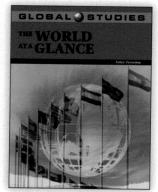

The **Taking Sides** volumes present current issues in a debate-style format designed to stimulate student interest and develop critical thinking skills. Each issue is thoughtfully framed with an issue summary, an issue introduction, and a postscript. The pro and con essays—selected for their liveliness and substance—represent the arguments of leading scholars and commentators in their fields. **Taking Sides** readers feature annotated listings of selected World Wide Web sites. An instructor's resource guide with testing materials is available with each volume. To help instructors incorporate this effective approach in the classroom, an excellent resource called "Using Taking Sides in the Classroom" is also offered.

Taking Sides: Clashing Views on African Issues, **second edition by Moseley (ISBN 978-0-07-351507-6; MHID 0-07-351507-8)**
Taking Sides: Clashing Views on Global Issues, **fourth edition by Harf/Lombardi (ISBN 978-0-07-352724-6; MHID 0-07-352724-6)**
Taking Sides: Clashing Views on Latin American Issues, **by DeGrave et al. (ISBN 978-0-07-351504-5; MHID 0-07-351504-3)**
Taking Sides: Clashing Views in World Politics, **thirteenth edition by Rourke (ISBN 978-0-07-339720-7; MHID 0-07-339720-2)**

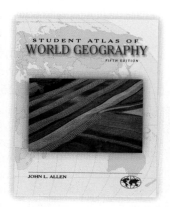

The **Student Atlas** series combines full-color maps and data sets to introduce students to the importance of the connections between geography and other areas of study, such as world politics, environmental issues, and economic development. In particular, the **Student Atlas of World Geography**, fifth edition, by John Allen combines over 100 full-color maps and data sets to give students a clear picture of the recent agricultural, industrial, demographic, environmental, economic, and political changes in every world region. This concise, affordable resource provides the most recent geographic data for geography students.

Student Atlas of World Geography, **fifth edition by Allen (ISBN 978-0-07-352757-4; MHID 0-07-352757-2)**
Student Atlas of World Politics, **seventh edition by Allen (ISBN 978-0-07-340146-1; MHID 0-07-340146-3)**

Global Studies is a unique series designed to provide comprehensive background information as well as vital current information regarding events that are shaping the cultures of the regions and countries of the world today. Each **Global Studies** volume features country reports in essay format and includes detailed maps and statistics. These essays examine the social, political, and economic significance of each country. In addition, relevant and carefully selected articles from worldwide newspapers and magazines are included to further foster international understanding.

Global Studies: Africa, eleventh edition updated by Edge (ISBN 978-0-07-337992-0; MHID 0-07-337992-1)
Global Studies: China, twelfth edition by Ogden (ISBN 978-0-07-337991-3; MHID 0-07-337991-3)
Global Studies: Europe, ninth edition by Frankland (ISBN 978-0-07-319874-3; MHID 0-07-319874-9)
Global Studies: India and South Asia, eighth edition by Norton (ISBN 978-0-07-337971-5; MHID 0-07-337971-9)
Global Studies: Islam and the Muslim World, by Husain (ISBN 978-0-07-352772-7; MHID 0-07-352772-6)
Global Studies: Japan and the Pacific Rim, ninth edition by Collinwood (ISBN 978-0-07-337990-6; MHID 0-07-337990-5)
Global Studies: Latin America, twelfth edition by Goodwin (ISBN 978-0-07-340406-6; MHID 0-07-340406-3)
Global Studies: Russia, the Eurasian Republics, and Central/Eastern Europe, eleventh edition by Goldman (ISBN 978-0-07-337989-0; MHID 0-07-337989-1)

Global Studies: The Middle East, eleventh edition by Spencer (ISBN 978-0-07-340405-9; MHID 0-07-340405-5)
Global Studies: The World at a Glance, by Tessema (ISBN 978-0-07-340408-0; MHID 0-07-340408-X)

Other supplemental titles include:
Afghanistan: Geographic Perspectives, by Palka (ISBN 978-0-07-294009-1; MHID 0-07-294009-3)
Iraq: Geographic Perspectives, by Malinowski (ISBN 978-0-07-294010-7; MHID 0-07-294010-7)
North Korea: Geographic Perspectives, by Palka and Galgano (ISBN 978-0-07-294011-4; MHID 0-07-294011-5)
Military Geography: From Peace to War, by Palka and Galgano (ISBN 978-0-07-353607-1; MHID 0-07-353607-5)

Instructors: Ask your sales representative about packaging options—special discounts may be available with some of the titles listed above!

GUIDED TOUR

Maps

A variety of helpful maps have been included in this edition. Special attention has been given to the new **population density maps.** They are featured in all ten chapters, providing an accurate illustration of the population distribution in the world. The first one opens Chapter 1. Other chapters open with a regional map of relief, country boundaries, and capital cities.

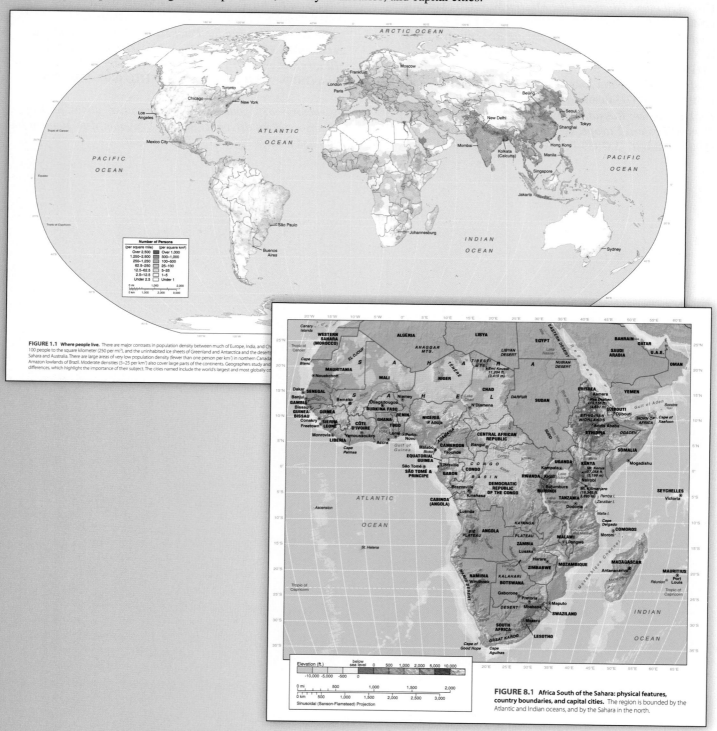

FIGURE 1.1 Where people live. There are major contrasts in population density between much of Europe, India, and China, with 100 people to the square kilometer (250 per mi²), and the uninhabited ice sheets of Greenland and Antarctica and the deserts of the Sahara and Australia. There are large areas of very low population density (fewer than one person per km²) in northern Canada and the Amazon lowlands of Brazil. Moderate densities (5–25 per km²) also cover large parts of the continents. Geographers study and explain such differences, which highlight the importance of their subject. The cities named include the world's largest and most globally connected.

FIGURE 8.1 Africa South of the Sahara: physical features, country boundaries, and capital cities. The region is bounded by the Atlantic and Indian oceans, and by the Sahara in the north.

Foldout World Maps

Foldout world maps from Klett-Perthes can be easily referenced from any chapter in the text.
The foldout section at the back of the text focuses on the following key areas:

- World climates
- Urbanization and migration
- There is also a guide to using and reading maps

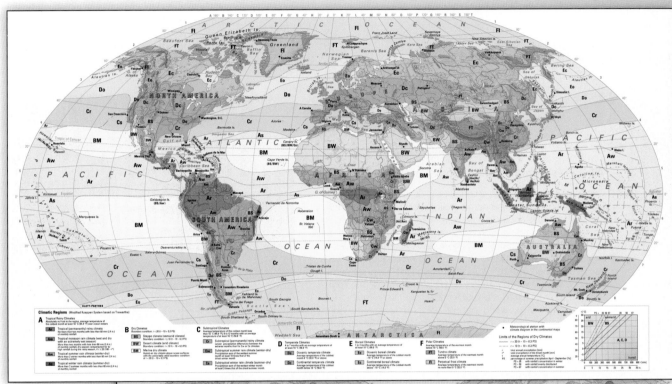

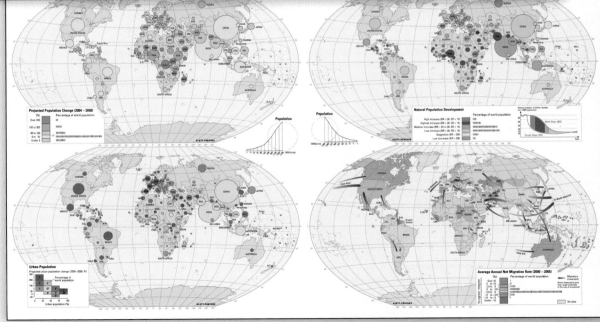

Geography At Work Boxes

Interesting **Geography at Work boxes**, found at the end of each chapter, provide valuable insights into the ways geographical understandings and skills are used in a variety of jobs and roles. **Geography at Work boxes** cover the following topics:

- Chapter 1: Alexander B. Murphy, *AAG President in Iran: Reconciling Differences*
- Chapter 2: Anne Hunderi, *Business Development*
- Chapter 3: Robert W. Hutton, *Wine Industry*
- Chapter 4: Erle Ellis, *China's Landscapes and Global Change*
- Chapter 5: *The Tsunami of December 2004*

- Chapter 6: Mohammed Ali, *Battling Infectious Diseases*
- Chapter 7: Mark Corson, *Operation Iraqi Freedom*
- Chapter 8: Elio Spinello and Steve Lackow, *Mapmakers and GIS Analysts*
- Chapter 9: Paul Hudson, *Floodplain Erosion in Mexico*
- Chapter 10: Brian Hannegan, *Using Geography to Aid Public Policy*

GEOGRAPHY AT WORK

AAG President in Iran: Reconciling Differences

Geography provides fundamental insights into the nuances of globalization, whether economic, political, or cultural. Geography can build bridges and help ease cultural tensions and misunderstandings. For example, Professor Alexander B. Murphy, from the Department of Geography at the University of Oregon, traveled to Iran in 2004 when he was president of the Association of American Geographers (Figure 1.26). He addressed the "Second International Congress of Geographers of the Islamic World," and spoke to students and faculty at two of Tehran's major universities.

Everywhere Alex went, people were extraordinarily nice, helpful, and friendly. He had traveled previously in other parts of the Muslim Middle East—especially Egypt, Jordan, and Palestine. Iran is clearly different—in language, in culture, in social norms, and much more. To visit Iran is to understand the fallacy

FIGURE 1.26 Professor Murphy, the tallest person, with colleagues at the Iranian geographer's conference.

...vernment news sources but from the Internet...and relatives in other parts of the world. He...an society has opened up in recent years,...with people and his media appearances...America that Iranians rarely see.

GEOGRAPHY AT WORK

Mapmakers and GIS Analysts

Elio Espinello and Steve Lackow are Californians who are the authors and managers of AtlasGIS, a software package that combines mapping with a geographic information systems approach. In the 1990s, the huge GIS software company, ESRI, took over AtlasGIS, but has recently placed marketing and development back with Elio Espinello and Steve Lackow's company, RPM Consulting.

As an example of their work, Elio completed a project that was an epidemiological assessment of river blindness (onchocerciasis) in Mozambique. He was commissioned by Aircare International, which provides small aircraft to help local groups in Mozambique and wanted to know the most important areas for delivering medical care. The disease does not kill, but is chronic and widespread and affects the lives of many people. The African blackfly inhabits areas with fast-flowing streams and the female carries the river blindness from an infected person to an uninfected one. Fibrous nodules form in the infected persons, producing microfilariae that attack skin pigmentation, causing skin atrophy and blindness. Some surgery is

needed to remove the nodules, but new medicines make it possible to treat many more people without bad side effects.

Elio began by mapping the factors that encourage a concentration of African blackflies: the density of population, the concentration of rivers, and the occurrence of steep slopes that give faster river flow. He put together a composite index of these factors, giving greater weight to the population and waterway density than slope steepness (Figure 8.29). The results highlighted districts of high risk (yellow), making it possible for Aircare International to concentrate its delivery of medical support.

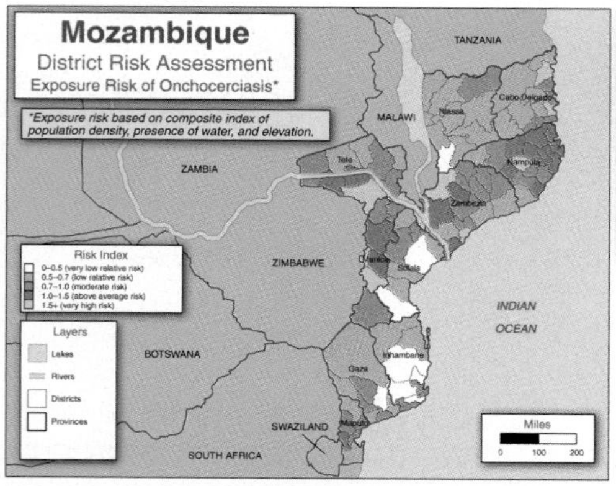

FIGURE 8.29 Mapmakers and GIS. (a) Mapping the risk of river blindness in Mozambique. (b) Elio Spinello. *Source: (a) Data from Elio Spinello, RPM Consulting.*

Debate Tables

Debate tables, found in virtually every chapter, provide contrasting views of controversial regional and world issues, engaging students and supporting critical analytical skills. Topics featured in the **debate tables** include:

- Table 2.3 Debate: The EU's Future
- Table 3.3 Debate: Is Russia Still a World Power?
- Table 4.3 Debate: Population Policies in China
- Table 5.3 Debate: Singapore
- Table 7.3 Debate: Israelis versus Palestinians
- Table 8.2 Debate: HIV/AIDS in Materially Wealthy and Poor Countries
- Table 8.4 Debate: The Future of Africa South of the Sahara
- Table 9.3 Debate: Tropical Deforestation
- Table 10.3 Debate: The Effects of NAFTA

TABLE 3.3   DEBATE: IS RUSSIA STILL A WORLD POWER?

Still a World Power	No Longer a World Power
Russia lost possession of 14 republics (now independent countries) and with them sizable populations and resources.	Russia's heartland remains together, and Russia is still the largest country in the world, with a sizable population and considerable resources.
Russia's international power i[s] presence of 25 million Russia[ns] as ethnic minorities in other [countries] careful not to offend these ot[her] endanger the Russians who l[ive] to make economic concessio[ns] protect Russian minorities in	
The Russian government has [difficulty] controlling independence m[ovements] republics. It has been conder[ned for] violations in Chechnya. Its m[ilitary is] unprepared, and demoralized[, unable to] control the conflict in the for[mer] expansion of NATO.	
With rampant corruption, Ru[ssia] and has floundered since the [collapse in] 1991. Russia is unable to capi[talize on its] resources on the world mark[et] oil is questionable. Russia has economic organizations.	

TABLE 8.4   DEBATE: THE FUTURE OF AFRICA SOUTH OF THE SAHARA

Africans Can Do It	No Hope for Africa's Future
There are signs that Africans are moving toward a better future. The peaceful transition in South Africa, Uganda's success with AIDS, fewer wars, and increasing democracy are examples of good trends.	Such signs are few and temporary. Most things get worse. Civil strife springs up in new places and most democracy is a façade. These are basic reasons today for crippling African development.
Younger "born frees" (since independence) are better educated and more inclined to expect good leadership from the current leaders rather than blaming the past.	The present problems stem from the colonizers and their racism. Whatever Africans do to put them right is futile. (This is a common complaint of older people born into colonial rule.)
African countries can produce goods in addition to the crops and minerals that others want to buy if trading terms with the wealthier countries are improved. Some countries are growing rapidly, some as a result of oil windfalls, and a few others such as Mozambique, Rwanda, and Uganda have seen a decade of economic growth after decades of strife and poverty.	Few leaders place much importance on sustained economic growth and their actions tend to reduce the ability of governments to develop education, health, and employment prospects. Few African leaders allow widespread involvement in government that reduces their powers of patronage and ability to make arbitrary decisions.
Smart businessfolk can do well, as in the oil companies on the west coast, the mining corporations, and those making cheap luxuries such as bottled drinks and soap powder prosper. Cell phones increased from almost none 10 years ago to nearly 100 million today. The business climate could improve with more privatization in areas such as utilities (telephones, electricity) and with shorter periods of business registration.	Too many countries place difficulties in the way of foreign corporations wishing to do business in African countries, including the continuation of the bribe culture and lawlessness. The past has left too many examples of squandered opportunities.
The rise in the numbers of urban Africans is leading to wider political participation and, hopefully, to demands for better government. Better government is particularly important at this time, but has to be widely wanted and supported. Power is based in country governments. But improved political involvement has seldom been linked to greater prosperity.	There is still too much rule and control by a few "big men" who turn the law and finances to their own ends. President Mugabe of Zimbabwe is a well-known example. His once fairly prosperous country is now among the poorest in the world. Too many governments are predatory and few are competent. There are still few leaders who leave after electoral defeat, compared with those overthrown by war or coup.
Aid agencies are now putting more research into pre-funding activities. More philanthropic donors are needed, overseeing their giving in relation to criteria such as "saving the maximum number of lives at minimum cost" (Bill Gates).	In the past, too many aid agencies left behind more harm than good. Dependency on outside resources, and spending aid such as World Bank grants in profligate ways, do not lead to local entrepreneurial actions.
Land reform is occurring, although there is a need to ensure that people who can farm get title to farming land and government favorites do not take over productive land.	Most countries suffer from a lack of security in property rights. People with communal or tenant rights cannot use that land to underpin bank financing.

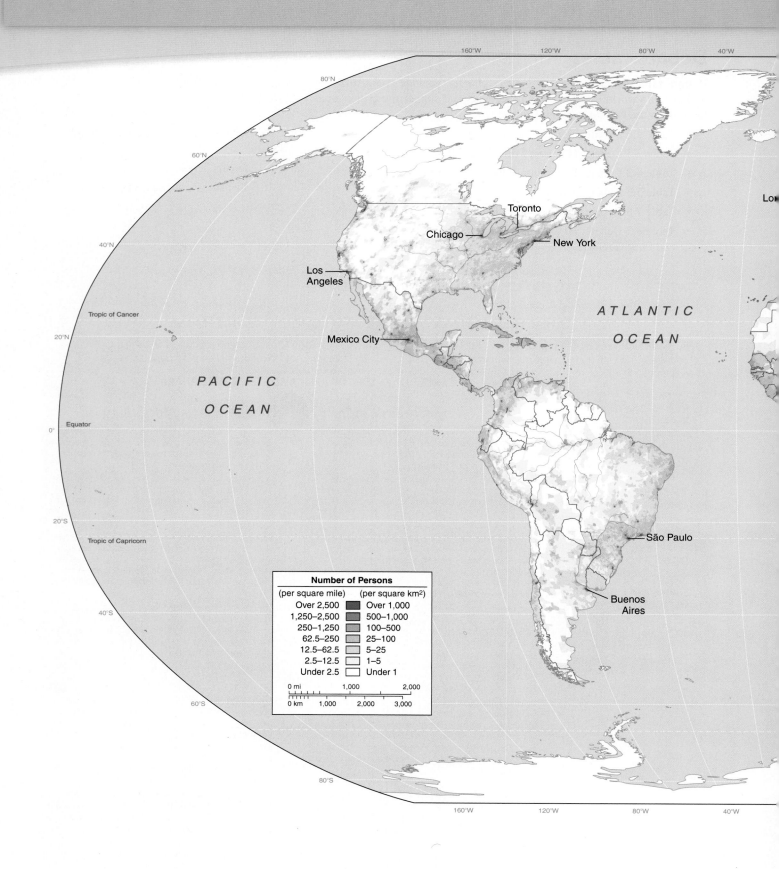

Number of Persons

(per square mile)		(per square km²)
Over 2,500		Over 1,000
1,250–2,500		500–1,000
250–1,250		100–500
62.5–250		25–100
12.5–62.5		5–25
2.5–12.5		1–5
Under 2.5		Under 1

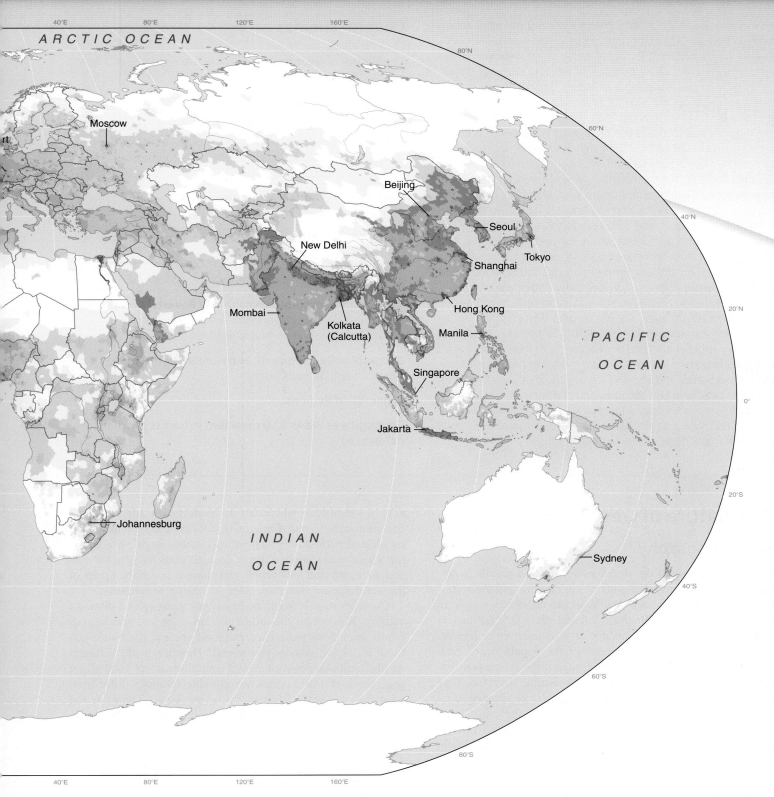

FIGURE 1.1 **Where people live.** There are major contrasts in population density between much of Europe, India, and China with over 100 people to the square kilometer (250 per mi.²), and the uninhabited ice sheets of Greenland and Antarctica and the deserts such as those in the Sahara and Australia. There are large areas of very low population density (fewer than one person per km²) in northern Canada and Russia and the Amazon lowlands of Brazil. Moderate densities (5–25 per km²) also cover large parts of the continents. Geographers study and try to explain these differences, which highlight the importance of their subject. The cities named include the world's largest and most globally connected.

As individuals we live in local places like small towns or neighborhoods, but our lives are affected by global trends in politics, culture, and economics. Two brief stories illustrate how our lives are connected to global processes. What would be your story?

"I am Miro and I was born in Mostar, Bosnia. It was a great place to grow up in, with its old buildings and spectacular bridge built in 1561. I trained as an auto mechanic. But in 1992 the fighting that followed the breakup of Yugoslavia forced my family to move to Serbia. Then my sister, who had stayed in Mostar, had her leg amputated after a shell exploded near her. Complications in her treatment led the U.S. Veterans for Peace organization to airlift her to Cumberland, Maryland. I joined her and after difficulties with learning English, having to take jobs with long commutes, and adjusting to the American way I am now working as a truck mechanic. I, my wife and children, and several friends from home are living happily here."

"I am Michiko and my home is on the southern Japanese island of Shikoku. While growing up, my family lived in a two-bedroom wooden house next to my grandmother, who taught me traditional Japanese social customs. My engineer father worked for the same company all his life and was seldom at home; my mother also had a weekday job. I worked very hard at school and then traveled abroad, spending time in Australia, the United States, and England. I enjoy the benefits of the consumer society, high tech facilities, and global travel. Gaining British qualifications for teaching English as a foreign language in Japan, I teach English alongside Japanese calligraphy. Recently my family moved into a new high tech house."

Chapter Themes

The Nature of Geography: why geography? places and flows; approaches to geography.

Globalization and Localization: globalization; localization.

Regions and Natural Environments: powerful natural systems; solid earth environments; atmosphere-ocean environments; earth surface environments; ecosystem environments; human impacts on natural environments; resources and hazards.

Regions and Cultural Geography: language; religion; cultural status: race, class, and gender; cultures and regions.

Regions and People: population distribution and dynamics.

Regions and Politics: countries still dominate; global governance; country groupings for trade or defense.

Regions and Economics: Wealth and Poverty: measuring wealth and poverty; creating material wealth; the global economy; regional economic emphases.

Geography, Development, and Human Rights: human development; issues of human rights.

Major World Regions

Geography at Work: AAG president in Iran: reconciling differences

The Nature of Geography

Why Geography?

We are all citizens of the world as well as of a specific country. What happens in other parts of our country and the rest of the world impacts us each day. What we do affects people in other places. Bombs in Baghdad kill and injure U.S. military as well as Iraqi citizens, perhaps one of your relatives or friends. Following Hurricane Katrina most of the population of New Orleans was spread out across the United States, some possibly in your classroom, while the price of gasoline rocketed. Your choices of clothes, food, and other goods may be manufactured elsewhere in your country, or abroad, increasingly in China. Closer contacts make us more conscious of the variety of conditions around the world, the differences that separate people, and the impacts of rapidly changing technology and political and cultural forces.

Geography is the study of spatial patterns in the human and physical worlds. Geographers examine where and how human and natural characteristics are distributed across Earth's surface, how they relate to each other, and how patterns change. The distribution of, for example, population (Figure 1.1), criminal activities, or hurricanes can be mapped and explanations offered as to how they occur where they do. Such studies have very practical outcomes, as our "Geography at Work" pieces at the end of each chapter illustrate. Geographical studies begin with knowing where places are and where certain events happen, but also seek to understand how people and cultures in different parts of the world interact with each other and their environments.

Consider what it would be like if you grew up in a different country. How might that affect the language you speak, your family's religious preference, the food you eat, the music you listen to, or the schools you attend? How might the different weather and other environmental conditions affect you? Might your views of world issues and possible solutions differ from what they are now? What might be the same? These are the sorts of questions geographers ask about people and the places where they live.

Places and Flows

Geographers study places on Earth's surface and flows between them. A **place** may be perceived as a point on a map or as a large area. It can be of different **scales** of geographic size. It might be an individual building (convenience store), small town (Freeport, Maine), large city (New York), rural area (western

Iowa), another state, or another country. Places are the environments or spaces where people live and through which they make life meaningful. Geographers draw maps to represent the features of places on Earth's surface and better understand the flows among these places. Geography thus provides a place- or space-related, **spatial view** of the human experience. **Flows** of people, goods, information, and ideas are dynamic aspects of geographical studies. They highlight the interconnections among places and resultant interactions.

Maps and Geographic Information Systems

In their approach to understanding and solving human problems, geographers use maps to present information about location, distance, direction, flows, and other characteristics of places. **Maps** are relatively small representations of much larger areas of Earth's surface. On maps, a scale implies a mathematical relationship between features on a map and on the ground area it represents. Map scales vary with the purpose of the map (Figure 1.2). Small-scale maps show areas at fractions of 1:250,000 or smaller (e.g., 1/1 million or a ratio of 1:1 million). Large-scale maps have map-to-ground ratios ranging from 1:10,000 to 1:250,000 (e.g., 1:50,000). The world maps used in this text (see, e.g., Figure 1.5b) are examples of small-scale maps, in which the scale along the equator is approximately 1:120 million. For the same size of map, large-scale maps cover smaller areas, as in town maps, but can include more details. Not everything can be drawn to scale on maps or features would be too small to be seen: for example, roads, rivers, and buildings are denoted by symbols.

Geographic information systems (**GIS**) combine maps and aerial and satellite images with data relevant to the area (Figure 1.3). Such systems bring together a range of information that can form the basis of finding answers to complex questions. Since about 1970, GIS has been a huge area of development in geography and most published large-scale maps now relate to satellite images.

Latitude and Longitude

Increasingly, the greater interconnectedness of people and places puts an emphasis on the movements, or flows, among places. Places often relate to other places by their position, or **location**, on the globe. **Absolute location** is the precise position of places on Earth's surface. Latitude and

longitude form the framework of the international reference system that pinpoints absolute location (Figure 1.4). Today radio beacons and satellites provide standard reference points by emitting radio pulses that are timed and interpreted rapidly by computerized navigation systems to give accurate position fixes in **global positioning system (GPS)** devices. Many cars are now equipped with such systems.

The **latitude** of a place is its position north or south of the equator, measured in degrees, as shown on Figure 1.4. The North Pole is at 90°N and the South Pole at 90°S. The equator is an imaginary line that encircles the globe midway between the North and South Poles. It is the 0° (zero degree) line of latitude. The almost spherical Earth's circumference is around 40,000 km (25,000 mi.) at the equator. An imaginary circle that joins places of the same latitude at Earth's surface is called a **parallel of latitude**. The ground distance from one degree of latitude to the next is approximately 110 km (69 mi.) on Earth's surface.

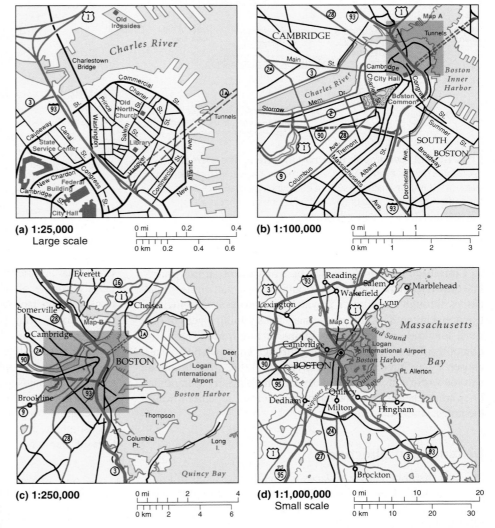

(a) 1:25,000 Large scale

(b) 1:100,000

(c) 1:250,000

(d) 1:1,000,000 Small scale

FIGURE 1.2 Map scale and detail. Maps of Boston from large scale (1:25,000) to small scale (1:1,000,000). The larger scale maps show more detail (streets, buildings) than the small scale maps (only major roads) for the same size of map.

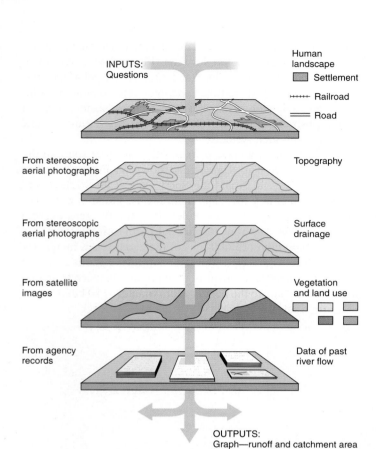

FIGURE 1.3   **Geographic information system.** The layers of information in this example are used to monitor a river system feeding a reservoir. Outputs from the system could include graphs of seasonal streamflow per unit of drainage basin area, or maps of dominant vegetation types or changing land use.

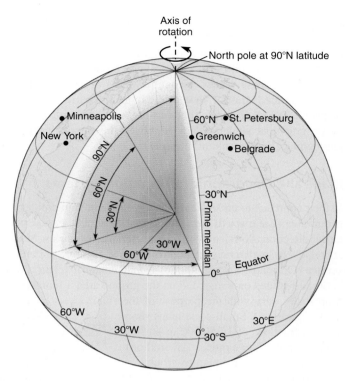

FIGURE 1.4   **Location: latitude and longitude.** Locating places on Earth's surface is made possible by the coordinate system of latitude and longitude. The degrees of latitude and longitude are formed by angles focused at Earth's center. Define the differences between 60°N and 30°N and between the 0° meridian and 60°W in terms of angles at Earth's center. Give the latitude and longitude of the cities marked on the globe.

Longitude measures position east or west of an imaginary line drawn from the North to the South Pole—a half circle—that passes through the former Royal Observatory at Greenwich, London, United Kingdom. Imaginary lines joining places of the same longitude are called **meridians of longitude**. The position of the Prime Meridian passing through Greenwich (0°) was chosen by an international conference in 1884, when London was the world's most powerful decision-making city. Longitude lines are farther apart at the equator and closer together toward the poles, where they intersect.

Distance and Direction

Physical **distance** between places is usually measured in kilometers or miles. However, travel time or travel cost (e.g., cost of gasoline) may be of the greatest significance in affecting the **relative locations** of places with respect to each other. The relative measures of time-distance and cost-distance are often substituted for measured distance in geographic studies. The increasing cost and time to cover distance between places gives rise to the idea of the **friction of distance**. As costs increase with distance, in-

teractions decrease. However, improving technology decreases the friction of distance. For example, the friction of distance between New York and Chicago was reduced in the 1800s, when time for the journey was cut from weeks to days by building the Erie Canal and railroads. Air travel between these places today takes a couple of hours. The increasing availability of rapid transportation facilities and the "global information highway" (the Internet) bring people into easier contact with each other, making them relatively—but not absolutely—closer. Friction of distance is still affected by other factors that slow or increase contacts among people. These factors include natural obstacles such as mountains and oceans, political factors such as country boundaries, and cultural factors such as language differences.

Geographers give **directions** by the cardinal points: north, east, south, west. Direction and distance help to define the locations of places relative to each other.

Approaches to Geography

Geographic studies have different emphases.

- **Physical geography** studies natural environmental processes across Earth's surface: the distributions of mountain building, climate varieties, river action, plant ecologies, or soil types, among other patterns. Physical geography increasingly examines the impacts of human actions on Earth's natural environments.

- **Human geography** is the study of the distribution of people and their activities (demography, cultures, politics, economies, and urban/rural changes).
- **Systematic, or thematic, geography** studies consider the spatial relationships and implications within a specific theme within physical or human geography, such as geographic aspects of climate, urban experience, industrial development, health, or migration.
- **Regional geography** focuses on specific areas of Earth's surface, or regions. Regional studies combine consideration of the systematic physical (rocks and landforms, climates, natural vegetation, soils) and human (culture, population, politics, economy) features that give a region geographic character. Some regions are defined formally as separate entities by political or marketing unit boundaries. Other regions are functional, exchanging flows of people, goods, information, and ideas with other regions. Regional linkages and boundaries change over time, or may be perceived and defined differently by their inhabitants or different geographers. The range of regional scales between the global and local includes world regions, subregions, countries, and smaller regions within countries.

This text adopts the regional geography approach. Regions form and change as different systematic aspects interact and influence one another in specific areas of the world.

- People create regions over time by interacting with natural environments and other regions. People in some regions view people in other regions as friendly or as different, perhaps hostile. Perceptions of the region's identity are often helped by propaganda.
- Regional character and identity influence the actions of people living there. There is thus a two-way interaction of people creating their region and of the region's character influencing the people.
- Regional character may change over time at varied rates as a result of introducing new people, products, cultural features, or political control. Individuals and small groups may be as responsible for change as governments. For example, powerful leaders, inner government groups, or local pressure groups may dominate the outcome of issues leading to changed or unchanged geographies.
- Regions are not isolated from one another. Ties among them became more intense over the last 200 years, causing more far-reaching and frequent changes of internal geographies and flows between regions.

Globalization and Localization

In our world of closer ties among peoples and places, together with more frequent and numerous movements of people, goods, and ideas in and out of regions, geographers study global trends and connections and their impacts as expressed by local voices.

Globalization

In its simplest form, **globalization** is the increasing level of interconnections among people and places throughout the world. Wide usage of the term, however, gives it a range of meanings. Globalization can be seen as the spreading influence of market economics through world political empires in the early 1900s or through the increasing trade and financial movements from the late 1900s. Although these trends began in the European empires before World War I, the United States and its corporations took the lead before and after World War II. Today, U.S. economic, political, and cultural ways influence the world. However, alternative economic and cultural competition from revived Japanese, Chinese, and European influences compete in the almost worldwide marketplaces and communities. Major countries such as Russia, China, and India, which kept apart from globalizing trends until the 1980s and 1990s, are now very much involved.

Global connections include the spread of ideas, technologies, crimes, and diseases; flows of goods and services; long-term migrations of people for work, seekers of political asylum, and family consolidation; short-term flows of people for business purposes or tourism; impacts of dominant ideologies, both religious and political; and the spread of images and messages through the media of film, TV, the Internet, and print. Today a few languages, notably English, Chinese, French, Spanish, and Arabic, act as a basis for global communication. Although many worldwide flows continue to be controlled by country-based legislation and policing, others, such as the trafficking of arms, drugs, or slaves, are difficult for individual countries to control.

Global economic growth since around 1990 has been the greatest in history. Overall, leading developing countries now accrue over 50 percent of world income, consume as much fossil fuels as the materially rich countries, produce over 40 percent of world exported goods (as compared with 20 percent in 1970), and possess 70 percent of the world's foreign exchange reserves. The fullest impacts of globalization, however, may be slow in coming to the majority of poorer countries. For example, the World Trade Organization (WTO) policies of increasing international trade by reducing restrictions on the movement of goods have made some progress. However, WTO's Doha Round instituted at the end of 2001 to allow greater access for materially poorer countries to markets in the richer countries, especially for agricultural produce, achieved little and was abandoned in 2006.

Localization

Localization is based on long-established local identities, economies, and cultures. For example, the lives and living conditions of ethnic groups of Papua New Guinea contrast with those in the commercial crop regions of Kenya and the urbanites of Paris, France.

As global exchanges and flows of information, ideas, people, money, and technology underlie increasing worldwide economic exchanges, political responses, cultural attitudes, and environmental concerns, people at local levels attempt to maintain their distinctive identities. Local voices express reactions to globalizing trends in such areas as regional identity in Europe, political nationalism in China, separatist groups in Indonesia, local customs and practices in Nigeria, economic marginalization in many parts of African countries, language usage in India, and religious differences in Iraq. These voices preserve local identities and may lead to the political fragmentation of larger entities. There is thus no total homogenization of societies under globalization. Ironically, globalization is encouraging a renewed pride in local identities. As a result, geographical diversity continues and even increases within world regions and countries.

Regions and Natural Environments

Physical geography is the study of natural environments and their world distribution. In world regional geography, the interactions between people and natural environments are important. Major physical features, such as oceans, seas, mountain ranges, and rivers often form boundaries between countries and world regions (see the maps that open each regional chapter) and frequently shape patterns of human movement.

For many centuries, human activities relied on natural resources. The length of growing season, amount of water available, soil types, and mineral-bearing rocks influenced the locations of people. Natural environments played important parts in the locations of early culture hearths and concentrations of people. From the 1800s, physical geography also played important roles in the locations of new human-engineered urban environments. Today, in a reversal of the idea that physical geography determines human affairs, human beings increasingly influence many of the ways in which natural processes function. For example, natural vegetation has been removed to make way for agriculture, rivers are straightened or diverted, slopes are graded, and mountains are tunneled. We now believe that human actions have increasing influences on our weather and climate.

Powerful Natural Systems

Four major natural environments, powered by energy from Earth's interior and the sun, comprise a dynamic and interacting system. First, in the **solid earth environment** (lithosphere), energy from the interior causes huge sections of the crust to collide with each other or move apart. These interactions produce earthquakes, volcanoes, mountain systems, and continental shifts. Second, incoming solar energy controls the circulations of the atmospheric gases and oceanic waters (atmosphere and hydrosphere), producing weather, climatic differences, and longer-term climate changes in the **atmosphere-ocean environment**.

Third, the interactions of atmosphere and hydrosphere with the lithosphere cause rain, glacier ice, wind, and ocean waves and currents to produce landforms such as hills, valleys, cliffs, and beaches on continental areas. Such **earth surface environments** provide the stages for human activities. Fourth, in the biosphere, solar energy is converted by plants to food for animals. Living organisms, including plants, animals, and humans, respond to and modify local climates, landforms, and soils in **ecosystem environments**. We take these environments in turn.

Solid Earth Environments

Earth is a multilayered planet with a hot molten core surrounded by the solid mantle and a relatively thin outer crust. Earth's interior provides the energy that forces large blocks of surface rock, known as **tectonic plates**, to crash into each other or move apart. There are six major and several smaller plates (Figure 1.5), each up to thousands of kilometers across and around 100 km (65 mi.) thick. Most earthquakes and volcanic outbursts of molten rock from beneath Earth's surface occur along plate boundaries. The volcanic "ring of fire" and earthquake zone around the Pacific Ocean is a consequence of plates clashing there.

Where plates move apart, or diverge, fissures open and rock erupts as molten lava, adding to the edge of a plate where it solidifies. Such **divergent plate margins** are responsible for widening oceans: the Mid-Atlantic Ridge marks one line of divergence. The ridge is mostly covered by water, but in the north sufficient lava has been erupted to form Iceland.

Plate collisions occur along **convergent plate margins** and often force one plate upward to form mountain systems such as the Andes Mountains of South America. A plate that is forced beneath another plate is said to be subducted. The subducted solid rock melts under high temperatures generated by burial and friction, and the molten rock produced rises toward the surface under pressure. It erupts to create volcanic mountains or piles of lava such as those forming the Columbia Plateau in the northwestern United States.

Where plates move parallel to each other but in opposite directions, without clashing or pulling apart, transform faults such as the San Andreas Fault in California create earthquakes and displace sections of crustal rocks (Figure 1.6). Few transform faults have volcanic eruptions.

Atmosphere-Ocean Environments

The **climate** of a place is based on long-term averages of the weather conditions (mainly temperature and precipitation). It is important for regional geography in indicating whether the place is more or less habitable and which crops can be grown. Differences in climatic conditions result from the transfers of heat and moisture through the atmosphere and oceans, and how they interact with the surface conditions. The transfers are powered by energy from the sun.

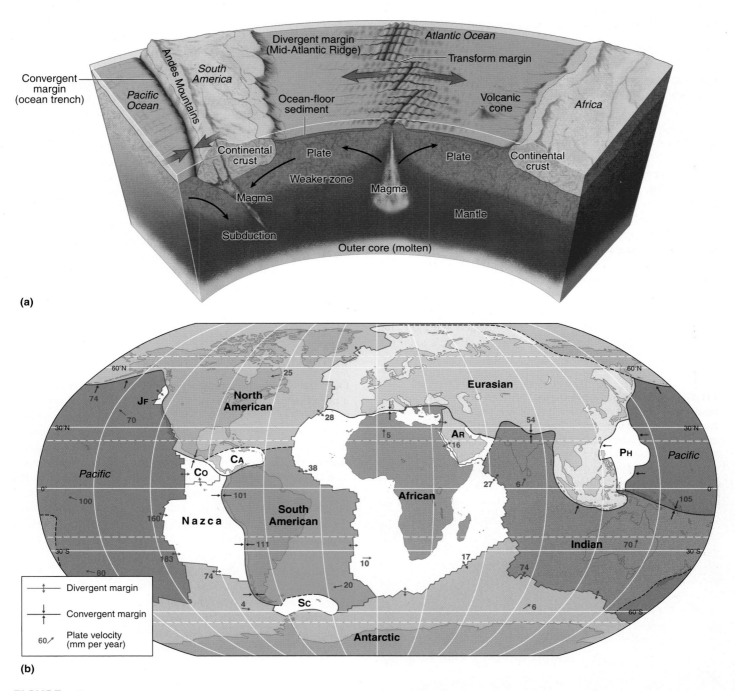

FIGURE 1.5 Plate tectonics, ocean basins, and mountain ranges. (a) A diagram that simplifies the plate tectonics relationships of the South Atlantic Ocean area. The large red arrows depict the plate movements: diverging from the Mid-Atlantic Ridge where molten rock (magma) rises and solidifies as new plate; converging where the Pacific Ocean and Andes Mountains meet, where plate material is destroyed and rises as molten magma to form volcanoes. A mid-ocean ridge is associated with diverging plates, and an ocean trench and high fold mountains with convergence. Where a plate is broken horizontally, a transform margin forms. All plate margins are associated with earthquakes and volcanic activity. (b) World map of major and the main minor plates. The minor plates include Nazca, Cocos (Co), Caribbean (Cᴀ), Juan de Fuca (Jғ), Arabian (Aʀ), Philippine (Pʜ), and Scotia (Sc). *Source: (b) Data from NASA.*

Heating the Atmosphere-Ocean System

Mostly visible light rays from the sun reach Earth's surface. Earth's atmosphere filters out other aspects of solar-ray energy that harm living organisms, including ultraviolet rays, x-rays, and gamma rays. Absorption of the light rays at Earth's surface causes rock, soil, and ocean water to be heated and to radiate heat upward. This heat is then absorbed in the lower atmosphere by water vapor and carbon gases, raising the temperature of the air. This natural process in Earth's atmosphere is known as the **greenhouse effect**. There is an approximate balance between the incoming solar radiation and radiation from Earth to space, reducing temperature fluctuations over time in Earth's atmosphere.

Important geographic differences in solar heating cause climatic differences from place to place. Earth rotates on its axis every 24 hours, creating day and night. It revolves in orbit around the sun once a year, producing seasonal changes. The seasonal progression of the overhead sun north and south of the equator brings summers of warmer weather and long days to each hemisphere, while the winter hemisphere with low sun angles has cooler weather and longer nights. Earth's axial tilt

FIGURE 1.6 Internal and external Earth forces mold the landscape. The gash across the Carrizo Plain in southern California caused by the transform San Andreas Fault splitting a section of Earth's crustal rocks. Two plates move against each other with the western side moving northward. Rain and rivers etched out the line of weakness along the fault. Continuing movement along the fault offsets the lines of stream valleys on either side.

causes the sun to be directly overhead at noon at the Tropic of Cancer (Northern Hemisphere summer) between June 19 to 23 and at the Tropic of Capricorn (Southern Hemisphere summer) between December 19 to 23.

Because the sun is more directly overhead for most of the year in tropical regions, its heating impact on the atmosphere there produces higher temperatures (Figure 1.7). Tropical areas have an excess of incoming energy over that which is radiated back to space. The polar regions, however, have a deficit of energy: in winter, they have several months of almost complete darkness, losing energy to space without any coming in. Flows of air and ocean water transfer the heat from the tropics toward the polar regions. Tropical oceans are huge reservoirs of heat, which ocean currents move poleward and heat the atmosphere of temperate and high latitudes. The air and waters cool by releasing heat in higher latitudes and then return as cool flows to the tropics, where they are reheated. This system makes human habitation possible outside the areas of greatest solar radiation.

Air and Water Circulating in the Atmosphere-Ocean System

Oceans are the major sources of water that evaporates into the atmosphere, condenses into clouds, and produces precipitation as rain, hail, or snow. Parts of the world with the highest rainfalls lie near the equator (Figure 1.8), where warm, humid airstreams collide, forcing the air to rise and produce frequent rainstorms. High precipitation totals also occur where moisture-laden winds from the ocean meet tropical islands or the coasts of temperate continents. Mountain ranges in the path of moisture-laden winds (as in Canada and southern Chile) force the air upward, a cloud and rain triggering process known as **orographic** ("mountain-related") **lifting**. Areas between these two rainy zones have less, and usually seasonal, rain.

Earth's rotation affects air and water movements—winds and ocean currents—across the surface. The effect of rotation increases away from the equator toward the poles, bending winds to form circulating weather systems, including cyclones (counterclockwise wind circulation in the Northern Hemisphere, clockwise in the Southern Hemisphere) and anticyclones (clockwise circulation in the Northern Hemisphere, counterclockwise in the Southern Hemisphere).

World Climate Regions

The transfers of energy and moisture and the circulation within the atmosphere produce seasonal changes and weather systems that distinguish climate regions (see foldout map inside back cover). **Tropical climates** experience high temperatures (around 30°C or 80–90°F) throughout the year. They have short winters, if any (no average monthly temperatures below 18°C, 64°F). The main tropical climates have a north-south distribution, from the equatorial climate with rain at all seasons (Ar), through wet-dry seasonal climates (Aw), to very dry, or arid

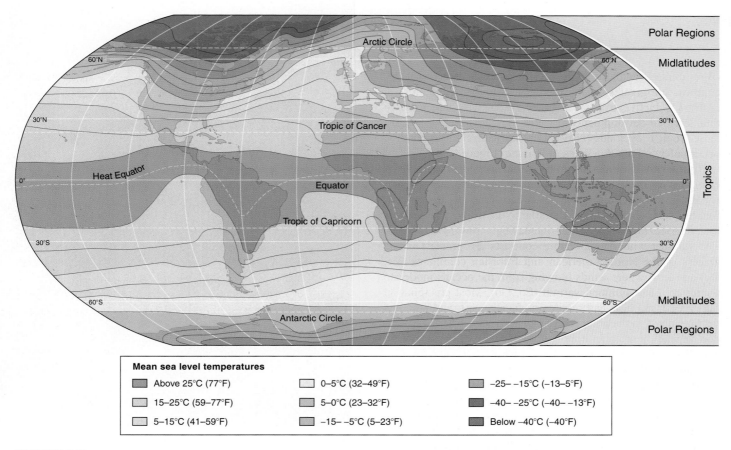

Mean sea level temperatures

▨ Above 25°C (77°F)	☐ 0–5°C (32–49°F)	▨ –25– –15°C (–13–5°F)
☐ 15–25°C (59–77°F)	▨ 5–0°C (23–32°F)	▨ –40– –25°C (–40– –13°F)
☐ 5–15°C (41–59°F)	▨ –15– –5°C (5–23°F)	▨ Below –40°C (–40°F)

FIGURE 1.7 Temperatures at ground level: January. Isotherms (lines joining places of equal temperature) during the Northern Hemisphere winter. The heat equator connects points of highest temperature at each meridian of longitude. Compare the Northern and Southern Hemispheres for the extent of very cold temperatures and the position of the warmest band of temperatures. What effect do the oceans and continents have on the air temperatures above them?

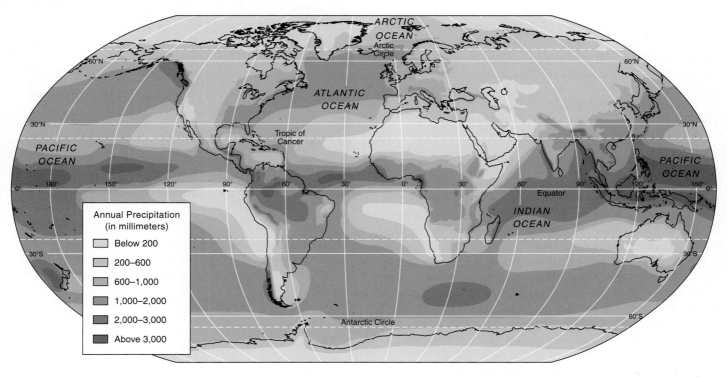

Annual Precipitation (in millimeters)

☐	Below 200
☐	200–600
▨	600–1,000
▨	1,000–2,000
▨	2,000–3,000
▨	Above 3,000

FIGURE 1.8 World precipitation (rain, hail, snow). Locate the main areas of low and high precipitation. Areas with less than 200 mm of precipitation in a year are classed as arid; those with more than 2,000 mm are nearly all in the tropical ocean areas. Warm air (which can hold more moisture than cold air) from the northern and southern tropics converges at the equator, rises, and forms clouds that precipitate rain.

(BW) climates without appreciable rainfall. Seasonal wet-dry contrasts are greatest in the monsoon climates of South Asia (Am). Distinctive tropical weather systems include massive thundercloud developments near the equator (Figure 1.9a). Intense tropical cyclones (also called hurricanes and typhoons) occur poleward of 10 degrees of latitude (Figure 1.9b).

Temperate climates in middle latitudes respond to the sun's changing overhead seasonal position. They have marked seasonal temperature differences between summer and winter. Summer temperatures may be similar to the tropics, but winters include freezing temperatures increasingly toward the poles. The summer-winter temperature contrasts are greatest in the centers of the North American and Eurasian continents (Ec, Dc) and least where winds blowing from the oceans moderate winter cold and summer warmth (Eo, Do). Most precipitation falls on coastal hills and mountains facing west, and amounts decline inland to the point where moisture is insufficient to support vegetation or agriculture. Temperate climates have a west-east distribution, with mild and moist west coasts (Do, Eo, Cs), dry continental interiors (BS, Dc, Ec), and east coasts with summer-winter temperature and rainfall contrasts (Cr, Do, Eo).

Temperate weather systems have circular wind patterns. Frontal cyclone systems with low atmospheric pressures bring rain and high winds. The oceanic temperate climates on the west coasts of North America and Europe benefit from the ocean currents that warm the air above. On the equatorward margins, transitional climates bring long, dry summers and wet, windy winters. They are typical of lands around the Mediterranean Sea and the central Californian coastlands. On eastern coasts air drawn in from the ocean produces summer rains, while cold winter winds from the interior bring frost and snow.

Polar climates are extremely cold throughout the year, seldom rising above freezing temperatures. Winter conditions dominate the Arctic climates (FT, FI), although short summer spells may melt some of the snow and ice. Truly polar climates are frozen all year. Temperate cyclones occasionally invade polar regions, bringing high winds and precipitation.

Global Climate Change

Changes in the path of Earth's solar orbit and the planet's axis angle alter the intensity of solar radiation and are major factors in climate changes over time. A cyclic progression repeats warmer and cooler phases with small changes over shorter periods, but larger over longer periods.

An example of a longer-term change was the intensive freeze that was the latest part of the Pleistocene Ice Age. It lasted for most of the last 100,000 years, ending around 10,000 years ago with a period of warming and ice melting that led to a sea surface rise of around 100 m (300 ft.) to its present level. During the freeze, huge ice sheets dominated the northern parts of North America and Europe and sea levels were lowered around the world.

After the ice cover retreated, smaller fluctuations of climate brought the warmest conditions around 5,000 years ago. The "Little Ice Age," from approximately AD 1430 to 1850, caused

(a)

(b)

FIGURE 1.9 Tropical thunderstorms and hurricanes.
(a) Thunderstorms over Africa, seen from a space shuttle. Several cloud tops amalgamate in fibrous masses of ice that spread outward and downwind. (b) Hurricane Katrina over the Gulf of Mexico, August 28, 2005. The hurricane passed over Florida (on the right) and narrowly missed Cuba and Mexico (foreground) before turning north toward the Mississippi river delta and New Orleans. Several hundred kilometers across, this hurricane shows the cloud-free eye of descending air in the center, surrounded by swirling clouds affected by strong winds. Once over the Gulf of Mexico, it picked up energy from the warm surface waters, intensifying the winds and surge of ocean water.

upland glaciers to advance several kilometers down valleys and cultivation to retreat from higher areas in temperate countries. Climatic warming since the early 1800s resulted in a reversal of those trends, with glacier melt and higher temperatures. This phase coincided with the industrialization spreading from Europe and North America, which releases greenhouse gases that reinforce the natural warming.

Earth Surface Environments

Earth's surface is 71 percent covered by oceans and only 29 percent occupied by the continents and islands on which people live. Weather and the action of the sea interact with Earth's internal continent-building forces to produce differences in the height and shape of the land, or its **relief**, in features such as mountains, valleys, and plains.

Once land emerges above sea level, atmospheric processes and ocean waves etch the details of surface relief. The temperature and chemical composition of the atmosphere, together with water from rain and snowmelt, react with the rocks exposed at the surface and dissolve them or break them into fragments. Such changes are called **weathering**. The broken and dissolved rock material forms the mineral basis for soil. On steep slopes, weathered material moves downhill under the influence of gravity. Such movement may be rapid in slides, flows of mud, or avalanches, or slow in local heaving and downslope creep of the surface. The mobile fraction of this broken rock material often enters rivers or glaciers and is moved toward the ocean.

The concentrated flows of water or ice and rock particles in rivers and glaciers gouge valleys in the rocks—a process called **erosion**. Glaciers formed of ice move slowly, helped by meltwater lubrication in the summer. When the flows reach a lowland, lake, or ocean, the rock particles drop to the valley, lake, or ocean floor in the process of **deposition**. Wind blows fine (dust-size) rock particles long distances, while sand-sized particles are moved around deserts or across beaches to form dunes. Along the coast, sea waves and tides fashion eroded cliffs and deposit beach features, often moving the rock particles supplied by rivers, glaciers, or wind (Figure 1.10).

In a particular region, the relief features of the land surface are determined by the combination of these internal and external forces, together with modifications added by human occupation (Figure 1.11). The landscape outcome depends on whether the region contains a plate margin, the nature of the rocks, which climatic elements fashioned the surface, how long the natural forces operated without catastrophic changes, and the extent to which people affected the surface processes.

Ecosystem Environments

An **ecosystem** is the total environment of plants and animals, which live in communities sharing the physical characteristics of solar energy, water availability, and soil nutrient chemicals.

FIGURE 1.10 Eroding the land. Ocean waves attack the cliffs and form beaches in Oregon. Inland, valleys are formed by a combination of river action and slope processes.

FIGURE 1.11 Landscape evolution in western Scotland. The rocks in northwestern Britain were formed over 400 million years ago. They were then crushed and raised into mountain ranges by colliding plates as an earlier Atlantic Ocean closed. Rivers wore down the mountains, exposing ancient and highly altered rocks. The Atlantic Ocean opened again beginning some 50 million years ago, raising this area with others along its margins. Within the last million years, the Ice Age covered the higher parts with thick ice that moved outward, carving deep valleys such as the one in this photo. On melting, the ice poured water back into the ocean, raising its level and drowning the lower parts of this valley as a sea loch. Rivers eroded the hills, depositing deltas along the edge of the deep water. People altered the rates of natural processes by clearing woodland or erecting buildings.

Most plants produce the food that animals require by capturing and storing the sun's energy in chemical form. Ecosystems exist at all geographic scales but for the purposes of this text are discussed in relation to the largest scale, or **biome**.

Forest biomes exist in climatic environments where sufficient water is available. Tropical types range from the dense rain forest of equatorial regions characterized by the crowding of great numbers of plant and animal species, to more open forests with fewer species in areas having marked dry seasons. Temperate forests include the deciduous forests, typical of the eastern United States, in which the leaves are shed in winter, and the evergreen forests typical of sandy soils, high latitudes in Canada and Russia and mountain slopes in western North America.

Grassland biomes occur where longer dry seasons or annual burning by humans restrict tree growth. In the tropics, savanna grasslands are characteristic of most of the seasonal climatic regions, as in Africa with the associated wide range of large animals from lions to elephants (Figure 1.12). Temperate grasslands once occurred in the prairies of North America, the steppes of southern Russia, the pampas of Argentina, and the veldt of Southern Africa. These environments are now dominated by grain farming and grazing.

With its low shrub vegetation, mountain and Arctic **tundras** occur where long cold seasons and strong winds prevent tree growth. They do not qualify as grassland but have some similar characteristics of that form and often support grazing animals.

Desert biomes occur in lands dominated by aridity, where evaporation exceeds water supply, supporting little or no vegetation. Any plants or animals that survive in such conditions have special means for storing water, as in the cactus, or short-flowering, seed-producing regimes.

Polar biomes, where ice cover combines with low inputs of solar energy, provide little sustenance for living organisms. Offshore marine resources around frozen coasts support a variety of swimming and flying animals.

Ocean biomes are differentiated by water temperatures and nutrient availabilities that produce zones of more or less abundant life. Microscopic plankton plants take advantage of sunlight in the surface waters and form the basis of living systems in the oceans. Smaller numbers of a wide range of different species occur in the tropics and greater numbers of fewer species in temperate oceans. The richest biomes with the most profuse numbers of organisms occur where rising cold water brings nutrients to the surface, as along the west coasts of Africa and North and South America.

Soils are types of ecosystems, which form as broken rock matter interacts with weather, plants, and animals. The rock materials supply or withhold nutrients. Water from rain and snowmelt makes nutrients present available to plants; decaying plant and animal matter releases the nutrients back to the soil in mineral form. Soil fertility, based on nutrient content and structure, together with climatic heat and moisture conditions, governs whether a region will produce good crops, support livestock farming, or have little farming potential.

Human Impacts on Natural Environments

The natural systems of physical geography interact with human activities. Throughout their existence, but increasingly over time and the history of technological developments, people changed from having their actions determined by nature to seeing their role as conquering nature. Human attitudes toward nature vary from the pragmatic utilization of natural resources to the romantic desire to return to an unspecified Eden.

Farming and Mining

The first farmers settled the lighter soils where there was not much vegetation to clear. From around 1000 BC, new iron implements made it possible to fell trees on a larger scale and extend farmland into heavy clay soil areas. Phases of woodland clearance increased soil erosion in uplands and deposition in lowlands, together with changes in the species composition of animals and plants. Drainage of marshes and protection of coastal lands also altered ecosystems. Human stripping of vegetation cover or soils continues to destroy the productive capacity of land. Environmental contamination, as when wastes from mining foul the land or nuclear or other toxic wastes leak into soil and water, makes land unusable for many years.

FIGURE 1.12 World biome types: savanna grassland. Tropical savanna grassland, Eastern Africa. The grasses grow in areas of moderate rainfall, supporting large numbers of grazing animals, such as zebras and wildebeest. These become food for lions and other carnivores at the top of the food chain.

Industry and Global Warming

From the 1800s, as factories poured their wastes into the atmosphere and rivers, environmental stresses built up rapidly over large areas of Europe and eastern North America. The carbon gases emitted by burning fossil fuels such as coal, oil, and natural gas exceeded the amounts that natural systems could absorb, and the carbon dioxide content of the atmosphere rose.

The present concern about **global warming** is that the carbon gases humans pour into the atmosphere from burning those fuels are adding to the gases that trap solar energy in the lower atmosphere, enhancing the greenhouse effect. Today emissions from burning fossil fuels are increasing in newly industrializing countries such as China and India. Rising temperatures melt ice sheet margins. As the meltwater pours into the ocean, its level is forecast to rise a meter or so within the next 50 years. The low-lying coasts of coral islands, wetlands, port areas, and millions of people will be at risk. This is by far the most important global environmental issue today and will increasingly affect people's lives around the world.

Increasingly confident scientific evidence for the future impact of human-controlled emissions on global atmospheric temperatures raised international concerns, expressed in conferences at Rio de Janeiro, Brazil, in 1992 and at Kyoto, Japan, in 1997. The **Kyoto Protocol** to the United Nations Convention on Climate Change addressed targets for reducing emissions of the six primary greenhouse gases: carbon dioxide, methane, nitrous oxide, hydrofluorocarbons, perfluorocarbons, and sulfur hexafluoride. By the end of 2004, enough countries had ratified this treaty for it to become UN policy by February 2005. The United States federal government persists in not ratifying this treaty, but some U.S. states and cities are adopting green agendas.

Although there is agreement about the future dangers of global warming, there are few adequate forecasts of its precise impacts on specific regions (apart from near coasts). Policies to reduce emissions are being slowly adopted by many countries that signed the Kyoto Protocol. However, greater use of natural gas (with lower carbon emissions) to generate electricity, the promotion of "greener" fuels for vehicles, and the use of more efficient engines for road vehicles and jet airliners scarcely offset the impact of our increasing demands for electricity and travel.

Ozone Depletion

The growth of the "ozone hole" over Antarctica poses another global problem. Ozone is a gas that forms a protective shield in the upper atmosphere, intercepting incoming ultraviolet rays from the sun. In the 1980s, it became clear that chlorine gases, including the human-made chlorofluorocarbons used in refrigeration systems, gradually permeate upward and destroy ozone by chemical reactions. The reactions are most intense during the polar winters over Antarctica, forming a hole in the ozone layer. The loss of the protective ozone shield increases the risk of skin cancer. Governments agreed to end the use of ozone-depleting gases, and the policy is reducing ozone destruction, although the potential dangers will last for several decades. People living in southern Chile and Argentina, Australia, and New Zealand take steps to mitigate the dangers, such as wearing hats and clothing cover in their summer months.

Acid Deposition

While global warming and ozone depletion may have worldwide effects, acid deposition affects areas up to several hundred kilometers downwind of major urban-industrial areas. It is mainly caused by sulfur and nitrogen gases from power stations and vehicle exhausts, which react with sunlight in the atmosphere and return to the ground as acids. Soils and lakes that are close to the pollution sources and are already somewhat low in plant nutrients suffer first. The phenomenon is a growing menace downwind of new industrial areas in developing countries.

Resources and Hazards

Natural resources and hazards are the "good" and "bad" of human interactions with natural environments. Both are distributed unevenly around the world.

Natural resources are naturally occurring materials that human societies use to maintain their living systems and built environment. They include fertile soils, water, and minerals in the rocks. However, resources valuable to one society or technology are not always rated highly by later groups. For example, Stone Age peoples used flint and other hard rocks that flaked with sharp edges to make tools and weapons, but such rocks have few uses today. The clay mineral bauxite was ignored until it was found that refining it produced the strong, lightweight metal aluminum. Among energy resources, emphasis shifted over time from wood to wind, running water, coal, oil, natural gas, and nuclear fuels.

Renewable resources are naturally replenished. Solar energy provides a constant stream of light and heat to Earth. Water is a renewable resource that is recycled from ocean to atmosphere and back to the ground and oceans. All renewable resources are, however, ultimately finite in quantity and quality or limited by human ability to exploit them. For example, the limits of water supply affect irrigation-based development in arid countries.

Nonrenewable resources are used up after extraction. They include the fuels and metallic minerals available in rocks. Technological advances or new and increased demands, however, assist in finding new sources, extracting sources that were once thought to be uneconomic, and recycling metal products. Such technologies extend the life spans of nonrenewable resources.

The natural environment poses difficulties and challenges for human settlement in zones of **natural hazards** such as volcanic eruptions, earthquakes, hurricanes and other storms, mudslides, river and coastal floods, and coastal erosion. Hazards interrupt human activities to a greater or lesser extent; they seldom deter humans from settling or developing a region if its resources are attractive. For example, people are drawn to living in California or the major cities of Japan despite the likely occurrence of earthquakes. Similarly, people continue to live and work in areas prone to hurricane damage or river flooding. In areas such as the Mississippi River valley in the United States and the lower Rhine River valley in the Netherlands, protective walls are designed to cope with all but the worst river floods. Extreme weather events may produce levels of flooding that overtop these walls, as in 1995 along the lower Rhine and in 2005 in New Orleans.

Hazards cause loss of life and destruction of property, but the costs of protection against hazards are also high and become technically, financially, or politically unfeasible in the face of the most devastating hazards. Most protection is provided in wealthier countries, where hazards cause the least loss of life but the greatest damage to property. Many poorer countries have few resources available to construct protective measures against natural hazards and often suffer major losses of life after floods, hurricanes, or earthquakes. For example, the few people killed when hurricanes strike the United States contrast with the high number killed in the earthquake and resultant tsunami that devastated coastal areas of Southeast Asia and around the Indian Ocean in December 2004.

Regions and Cultural Geography

One of the ways geographers identify and examine world regions is by analyzing the elements of culture present in the landscape and lives of people, including language, religion, community and social norms, and popular culture (among many others). Many of these elements emerge through interactions of people with their natural environments over time. Historical events such as wars and movements of people may lead to the mixing of cultures, to the overlaying of several cultural patterns in the same place, or to territorial claims and rivalries.

Cultural geography is the study of spatial variations in cultural features. The culture of a group of people is basic to how they create and recreate the regional distinctiveness that in turn affects their lives and those of their descendants. People often explain their behavior and find comfort and security through identification with a group who share common cultural characteristics.

The most important thing to understand about culture is that it is learned behavior. A combination of traditions and behavior practices, as in language or religion, is transmitted from generation to generation along with adaptations, variations, and new ideas or innovations more recently acquired or accepted by groups of people. Such cultural history provides an important understanding of the human geography of each world region.

Language

A **language** is a means of communication among people, including speech, writing, and signing. It is an important factor in geographic diversity. Languages grow out of historic experiences and traditions and often provide a shared identity for a cultural group. Some groups make a point of using their language to enhance their identity. For example, the French-speaking Québecois people in Québec, Canada, increased their focus on language when they wanted to achieve greater political recognition.

Thousands of languages are spoken around the world, many by small groups of people in isolated environments such as South America's Amazon River basin. Twelve dominant languages each have over 100 million people speaking them. Six of these (English, French, Spanish, Russian, Arabic, and Mandarin Chinese) are official languages of the United Nations, mostly, apart from Arabic, chosen from the victorious allies at the end of World War II.

Related languages can be grouped in families (Figure 1.13). For example, the Indo-European family includes most of the languages of South Asia (e.g., Hindi), the Slavic languages of eastern Europe and Russia (e.g., Polish and Russian), and the languages of southern, western, and northern Europe. Some of these languages, especially Spanish, Portuguese, English, French, and Dutch, disseminated with European colonization from around 1450, dominating the Americas and becoming important in many parts of Africa and Asia.

Globalization leads to the strengthening of some languages and the marginalization and extinction of others. English is increasingly used in communications among scientists throughout the world, in international air traffic control, in computer software, and in the entertainment industry where films and TV programs in English dominate. It remains a common language in the former colonies of Britain where there are rivalries among local languages.

Religion

Religions also generate geographic cultural variations through strong group loyalties and exclusive attitudes. Each **religion** is an organized set of practices that professes to explain our existence and purpose on Earth. Some also project a system of values and faith in and worship of a divine being or beings. Regional emphases, such as social and legal practices, and visual features, such as building designs, often reflect religious allegiance (Figure 1.14). Religions play a significant role in transferring cultural values and practices from one generation to the next.

Major World Religions

The religions claiming the largest numbers of adherents are Christianity, Islam, Hinduism, and Buddhism; each has around one-fifth of the world's population. Judaism has a smaller number of adherents but an important place among world religions. Christianity, Islam, and Buddhism are religions that can be joined by anyone in any country and actively seek to extend their membership: they are **universalizing**, or global, **religions**. Hinduism and Judaism are considered to be a matter of birth by adherents, closely tied to family and region: they are **ethnic religions**. People practicing ethnic religions do not actively seek converts because they see their religions as only appropriate for

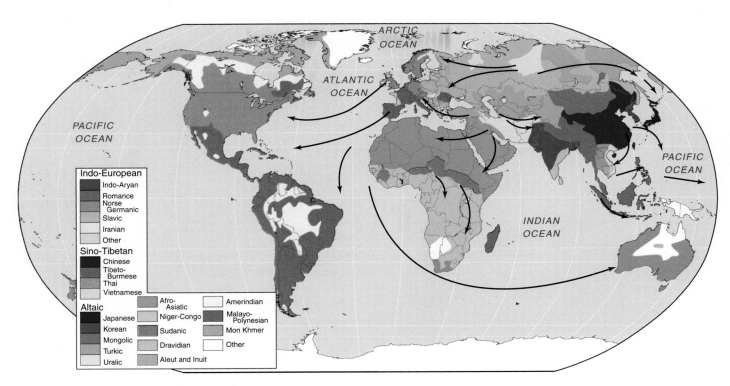

FIGURE 1.13 Cultural geography: world language families. The arrows show some diffusion routes. The map records the majority languages; there are often many different languages, including those of minority groups, in each area.

their own ethnic groups. Although India and Israel are officially secular countries, there are movements to proclaim India as a Hindu country, while religious Jews seek to extend their control in Israel. Buddhism, often combined with older local religious elements, is the main religion of East Asia, while Hinduism is the predominant religion of South Asia.

Major religions are often associated with specific world regions, although many spread out from their earlier centers. Internal splits are also geographically significant. In Christianity, for example, the rift between Roman Catholicism and the Orthodox churches set apart the cultures of western and eastern Europe. Protestantism grew out of Roman Catholicism in northern Europe after 1500. Overseas explorations from the 1400s led to Christian missionary expansion outside Europe. Today, some 60 percent of world Christians are Roman Catholics, 25 percent belong to Protestant groups, and 10 percent are Eastern Orthodox. Islam spread out from Arabia from the AD 600s, reaching northern Africa, Central Asia, South and Southeast Asia, and the southern margins of Europe. Today its adherents are found worldwide.

Religion and Society

Religious adherence often determines responses to issues affecting society. Many Roman Catholic and Muslim leaders, for example, officially resist policies aimed at reducing births—although individual families increasingly make their own choices. Most religions profess a care for the natural environment, although the practice of societies has often not matched

such statements. Religious adherence also motivates believers to advocate certain social, economic, and political policies.

Religious differences among peoples may result in strong and conflicting views being held over different social, economic, or political ends. Such conflicts often arise from disregard (or ignorance) of the views and needs of other groups and from exclusive claims that are fanned for their own ends by those with political power. Groups labeled Christians or Jews, Muslims or Hindus, Christians or Muslims fight each other. Conflicts extend to Catholic against Orthodox (former Yugoslavia) or Protestant (Northern Ireland) Christians and to Shia against Sunni Muslims. Saddam Hussein in Iraq favored the Sunni Muslims and the areas where they live against the dominant Shia Muslims; today the Shias dominate the Iraqi government.

It is likely that religion will continue to influence culture, perhaps in new ways. In the United States, religious groups increasingly get involved in politics, arguing that the country is getting less religious and needs such interventions. From the 1990s, the resurgence of Christian, Hindu, Jewish, Islamic, and Buddhist extremists in some countries is one sign that religious motivations are again significant after the Cold War period of largely political conflicts.

Cultural Status: Race, Class, and Gender

Personal or group status—and the ability to effect or suppress changes—varies greatly throughout world regions. In many cultures, status is inherited through the family, as in the caste

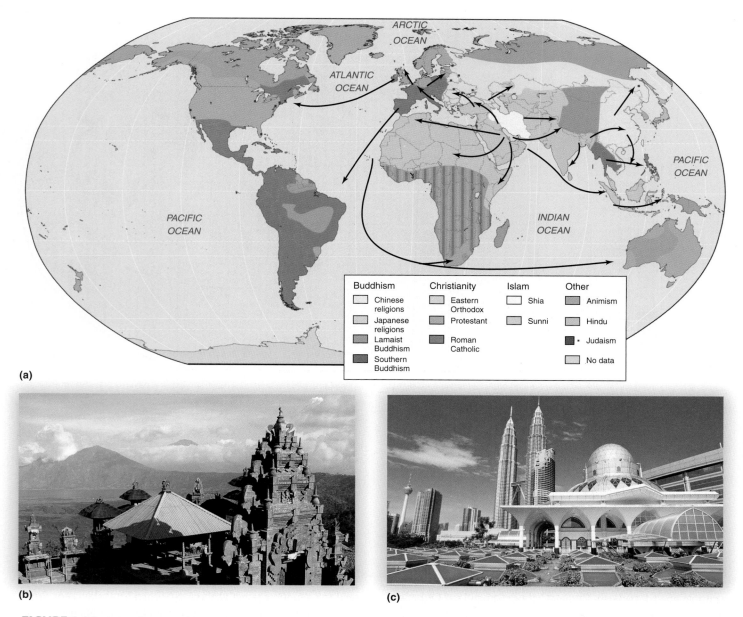

(a)

(b)

(c)

FIGURE 1.14 Cultural geography: world religions. (a) The map shows the geographical dominance of Christianity in its various forms, Islam, Hinduism, and Buddhism. These also have the highest numbers of adherents. The Jewish religion is spread around the world, mainly in cities. Older and local religious practices remain in localized areas. Many places have mixtures of allegiances to different religions – sometimes working together, other times in conflict. (b) and (c) Religious and secular symbols: (b) Hindu temple, Bali, Indonesia; (c) Asy-Syakirin mosque with the Petronas Towers, Kuala Lumpur, Malaysia.

groups of India or the royal families and linked aristocracies of Europe. In other cultures, status may be achieved by creating wealth, political allegiance, media prominence, or sporting performance. Throughout the world, race, class, and gender influence access to position in society; they vary geographically and are particularly significant in local contexts.

Race and Ethnicity

An **ethnic group** is a cultural group. Its members identify a common origin (real or imaginary) and are set apart by religion, language, national origin, or physical attributes. Such groups may be called tribes, clans, or segregated minorities.

The concept of **race** is assumed by many to be based on essential biological differences, but characteristics such as skin color, eye shape, or hair type vary as much within identified "races" as between them. Although race and racism are often at the center of human conflicts, the most basic human biological features, DNA and blood type, demonstrate little variation across the human species, which is a single reproducing group. However, cultural status is often given to bodily features as a means of defining what are essentially ethnic "them-us" differences.

Examples of cultural racism include South Africa, where the minority white population operated a supremacist apartheid policy until 1994, separating whites, blacks, and other "colored" peoples into segregated neighborhoods. The country is now reversing this policy (see Chapter 8). In the United States, most African Americans, a minority ethnic group identified mostly by skin color, still struggle against discrimination from people of European origin. Even in Brazil, where Native American, European, African, and Asian peoples mix freely, most wealthy people are of European origin and most people of African or Native American origin are poor.

In many countries, ethnicity decides the membership of opposing political groups and may lead to armed conflict. The promotion of ethnic groups as "superior nations" had much to do with World War II holocausts, ethnic cleansings, and "shoot-to-kill" mentalities in Europe and Asia. Before that, ethnic, or tribal, differences were often highlighted and exploited by colonial powers, playing one group off against another. After the independence of former colonial countries in Africa, Asia, and Latin America, however, many new rulers brought in one-party rule and often increased the perceived differences among ethnic groups by rewarding those in their own ethnic group and penalizing others.

Class Distinctions

Class in society is imposed through combinations of religious, economic, and social criteria. Wealth, education, and perceived birthright are common bases for class distinctions in all countries. In the United Kingdom, the royal family is linked to a hierarchy of hereditary dukes, earls, and knights, with commoners as "other." In India, the caste system and, increasingly, education define position in society. In Communist countries, the expectation of a classless society is contradicted by the status and privileges given to members of the Communist Party. Most Americans identify general class differences on the basis of material possessions and appearances: a lower-middle class of factory and shop workers, an upper-middle class of managers and professionals, and a group of exceptionally wealthy financiers, property owners, and sports and media stars.

Gender Inequalities

Gender—the cultural implications of one's sex—also highlights differences and inequalities of opportunity within and among societies. Males dominate most societies and have a history of denying full rights to women. Although major changes occurred in the 1900s, particularly in extending voting franchises to women, some countries still deny women the human rights defined by the United Nations. There may be as many as 100 million fewer women than men in the world due in large part to cultural preferences and practices favoring male children.

In the late 1990s, the Taliban rule in Afghanistan became notorious for taking women out of education, confining them to the home, and selling some of them into slavery. In Africa, female genital mutilation, or "cutting," continues in some societies. In Iran and northern Nigeria, women—but not men—can be stoned to death for adultery. Illiteracy among women is still much higher than among men in most of Africa and Asia. Few European countries allowed women to vote in elections until the 1900s, with Switzerland delaying it until 1971 and one of its cantons until 1990.

Women, even in the world's wealthier countries, commonly receive lower wages than men for the same job and constitute a minority of doctors, engineers, corporate executives, and elected politicians. Some jobs such as nursing, secretarial work, elementary school teaching, and sales clerking have been widely regarded as "women's work" and often have lower status. Further, women commonly take a major role in home management and child care, which often affects their career prospects. In some cases, marriage breakups leave women at a financial and social disadvantage.

Some aspects of women's inequality are gradually being tackled. The high rates of female illiteracy in poorer countries have major implications for future population-resource issues. Better education gives women confidence, enables them to take jobs, improves their self-esteem, and increases their roles in decision-making. It also impacts such matters as how many children are wanted in a family, since their education is perceived as vital but expensive. In many places, male attitudes and cultural traditions still prevent women from controlling how many children they have. It is clear that many women prefer smaller families, but cultural beliefs and fears raised in earlier periods of higher infant and child mortality still encourage families to have more children.

Despite their low status in most societies, women now play major roles in the expanding world economy. The rising number of women millionaires in the West in 2005 was a consequence of better opportunities, often self-made, in business services. The proportion of women in manufacturing jobs in the United States and Canada grew. Worldwide, female employment in the electronics industry expanded, but men were laid off when "male" jobs were taken over by machine tools. In materially poor countries, women comprise around 80 percent of the labor force in export-oriented electronics, apparel, and textile industries. Expanding numbers of jobs in services such as retailing, health care, teaching, banking, and tourism further increased female employment.

The materially wealthier countries, including the industrializing countries of East Asia, experienced greater household prosperity through dual incomes as females entered the paid labor force. In Russia, however, the wide involvement of female labor under the former Soviet Union's Communist regime gave way in the 1990s to a greater focus on male employment.

Cultures and Regions

Cultural differences bring character and identity to the world's major regions, countries, and smaller regions within them. Many cultural differences can be traced back to early human history and the development of agricultural technology, religion, and language in small regions with specific environmental conditions. Regions that are termed **culture hearths** were centers from which techniques, useful materials, and social norms diffused to other regions. Mesopotamia, the Indus River valley, north-central China, and central Mexico are examples of major cultural hearths.

Current world geography includes the imprint of many newer cultural overlays on older ones. Regions are constantly being recreated. For example, remnants of the Maya and Aztec empires, as well as of Spanish colonization, still exist in Mexico. The Mayan remains in Yucatán are being revealed from beneath the forest, while modern urban-industrial buildings are replacing Aztec and Spanish relics around Mexico City. In Southwest Asia, the Muslim culture removed most non-Muslim relics, but some, such as the ruins of medieval Crusader castles, still exist. In Communist China, the palaces of former imperial dignitaries occupy large areas of Beijing, while the new cities of the eastern coastlands impose modern urban-industrial features on the rural landscape that evolved over centuries.

Just as culture hearths identify areas that had formative influences on cultures in wider areas, **culture fault lines** exist between cultural regions at different geographic scales. Tensions along these lines may lead to conflict. One prominent cultural fault line lies between the Muslim and Christian cultural regions from southern Europe through southern Russia. Although the current disputes are not merely religious, the antagonisms between Muslim and Christian cultures along a line through Bosnia, Kosovo, Macedonia, Bulgaria, Armenia, and the Russian republics such as Chechnya bordering on the Caucasus Mountains have led to conflicts since the 1990s.

Regions and People

People are central to geography. They create, live in the context of, and reshape geographic regions. Population dynamics such as growth, settlement patterns, and migration account for major differences among regions. In late 1999, the world population passed 6 billion, rising at nearly 80 million people per year; by 2006 it reached 6.5 billion (Figure 1.15). After 50 years of rapid population growth, the rate is slowing, but the world's total population will continue to expand and put pressures on resources. When population growth is greater than economic productivity, it challenges the ability to improve education, children's rights, and the roles of women in society.

We all have choices over the future of population growth and the conservation of resources. Do we belong to the "bigger pie" school, which proposes expanding production through applying technology in such areas as genetically modified foods, additives, plastics, synthetics, and alternative fuels? Or to the "fewer forks" school, which emphasizes environmental protection and the slowing, stopping, or reversing of population growth? Or to the "better manners" school, which highlights cultural values as a source of improving the terms on which people live in com-

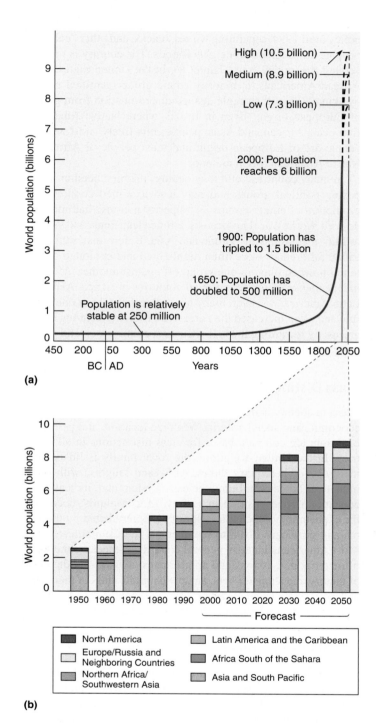

(a)

(b)

FIGURE 1.15 World population growth. (a) For most of the human occupation of Earth, population growth was slow compared to the last 300 years. The total world population took 1,300 years to double from 250 to 500 million, then doubled again in 200 years. In the 1900s, world population quadrupled. The high, medium, and low projections are from the United Nations. (b) When the population increase is broken down by region, all major regions outside North America and Europe show big increases. *Sources: (a) United Nations; (b) Data from* The Economist, *1999.*

munities? Our lifestyle choices concerning transportation, food, and clothing, among many others, will impact future patterns of human and physical geography.

Population Distribution and Dynamics

Where People Are

Population densities—the numbers of people per given area (e.g., square kilometer or square mile)—vary greatly around the world (see Figure 1.1). Density is a common measure of **population distribution** because it takes into account that each country is a different size. The highest densities developed in fertile lowlands over the centuries and the lowest densities in physically challenging regions such as deserts and mountains. Urban areas today expand over good farmland and along coastal areas.

Urban Growth

One of the most basic geographic differences is between **urban** and **rural areas**. At present, half the world's people are concentrated with very high densities in expanding urban areas. In addition to high population densities, urban areas are marked by such characteristics as large proportions of offices, housing, and factories, a concentration of economic activities, and political boundaries of legally incorporated places.

Although the materially wealthy countries have over 70 percent urbanites, it is estimated that 97 percent of world urban population growth from 2000 to 2050 will be in the larger poorer countries, increasing the contrasts among groups of people and regions. Thirty-five percent of the growth is expected to be in China and India—already the world's most populous countries; Pakistan, Indonesia, Nigeria, Brazil, Bangladesh, Mexico, the Philippines, and the United States (the only materially wealthy country in the list) are likely to contribute another 25 percent.

The largest cities grow larger. Estimates for 2015 of total metropolitan area populations over 20 million raise metropolitan Tokyo to 29 million, followed by Mumbai (Bombay, India, 27 million), Lagos (Nigeria, 24 million), and Shanghai (China, 23 million). Jakarta (Indonesia), Karachi (Pakistan), and São Paulo (Brazil) may each have 21 million. By comparison, New York's population, over 18 million in 2003, is increasing slowly and is unlikely to reach 20 million. In each regional chapter there is a listing of major city populations and their projected growth. Where a country has a single very large city, often several times the population of its other cities, it is known as a **primate city**.

Rural Poverty

Rural areas still contain most of the populations of materially poorer countries—around 70 percent in India, for example. Many rural areas are ruled by traditional regimes that resist change and are often dominated by low-wage jobs in farming, mining, and basic manufacturing. They do not have good access to educational, health care and other services. However, distinctions between urban and rural are becoming less clear in many countries as new industries and housing are built outside the established urban limits, as in eastern and southern China.

Population Ups and Downs

Growth or decline in the population of an area are indicators of the overall state of population: whether births exceed deaths (natural change); whether immigration exceeds emigration; and the balance between natural change and migration. **Demography** is the study of population structure and change.

Natural change elements include:

- The crude **birth rate**, the number of live births per 1,000 habitants per year in a given population. It is related closely to the **total fertility rate**—the average number of births per woman in her lifetime. Total fertility rates of 6 to 7 are typical of many poorer countries, while wealthier countries have rates of 2 or below.
- The crude **death rate**, the number of deaths per 1,000 inhabitants per year in a given population. It is often broken down into age groups. **Infant mortality** (deaths per 1,000 live births in the first year of life) and child mortality (deaths per 1,000 live births in the first five years of life) are examples. Infant mortality rates below 10 (i.e., 10 infant deaths per 1,000 live births) in wealthier countries compare with those above 100 in many poorer countries. The crude birth and death rates are labeled "crude" because they do not account for age and gender structure of a population.
- The crude birth rate minus the crude death rate equals the rate of natural population increase or decrease.

The Migration Factor

Migration is the long-term movement of people into or out of an area. When immigration to a country or region exceeds emigration, and is not matched by natural decrease, population numbers increase. Immigration is a major source of population growth in the United States and Europe.

Major people migrations have peaks and troughs. As trade and European colonization spread in the late 1800s, and as refugees from political persecution could escape to new lands, millions of people moved mainly from Europe to the Americas and Australia. Flows slowed in the period between 1930 and 1960, after which new migrations occurred.

By 2002, 150 million people worldwide lived outside their country of birth, making up just 2.5 percent of the world population but often playing important roles in their host and home countries. For example, low and moderately skilled workers moved from South and Southeast Asia to work in the Persian Gulf oil countries and Europe, sending their paychecks home. Such sources may provide a high proportion of national income in some poorer countries. Remittances from abroad make up over half the national income of Haiti. At the same time, large numbers of refugees from war-torn countries in Africa move into neighboring countries, often leading to new tensions and problems of accommodating these migrants.

Migrants to urban centers such as London, Paris, Washington, D.C., and Toronto are dramatically changing the human geography of these areas. Educated men and women from both materially wealthy and poor countries around the world move to the cities of the United States, Europe, and Japan for better-paying jobs.

The specific reasons for migration and the policies adopted by receiving countries vary and change over time. For example in the 1990s, materially wealthier countries that previously offered permanent residence and full citizenship rights to immigrants began to examine more closely the increasing numbers of migrants and asylum claimants. Conflicts and repression in Afghanistan, Iraq, and many African countries bring asylum-seekers to Europe and the United States. However, internal political problems arise from the difficulties of integrating immigrant groups with established communities, causing some countries to reduce official intake numbers and offer only limited-term work permits. In a contrary trend, aging populations and declining numbers of workers in European countries lead to demands for more immigrants to fill jobs.

Overall Population Change

When natural change is considered with migration changes, an overall population growth (due to natural change or migration alone or the two together) of 1 percent will lead to a **population doubling time** of 70 years. An increase of 2 percent means a doubling in 35 years; one of 3 percent means a doubling in 23 years. Wealthier countries today commonly have below 0.5 percent population increase, while poorer countries have rates of 2 to 3 percent. Countries with high emigration, low birth rates, or high death rates may experience population losses.

The composition and recent history of a country's population characteristics are often summarized in an age-sex diagram, also termed a "population pyramid" (Figure 1.16). Migrations into the country or baby booms show up as expansions in particular age and gender groups; deaths in major wars may be reflected in a narrowing of specific cohorts. Long life expectancy results in larger groups of older aged people, who impose social stresses on their societies through health care and housing demands, but also provide an economic boost when they have good pensions.

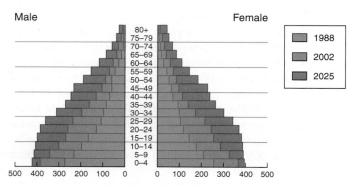

(a) Papua New Guinea

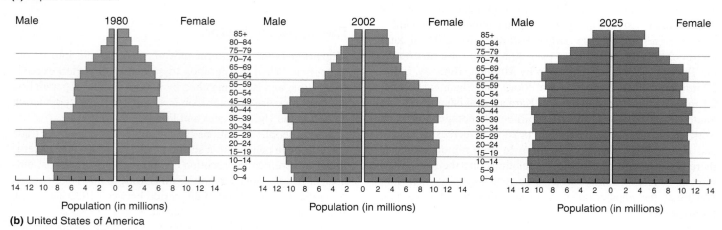

(b) United States of America

FIGURE 1.16 Age-sex diagrams (population pyramids). Diagrams for three years are overlaid or set side-by-side to show changes. In each case, the bars represent a five-year age group (male and female). Total numbers of people in each group are preferred here to percentages of the total population to allow comparisons of places. (a) Papua New Guinea, a materially poor country with large numbers of young people and fewer old; middle-age groups will increase by 2025. The progressive increases allow the three years to be superimposed. (b) The United States, a typical wealthier country with a more even spread of numbers in each age group. The baby boom of 1950-1965 is seen to move upward over time. The three years are separated for clarity. These diagrams can be found in each of Chapters 2 through 10. *Source: U.S. Census Bureau, International Data Bank.*

Regions and Politics

Political geography is the study of how governments and political movements (e.g., nongovernmental organizations, labor unions, political parties) influence world regional geography.

Countries Still Dominate

Self-governing **countries** are the basic political units: within their borders, country governments are assumed to have total political control, or sovereignty, over the country's inhabitants. The numbers of self-governing countries increased from 62 in 1914 to 74 in 1946 and 193 in 1998. However, not all units shown on maps are recognized as independent countries. A few are colonies, while others, such as Taiwan, which the People's Republic of China claims as its territory, lack recognition from other countries. In world regional geography, countries provide the main subunits of study and the basis of the data tables within each world region.

Countries tax their citizens to provide public services, including military capabilities and economic and social welfare. During the 1900s, the average proportion of a country's wealth taken and used by its government increased from under 10 to over 40 percent and continues to rise. Countries often have systems of regional, state, or local government that carry out some of the governmental responsibilities at different geographic levels. Country governments promote and protect their peoples in world affairs and may join other country governments in mutual trading or defense agreements. However, some aspects of country sovereignty are now contested by globalizing trends.

Nations and Nationalism

A country is rarely, if ever, the same as a nation. A **nation** is an "imagined community," or politicized ethnic group of people, who believe themselves to share common cultural features, often associated with a specific area of land. The cultural features may be language, religion, or other characteristics such as a shared history. In the country known as the United Kingdom, the English, Scottish, Welsh, and Northern Irish consider themselves distinctive nations. Each fosters separatist groups, and they often enter separate teams in world sporting events. The Scots, Welsh, and Northern Irish—but not the English—now have separate home parliaments with some limited powers, although their people still elect members of the United Kingdom's parliament. Thus, a nation may not have sovereign government control over a country.

Nationalism is the desire of a people to have their own self-governing country. The idea was basic to the formation of countries in Europe since around 1800. Increased levels of communication brought about by technical advances bolstered nationalism through printed books and newspapers, the telegraph, telecommunications, radio, and television. Education provided by governments supported nationalist themes through selective views of history glorifying the national experience. The idea

of a "nation-state" emerged—with every nation having its own state. The term "nation-state" is used widely in the social sciences as equivalent to country.

In a country, stronger nations often dominate other peoples, while smaller nations sometimes seek to increase their role or work for independence. Thus, the country of France includes culturally distinct local ethnic groups, the Bretons, Basques, and Alsatians, as well as the French and groups of immigrants from Algeria and other former French colonies. The Basque nation seeks to establish its own country straddling the French-Spanish border, but the Bretons and Alsatians are more content to remain part of France. Germany emerged as a country in the later 1800s, when a group of smaller states united under a nationalist linguistic banner and Prussia's military leadership. In the 1900s, Germany used the idea of uniting separated German-speaking minorities in neighboring countries as a reason for acquiring further territory they called "living space." This led to world wars when Germany discovered limits to the expansionism its neighboring countries would tolerate.

Indigenous Peoples

Indigenous peoples are the first inhabitants (or nations) of any modern country. Today indigenous groups often remain as minority populations within individual countries, subject to some denial of human rights and development opportunities. The numbers of indigenous peoples often dramatically declined during the colonial period and world wars, and some groups were exterminated. Other groups evolved their own cultural and political aspirations as minority "nations." For example, the Native Americans in Canada initially disputed land ownership with European colonists and recently won major rights as "First Nations" in Canada.

Governments

Some countries have a **unitary government** structure, administering all parts from the center for all aspects of government. Other countries, including most of the world's largest (Russia, India, Brazil, Nigeria, the United States, and Canada), have a **federal government** structure, dividing the authority for various activities between a central government and subunits commonly called states or provinces. Federal governments have responsibility for external relations, defense, and interstate relations, while the states or provinces generally have authority for education, local roads, and physical planning within their borders.

Government functions are concentrated in **capital cities**, where the heads of state live and administrative and government offices are situated. Many capital cities are the largest cities in the country, as, for example, London (United Kingdom), Tokyo (Japan), and Nairobi (Kenya). In many federal countries, new capital cities were built as a gesture to replace colonial choices, to expand economic development of the interior, or to provide a more central location. Washington, D.C., Brasília (Brazil, Figure 1.17), Abuja (Nigeria), and Canberra (Australia) are examples.

FIGURE 1.17 Brasília, the planned capital city of Brazil. The central avenue is lined by public buildings and government offices with the Parliament at the far end. There are residential "wings" on either side. Affluent residential districts surround the artificial lake in the distance. Many poorer workers live in satellite towns and bus into work.

Global Governance

No global government exists with the same powers as country governments. Any vision of a worldwide government remains a long way off. However, the term **governance** is increasingly used to include organizations, which seek to legislate for and regulate aspects of human activities outside the powers of sovereign countries, or have cooperative agreements to work with intergovernmental or supranational bodies. An evolving global governance complex brings together countries, international institutions such as the United Nations, and public and private networks of transnational agencies. Much of this complex forms an extension of country government activities into the wider world, but there are also networks that function across country borders, some with minimal government consultation. The process has been called "the stretching of politics."

United Nations and Linked Agencies

The United Nations (UN) is the largest and most influential institution attempting global governance. Countries pay dues to the United Nations and these are used in its various programs and specialized agencies. The United Nations has few specific programs in the security field apart from the Security Council and the groups of peacekeeping military that are drawn from member countries. Although the United Nations' members include almost all of the world's countries, it fails to prevent all civil wars, nuclear testing, or drug, weapons, and slave trafficking. The difficulties in obtaining agreement from all or a majority of UN members resulted in the United States, supported by other wealthier countries, intervening directly in the affairs of Caribbean islands, Kuwait, Bosnia, Kosovo, Afghanistan, and Iraq.

The UN's greatest strengths are in coordinating aspects of human welfare, economic development, and care for the environment. Worldwide organizations associated with the United Nations are designed to promote the economic development of poorer countries (International Monetary Fund, World Bank) or to liberalize trade among countries (World Trade Organization).

Nongovernmental Organizations

Increasingly, nongovernment organizations assume responsibilities for government-like activities. **Nongovernmental organizations** (NGOs) are also called "private voluntary organizations" or "civil society organizations." They include any institution engaging in collective action of a noncommercial, nonviolent manner that is not on behalf of a government. Some NGOs have local or country bases, but the largest engage in international activities.

Many international NGOs are contracted by governments and international agencies to supply aid irrespective of country borders. NGOs such as the International Red Cross, Green Crescent, Oxfam, Save the Children, Greenpeace, and Médécins Sans Frontières (Doctors Without Borders) are better known than many smaller countries. NGOs often work in fields similar to those of the United Nations, which distinguishes them from governments by giving a consultative status.

NGOs such as Amnesty International and Greenpeace assume advocacy roles, campaigning against human rights and environmental abuses respectively. However, when some NGOs oppose aid agency projects, they may set back attempts to bring improved conditions to materially poor people. Their use of the media and politicians may run counter to the needs and wishes of those they profess to protect.

Country Groupings for Trade or Defense

Few sovereign countries have joined each other and remained together in a federated country. Examples include Yemen (1972), East and West Germany (1990), and the United Arab Emirates (1990). While closer political integration among countries is difficult to achieve, many countries make agreements with other countries, often on a world regional basis. Specific agreements of this type foster security through common trading and defense interests.

During the later 1900s, there was a widespread political will among countries to free world trade from barriers. The General Agreement on Tariffs and Trade (GATT) was established in 1948 to encourage countries to lower their tariffs. After 1993, the World Trade Organization took over GATT's role of trying to prevent discrimination among trading partners, but its activities appear to many to support the wealthier countries at the expense of the weaker.

Most progress on liberalizing trade has been made at the world regional level in free-trade areas, the members of which impose common tariff rates on imports. The largest trading group at present is the European Union (EU), which is adding the former Communist countries of east and central Europe. The North American Free Trade Agreement (NAFTA) may extend to much of Latin America. Even more ambitious is the prospect of the Asia-Pacific Economic Cooperation Forum (APEC) that pledges to free internal trade among its wealthier countries by 2010 and among other countries by 2020. Such groupings of regional interests are considered in each chapter of this text.

The Cold War period that began after World War II generated defense agreements on both sides. The North Atlantic Treaty Organization (NATO) linked North America and western Europe in a common response to a perceived military threat from the Soviet Union. The Soviet Union established the Warsaw Pact to unite it with the countries of Eastern Europe in the Soviet bloc. In the 1990s, the end of the Cold War led to shifting emphases among regional groupings of countries. The Warsaw Pact ended after 1991, and NATO extended its agreements to some of the former Soviet bloc countries, despite resistance from Russia. The Association of South East Asian Nations (ASEAN) began with political objectives during the Cold War, opposing Communist countries such as Vietnam, but later shifted to increasingly economic objectives and admitted Vietnam as a member. Through the 1990s and early 2000s, countries and regional groups of countries refocused their policies and rhetoric, becoming less confrontational.

Regions and Economics: Wealth and Poverty

Material wealth and poverty and their distribution are central concerns of regional geographers. The huge numbers of poor people constitute the greatest economic problem facing our world today. **Economic geographers** study how human and natural resources are utilized and scarce goods are produced, distributed, and consumed around the world. The spatial distribution of material wealth is depicted in the World Bank map (Figure 1.18). In contrast to the large numbers of poor people in the world, extremely wealthy people are fewer than 1 percent of the total, but they control more than one-third of the world's wealth and their numbers are growing. Dollar billionaires increased from 140 in 1985 to 476 in 2003 and 793 in 2006; 56 percent are North Americans and 26 percent are Europeans.

Measuring Wealth and Poverty

Specific indicators that give more precise meanings to material wealth and poverty enable comparisons of differences among groups of people. The ownership of consumer goods and access to piped water and energy resources are vivid indicators of differences in material wealth among countries (Figure 1.19). Poor people's luxuries such as better drinking water, food, clothing, and shelter are often wealthier people's normal expectations.

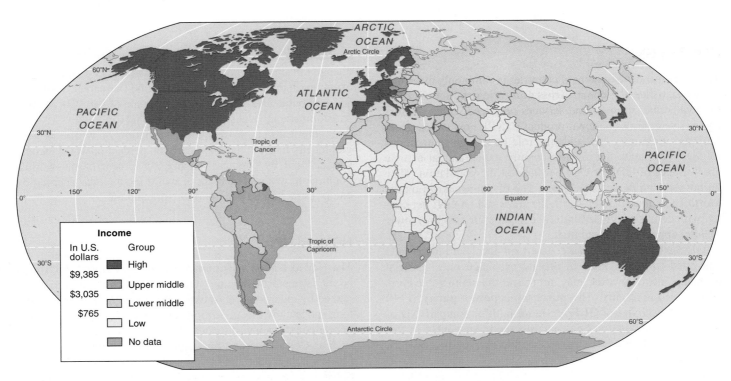

FIGURE 1.18 Major income groups of countries. A World Bank division based on GNI per capita for each country. *Source: Data from* World Bank Atlas, *World Bank, 2006.*

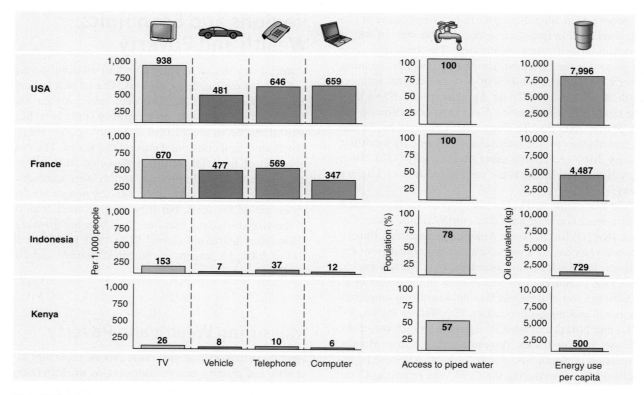

FIGURE 1.19 **Consumer goods, access to piped water, and energy use.** As world societies see consumer goods as increasingly important, differences among countries in consumer goods ownership have significant geographic implications. Access to piped water and energy are also important to the well-being of people. The ownership of consumer goods is shown as the number per thousand people (e.g., 938 people in each 1,000 in the United States have a TV). Access to piped water is shown as a percentage of the total population. Energy use is in kilograms of oil equivalent per capita. This diagram can be found in each of chapters 2 through 10. *Source: Data (for 2002) from* World Development Indicators, *World Bank, 2004.*

GDP and GNI

The economic development of countries is commonly measured by two statistics of income. **Gross domestic product** (GDP) is the total value of goods and services produced within a country in a year. Gross national product (GNP), now called **gross national income** (GNI), adds the role of foreign transactions to GDP. Per capita figures of a country's total annual income are averages of GDP or GNI per head of the population. They do not indicate personal incomes. The divisions shown on the World Bank map (see Figure 1.18) are based on GNI per capita, with countries divided into four income groups: low, lower middle, upper middle, and high.

GDP and GNI values are based on local currencies and converted to U.S. dollars at official exchange rates. Official exchange rates, however, may not reflect the comparable costs of living in a country. The **purchasing power parity** (PPP) estimates of GNI and GDP are more faithful comparisons of living costs among countries and are used extensively in this text. Prices in India, for example, are much lower for equivalent items you might buy in the United States. In 2003, US$440 bought as much in India as US$2,230 did in the United States. Countries with high incomes and high living costs have a lower

PPP estimate of income than the GDP or GNI based on exchange rates; poorer countries often have higher estimates. For example, in 2001 Switzerland had a GNI per capita income of $38,330 but a GNI PPP per capita estimate of $30,970; Mexico had comparable values of $5,530 and $8,240.

Figure 1.20 shows that the distribution of wealth is geographically uneven: in 2005 nearly 60 percent of world income was accrued by countries having 20 percent of the world's population, while 26 percent of world income was accrued by countries having 60 percent of the population.

HDI and HPI

The United Nation's **Human Development Index** (HDI) is a broad estimate of human well-being, in which countries are ranked by incorporating statistics calculated from averages of a long and healthy life, education, and a decent standard of living. Poorer countries investing heavily in education and health care, such as Costa Rica and Sri Lanka, provide a better quality of life for their people and have a higher HDI than GDP (merely based on a country's income) rank. By contrast, many of the oil-rich Persian Gulf countries have high income rankings based on

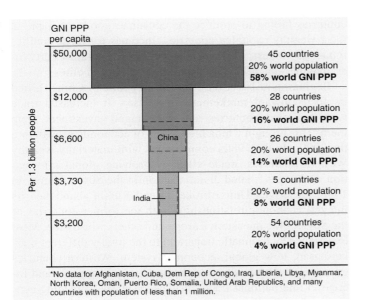

FIGURE 1.20 Distribution of world incomes. The 2005 world population was divided into five groups of 1.3 billion people each. Gross National Income Purchasing Power Parity (GNI PPP) forms the basis of comparison of these groups. Each country's position is shown in the regional chapters, 2 through 10. The United States is in the top group and has 5 percent of world population and 23 percent of GNI PPP. *Source: Data (for 2005) from* World Development Indicators, *World Bank and Population Reference Bureau.*

oil exports but lower HDI rankings because of slow improvements in schooling, especially for girls, and health care.

As the emphasis in thinking about economies shifted toward the needs of the poorest people, the United Nations introduced the **Human Poverty Index** (HPI). Linked to the HDI, it is another composite index that measures deprivation in HDI criteria. For example, the percentage of people expected to die before age 40 indicates health vulnerability; the percentage of illiterate adults indicates restrictions on entry to better jobs and full community life; and a combination of percentages of people without access to health care or piped water and of malnourished children under age five indicates a lack of decent living standards. The HPI ranks countries by the proportions of populations affected by such deprivations. Values range from around 10 percent in Cuba, Chile, and Costa Rica to over 50 percent in many African countries and Cambodia in Southeast Asia. Contrasts also occur within countries: in China the coastal regions have HPI values of 18 percent, while areas in the remote interior have values of 44 percent.

Creating Material Wealth

Many different industries produce and distribute goods and services. There are four major sectors of a country's economy.

The **primary sector** produces raw materials from natural sources including minerals, oil, gas, timber, and fish. Farm products come from domesticated plants and animals, subject to local soil and climate conditions. The industries producing these goods

provide livelihoods for many people, albeit commonly with low wages. In materially poor countries the greatest proportion of the workforce is employed in such activities. Over the last 200 years, output has been increased by the use of machinery in mines, forests, and farms and by using manufactured chemical nutrients and insecticides to improve agricultural yields.

The modern materially wealthy countries achieved their prominence from the **secondary sector** of production in manufacturing and construction. Extra value and profit often comes from processing raw materials to make clothes, furniture, food and drink products, pharmaceuticals, railroads, engines, road and air vehicles, consumer electrical goods, and many other products. The cost of raw materials is a relatively small part of total product costs. The expansion of manufacturing after 1800 in Europe and North America created huge industrial areas dominated by towns, factories, and railroad transportation. As railroads and larger merchant ships enabled worldwide sales, products became more complex, with the assembly of many components into the finished goods. Component manufacture was **outsourced** to other companies in the local area and later to other industrial countries around the world. Today, for example, Boeing aircraft are assembled in Seattle and Airbus aircraft in Toulouse, France, but engines, wings, and tailplanes for both are made elsewhere in the United States and Europe. From the 1990s there has been an increase in the relocation of manufacturing operations that can be carried out by low-cost labor to materially poorer countries in Asia and Latin America.

Many service industries of the **tertiary sector** began as support for the manufacturing companies. Their numbers expanded when governments decided to provide health services, education and a wide range of other services. This sector also includes retail and wholesale trade, and is mainly present in urban areas. During the second half of the 1900s, all service industries experienced huge employment increases.

The **quaternary sector** of information-based services (legal, financial, media, Internet) emerged as a specialized and highly sophisticated branch of the services sector, and was at first concentrated in the cities of materially wealthy countries. However, as these services became more complex, the headquarters of many large corporations first outsourced routine paperwork to local "office factories." They then sent many aspects of the work to low-cost countries, encouraged and enabled by information technology, falling telecommunications costs, and low wages. The growth of the call center industry is one of the major outgrowths of this trend, but higher-level business services from logging insurance claims to making payments are also moving to these global centers.

Marketplace Economy

For much of the 1900s two systems closely related to political systems competed with each other in strategies for economic organization. The **free market enterprise**, or **capitalist, system** is based on accumulating capital and investing it to accumulate more. Individual purchasers of goods may choose what they want from a range of products. Those who provide for these wants—and sometimes create them through advertising—invest financial capital with the aim of making profits. They "buy" labor and machinery to produce saleable goods at the lowest cost. Competition among small firms, large corporations, and countries is an essential feature of the system.

Capitalism developed in Western countries from the 1700s, and especially in the United States from the late 1800s. It involves the private and corporate organization of investment, production, and marketing. Western countries produce and market sophisticated goods at high prices and buy low-priced raw materials and cheap products from the materially poorer countries.

In theory, governments intervene in free-market economies mainly to regulate the terms of trade and ensure "fair play" among producers. In practice, decisions made by governments and the World Trade Organization on what trade should take place and what is fair are not always framed or carried out in the interests of all countries or those of ordinary people. Political considerations favoring some groups backed by powerful lobbies influence legislative actions. For example, a stated goal of open borders to trade may be contradicted by placing protective taxes on imported goods and subsidizing internal products. Furthermore, the governments of many wealthier countries provide the social services and build infrastructure (roads, airports, harbors, water supplies, waste disposal) that give businesses and people in those countries many cost advantages compared to the poorer countries. As a result of increased government intervention, capitalist countries become less "free" market.

Central Planning Economy

The Communist **centrally planned economic system** provided the political and economic opposition to the Western marketplace countries during the Cold War. It was practiced by the former Soviet Union, its satellite countries, the People's Republic of China, and linked countries such as Cuba. This system centralized planning and decision-making responsibilities in the government on the grounds that the whole country's interests came first rather than the interests of enterprising individuals. The central ministries knew what was best for the people and planned the production of goods considered essential—whatever the cost and whether or not the goods met actual consumer demands. Central governments provided welcome medical care and education to support a fit and able workforce. These governments spent highly on strong military defenses.

Those in command of centralized policymaking, however, often made large-scale mistakes. They were handicapped even more than the marketplace countries by a lack of information, or just as much by personal bias or interest. Many leaders were reluctant to change past policies, even if inefficient or oppressive, while regional bureaucrats often obeyed central commands despite knowing the policies would fail. Overproduction of some goods and underproduction of others led to these

countries failing to produce the consumer goods available in most wealthy capitalist countries. Incomes for most families remained modest, while members of the Communist Party hierarchy became relatively wealthy or privileged elites.

Beginning in 1978, the People's Republic of China shifted its focus toward marketplace processes. It increased trade with capitalist countries and encouraged investment from them. This brought China high levels of economic growth into the 2000s. In the 1980s comparisons with materially wealthy Western countries, made clear by growing global information exchanges, fueled dissatisfaction in the Soviet Union. In 1991, the Soviet Union broke up and Russia abandoned its integral economic relationships with the former countries of the Soviet bloc in eastern Europe and others worldwide. Most experienced a traumatic transition to the totally different market-based, now global, economic system. Western countries encouraged the transition to democratic governments and increased international trade on Western terms, spreading economic globalization.

The Global Economy

The collapse of the Soviet Union in 1991 brought the end of the Cold War and the dissolving of Soviet economic and defense associations. It left the marketplace society dominant. While most political and economic forces operate at the country level, global exchanges and institutions support much of the growing wealth. The roles of multinational corporations, global financial and information developments, and the rise of global city-regions are particularly important.

Multinational Corporations

Multinational corporations (MNCs) make goods or provide services for profit in several countries but direct operations from a headquarters in one country. Strictly, the term "transnational corporation (TNC)" refers to corporations that are no longer headquartered in a single country, but it is sometimes used instead of MNC. The greater ease of travel and telecommunications contacts, together with the Internet transfer of information, encouraged MNCs to expand in the later 1900s. They are a major force in globalization trends.

Multinational and transnational corporations place production facilities in countries outside their homelands to take advantage of cheaper labor, land, energy and access to local markets. Some look to less stringent worker safety and environmental laws in other countries. For example, auto manufacturers spread the manufacture of components across several countries to respond to local design preferences and ensure supplies during labor strikes (Figure 1.21). Most early multinational corporations were based in the United States, but those of European, Japanese, and South Korean origin are increasingly significant. Of the top 100 MNCs with the highest level of assets outside their home country in the early

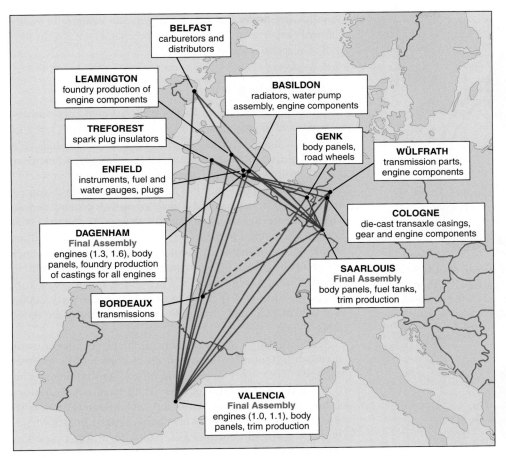

FIGURE 1.21 Multinational corporation's international linkages. The European distribution of Ford Company factories includes makers of parts and final assembly locations for the Ford Fiesta car in the late 1980s. Model and other company changes altered the roles of these places in subsequent years, but the international associations remained. *Source: Data from Dicken, 2003.*

better wages and provide better employee benefits and prospects than local companies.

Global Financial Services

The growth of financial services to the global scale in the later 1900s was both caused by global economic activity and enabled its expansion. The breakdown of the system of fixed exchange rates after 1970 started more frequent flows among currencies. Oil producers raised prices in the mid- and late 1970s and U.S. dollars accumulated in foreign banks, which created new financial markets. Materially poor countries such as Brazil borrowed large sums at low interest rates to develop roads and power dams. Money flows in the 1970s and early 1980s were mainly from materially wealthier to poorer countries.

Although it is the world's largest economy, the United States ran huge budget deficits in the 1980s and early 2000s, financed by borrowing in dollars. In the 1980s this raised world interest rates, left huge debts for posterity, and slowed economic development worldwide. The rising interest rates prevented the poorer countries from repaying their debts. During the late 1980s and 1990s, foreign investment supplied by the wealthier countries such as the United States, Japan, and the countries of western Europe was nearly all used in those countries (Figure 1.22).

2000s, 50 were based in Europe, 27 in the United States, and 17 in Japan. They produce a huge range of brands sold worldwide, including Coca-Cola, PepsiCo, Ford, General Motors, Volkswagen, Mercedes, Exxon, Shell, Toyota, Sharp, Samsung, Kellogg, Nestlé, Hyundai, and IBM. By the early 2000s, multinational corporations accounted for 40 percent of all international movements of goods.

Multinational corporations are not only manufacturers. MNCs in service industries have expanded since the 1970s. By the early 2000s over 40 percent of foreign direct investments to countries went to corporations involved in tourism and travel, data processing, advertising, market research, banking, and insurance. Some manufacturing corporations, such as the Ford Motor Company and General Motors, diversified into financial loans and credit cards.

MNCs wield considerable power in the countries where they operate. Some MNCs are perceived as uncaring monolithic institutions without concern for the best interests of the people they employ in either home or adopted countries. However, other MNCs transfer wealth and technology to poorer countries, provide jobs where none existed in rural areas, and pay

Global Information Services

The rapid expansion of Internet telephone-computer linked services in the 1990s fueled the growth of information services. E-commerce (electronic commerce), the trade-based sector of such services, is mainly business-to-business ("B2B") transactions. Large corporations such as General Motors work with their suppliers over the World Wide Web. Business-to-consumer ("B2C") facilities include retail sales, bidding (e.g., for airline tickets), and auctioning. In the initial stages of B2C, few companies made rapid trading profits, and many went out of business after initial high share valuations. Although some new car sales, for example, take place over the Internet, many people with Internet access in the United States use it as a source of information before going to their local auto outlet to buy.

Just as some multinational corporations moved manufacturing production facilities to places with lower labor costs, so

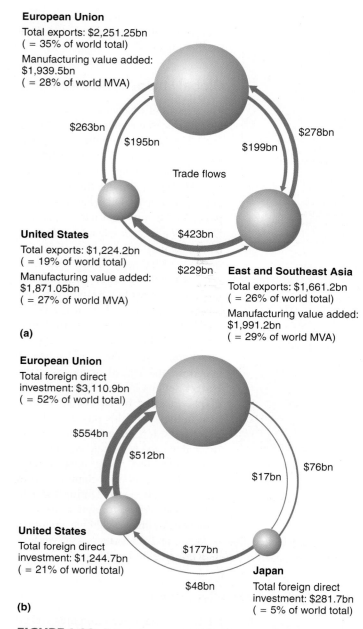

European Union

Total exports: $2,251.25bn
(= 35% of world total)

Manufacturing value added:
$1,939.5bn
(= 28% of world MVA)

$263bn

$195bn

$278bn

$199bn

Trade flows

United States

Total exports: $1,224.2bn
(= 19% of world total)

Manufacturing value added:
$1,871.05bn
(= 27% of world MVA)

(a)

$423bn

$229bn

East and Southeast Asia

Total exports: $1,661.2bn
(= 26% of world total)

Manufacturing value added:
$1,991.2bn
(= 29% of world MVA)

European Union

Total foreign direct
investment: $3,110.9bn
(= 52% of world total)

$554bn

$512bn

$17bn

$76bn

United States

Total foreign direct
investment: $1,244.7bn
(= 21% of world total)

$177bn

Japan

(b)

$48bn

Total foreign direct
investment: $281.7bn
(= 5% of world total)

FIGURE 1.22 Concentrations of manufacturing, trade, and foreign direct investment in the late 1990s. The United States, European Union countries, and East Asia (Japan and China) dominate the world in these important areas: 84 percent of world value added in manufacturing, 80 percent of world export value, and 78 percent of FDI (Japan only for East Asia). (a) Trade and value added in manufacturing. (b) Foreign direct investment. *Source: Data from Dicken, 2003.*

others moved information handling to such places. For example, in 1983 American Airlines established Caribbean Data Services in Bridgetown, Barbados, to process the paperwork related to its tickets and boarding passes. It became the largest single employer in Barbados. U.S. insurance companies process claims in Ireland. India, with its large population of English speakers,

now has several of the world's largest call centers for multinational corporations. Some Indian companies even train their staff to respond in American accents.

Global City-Regions

The multiplication and growth of multinational corporations, international financial institutions, dense networks of telecommunications, information processing facilities, and international airline routes, together with the rising significance of quality business services, placed a new focus on the world's largest cities. Those cities with an increasing involvement in the global economy often have as much or more contact with foreign cities as with those in their own country. New York, for example, is the center of a region with 18 million people and an economic product greater than countries such as Canada or Brazil; its businesses receive 40 percent of their revenues from foreign sources. Foreign banks with New York offices rose from 47 in 1970 to over 200 in the early 2000s; over half of U.S. law firms with overseas business are based in New York.

Such cities have major impacts on the places immediately surrounding them as well as on cross-border links to other countries. Their geographic scale of size and influence merits the term **global city-regions**. They have concentrations of high-salaried people, high-end technological and business services, specialized workplaces, hotels, homes, major sports stadia, and concert halls. They have high-rise office and apartment blocks and a wide range of art and sports facilities. At the same time, their corporations employ increasing numbers of foreign experts and a growing underclass of poorly paid support workers, often migrants from materially poorer countries.

One approach to identifying and classifying global city-regions focuses on the importance of four categories of global corporate services (accounting, advertising, banking, and law). Prime centers in all categories are New York, London, Paris, and Tokyo, closely followed by Chicago, Los Angeles, Frankfurt, Milan, Hong Kong, and Singapore (Figure 1.23). The uneven distribution of these cities—the "control centers" of the global economy—highlights the diversity of today's world.

Regional Economic Emphases

Although the marketplace capitalist economic system prevailed globally after 1991, its geographic influences and benefits are not evenly distributed or applied. Instead of a single path toward a global economy, distinctive regional variants of capitalist economies developed. The "Asian Way" builds on family linkages connected to government-business liaisons, rather than on the independently verified banking and legal systems that are basic to capitalist economies in Europe and the United States. The "Chinese Way" merges Asian features

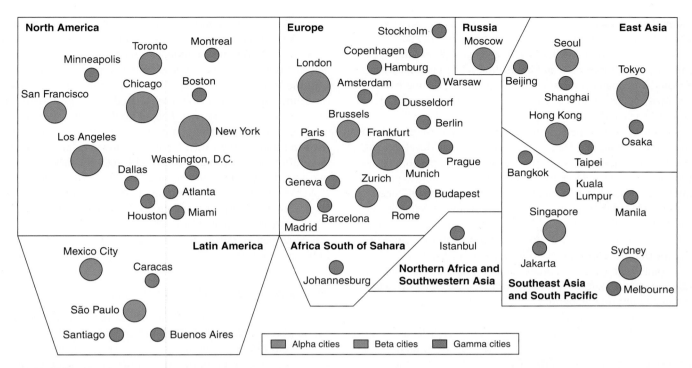

FIGURE 1.23 Global city-regions. Major world cities and their immediate regions are centers of global economic activities. The threefold classification is based on four categories of international business services: advertising, accounting, banking, and law. Each of the four is given a 1–3 value, making a maximum score of 12 for the city regions. "Alpha" cities have scores of 10–12; "beta" cities of 7–9; "gamma" cities of 4–6. *Source: Globalization World Cities Study Group and Network, University of Loughborough, U.K.*

with the continuing dominance of central government planning alongside attracting foreign direct investment to expand its export industries. The "European Way" makes much of providing social welfare to support those who are not able to benefit from or exploit the capitalist system. Countries in Latin America are moving out of self-sufficient and inwardly focused economic systems and into the world system with some pain. The former Communist countries of the Soviet bloc struggle to construct new economic identities. And, on a much smaller scale, groups of peoples living in such isolated areas as the Amazon rain forest and Papua New Guinea, as well as increasing numbers in African rural areas, engage little, if at all, with the global economy. Some local voices are not heard.

Geography, Development, and Human Rights

Geographers consider how some places experience greater material and personal well-being than others, and how improvements, or **development**, of poorer regions may occur. For each place or region they build a knowledge of:

- the complex interactions of people with the natural environment,

- the historic growth of population numbers and cultural expressions,
- the evolution of political systems and their present operation,
- the growth of economic output, and
- the effect that different views of human rights have on those living in different parts of the world.

In this chapter we highlighted geographically related analyses of natural environmental concerns, resources, cultural features, population, political governance, wealth and poverty, and the global economy. They affect human development and human rights in different ways in different parts of the world.

Human Development

Throughout history, the differences among places led some countries that were well endowed with natural resources and strong leadership to assume positions of dominance and superiority. For example, the Chinese long believed they were the only civilized people and all others were barbarians. They spread their culture widely in Asia, but withdrew from wider geographic control beginning in the 1400s. During the 1800s, European countries claimed they were taking their form of civilization and modern ways to parts of the world that they often colonized. In both cases, assumptions of superiority and dependence changed perceptions of well-being among places. Despite (or because of) the trends toward globalization in the late 1900s and early 2000s, the world remains full of differences and inequalities. The material wealth of many Americans contrasts with the poverty of some parts of their own country as well as the extreme material poverty of millions in Africa and Asia.

Studies of how some regions and countries move ahead and others fall behind are based on the concept of development. **Human development** is the process of enhancing human capabilities and improving life quality by providing access to better incomes, education, health care, piped water, and energy supplies. Concern is given to the possibilities of helping materially poor, "less developed" or "underdeveloped" countries and regions to catch up with the wealthier countries. **Sustainable human development** involves economic growth that does not deplete renewable resources for the future. It thus links to both human and natural resources, drawing together studies of human and physical geography.

The United Nations Human Development Program and recent World Bank publications focus on the need to eradicate material poverty. The last fifty years saw major reductions of income poverty in large parts of the world, improvements in human development indicators—particularly in health and education—and the wider spread of law and fair administration of justice. However, the fact that so many people in the world remain materially poor is a challenge. In 2000, the United Nations and other global organizations formulated the Millennium Development Goals, to be achieved by 2010–2020 (Table 1.1). By 2006 the lack of progress in achieving the goals suggested that these would take longer to fulfill.

Three Worlds

In the 1950s and 1960s, the combination of many former colonial countries becoming independent and the growth of the Cold War (perceived at the time as "**First World**" vs. "**Second World**") set some economists thinking about development issues. Western economists assumed that the materially wealthy Western countries understood how economic growth worked and that their experiences could be transferred. The Soviet and Chinese Communist regimes offered alternative patterns based on central government dominance. Both sides promoted their ideas in the new countries. The "**Third World**" was born as many newly independent countries outside the two main blocs tried to establish their own identities. However, such countries were often drawn into alignment with one or other of the first two worlds. Although the clash of the first two worlds in the Cold War did not end in direct nuclear destruction, World War III is seen as having taken place in Third World countries such as the Koreas, Vietnam, Cambodia, and many African and Central American countries. Such conflicts held back development in these countries.

Modernization and Growth Poles

Two main approaches to development were adopted. First, it was suggested that **modernization** from traditional (i.e., underdeveloped) to new economic processes (i.e., developed) would take place as agriculture-based societies gave way to industrial production and mass consumption of the manufactured goods. In the second approach, a movement from dispersed rural to urban "**growth pole**" living in an industrializing world would bring improved access to waged jobs, education, health, and other services.

The marketplace capitalist countries and the centrally planned Communist countries both encouraged modernization in developing countries. However, outcomes were patchy, resulting in geographically uneven development, both among and also within countries. The formula worked in a few countries, particularly those, such as Japan and South Korea, which had U.S. help in war recovery investment and heavy state intervention. Other Asian countries such as Taiwan, Malaysia, Thailand, and Singapore also invested in manufacturing and service industries and experienced rapid economic growth. Larger countries, such as Mexico, Brazil, India, and China possessed internal resources and potentially large markets for home-produced goods. Small countries often lacked both resources and large markets.

Competing Systems

From the 1800s European colonial powers traded low-value raw materials with their own high-value manufactured goods. Colonial domination and other historical conditions produced a set of materially wealthier **core countries** and a set of dependent **peripheral countries**, making it difficult for the poorer countries to follow the same path. By the 1950s it became clear that although most materially wealthier places got wealthier, few poorer places experienced improvements.

In the 1960s and 1970s concerns of those living in Third World countries included the lag in economic development, resentment of previous colonial domination, and the apparent

---

TABLE 1.1	**Millennium Development Goals**

In 2000, many target dates were set between 2010 and 2020, but lack of progress by 2006 suggested that these goals will take much longer to achieve.

Goal 1: Eradicate extreme poverty and hunger.

Goal 2: Achieve universal primary education.

Goal 3: Promote gender equality and empower women.

Goal 4: Reduce by two-thirds the under five mortality rate.

Goal 5: Improve maternal health.

Goal 6: Combat HIV/AIDS, malaria, and other diseases.

Goal 7: Ensure environmental sustainability.

Goal 8: Develop a global partnership for development.

success of Communist countries at the time. A chain of dependency from the poorer to the wealthier countries was identified, with rural peasant farmers depending on local market towns, and those on international ports and centers of global trade in wealthier capitalist countries. The wish of many Third World countries to be self-sufficient, and released from involvement in the West-dominated world economy led to political independence, development of home-based manufacturing (**import substitution**), local trade barriers, restrictions on foreign corporations, and the formation of trading groups of countries with similar concerns. By the 1990s, however, it became clear that isolation from the world economy might have some internal cultural and political value in the larger countries, but even there did not result in economic equality with the Western countries.

"Bottom-Up, Not Top-Down"

From the 1970s, greater emphasis was placed on self-reliance and the local characteristics of places. Rural development became a focus for grassroots and populist movements to re-establish local communities and strengthen their abilities to maintain themselves and combat the external pressures. However, basic needs programs to increase the availability of food, clothing, and housing worked slowly. Separation of communities from urban influences often left them in a time warp. In some countries such as China, Cuba, and Tanzania, socialist regimes vowed to integrate culture, history, and local institutions to a self-sufficient approach mobilizing human and natural resources, but seldom succeeded.

In many of the materially poor countries today, few jobs in the **formal economy** pay taxable salaries. The majority of people have to gain income as best they can in the **informal economy**. They cannot use their homes or land as collateral for loans, and this failure to extend formal property rights to most of the population is a major cause of poverty and a lack of enterprise opportunities.

In response to the lack of local opportunities despite a positive view of the potential of individuals, groups of people set up microcredit banks, building on the experience of the Grameen ("Village") Bank in Bangladesh, which grew out of a program of small individual loans by Nobel Peace Prize winner Professor Mohammad Yunus. Small amounts of money enabled craft workers and others to establish small businesses that helped them emerge from poverty. These methods are now applied in 58 countries, including materially wealthier countries such as the United States and Canada. As the movement has grown and extended its influence, so have the critics, who suggest, for example, that such loans add to the oppression of women, who have been more successful in such businesses than men and so take on more work as well as running their homes.

Structural Adjustment

In the 1990s, the uneven and often unsatisfactory results of development led to new approaches. The experience of Asian countries in achieving economic growth through export-based manufacturing led the major global agencies, such as the World Bank, United Nations, and many non-governmental organizations to insist that countries applying for aid should open their trade by exporting manufactured goods and allowing foreign investment—a process known as **structural adjustment**. Such policies involved downsizing internal bureaucracies and central government planning. Copying the Asian Way, however, proved no easier than copying Western experiences.

Globalization, Technology, and Development

Globalizing trends challenged whatever "underdeveloped" countries tried to do for themselves. Friction of distance ceased to determine the economic costs of locating many production facilities; low labor costs increased in significance for manufacturers and some service industries. Fewer controls on trade and money movements and more open local governments allowed MNCs to choose their places of investment. Better communications by satellite TV, the Internet, and cell phones enabled wider global information transfer. Financial markets based on New York, Tokyo, and London but expanding to regional global city-regions in countries experiencing economic growth increased the speed and numbers of transactions.

A new division of the world emerged from technological innovation and transfer (Figure 1.24). Around 15 percent of Earth's population living in Canada, the United States, western and northern Europe, Japan, South Korea, Taiwan, and Australia is responsible for nearly all the technological innovations and their applications. A further 50 percent of the world's population is able to adopt some technologies in production and consumption. The remaining 35 percent of the world's population lies outside these two zones, is "technologically disconnected," and often caught in a trap of poverty, disease, low agricultural productivity, and environmental decline. These people need technological solutions they cannot afford.

Globalization and Diversity

Globalization does not result in uniformity. The greater flexibility, openness, and complexity of its processes lead to new differences among places and the spatial inequalities of uneven development. Local responses and fragmentation result from the impacts of global forces.

The combination of growing multinational corporation influences and the increasing focus of economic activity in global city-regions highlighted new geographic patterns of flow among people, finances, goods, and ideas. Any global convergence of economy and culture is paralleled by the divergence caused by dispersed manufacturing and local responses. The predominance of Western tastes in consumer preferences, food outlets, and mass media offerings is most noticeable in the world's major cities and among the urban elites. As the better-off urbanites in materially poor countries adopt Western lifestyles, however, they often combine them with local cultural elements, as demonstrated in popular music, movies, and their own fast food

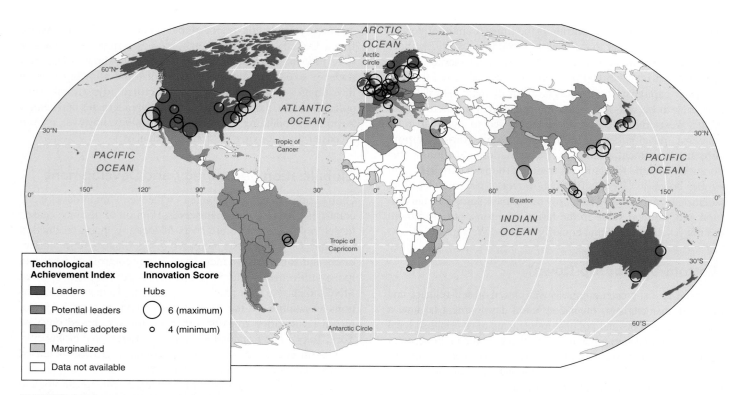

FIGURE 1.24 Global hubs of technological innovation. In 2000, *Wired* magazine consulted government, industry, and media sources to find the locations that matter most in the new digital geography. Each was rated 1–4 in four categories, giving maximum scores of 16 and minimum of 4: area universities and research facilities training skilled workers and developing new technologies; established companies and MNCs providing expertise and economic stability; local entrepreneurial drive starting new ventures; and availability of venture capital. Forty-six technology hubs, shown by circles on the map, were identified. The highest scores were achieved by Silicon Valley (California), Boston (Massachusetts), Stockholm (Sweden), Israel, Durham-Chapel Hill (North Carolina), and London (UK). Others in the top 40 are found in the United States, 12 in Europe, and 9 in South, East, and Southeast Asia. *Source:* United Nations Human Development Report, *2001.*

outlets. Such developments, by diversifying the cultural environment, often set off return influences taken by immigrants to Western countries.

The world's leading industry of the early 2000s, tourism, provides one of the most significant associations between development, globalization, and diverse geographies. Many "underdeveloped" countries adopt tourism as a strategy for modernization. Closely related to conspicuous consumption in materially wealthier countries, tourist venues based on jet aircraft and ocean liner access enhance the differences among places. Although tourism aids a greater understanding of conditions in other parts of the world, its negative side is often a low local return. Some countries use local food and market locally produced goods, but others import food and trinkets. International corporations build the hotels and cruise ships. Historical reconstructions for tourists mute the unpleasant aspects and guilt arising from, for example, slavery or wartime oppression in favor of inaccurate stereotypes.

Responsible Growth and Emerging Countries

By 2004, "**responsible growth**" was replacing earlier terminology. Arising from the Johannesburg World Summit on Sustainable Development held in August 2002, a consensus emerged

directing world leaders to new development paths that build on the UN Millennium Development Goals and link economic growth, environmental sustainability, and social equity. In this view, directed at the initial period to 2020 and longer term toward 2050, poverty reduction is not an end in itself but a precondition for peaceful coexistence and ecological survival.

A group of **emerging countries** have growing economies that challenge the wealthiest countries. They include Brazil, Russia, India, and China (the largest four "BRIC" group), along with Mexico, South Korea, Poland, Taiwan, Singapore, Hong Kong, Saudi Arabia, Turkey, South Africa, and Chile. These countries drive the advancing global economy, playing important roles by reducing inflation, interest rates, wages, and profits in wealthier countries. For example, although the United States has record high current account deficits, the dollar remains strong as the emerging countries invest their trading surpluses in low-yield U.S. bonds. Furthermore, although oil prices tripled from 2002 through 2006, due to a combination of increasing demand from emerging countries, Hurricane Katrina damage, and Middle East conflicts, global growth continued with any inflationary impacts offset by the falling prices of goods and services.

The high levels of global economic growth are expected to continue, with the emerging countries' contribution rising from half of world GNI PPP today to two-thirds by 2025. However, the gap between the incomes of the emerging economies and those of the rest of the developing world is increasing.

China is the main emerging country in economic terms, a result of its increasing openness and trade with the rest of the world, forming 70 percent of its GDP in 2005 (compared to up to 30 percent in the United States and India). It has a large educated and spending middle class and rapidly improving infrastructure of roads, airports, and electricity generation. With other emerging economies it has doubled the labor force engaged in global industries and achieved real productivity gains. Although the emerging economies remain volatile, as financial crises in the late 1990s showed, the increasing spread of economic growth brings greater stability.

The emerging economies involved heavily in global trade and investments help to keep prices and inflation low in the materially wealthy countries. However, competition with emerging countries imposes lower wage rates for manufacturing and many clerical jobs, particularly affecting low-income groups in wealthier countries.

Thus our world and its different regions and countries should not be seen as an unchanging set of rich and poor countries. Shifts of economic and political power are accompanied by cultural changes and challenges. In early 2007 further diversity is promised as some Islamic voices become louder in rejecting not only the Western-based globalizing influences, but also democratic politics and cultural trends that highlight, for example, the nature of women's clothing and their place in society. World regional geography provides an enlightening approach to all these cross currents and their impacts on people and places.

Issues of Human Rights

The concept of **human rights** emerged in the late 1700s, in part from revolutions in Europe and the United States. "Liberté, Égalité, Fraternité" (freedom, equality, brotherhood) was the slogan in France during the 1789 Revolution, and it was enshrined in the U.S. Bill of Rights. Stressed again after World War II by the United Nation's Declaration of Human Rights (Table 1.2), international agreements and actions often refer to these rights. However, few countries fully implement the UN list and some groups claim that the imposition of human rights legislation impinges on their traditional rights. Muslim countries in particular debate the largely secular basis of the UN list, which conflicts with some of their religious beliefs and fierce legal provisions. Saudi Arabia refused to attend UN conferences on such topics on the grounds that it has a God-given right to gender discrimination.

Whereas pressure for human development emerged from mainly economic considerations that were later broadened to cultural, political, and environmental concerns, pursuing human rights has often been a special objective of lawyers, philosophers, and political pressure groups. The United States and Western Europe have long argued for **political rights** (the right to vote and participate in one's own government). Oth-

TABLE 1.2	United Nations Declaration of Human Rights

Freedom from discrimination because of gender, race, ethnicity, national origin, or religion.

Freedom from want and a decent standard of living.

Freedom to develop and realize one's human potential.

Freedom from fear of threats to personal security in arbitrary arrest or violence.

Freedom from injustice.

Freedom of thought and speech to participate in decision making and forming associations.

Freedom for decent work without exploitation.

ers, including the former Communist governments, argued for **social rights** (the right to have a job and earn a living with basic material standards). People in materially poorer countries often argue for **cultural rights** (the right to protect one's cultural traditions, often developed in response to Western ideals and practices). An example of cultural rights involves Western medicine patents, where people in other countries must pay royalties to a Western corporation for a medicine they had discovered hundreds of years ago before the institution of patenting systems.

There are considerable debates as to what should be included or is feasible as human rights. For example, the world's materially wealthier countries often argue against trading with poorer countries because their factories violate human rights by employing children or ignoring environmental protection. Such arguments can be seen as a disguise for protectionist policies designed to secure low-wage, low-skill jobs in the materially wealthier countries.

After 9/11, there was much criticism in Western countries of the Taliban government in Afghanistan and other strict Muslim regimes for oppressing women and carrying out amputation punishments for "minor" crimes. And yet, Muslims claim that their strict rules protect women, while many Western governments allow men the freedom to mistreat and degrade women. Different cultural definitions of human rights contribute to differences among people's expectations in different places.

Human Development and Human Rights

Although having different emphases and emerging from different sources, the two strands of human development and human rights reinforce each other. Human rights add to the development agenda by drawing attention to those who are accountable for respecting rights and adding social justice to economic and cultural principles. An emphasis on human rights shifts priorities toward the most deprived and excluded. At the same time, human development brings a long-term perspective to fulfilling the rights by assessing the workings of socioeconomic contexts and institutional constraints.

In the early 2000s, geographic differences continue to affect many areas of discrimination, poverty, personal insecurity,

injustice, and abuses of free speech. For example, internal armed conflicts in many parts of the world hold back human development, abuse human rights, and create increasing numbers of dispossessed refugees. The move toward democracy in Africa and eastern Europe with multi-party elections brought some advances in human development and human rights. These trends, however, also led to new conflicts in some countries over previously suppressed ethnic demands. In contrast to decades of authoritarianism, more open government appears weak.

Major World Regions

In this text we identify nine world regions (Figure 1.25a,b) that are the subjects of Chapters 2 through 10. Each **world region** contains a group of countries having distinctive physical and human geographies and incorporating internal diversity in subregions, countries, and local areas. Each world region faces issues of significance that demonstrate the importance of a geographical understanding.

Europe

Europe (Chapter 2) is the source of many Western trends in politics, culture, and economics. The region is defined as those countries that are members of the European Union or likely to be in the next few years. It is marked by temperate natural environments and advanced capitalist economies. Its cultures continue to be affected by the past dominance of Roman Catholic, Protestant, or Orthodox Christian groups, increasingly challenged by secular ideologies and the religious demands of immigrants from other world regions. Europe is marked by a variety of spoken languages, while increased immigration from the late 1900s makes it home to a growing number of people speaking other languages from diverse world regions, contesting what it means to be European.

Russia and Neighboring Countries

Russia and Neighboring Countries (Chapter 3) includes all the countries that emerged from the breakup of the Soviet Union apart from the three small Baltic countries of Estonia, Latvia, and Lithuania (now part of Europe, Chapter 2). The region extends from easternmost Europe across northern Asia—a huge area that resulted from the expansion of the Russian Empire in the last four hundred years. European cultures extended into Asia, incorporating and interacting with a wide range of peoples, cultures, and languages in varied environments to produce today's geographic diversity. Through much of the 1900s, the centralizing Communist government of the former Soviet Union kept under control the historic clashes between the Orthodox Christians and Muslims and between the Russian and other peoples. These conflicts emerged again after the Soviet Union breakup, and they are part of the difficult

transition from Communist to marketplace economies and open religious allegiance.

East Asia

East Asia (Chapter 4) includes Japan, the Koreas, Mongolia, Taiwan, and a resurgent China. The region includes the world's most successful countries in recent economic growth, challenging the economic dominance of the United States and Europe. In economic terms, China now has the second highest total output by value (GNI PPP) in the world after the United States, and Japan is third. Neither North Korea nor Mongolia share the economic growth. The historic and modern influences of the Chinese kingdoms and related cultures were less affected by European colonization than other world regions and bring special character to this region and to surrounding countries, where many Chinese families and business personnel live. Geographic diversity occurs in all countries.

Southeast Asia, Australia, and Oceania

Diverse terrains in *Southeast Asia, Australia, and Oceania* (Chapter 5) include part of continental Asia and a series of major islands, thousands of small islands extending out into the Pacific Ocean, and the continents of Australia and Antarctica. The countries of the region are tied by being neighbors, by histories of ancient people movements and more recent colonial occupation and influence, by mainly maritime orientations, and by increasing trade with each other. Having lost many of their links with Europe, the former colonies increasingly look to South and East Asia, the Americas, and each other for trading opportunities. Their essentially European outlooks at times bring Australia and New Zealand into conflict with Southeast Asian and Pacific island attitudes, but necessity draws them together. We also include Antarctica in this world region, inhabited solely by scientific colonies. It is the only part of Earth's land surface that is not divided into countries, although it has been divided by countries into zones for study.

South Asia

Diversity within *South Asia* (Chapter 6) is based on the major religions including Islam, Hinduism, Sikhism, Jainism, and Buddhism. Today, Hinduism (the dominant religion in India) counts 80 percent of India's population as its adherents; Islam is the national religion of Bangladesh and Pakistan; Buddhism is dominant in some smaller countries. The British Indian Empire, which included the whole region, ended with the partition into separate countries at independence in 1947. Since independence, partly successful policies of self-sufficiency and alignment with the Soviet bloc brought moderate economic growth before the rise in foreign trade and investment beginning in the 1990s. Growing middle classes and technocratic and wealthy elites exist alongside large numbers of poor people.

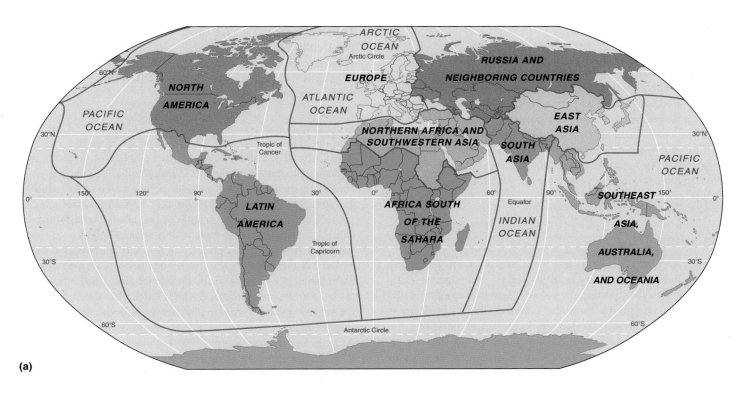

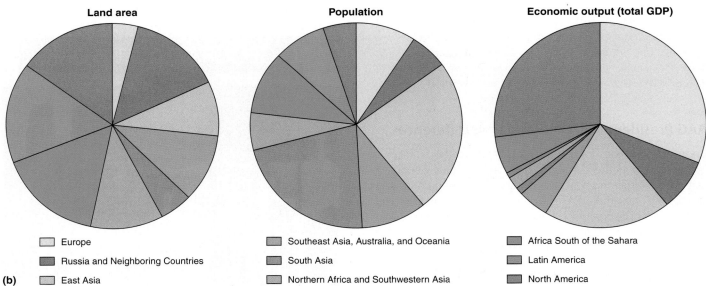

FIGURE 1.25 Major world regions. (a) Map of the world regions — the subjects of Chapters 2 through 10. (b) Comparisons of area, populations, and economic output (gross domestic product, GDP). Which regions have greater proportions of the world's land areas than its population? Which major regions have higher proportions of the world's economic output than its population?

Northern Africa and Southwestern Asia

Northern Africa and Southwestern Asia (Chapter 7) has a distinctive and strategic position at the junction of Europe, Africa, and Asia. It includes the birthplaces of the world's three monotheistic (believing in one God) religions, Judaism, Christianity, and Islam. The Islamic religion is dominant today, partly paralleled in its extent by the Arabic language—although the two Muslim countries with the largest populations in this region, Iran and Turkey, have their own languages. The presence of Jewish Israel in the predominantly Muslim region raises cultural, economic, and political tensions that result in almost continuous hostilities. The location of the world's largest oil reserves and water shortages in the largely arid natural environment of much of this region pose internal problems of uneven resource availability.

Africa South of the Sahara

Africa South of the Sahara (Chapter 8) was the cradle of the human race and most of its current population is of indigenous peoples and their many ethnic groups. It remains one of the most culturally diverse world regions. The region contains great mineral riches and has underdeveloped economic potential. European traders established commercial farming and mining, but colonial settlement by Europeans occurred much later and on a smaller scale than in the Americas or India. Decolonization began in the 1950s, but was often followed by repressive dictatorship governments. The people of this region are among the world's most materially poor.

Latin America

Latin America (Chapter 9) has a diverse physical geography of mountains, plateaus, major river basins, and islands. Many current political and economic issues relate to the region's history of colonization by European countries and its position as the nearest neighbor of the United States. From the 1500s, the Spanish and Portuguese invasions reduced the numbers of indigenous peoples, but those peoples still form major portions of the population in the Andes and Central American highlands.

The Latin-based Romance languages and Roman Catholic religious culture brought by Spanish and Portuguese settlers influence most of this region. There are enclaves of other European languages including French, Dutch, and English, particularly in the Caribbean, along with local centers of Protestant Christianity.

North America

North America, comprising the United States of America and Canada (Chapter 10), is the world's materially wealthiest region, containing the only current world superpower. The two countries are dominated by cultures first brought by European settlers beginning in the 1500s. French and Spanish are spoken in some localities, but English is the most widely spoken language in North America. Almost wiped out by disease and warfare as a result of the European influx, the indigenous peoples formed smaller and smaller proportions of the population, but some lands were restored to them in the later 1900s. Geographic diversity within the region results from the physical geography, the history of settlement and development, and the outlooks to the Pacific, the Atlantic, and Latin America. The high urban concentrations of people and the wish of increasing numbers to immigrate there are current features of its geographical changes.

GEOGRAPHY AT WORK

AAG President in Iran: Reconciling Differences

Geography provides fundamental insights into the nuances of globalization, whether economic, political, or cultural. Geography can build bridges and help ease cultural tensions and misunderstandings. For example, Professor Alexander B. Murphy, from the Department of Geography at the University of Oregon, traveled to Iran in 2004 when he was president of the Association of American Geographers (Figure 1.26). He addressed the "Second International Congress of Geographers of the Islamic World," and spoke to students and faculty at two of Tehran's major universities.

Everywhere Alex went, people were extraordinarily nice, helpful, and friendly. He had traveled previously in other parts of the Muslim Middle East—especially Egypt, Jordan, and Palestine. Iran is clearly different—in language, in culture, in social norms, and much more. To visit Iran is to understand the fallacy of treating the "Islamic World" as a monolith.

Very few Americans go to Iran these days, and this opportunity allowed Alex to give Americans a human face in Iran. At the same time, he gained insight into the diversity of opinion in Iran on political and social issues. Iranians express views that span the political spectrum. What particularly impressed Alex, however, was how well informed Iranians seem to be—gaining informa-

FIGURE 1.26 Professor Murphy, the tallest person, with colleagues at the Iranian geographer's conference.

tion not just from government news sources but from the Internet and from friends and relatives in other parts of the world. He saw how much Iranian society has opened up in recent years, and his discussions with people and his media appearances highlighted a side of America that Iranians rarely see.

FIGURE 2.1 **Europe**. Physical features, country boundaries, and capital cities. The region is defined by the Atlantic Ocean to the west, the Arctic Ocean to the north, the Mediterranean Sea to the south, and the land boundary with Moldova, Ukraine, Belarus, and Russia to the east.

Many Europeans are so passionate about soccer (which they call football) that the sport is a source of both national pride and national competition within Europe. All semifinalists in the 2006 World Cup were European countries. As one of the world's best soccer players of all time, Zinedine Yazid Zidane is one of Europe's biggest celebrities and greatest heroes. However, a close look at Zidane and other soccer players forces one to ask, hero for which country? Zidane has played for the French national team, as well as top clubs such as Juventus of Turin, Italy, and Real Madrid, Spain, helping each team to win championships. A closer look reveals that Zidane is a Muslim Algerian Kabyle Berber born in Marseille, France. Zidane's wife, Veronique, is a Christian Frenchwoman of Spanish ancestry. Although Zidane was ejected for head-butting an Italian opponent in the 2006 World Cup final, and France lost, he received a hero's welcome from the French president and media support. Such is typical of European soccer players, who are only required to be born in a country or have parents born there to play on a national team. It means that many players are children of non-European immigrants and frequently play for clubs in other countries. In many ways, soccer is the epitome of Europe today in that teams are a source of nationalist fervor but team members are much more multicultural.

Chapter Themes

European Influences

Distinctive Physical Geography: geologic variety; long coastlines and navigable rivers; temperate climates; forests, fertile soils, and marine resources; environmental impacts.

Distinctive Human Geography: cultural diversity, aging populations; urban pressures; evolving politics; global economics; growing economies.

Geographic Diversity
- Western Europe.
- Northern Europe.
- Mediterranean Europe.
- East Central Europe.

Contemporary Geographic Issues
- Political changes: European Union (EU).
- Multicultural societies.

Geography At Work: Business development.

European Influences

European influences dominate studies of world regional geography in the twenty-first century. For almost 500 years, from the late 1400s to the mid 1900s, European empires colonized or forced the other world regions into trading relations with European countries. In doing so, they also forced their cultures on other peoples. For example, the English, French, Spanish, and Portuguese languages and the Roman Catholic and Protestant variants of Christianity spread around the world through trade, religious missions, and political control. Other ideas that originated in Europe and diffused include democracy, colonialism, imperialism, capitalism, the basis of scientific research, nationalism, fascism, communism, socialism, and genocide. At the same time, many local ideas, technologies, and faiths from cultures in Asia, Africa, and the Americas were imported into various European cultures.

European discoveries and interactions abroad stimulated an interest in scientific and technological advancement at home. The scientific discoveries include the Earth being spherical and revolving around the sun, the laws of motion and gravity, the theory of evolution, genetic science, the causes of disease and pharmacology, radioactivity, and the theory of relativity. New technologies include movable type for the printing press, the microscope, telescope, steam engine, railroad, internal combustion engine, automobile, radio, antibiotics, orbital satellite, and digital computer. Many of these early discoveries led to and propelled the Industrial Revolution, which in turn increased the economic and political power of Europe.

As many Europeans used their technologies to explore and gain more material wealth from abroad, they also used their technology to transform Europe's natural environments (Figures 2.1 and 2.2). For 2,000 years farms and market towns replaced forests; in the last 200 years factories and cities replaced many farms. Rivers were straightened for navigation and dammed to prevent flooding and provide power. The earth was mined for coal and ore and then drilled for oil. These activities improved the material standard of living and created new landscapes, although some had negative environmental impacts.

For better or worse, over the last 500 years, many Europeans used their local ideas to shape the modern global economy, structure the world's political system, and spread their culture around the world. Though they forced their ways on other cultures for most of that time, they became less aggressive in their interactions with other cultures after the trauma of World War II. Since then, many Europeans have been major proponents of human rights and human development. European cultures continue to interact with and react to cultures in other areas of the world, illustrating that globalization is not destroying geographic differences but is often resulting in different and alternative cultural practices. Indeed, though many Europeans are cooperating both economically and politically more than previously, Europe is as culturally diverse as ever.

(a)

(b)

FIGURE 2.2 **Diverse transformed landscapes in Europe.** (a) Urban Leiden, low-lying Netherlands. (b) Rural Monterchi in Italy.

Distinctive Physical Geography

The regional physical geography of Europe is generally noted for closeness to the ocean and abrupt changes in the physical landscape over short distances (see Figure 2.1). Europe's natural environments have small-scale geologic provinces, mineral resources, extensive plains and valley routeways, long, indented coastlines, temperate climates, woodlands, and areas of fertile soil. These characteristics influenced human actions and have been shaped by human actions over time, contributing to the development of diverse modern geographies.

Geologic Variety

Within its relatively small area, Europe includes almost the world's entire range of geologic features. There are ancient shield areas around the Baltic Sea, the uplands of central Europe, the young folded mountains of the Alps, and the extensive plains in countries around the North and Baltic seas (see Figure 2.1). Volcanoes and earthquakes are active along the Mediterranean Sea.

The Mediterranean Sea is the remnant of a larger ocean that occupied the area between Africa and Europe but closed as the two continents clashed along a convergent tectonic plate margin (see Figure 1.5). Within this zone, the young folded mountains of the Alps form the highest ranges, with many peaks rising above 4,000 m (13,000 ft.). The highest point is Mount Blanc (4,807 m, 15,771 ft.), found on the French-Italian border. Farther east, the ranges are not so high in Austria, where few peaks exceed 3,000 m (10,000 ft.). The high Alps are part of a series of ranges that includes the Sierra Nevada in southern Spain, the Pyrenees between France and Spain, the Apennines that form the Italian peninsula, the coastal ranges of the Dinaric Alps in Croatia, Bosnia, Macedonia, and Albania, and the Pindus in the Greek peninsula. The curve of the Carpathian Mountains in Slovakia and northern Romania, which continues in the Balkan Mountains of Bulgaria, forms a further extension.

Around the young folded mountains are hilly plateaus (1,500 m, 4,500 ft.) of older rocks that once formed mountain ranges. The ranges were worn down by erosion and then raised again by faulting. Such areas include much of Spain and Portugal (the Meseta), France's Massif Central and Brittany areas, the Rhine Highlands of southern Germany, the Bohemian Massif of the Czech Republic, the uplands of western and northern Britain, and those of Norway. As the Atlantic Ocean opened along a divergent plate margin, uplift occurred along continental margins, together with volcanic activity. Rifting extended through the North Sea area, providing a downfaulted block of rocks that became oil reservoirs.

Lowlands with extensive areas less than 300 m (1,000 ft.) above sea level dominate southeastern England, northeastern France, the Low Countries, northern Germany, Poland, and the Baltic countries. The North European Plain extends from northern Germany eastward into Russia.

Long Coastlines and Navigable Rivers

Peninsulas such as Scandinavia (Norway and Sweden), Jutland (Denmark), and Brittany (France) in the north, and Iberia (Spain and Portugal), Italy, and Greece in the south allow arms of the ocean, such as the Baltic, North, Mediterranean, Adriatic, and Aegean seas, to reach far inland. Islands in the Baltic and Mediterranean seas add to the length of coastline that encouraged many groups to engage in trade and develop ship technology at a time in history when water transportation was easier than land transportation. Many **estuaries,** where rivers meet the seas, facilitated port building. During the most recent Ice Age, glaciers excavated deep valleys of Norway's Kjöllen Range as they cut their way to the sea. When the ice melted the sea rose and flooded inlets that are known as **fjords.**

On land, connections between places and world trade routes were eased by the existence of the North European Plain, covered in places by glacial moraine deposits dumped by the melting ice, and also by major river valleys that were also altered during the Ice Age. The Rhine and Elbe rivers in Germany and the Danube River flowing from Germany through Austria and the Balkans were particularly important. The Rhône, Seine, and Loire rivers in France, the Thames River in England, the Vistula River in Poland, and the Po River in Italy were also significant in movements of people and goods from early times. River valleys large and small also provided good soils. Sites for towns at river crossings and at the highest navigation points attracted and concentrated the settlement of growing populations. Modern investments in the Euroports at the mouths of the Rhine and Rhône rivers are important for maintaining Europe's continuing role in world trade.

The Danube River is longer than the Rhine, but the latter is used heavily for transportation and is the world's busiest waterway. The Rhine's watershed includes significant parts of four countries (Switzerland, Germany, France, and the Netherlands). Some of the main efforts at managing the Rhine waterway were devoted to making navigation possible for large barges from the international port of Rotterdam at its mouth up to Basel, Switzerland. The river was canalized to that point. Canalization straightened the rivers, producing faster flow with greater channel bed erosion in some sections and silt deposition in others. Channel deposition partly filled some sectors and caused flooding. Higher levees were constructed to protect from flooding the lands on either side that were used more and more intensively. At the Rhine mouth in the Netherlands, greater efforts were put into keeping out the sea than into maintaining the levees, which have been breached in places during times of high river levels.

The transportation uses of the Rhine River are linked to the growing industrialization of its watershed since the late 1800s. The coalfield and steelmaking areas of the Ruhr and Saar are now less productive and polluting, but some 20 percent of the world's chemical industry output occurs along the river with major centers around Basel, Mannheim, and the Ruhr area. Pollution was at its worst in the 1970s. Since then, an international agreement has been reached to reduce the problem, helped by improved water treatment technology. Concerns remain over nondegradable chemicals and metals that are still at high levels in the river.

Temperate Climates

Europe's climates are mostly temperate and humid, but include long freezing winters in the north and dry summers in the south. Most of the region enjoys environments that support farm production and reduce modern costs of both heating and air conditioning (see foldout world climate map inside back cover). No part of Europe is more than 500 km (320 mi.) from the coast, and mild and humid oceanic atmospheric influences affect the whole region.

Oceanic temperate climates extend from southern Norway to Northern Spain. The southern coasts have mild winters (0–10°C, 30–50°F) and warm summers (18–25°C, 60–80°F) with precipitation through the year. The North Atlantic Drift current, an extension of the Gulf Stream, has a major impact, bringing warmer ocean waters across the North Atlantic Ocean. Inland, the far north of Norway, Sweden, and Finland and the eastern parts of the region have climates with long, very cold winters and months of snow cover.

In the far south along the Mediterranean Sea coastlands, the **subtropical winter rain climate** (Mediterranean climate) prevails. Summers are hot and dry with drought-like conditions. In the winter, the midlatitude belt of cyclones moves southward, bringing rain and wind. In Central and Eastern Europe the **continental temperate climate** has severe winters as cold winds bring freezing temperatures from Russia and the northern countries. In summer, air over the land brings higher temperatures than on the coasts and thunderstorm rains.

Since the mid-1800s, Europe's climates have been getting warmer. Retreating glaciers in alpine valleys (Figure 2.3) have

FIGURE 2.3 Europe: changing climate. One hundred years has made a difference in Zermatt, Switzerland. (a) The glacier on the left side of the 1880 painting disappeared in the modern photo, (b), having retreated over a kilometer. Bare rock and icemelt deposits are exposed on the valley floor. Locate the church in both views and compare other 100-year differences. How might the changing climate have affected tourism?

(a)

(b)

influenced the tourist industry and the generation of hydroelectricity. This retreat and decreasing extent of snow cover shortened the winter sports season but added warm days for summer vacations. Inlets to hydroelectricity projects are now sited higher up the valleys, increasing the elevation (and energy) of water falling on the electricity-generating turbines.

Many Europeans are concerned about the potential impacts of global warming. Rising sea levels could drown the extensive low-lying parts of Europe, including the many areas of former coastal wetland that have been reclaimed and intensively populated. The Netherlands has particular worries in this area, but attention has also focused on the plight of Venice in northern Italy. Venice, a medieval trading city, is one of Europe's greatest architectural treasures and tourist attractions, but it is gradually subsiding (Figure 2.4). Increasingly frequent high tides weaken its foundations. The worst flood in Venice's history occurred in 1966. A south wind raised the tide level by 2 meters (6 ft.), covering St. Mark's Square with oil-polluted water that left stains on all the buildings. In 1990, St. Mark's Square was flooded 9 times, and in 2000, it was flooded 90 times. The industries and oil refineries of the lagoon that backs Venice pollute the waters and demand deepwater channel access, making protection of Venice difficult. As Venice has become increasingly less secure, the city's population has fallen from 175,000 in 1950 to 60,000 in 2005.

Forests, Fertile Soils, and Marine Resources

The natural vegetation of Europe is temperate forest. Deciduous trees (oak, elm, chestnut, beech) and associated brown earth soils dominated most lowland environments. Evergreen trees (mainly firs and pines) grew on the thin soils of the uplands and on sandy soils elsewhere, lowering their quality by acidification.

Because clearance of light woodland was easier, initial settlement (which occurred around 5,000 years ago by early Neolithic farmers) favored thinly vegetated uplands on limestone, sandstone, and some granite rocks. Also favored were the easily cultivated **loess** (windblown glacial debris) soils found along the southern edge of the North European Plain from Poland to northern France and southern Britain. The loess soils also provided an easy route of diffusion for farming technology originating around the Mediterranean Sea, via the Danube River valley.

After the introduction of Iron Age tools, humans rapidly cut into the denser forest on the lowlands, farming the exposed fertile brown earths. By the time of the Roman occupation of France and lowland Britain, a large proportion of the forest on the lighter lime-rich soils had been cut and some inroads made to the denser forest on heavier soils. Further expansion of the cultivated area onto the clay soils occurred during the Middle Ages. Huge demands for wood to construct ships and for charcoal used in smelting iron resulted in the removal of most natural woodland by around 1600. Later reforestation and the importing of exotic species added diversity to European woodlands.

As soon as humans cut the forest, soil washed down hillsides more rapidly and added silt to rivers. In the Middle Ages, for example, a combination of growing populations, rigid political systems, close grazing by sheep and goats, and climate change caused intense soil erosion on the hills of the Mediterranean peninsulas. In much of Europe outside the alpine area, however, the slopes are less steep, and cultivation methods were adopted that maintained the soils and their productivity over many centuries. In the 1800s, competition from cheap grain imported from newly opened and settled lands in North and South America resulted in once-plowed lands in Europe being sowed with grass for livestock production, further reducing soil erosion and helping to maintain soil quality.

In the Mediterranean, Baltic, and North seas and the North Atlantic Ocean fishing, related ports, and ships grew in significance in the later medieval period. In the 1900s and 2000s, overfishing led to decreasing supplies and made ocean fishing allocations a source of contention and conflict among European countries.

Environmental Impacts

Today, European countries are concerned about maintaining environmental quality, spending billions of dollars protecting the natural environment.

Impacts of Industrialization and Road Traffic

The Industrial Revolution and the spread of factory-concentrated production led to widespread pollution of the rivers and air. Many rivers lost their fish stocks. Occasional major pollution incidents still result in fish kills in major rivers such as the Rhine. Various

FIGURE 2.4 Venice. The medieval port city accumulated huge wealth on its unique site in a lagoon. Modern Venice is prone to flooding, as seen here.

(a)

(b)

FIGURE 2.5 **Environmental degradation in East Central Europe**. (a) Kraków-Nowa Huta, Poland. Sędzimir Steel Mill, formerly the Lenin Steel Mill. Communist governments rarely required any kind of pollution controls, and little is different for this factory today. The cemetery in the foreground is a telling commentary. (b) Pollution levels dropped in many areas of East Central Europe with the end of communism in 1991, after many inefficient factories closed because they could not compete in the global market. UN funds were provided to shut down this heavily polluting factory in Copsa Mica, Transylvania, Romania. Considered the most polluted place in Europe, local people called it "black town" because everything was black, from the air and sky to laundry on clothes lines and children's faces. Shown a few years after operations ceased, the sky is now blue and the grass green again.

governmental agencies are making great efforts to improve the quality of river water in European countries. In the Thames River of England, the reduction of pollution was so successful that fish stocks revived in the 1990s after decades of absence.

Winter smoke fogs (smog) that blighted the major industrial and urban areas of Europe in the 1950s have greatly declined due to legislation that outlawed domestic and industrial coal burning. Emissions of sulfur compounds from thermal power stations burning coal created **acid deposition**—of dry particles near the source or of wet "acid rain" farther downwind. It affected coniferous forest on thin soils and shallow lakes in the Alps and Scandinavia. The highest levels of acid pollution occurred during the Communist era along the Czech, Polish, and East German borders in an area known as the **Black Triangle** (Figure 2.5). Since the 1990s, acid rain and its effects were greatly reduced and alpine forests recovered.

As industrial pollution has declined, the increased number of automobiles and trucks pour sulfur, nitrogen oxides, and carbon particles into the air; these react with sunlight and lower air quality. Under anticyclonic meteorological conditions of slow-moving air, pollutants may accumulate to dangerous levels in broad valleys such those of the Po (northern Italy), Seine (France), and Tagus (Portugal). Although road congestion has increased, new cars from the 1990s emitted 93 percent less carbon dioxide and 85 percent less hydrocarbons and nitrogen gases than in 1970.

Global Environmental Action

The negative environmental effects of industrialization have motivated many Europeans to engage in global efforts to reduce pollution. Representatives from European governments were active participants in the Rio Earth Summit in 1992 and the

Kyoto conference in 1997. European countries comprise most of the industrialized countries on Annex I of the Kyoto Protocol (see discussion in Chapter 1, p. 13), accounting for more than 32 percent of global carbon dioxide (CO_2) emissions. Their ratification was key to making the Protocol go into effect. Romania was the first Annex I country to ratify the Protocol (32nd country overall to ratify). Iceland was the 55th country overall to ratify the Protocol, thus satisfying the first requirement for the Protocol to go into effect. European countries were also instrumental in motivating the Russian Federation to ratify, which it did in November 2004, likewise effectuating the Protocol by satisfying the second requirement that Annex I countries representing 55 percent of CO_2 emissions ratify the Protocol. However, although European countries support the Protocol in principle, most have been slow to meet the agreed emission reductions.

Distinctive Human Geography

Though Europe is small in area compared to other world regions, it has a complex mosaic of languages and religions. In comparison to much of the rest of the world, Europe's populations are rapidly aging and very urbanized, with major European cities serving as important nodes in the global economy. Over the last 500 years, Europeans developed new forms of identities and governance, in turn creating a template for modern nations and states. Intense competition between European peoples and governments led to two disastrous wars in the early 1900s and ended European imperialism. Economically, Europeans began creating new global trade routes in the late 1400s, garnering them great material wealth and technological advantages. Europe became the hearth of the Industrial Revolution and today continues at the forefront of modern agricultural practices, sophisticated manufacturing industries, and service industries such as tourism.

Cultural Diversity

Patterns of Language

Most Europeans speak languages in the Indo-European language family, which fall into one of three major groupings: Romance, Germanic, and Slavic (Figure 2.6a). The Romance languages stem from Latin, the language of the ancient Romans, and generally are still found in many of the lands that the Romans controlled longest. After Rome fell and its unifying influence was destroyed, Latin splintered and evolved into the differing Romance languages spoken today. The major ones are Portuguese, Spanish, French, Italian, and Romanian.

Spoken to the north and west of the Romance languages are the Germanic languages, which include English, Flemish, Dutch, German, Danish, Norwegian, and Swedish. English was heavily influenced by French after the Normans invaded England in AD 1066. Thus English has many Romance words, causing many to think that it is a Romance language though it is fundamentally a Germanic language. The influence of French has also given the English language a comparatively large vocabulary.

The Slavic languages dominate in East Central Europe and are divided into three major groups: western, southern, and eastern. The western group includes Polish, Czech, and Slovak. The southern group is comprised of Slovenian, Croatian, Bosnian, Serbian, and Bulgarian. The Slavic word for "south" is "yug," hence the name "Yugoslavia" ("land of the south Slavs"). The eastern Slavic languages are Russian, Ukrainian, and Belarussian and are spoken in Russia and Neighboring Countries (see Chapter 3).

In addition to these major language groupings, other smaller groupings and individual languages are noteworthy. The Celtic language group once was spoken in large areas of Central and Western Europe but was driven to the northwestern fringes of Europe by the Germanic peoples. Today, Irish Gaelic (Erse), Scottish Gaelic, Welsh, and Breton are the most commonly spoken of this group. In southeastern Europe, Greek and Albanian are spoken. Other languages found in Europe are Latvian, Lithuanian, Estonian, Finnish, Hungarian, Rom (Gypsy), and Basque.

Patterns of Religion

Christianity predominates in Europe though Islam and Judaism are important religions in the region (Figure 2.6b). Christianity became significant in Europe after the Romans adopted it as their empire's official religion in AD 381. The empire broke in two, leaving a Roman Catholic, Western Europe centered in Rome and an Eastern Orthodox, Eastern Europe centered in Constantinople (now Istanbul). The religious split became official in AD 1054. Eastern Orthodoxy organized along national lines and includes Greek Orthodoxy, Bulgarian Orthodoxy, and others. Roman Catholicism remained more homogeneous until the Protestant Reformation in the 1500s. At that time, the northern, mostly Germanic areas outside of or on the fringes of the former Roman Empire broke from Rome.

Jews have lived in Europe since Roman times, first along the Mediterranean and then eventually in small groups everywhere else, especially East Central Europe. Jews contributed greatly to European culture. Persecution of the Jews occurred through the centuries and culminated in the Nazi Holocaust in the 1930s and 1940s that sharply reduced their numbers. The *American Jewish Yearbook* estimates that in 1933, 7 million Jews lived in Europe as defined in this chapter. Poland, Romania, and Germany had the largest populations with 3 million, 980,000, and 565,000 respectively. In 2002, it was estimated that these countries had 3,000, 10,800, and 103,000 Jews respectively. Of the approximately 1.2 million Jews living in Europe in 2002, France had the greatest number with more than half a million.

In the Middle Ages, Islam spread into the Balkan peninsula with the Ottomans and into the Iberian peninsula with the Moors. After 1492, Muslims were driven out of the Iberian peninsula. Islam is still a significant religion in the Balkans. Following World War II, millions of Muslim migrants settled in Europe, especially in France (5–6 million), Germany (3 million), and the United Kingdom (1.6 million). Mosques and other expressions of Islamic culture are increasingly seen in the landscape.

Aging Populations

Dynamics

European countries are notable for their zero population growth rates. Europe's population is decreasing in most Mediterranean and East Central European countries and increasing only a little in most Western and Northern European countries (Table 2.1). Rates of annual natural population change ranged between –0.5 and 0.8 percent in 2006. Some countries, such as Germany and many in East Central Europe, have death rates that are above birth rates. Throughout Europe, total fertility rates declined from as high as 3 births per female in 1965 to commonly less than 2 in 2006. Italy and Spain had the lowest fertility rates (1.3) in the world. Though these two countries and Portugal are predominantly Roman Catholic in religion, the church's opposition to birth control is clearly having little effect. In Italy, few babies are born outside marriage, but greater access to careers for women, young people continuing to live with parents, and the end of pressures to have children tend to defer marriage and reduce the numbers of children. Western Europe has also experienced a greater number of divorces, later marriages, and increased numbers of widows due to longer life expectancy for women (80 years), creating more and smaller households so that more housing units were required. The weak economies of the former Communist countries of East Central Europe also have dampened population growth in that subregion.

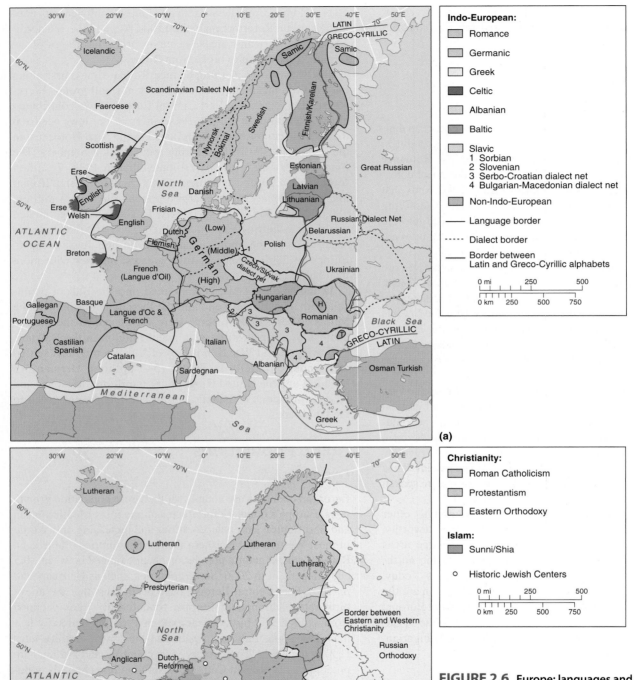

(a)

Indo-European:
- Romance
- Germanic
- Greek
- Celtic
- Albanian
- Baltic
- Slavic
  1 Sorbian
  2 Slovenian
  3 Serbo-Croatian dialect net
  4 Bulgarian-Macedonian dialect net
- Non-Indo-European
- —— Language border
- ---- Dialect border
- —— Border between Latin and Greco-Cyrillic alphabets

0 mi   250   500
0 km 250 500 750

(b)

Christianity:
- Roman Catholicism
- Protestantism
- Eastern Orthodoxy

Islam:
- Sunni/Shia

○ Historic Jewish Centers

0 mi   250   500
0 km 250 500 750

FIGURE 2.6 Europe: languages and religious groups. (a) Most European languages are in the Indo-European language family. The Latin alphabet, such as that used for English, is used by languages to the west of red line, and generally in Roman Catholic and Protestant Europe. The Greco-Cyrillic alphabets are generally used by Eastern Orthodox Christians to the east of the line. (b) Most Europeans are members of a Christian denomination though secularization is strong and Islam is growing. Are there any spatial correlations between language and religious groups? *Source: Data from Jordon-Bychkov.*

TABLE 2.1 — **EUROPE: Data by country, area, population, urbanization, income (Gross National Income Purchasing Power Parity), ethnic groups**

Country	Land Area (km²)	Population (millions) mid 2006 Total	Population (millions) 2025 est. Total	% Urban 2006	GNI PPP 2005 Total (US$ billions)	2005 Per Capita (US$)	Ethnic Groups (%)
WESTERN EUROPE							
Austria, Republic of	83,850	8.3	8.7	54.0	274.7	33,140	German 99%
Belgium, Kingdom of	33,100	10.5	10.8	97.1	343.7	32,640	Flemish 55%, Walloon 33%
French Republic	551,500	61.2	63.4	75.7	1,869.6	30,540	Celtic/Latin with Teutonic, Slavic, Nordic, African, Indochinese, Basque
Germany, Federal Republic of	356,910	82.4	82.0	87.5	2,406.5	29,210	German 92%, Turkish 2%
Ireland, Republic of	70,280	4.2	4.5	59.6	147.0	34,720	Celtic, English
Luxembourg, Grand Duchy of	2,586	0.5	0.5	91.0	30.1	65,340	Celtic/French/German 75%, some Portuguese, Italian
Netherlands, Kingdom of	37,330	16.4	16.9	64.9	531.2	32,480	Dutch 96%, Moroccan, Indonesian, Turkish
Swiss Confederation	41,290	7.5	7.4	67.8	277.5	37,080	German 65%, French 18%, Italian 10%
United Kingdom of Great Britain and Northern Ireland	244,880	60.5	65.8	88.9	1,976.9	32,690	Celtic, English 94%, South Asian, black, other 6%
Totals/Averages	**1,421,726**	**251**	**260**	**76**	**7,857**	**36,427**	
NORTHERN EUROPE							
Denmark, Kingdom of	43,090	5.4	5.6	71.8	182.5	33,570	Danish, Inuit (Eskimo), Faroese, Greenlander
Finland, Republic of	338,100	5.3	5.4	62.1	164.1	31,170	Finn 93%, Swede 6%
Iceland, Republic of	103,000	0.3	0.3	92.5	10.5	34,760	Norwegian/Celtic descendants
Norway Kingdom of	323,000	4.7	5.2	77.6	188.2	40,420	Germanic (Nordic, alpine, Baltic)
Sweden, Kingdom of	449,960	9.1	9.9	83.9	285.3	31,420	Swedes, Finns, Danes, Norwegians, Greeks, Turks
Totals/Averages	**1,257,150**	**25**	**26**	**78**	**831**	**34,268**	
MEDITERRANEAN EUROPE							
Hellenic Republic (Greece)	131,990	11.1	11.4	60.1	262.8	23,620	Greeks 98%
Italian Republic	301,270	59.0	58.7	89.8	1,701.3	28,840	Italian, Sicilian, Sardinian, German, French
Portuguese Republic	92,390	10.6	10.4	53.0	209.4	19,730	Mediterranean
Spain, Kingdom of	504,780	45.5	46.2	76.3	1,175.1	25,820	Spanish 74%, Catalan 16%, Basque 2%
Totals/Averages	**1,030,430**	**126**	**127**	**70**	**3,349**	**24,503**	
EAST CENTRAL EUROPE							
Albania, Republic of	28,750	3.2	3.5	45.0	17.1	5,420	Albanian 95%, Greek 3%
Bosnia-Herzegovina, Republic of	51,130	3.9	3.7	43.0	30.1	7,790	Muslim 40%, Serb 38%, Croat 22%
Bulgaria, Republic of	110,910	7.7	6.6	70.2	66.4	8,630	Bulgarian 85%, Turkish 9%
Croatia, Republic of	56,540	4.4	4.3	55.7	56.7	12,750	Croat 78%, Serb 12%
Czech Republic	78,860	10.3	10.2	77.0	206.8	20,140	Czech 81%, Moravian 13%, Slovak 3%
Estonia, Republic of	45,100	1.3	1.2	69.3	20.7	15,420	Estonian 64%, Russian 29%
Hungary, Republic of	93,030	10.1	9.6	65.1	170.5	16,940	Hungarian (Magyar) 90%, Romany (gypsy) 4%
Latvia, Republic of	64,500	10.1	9.6	65.1	170.5	16,940	Latvian 55%, Russian 32%
Lithuania, Republic of	65,200	3.4	3.1	66.7	48.2	14,220	Lithuanian 80%, Russian 8%, Polish 8%
Macedonia, Former Yugoslav Rep. of	25,710	2.0	2.1	59.0	14.5	7,080	Slav 65%, Albanian 21%
Poland, Republic of	312,680	38.1	36.7	61.6	514.6	13,490	Polish 98%
Romania	237,500	21.6	18.1	54.8	192.9	8,940	Romanian 89%, Hungarian 7%, Roma (gypsy) 2%
Slovak Republic	49,010	5.4	5.2	55.6	85.0	15,760	Slovak 86%, Hungarian 11%, Romany (gypsy) 2%
Slovenia, Republic of	20,050	2.0	2.0	48.9	44.4	22,160	Slovene 88%, Croat 3%, Serb 2%
Yugoslavia, Fed. Rep. of (Serbia-Mont.)	102,170	9.5	9.2	51.6	—	—	Serb 62%, Albanian 17%, Montenegrin 5%
Totals/Averages	**1,341,140**	**133**	**125**	**59**	**1,638**	**13,263**	
Europe Totals/Averages	**5,050,446**	**535**	**538**	**71**	**13,675**	**27,115**	

Source: World Population Data Sheet 2006, Population Reference Bureau. Microsoft Encarta 2005.

European countries are approaching or are in the fourth stage of the demographic transition, where birth rates and death rates are almost equal. The age-sex diagrams for Italy (Figure 2.7) clearly show where these countries are in the demographic transition. The exceptions to population decline are countries (e.g., Albania, Bosnia-Herzegovina, Macedonia) with significant Muslim populations. The traditional way of life still practiced in these countries is just as likely an explanation for population growth as religious belief. Europe's population is clearly aging with those over 65 years making up increasing percentages of the total population. Consequently, the burden on the welfare systems is likewise increasing and resulting in higher taxes and the reduction of some welfare programs, such as health and education, after years of wide-ranging coverage. Many countries in the region have had to cut back on the services they provide.

Densities

The highest densities of population in Europe are in the urbanized industrial belt that runs southeastward from central Britain, through northern France, Belgium, and the Netherlands, and into Germany (Figure 2.8). In Northern Europe, most people live toward the southern part of the subregion where the climate is relatively warmer. In Mediterranean Europe most people live in the lower parts of major river valleys, such as the Po River valley of northern Italy, and along the coasts. In East Central Europe, the main concentrations of people are in the urban-industrial areas on either side of the borders between the Czech and Slovak republics and southern Poland. Farther south, the Danube River valley has the main population centers in the cities of Belgrade (Serbia), Budapest (Hungary), and Bucharest (Romania) along the river's length.

Low population densities in Europe are typically in the mountainous areas such as the Alps, Apennines, Greek mountains, the Pyrenees, the Carpathians and the Dinaric Alps or in the areas of extreme climates such the colder areas of northern Europe or the hot, dry interior of central Spain

(compare Figure 2.1 and the foldout world climate map with Figure 2.8).

Urban Pressures

European countries are among the most highly urbanized countries in the world. The high proportions reflect the economic focus on urban-based manufacturing and service industries. Europe has few extremely large cities, since functions are often spread among several cities (Table 2.2). Paris (France) and London (United Kingdom) are by far the largest urban centers. Many cities have populations in the 1–3 million range.

The geographic characteristics of the city landscapes of Europe result from centuries of making and remaking built environments. They record the consequences of change (Figure 2.9). A number of Western and Mediterranean European cities contain within their centers relics of historic cultures such as ancient Greek, Roman, and Moorish. In medieval times, a network of towns grew all over Europe with walled defenses and market and administrative functions. Internal spatial differentiation by class occurred within buildings (with servants living in attics and basements and having to leave and enter from the back of buildings) rather than between central cities and suburbs.

With the Industrial Revolution of the 1800s, cities grew tremendously and land uses differentiated into central business districts of shops and offices, industrial areas of large factories and worker housing, and suburbs in which the growing management classes lived. In the early 1900s, the development of public transportation in cities led to further differentiation of land use and the building of suburban housing linked by streetcar or bus to the city centers where retailing, commercial, and manufacturing activities concentrated.

During World War II, bombing and ground fighting reduced parts of many European cities to rubble, destroying older areas including medieval and industrial buildings. The period from 1945 to 1970 was one of rehabilitation, expansion, and restructuring of cities. Nearly all cities expanded their functions and populations. City centers were rebuilt with utilitarian buildings

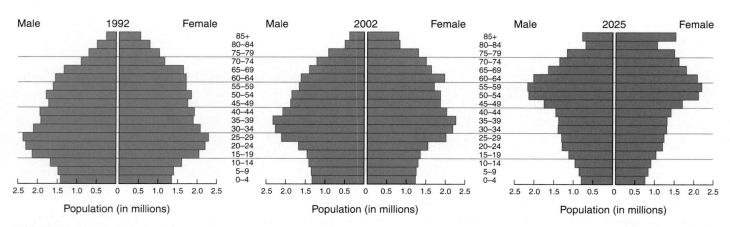

FIGURE 2.7 Age-sex diagram for Italy. The narrowing base of each successive pyramid is a feature of countries with declining birth and fertility rates.
Source: U.S. Census Bureau; International Data Bank.

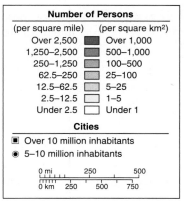

Number of Persons	
(per square mile)	(per square km²)
Over 2,500	Over 1,000
1,250–2,500	500–1,000
250–1,250	100–500
62.5–250	25–100
12.5–62.5	5–25
2.5–12.5	1–5
Under 2.5	Under 1

Cities

■ Over 10 million inhabitants
◉ 5–10 million inhabitants

0 mi   250   500
0 km 250 500 750

FIGURE 2.8 Europe: population distribution. Explain the heaviest and lightest concentrations within and among subregions.

(Figure 2.10), while large tracts of public housing catered to lower-income groups. New towns were built at a distance from the largest cities, separated by "green belts" of rural land in which new building was seldom allowed. A major result was decentralization—a movement away from the previous focus on the central business district and toward more economic activity in the extensive suburbs built since 1945.

Beginning in the 1970s, disillusionment grew in non-Communist Europe over the outcome of postwar reconstruction and the bleak nature of low-cost public housing. Large public housing tracts, especially where high-rise buildings were common, proved unpopular and often became centers of unemployment and crime. A significant move began toward public and private cooperation, particularly in efforts to make city centers and old dockland waterfronts more livable. Areas with buildings of historic interest were declared conservation areas, but this time the finances for pursuing such a policy came from pri-

vate investors who saw the potential returns from encouraging tourism and building new accommodations for those wishing to move back near the city center.

Older couples whose children had become independent and younger professionals moved to neighborhoods near city centers. They bought and renovated poor and dilapidated housing, increasing the value of the housing stock and raising neighborhoods from low income to higher income. This process is known as **gentrification.** Other changes affected city centers, bringing pedestrian-only traffic, arcades of specialty shops and the congregation of high-order services requiring support from people living in a wider area—theaters, opera houses, cinemas, and clubs. Areas around the city center that had become derelict, such as old railroad yards, factories, workshops, and worker housing, became sites for the new facilities.

In the 1980s, the relaxation of planning constraints led to a further dispersal of economic activity from city centers to the

TABLE 2.2	Populations of Major Urban Centers in Europe (in millions)	
City, Country	2003 Population	2015* Projection
WESTERN EUROPE		
Paris, France	9.8	10.0
London, UK	7.6	7.6
Essen (Rhein-Ruhr North), Germany	6.6	6.6
Frankfurt (Rhein-Main), Germany	3.7	3.7
Berlin, Germany	3.3	3.3
Düsseldorf (Rhein-Ruhr Middle), Germany	3.3	3.3
Cologne (Rhein-Ruhr South, Germany	3.1	3.1
Hamburg, Germany	2.7	2.7
Stuttgart, Germany	2.7	2.7
Munich, Germany	2.3	2.3
Birmingham, UK	2.2	2.2
Manchester, UK	2.2	2.2
Vienna, Austria	2.2	2.2
Mannheim (Rhein-Neckar), Germany	1.6	1.6
Leeds, UK	1.4	1.4
Lyon, France	1.4	1.5
Marseilles-Aix-en-Provence, France	1.4	1.4
Bielefeld, Germany	1.3	1.3
Hannover, Germany	1.3	1.3
Nuremburg, Germany	1.2	1.2
Aachen, Germany	1.1	1.1
Amsterdam, The Netherlands	1.1	1.2
Rotterdam, The Netherlands	1.1	1.2
Dublin, Republic of Ireland	1.0	1.1
Lille, France	1.0	1.1
Tyneside (Newcastle), UK	1.0	1.1
NORTHERN EUROPE		
Stockholm, Sweden	1.7	1.8
Copenhagen, Denmark	1.1	1.1
Helsinki, Finland	1.1	1.1
MEDITERRANEAN EUROPE		
Madrid, Spain	5.1	5.3
Barcelona, Spain	4.4	4.5
Milan, Italy	4.1	4.0
Athens, Greece	3.2	3.3
Naples, Italy	2.9	2.9
Rome, Italy	2.7	2.6
Lisbon, Portugal	2.0	2.1
Porto, Portugal	1.3	1.4
Turin, Italy	1.2	1.2
EAST CENTRAL EUROPE		
Katowice, Poland	3.0	3.0
Warsaw, Poland	2.2	2.2
Bucharest, Romania	1.9	1.8
Budapest, Hungary	1.7	1.7
Prague, Czech Republic	1.2	1.2
Belgrade, Serbia and Montenegro	1.1	1.1
Sofia, Bulgaria	1.1	1.0
*estimated		

Source: United Nations Urban Agglomerations 2003, with estimates for 2015 (2003).

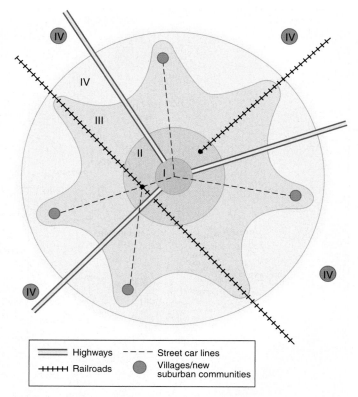

══════ Highways	– – – Street car lines	
╁╁╁╁╁ Railroads	⬤ Villages/new suburban communities	

FIGURE 2.9 **Generalized morphology of non-Communist European cities.** I = medieval core; II = Early growth area later penetrated by railroads and the location of early industrial factories and worker housing; III = Star-shaped pattern created by suburbanization along streetcar lines in the early 1900s; IV = Post World War II suburbanization with automobiles.

suburbs and beyond. Manufacturing had already moved out from cramped inner city sites to suburban industrial estates. Now retail and office facilities followed to become suburban warehouse shopping, hypermarkets, shopping malls, and office parks. After communism ended in East Central Europe around 1990, East European cities began experiencing this same phenomenon. Though these changes parallel what has been happening in the United States, densities of European cities are still much higher than in the United States and suburbanization is much less pronounced.

Global City-Regions

One way that Europe plays a leading role in the global economic system is through its cities, where multinational corporations site factories and where sophisticated service industries play important global roles. Of the global cities identified in Figure 1.16, Europe has by far the most with well over one-third. These European cities are no longer at the top of the list of the most populous cities in the world, but they are among the most globally connected cities. London and Paris, along with only New York and Tokyo, are the most important cities

(a)

(b)

FIGURE 2.10 Europe: urban landscapes. (a) Plymouth, England. The central shopping area was rebuilt after being bombed in World War II. It remains the largest retail district in the city despite the growth of out-of-town supermarket and warehouse shopping facilities since the 1980s. Some streets in this shopping area were made into pedestrian zones in the early 1990s. (b) Warsaw, Poland. The 38-story Palace of Culture was built with Soviet Union funds in the 1950s. It houses scientific and cultural institutions, theaters, and Congress Hall.

for accounting, advertising, banking, and law. When we think of cities such as London and Paris, we should not simply consider their businesses as only serving Londoners and Parisians, respectively. The businesses within them serve people all over the world, whether in the Americas, Africa, or Asia. Many millions of dollars flow in and out of these cities from and to other countries around the world. After London and Paris, cities such as Frankfurt, Milan, Zürich, Brussels, and Madrid rank high in their global importance.

Evolving Politics

Modern Countries: Nation-States

Beginning shortly after AD 1000, the identities of Europeans changed as people began to shift their loyalties from more local feudal leaders to large social groups and emerging countries. With the French Revolution in 1789, the idea of the nation emerged. As noted in Chapter 1, a nation is an "imagined community" of people who believe themselves to share common cultural characteristics (e.g., language, religion, ideology). Soon many Europeans believed that each nation should be free to govern itself and can only do so if it has its own **state** (i.e., country), which is a politically organized territory with an independent government. Thus, nations become linked with states in what is known as the **nation-state.**

Many Europeans spread their ideas of the nation and the nation-state around the world. Today each country in the world is also called a nation-state, that is, the home of a single people (e.g., France as the home of the French, Japan as the home of the Japanese, etc.). However, few countries truly live up to the definition of the term. Even France, where the nation-state concept originated, contains peoples such as the Basques who regard themselves as a nation separate from the French. Thus, the European belief in linking nation to state has created more of a **nation-state ideal** than a reality. Nevertheless, since the age of nationalism began in the late 1700s, many dominant nations imposed their national cultures on other nations living within their borders in an attempt to make the nation-state ideal a reality. In response, many minority peoples resisted, and some have tried to establish their own nation-states. The aforementioned Basques are an example and so are the Palestinians and Kurds (see Chapter 7). The nation-building projects of dominant peoples who try to homogenize their states and of minority peoples who try to declare their own nation-states have caused much violence, even war (see "Human Rights," p. 33).

Not all nations have been able to incorporate their nations' territories into their nation-states. Stronger nations hold onto some territory that other nations consider their own. These situations have caused the phenomenon of **irredentism,** which is the desire to gain control over lost territories or territories perceived to belong rightfully to one's group. With boundaries having changed so frequently in European history, irredentism is a major issue and a source of much conflict, more so than conflicts arising from simple cultural differences among groups.

Nationalism and World Wars

As capitalism ad nationalism grew in Europe, so did competition. European armies that were created to conquer and colonize the rest of the world were turned toward one another in 1914 as war erupted between the European powers, later to be known as World War I. Though Germany and Austria-Hungary were decisively defeated in 1918, the trouble was not over. War costs and protectionism caused European economies to slump in the postwar 1920s and 1930s, bringing hardship to millions of individuals and families. Discontent and resentment grew in the defeated countries. The nationalist competition became more bitter and fed the more extreme but opposing ideologies of fascism and communism, two other concepts Europe gave to the world. A European war engulfed the rest of the world once again between 1939 and 1945, known as World War II. The intolerant side of nationalism under fascism led to the extermination of millions of people of specific groups such as Jews (see "Patterns of Religion," p. 43) and Roma (Gypsies), a phenomenon called **genocide.**

Europe After 1945

After World War II, many Europeans seriously reevaluated their role in the world and their relationships with one another. The war had been so devastating that even the winners suffered destruction and huge financial, political, and cultural losses. In Western Europe, the United Kingdom, France, and the Netherlands were confronted with independence movements in their colonies at a time of weakness. Fueled by the ideologies of nationalism and communism that originally came from Europe, these movements ended four centuries of building colonial empires on which the "sun would never set."

At the same time, the politically and economically weak European countries were faced with the United States and the Soviet Union, which emerged as new world powers. At the end of World War II, the Soviet Union showed its strength when the Red Army moved into most of the countries of East Central Europe and fostered the establishment of **communism.** The Communists believed that capitalists used their riches to manipulate their governments in order to protect, even increase, their privileged positions in society and keep the majority of society, especially the working classes, powerless and in relative poverty. Instead, Communists argued for **democratic centralism:** the belief that the Communist Party, claiming to be the political party of the working class, was the only true representative of the people and, therefore, the only party with the right to govern. To keep capitalists and others from taking advantage of the people, Communists also believed in **state socialism:** governance by the Communist Party, which actively runs the political, social, and economic activities of the people. The state owned all the businesses and decided what was produced. The capitalist practice of competing companies producing similar products was seen as wasteful. Rather, large corporations owned by the state made each product. The state, not the free market of consumers, decided what needed to be produced through a **planned economy.**

As the Soviet Union imposed its Communist political and economic systems on East Central Europe, the rest of Europe (i.e., Western, Northern, and Mediterranean) immediately began to cooperate to counter further Soviet moves. In 1949, they formed the **North Atlantic Treaty Organization (NATO)** with the United States, which was seen as an ally against the Soviet military threat (Figure 2.11). To counter NATO, in 1955 the Soviet Union created a similar military alliance, known as the Warsaw Pact, among East Central European countries.

After the breakup of the Soviet Union in 1991, many questioned the need for NATO. Others feared that Russia would eventually become a formidable power again, though the Soviet Union no longer existed and Russia was weak. Having emerged from more than 40 years of Soviet domination, East Central European countries began joining NATO. In 2002, NATO formed a partnership with Russia, the country that NATO was created to defend against! Post-Cold War NATO is actually more focused on resolving or policing disputes within the expanded Europe and its immediate neighbors—as in Bosnia-Herzegovina, Kosovo, and Macedonia.

Global Economics

Capitalism and Imperialism

Of the European influences mentioned in the beginning of the chapter, capitalism has been very important and characterizes today's global economy. Capitalism, the practice of individuals and corporations owning businesses and keeping profits, began in the late 1400s when merchants invested in trade expeditions that brought in precious metals (gold and silver). The Portuguese and Spanish launched the first expeditions but were later followed by other Europeans. They all discovered and conquered new lands, radically altering and frequently destroying many local economies and cultures of indigenous peoples as they began a new era of **imperialism.** They also frequently pursued imperialism through **colonialism,** the settlement of their peoples in colonies in the new lands they conquered (see Figures 5.6 and 8.8).

The best known of the new voyages was made by Christopher Columbus in 1492. Funded by the Spanish crown, Columbus sailed westward in the belief he would get to India by a quicker route. He did not know that the Americas (supposedly later named after another Italian sailor, Amerigo Vespucci) lay in his path. At the same time, Vasco da Gama led the Portuguese explorations around the southern tip of Africa to India. The French, Dutch, and British followed the Spanish and Portuguese in the 1500s and 1600s.

Industrial Revolution

European technological innovations in the mid-1700s led to the **Industrial Revolution,** which was characterized by the use of machines for the mass production of goods. The resulting need for workers led to the growth of large urban centers devoted

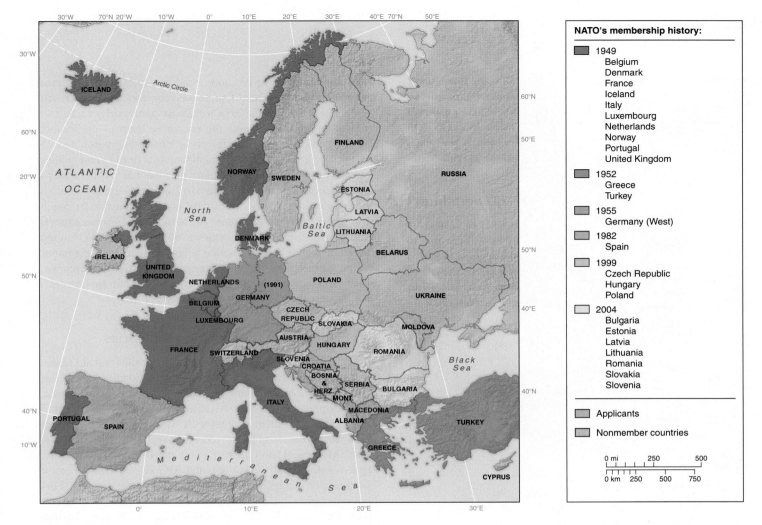

FIGURE 2.11 **Europe: NATO**. The North Atlantic Treaty Organization: member countries and those applying for membership. The United States and Canada are also original members. In 2002, Russia formed a special partnership with NATO.

to industrial work. Though machinery was first driven by waterpower, coal soon became the major energy source, leading to the growth of many urban-industrial centers near coalfields (Figure 2.12). With rivers, canals, and the sea as the initial forms of transportation used to transport raw materials and finished products, huge port facilities grew in the estuaries of Europe's major rivers. During the 1800s, railroads became more important and connected ports and coalfield industries.

The Industrial Revolution began in Great Britain and then spread across the English Channel to the Netherlands, Belgium, northern France, and the western areas of Germany. In the late 1800s and 1900s, the Industrial Revolution diffused further to central and eastern Europe and to other areas of the world. The European empires used their colonies to produce the raw materials needed in Western European factories. Cotton, wool, indigo, tobacco, and foodstuffs are a few examples. Forced to supply Western Europe with raw materials and having Europe as their only source of finished products, the colonies were pushed into economic dependency. Local economies and traditional ways of life were brought to an end as peoples in the colonies had to change their lives radically to produce exports for Europe.

Growing Economies

Two world wars and intervening economic depressions shattered the economies of European countries and their economic relationships with the rest of the world. After 1945, greater governmental intervention restored **productive capacity** (the amount of goods a country's businesses can produce) and now ensures education, health care, unemployment benefits, and pensions for all. In non-Communist Europe, older, "heavy" (**producer goods**) industries, such as steelmaking, heavy engineering, and chemicals located on coalfields, were replaced in value of output and employment by motor vehicles, consumer goods, and light engineering products. The availability of electricity spread, so that new products could be manufactured wherever there was plentiful semiskilled labor, often in the larger cities that were also large markets for the consumer goods. This process led to greater material wealth in the cities compared to the older industrial and rural areas. Unemployment increased in the polluted old industrial centers.

New industrial areas grew at locations more suited to the needs of developing technologies and industries, although

51

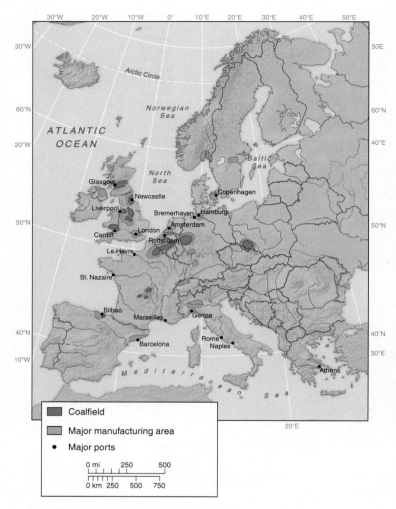

FIGURE 2.12 Europe: major manufacturing areas in the early 1900s. Note: Contemporary boundaries are drawn for reference.

long-established production continued in the older areas. It was too costly to move old production facilities. The advantages of producing goods in an area having a trained labor force, assembly and distribution systems, and financial and other services, build up and reinforce the original locational advantages as **agglomeration economies.** Keeping production in an area despite higher costs than in possible competing areas creates **geographic inertia.** It is only when a substantial change in costs occurs, new products emerge, or the demand for the original product is reduced that new areas develop and older manufacturing centers decline.

Deindustrialization occurred when the numbers of jobs in manufacturing fell rapidly and factories became derelict in older industrial areas. A decline of 20 percent in European manufacturing jobs between 1970 and 1985 was more than balanced by a 40 percent rise in tertiary sector jobs. However, the skills of blue-collar miners and factory production-line workers were seldom convertible into the new white-collar office, hospital, or classroom jobs that were taken by younger and better-educated people. Those workers

unable to retrain for the new jobs faced long-term unemployment. Older production workers, female workers, and poorly educated young people entering the labor force were particularly at risk.

Western Europe remains the economic heart of the region, with 60 percent of all manufacturing jobs in the region and 75 percent of the research and development that devises and applies new technology. Germany in particular has one of the world's largest economies and the largest in Europe. Germany is often referred to as the "motor" of Europe and is responsible for much of Western Europe's economic strength. Total German exports almost meet the combined exports of France and the United Kingdom, and its central position helps Germany to be the major trading partner for almost 20 European countries. By contrast, France and the United Kingdom are each the main trading partners for only three to four other European countries.

Sophisticated Manufacturing Industries

Two of Europe's major industries are automobiles and airplanes. Originally, the automobile industry was nationally based. For example, French companies were Renault and Peugeot-Citroën, Italian producers were Fiat, Alfa Romeo, Ferrari, Maserati, and Lamborghini, Swedish makers were Volvo and Saab, German firms were Volkswagen, Porsche, BMW, Audi, and Daimler-Benz (Mercedes), and British companies were Rolls Royce, Jaguar, Aston Martin, Bentley, and MG Rover. Globalization has radically changed the European automobile industry. Foreign companies have made great inroads in Europe in recent years. For example, the Japanese carmakers Nissan, Toyota, and Honda all built assembly plants in the United Kingdom to gain a foothold in the EU market. However, European carmakers have extended their operations abroad and into one another's countries (see Figure 1.21). As automobile companies are purchasing and investing in one another, it is no longer accurate to associate a company with a single country anymore. For example, in 1998, Daimler-Benz (Mercedes) merged with Chrysler, and Audi purchased Lamborghini. Volkswagen now owns Seat (Spain) and Škoda (Czech Republic). Renault bought a large share of Nissan in 1998. For a time in the 1990s, BMW owned MG Rover and then sold it again but kept the Mini. In 2005, the parent company of MG Rover sold MG Rover to a Chinese company. No major British automobile companies exist anymore though their names live on. Nevertheless, other European automakers are globally active by building auto-making plants in Poland, Turkey, Brazil, India, and China.

The aerospace industry is another key sector with aeronautical research and manufacturing well developed. Defense was a reason for the national support of this industry, but commercial production receives more attention. The small size of European countries makes it difficult for them to compete commercially with a country as large as the United States and its airline manufacturers such as Boeing. To compete, a group of European airplane manufacturers formed the Airbus consortium in the late 1960s to pool the resources of several countries. Companies in France, Germany, the United Kingdom, and Spain all manufacture various components that are then shipped to assembly plants in either Toulouse or Hamburg (Figure 2.13). Toulouse is the bigger assembly plant and headquarters for Airbus. The first airplane rolled off the assembly line in 1972, and Airbus soon captured 10 percent of market share. Growth has been steady and rapid. Today, Airbus employs 44,000 and captured the biggest share of new passenger-jet orders worldwide in the early 2000s.

Competition between Airbus and Boeing is now fierce. With commercial airline traffic at a high and likewise airport congestion, the two companies are staking their market positions on two different scenarios of what the future may bring. Airbus believes that continued congestion will result in greater difficulties for airlines in obtaining gates at major airports such as London's Heathrow, Tokyo's Narita, and New York's John F. Kennedy. Therefore, Airbus launched the A380, a plane that has 50 percent more floor space and at least one-third more seats than Boeing's 747. The plane flies greater distances, and airlines can have versions of the plane built to offer more comforts such as shower facilities, a piano bar, an exercise room, and a barber shop. The plane was scheduled to make its first commercial flight in 2007. In contrast, Boeing believes that greater airport congestion will lead to both new airport construction and the expansion of current facilities, calling for more mid-size and faster jets. Thus, Boeing is developing a modest-sized but very fast subsonic cruiser. It is difficult to predict the future, but European cooperation in aeronautics may make Europe the world's leader in the commercial airline industry.

Energy Sources

Domestic coal was the main source of energy in the region until the 1950s when cheaper imported oil, especially from Southwestern Asia, took over. Discoveries of natural gas in the Netherlands in 1963 led to the development of the North Sea basin (Figure 2.14). Today, Norway is one of the world's largest crude oil exporters. By the mid-1980s, natural gas became the cheapest and cleanest fuel for electricity generators. After the fall of communism in the Soviet Union in 1991, Europe began receiving oil and natural gas via pipeline from Russia's vast supplies, although its countries are concerned about the future level of control this might bring.

Of other sources of energy, nuclear-powered electricity generation is the most important, especially in France and Belgium (76 and 54 percent of the national totals, respectively, in 2002). Hydroelectricity is important in Alpine Europe. Wind power is found in northern Germany and the United Kingdom and tidal power in northern France, but these and other alternative energy sources provide a tiny fraction of current needs.

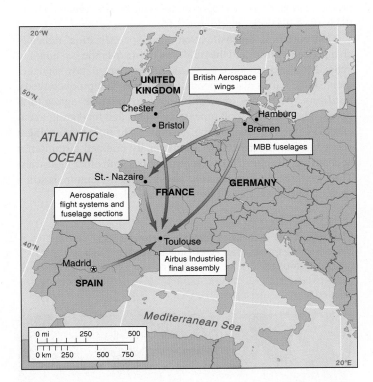

FIGURE 2.13 **Western Europe: cooperative European manufacture of Airbus aircraft.** The high levels of capital inputs and technological complexity required for constructing passenger aircraft make it impossible for the aerospace industry in one country of Europe to support the whole process. What are the political and economic implications of these movements?

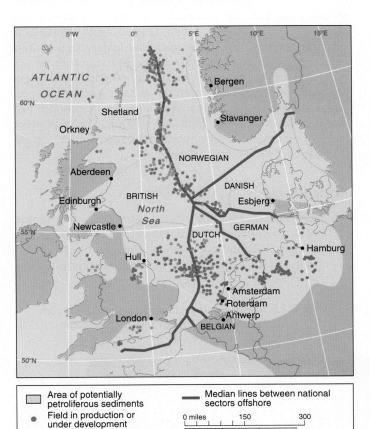

FIGURE 2.14 **North Sea oil and natural gas fields, 1995.**
Source: Data from Pinder 1998.

Service Industries

After World War II, service jobs grew in importance as the European countries increased their populations, became richer, and instituted strong social welfare programs. Jobs in retailing, wholesaling, education, health care, and government employment relate closely to numbers of people and population distribution. The providers range from private corporations, such as food and drink retailing, to state-controlled institutions in health care and education. In the mid-2000s, employment in the service sector made up over 75 percent of the total work force in most Western European countries and somewhat less in the other subregions of Europe.

The main growth areas in the service sector are in producer services and tourism. **Producer services** are involved in the output of goods and services, including market research, advertising, accounting, legal, banking, and insurance. They serve other businesses rather than consumers directly. Greater use of computers and information technology increased **productivity,** which is frequently measured by the amount of product generated or work completed per hour of labor. Producer services are closely related to global commercial developments and concentrate in the centers of major cities, where agglomeration economies are significant. As stated previously, agglomeration economies involve the clustering of businesses in a location, often near governmental agencies in the case of services, to save costs from the sharing of infrastructure, labor pools, market access, and transportation. London, Paris, Amsterdam, Frankfurt (Figure 2.15), and Munich have major shares of these industries. While some functions, such as back-office routine processes, are decentralized in suburban and small-town "paper factories"—often in landscaped office parks—major city centers retain the high-order functions in which face-to-face personal contact is important. In the early 2000s, many back-office and call-center jobs went to cheaper locations in Ireland and overseas to India and the Caribbean.

Agriculture

Beginning in mid 1700s, the Industrial Revolution transformed traditional agriculture in which peasants tilled the land with draft animals. Scientific crop rotation, which involved the planting of soil-enriching crops such as clover and turnips every four years, made it unnecessary to leave half or one-third of the land fallow every year. The use of tractors, fertilizers, and pesticides required fewer people to work larger farms. Particularly in the 1800s, people moved from the countryside to the cities, where they found jobs in industry and services. Although small farms of a few hectares continue to exist, many were taken over by more successful farmers, resulting in fewer but much larger farms, a trend known as **concentration.** Tractors, fertilizers, and pesticides also resulted in greater agricultural productivity per hectare, known as **intensification.** Today, agriculture still occupies almost as much land as before but accounts for only 2 to 7 percent of total employment in Europe.

Modern agriculture motivated farmers to switch from producing a variety of crops for their families to single crops that brought high profit, such as sugar beets or oil seeds. This is known as **specialization.** Specialization first occurred in basic grain and root crops. Then specialization developed in certain fruits and vegetables, more perishable and expensive to transport. These foods are associated with gardens, hence specialization in them is called **market gardening** (truck farming in the United States). For example, warm dry summers and cool winters that seldom drop below freezing allow the Mediterranean countries to produce crops that are difficult to grow in the other subregions of Europe, such as olives, table and wine grapes, citrus fruits, figs, and specialized cereal grains for pasta. In addition, Portugal produces most of the world's cork for wine bottles, obtained from the bark of the cork oak (Figure 2.16).

Concentration, intensification, and specialization radically changed agriculture and the rural way of life over the last hundred years. Most large modern farms require farmers to have management skills. Farms also depend on industries that produce seeds, fertilizers, pesticides, and machinery to produce agricultural goods. They rely on food processing plants and marketing agencies to get their products to the consumer. The term **agribusiness** was coined to describe how commercially oriented farming has become and the close links that farming has with other industries. Thus, while fewer Europeans work on farms, agriculture is still a very significant sector of European economies.

FIGURE 2.16 A cork oak forest in Portugal. Bottle corks are made from the bark of these trees. Little processing is required. Once the bark is stripped from the tree, bottle corks are cut. Numbers on the trees refer to a calendar year and indicate when the bark can be stripped from the tree again.

FIGURE 2.15 Western Europe: financial services. Frankfurt, Germany, is one of Europe's financial centers. Notice the preserved medieval buildings among the newer office complexes.

Despite the rise of agribusiness, farming is still a risky prospect with bad weather and pests causing bankruptcy. To provide a stable food supply, European governments subsidize farmers. Within the European Union (EU) (see "Political Changes: European Union (EU)," p. 63), the Common Agricultural Policy (CAP) was created. It first provided a set of price supports that made farming almost risk-free. As technology advanced, it led to increased output. Mountains of grain and butter, and lakes of milk and wine resulted. Farming is now less risky but agricultural subsidies to farmers remain the largest expense of the EU's funds.

During the 1980s, the EU began to change CAP to reduce the costly and unwanted surpluses. One result was to encourage live-stock farmers to produce less from the same area, a process known as **extensification.** Overall, the results of the attempts to reduce farm output were still not clear in the mid-2000s. With agriculture a politically sensitive topic, cutting subsidies has been difficult. Countries like France, Spain, Ireland, and Portugal benefit greatly from agriculture and resist change. Moreover, despite cutbacks, the agricultural sector continues to produce surpluses. The growing global economy in the 2000s led to greater complaints from other countries such as the United States that EU agricultural subsidies gave European farmers unfair advantages in international trade.

Tourism

Europe dominates the international tourist market as it received around 350 million, or 47.5 percent of 736 million world total, in 2005. European countries held six of the top ten places in terms of tourist arrivals: France was first and Spain second (the United States was third and China fourth) with Italy fifth (Figure 2.17a; see Figure 2.4), United Kingdom sixth, Germany

ninth, and Austria tenth. Germany and the United Kingdom provided most of tourists in other European countries that are also magnets for American and Japanese visitors.

Most tourists went to the Mediterranean countries with their warm sunny beaches, but the mountainous, rural, coastal, and historic urban areas elsewhere in Europe also increased their tourist trade (Figure 2.17b). While tourism is increasingly significant in Europe, tourist industries in other world regions have developed more rapidly. Europe still dominates the world in tourism, though its share of tourism receipts declined from 69 percent of the world total in 1975 to 47.5 percent in 2005.

Tourism is encouraged by the governments of each country, EU regional investment in infrastructure (roads, airports), and agreements on the sharing of health facilities, currency regulations, and customs. Individual countries promote their facilities through tourist boards. In the mid-2000s, tourism generated one EU job in eight, making it the largest industry in Europe. Much tourism employment, however, is seasonal, poorly paid, and female-dominated. The unskilled and low-paid nature of many jobs lead to them being taken up by migrants and thus having less impact on local economies than anticipated.

Geographic Diversity

Though Europe is smaller than many other world regions, it has considerable internal political, economic, and cultural diversity which can be generalized into four distinct subregions (Figure 2.18 and see Table 2.1).

(a)

(b)

FIGURE 2.17 Europe: tourist attractions. (a) Florence, Italy, has cathedrals, art galleries, palaces, and the Arno River. (b) Prague, Czech Republic, escaped much World War II destruction and is one of the best-preserved cities in East Central Europe. The tourists in this photo are on the Charles Bridge with Prague's castle on the hill.

FIGURE 2.18 Europe: subregions.

Subregion: Western Europe

Western European countries have been most significant in creating Europe's image as a global leader. The United Kingdom, France, and the Netherlands were three of Europe's most powerful colonial powers (Figure 2.19). They spread many European cultural characteristics around the world. Still today, English and French are two of the most commonly spoken world languages. Many Western Europeans were also the driving forces behind the spread of Protestant Christianity. The British, French, and Dutch in particular carried Protestantism to North America, South Africa, Australia, New Zealand, and parts of Asia.

Western European countries were also among the first to experience the Industrial Revolution. The combination of colonialism and industrialization gave Western Europe the ability to establish many of the global trade flows that are still in effect today. The United Kingdom maintains ties through the British Commonwealth. In 1945, France created a currency known as the CFA franc (franc of the French Colonies of Africa) for use in its western and central African colonies. These colonies are now independent countries, but the CFA franc still exists and ties their economies closely to France. France also maintains the French Foreign Legion, a military unit of non-French citizens that fights for French interests. France and Canada are the main supporters of the Agency for Francophony, an organization that promotes the French language and culture around the world. Western European countries also still import raw materials from countries that were their former colonies and then sell finished products back to them. Colonialism helped to make

Dutch cities such as Amsterdam and Rotterdam great trading centers, the latter being one of world's largest ports.

Today, Germany, United Kingdom, and France rank among the top ten largest economies in the world. These three countries are members of Group of Eight (G8), an informal organization representing the world's most materially wealthy countries. (Other G8 countries are the United States, Canada, Japan, Italy, and Russia.) The political power of Western Europe is underscored by the UK and France occupying two of the five permanent seats of the UN Security Council, the most powerful organ of the United Nations. The human environment of this subregion ranks high in health, education, and income standards with all countries ranking within the top 20 on the Human Development Index.

German Reunification

Germany is Europe's largest economy but has not had the same political influence as France and the United Kingdom since the end of World War II. Following Germany's defeat, the Allies (the United States, the United Kingdom, France, and the Soviet Union) divided Germany into occupation zones (Figure 2.20a). Though allies during the war, the Soviet Union disagreed with the Western Allies over Germany's fate. In 1949, the Western Allies formed their three occupation zones into the Federal Republic of Germany (West Germany), a democratic/capitalist country. The Soviet Union then created out of its occupation zone the German Democratic Republic (East Germany), a Communist country. With Berlin similarly divided into four occupation zones, the three western zones became West Berlin and politically associated with West Germany, though it was deep inside East Germany. Likewise, the Soviet zone became East Berlin and part of East Germany.

The Cold War division of Europe between non-Communist West and Soviet Communist East ran through Germany and its capital Berlin. As the Cold War intensified, the division of Germany and Berlin deepened. For example, in 1955, West Germany joined the North Atlantic Treaty Organization (NATO) and East Germany the opposing Warsaw Pact. At first, the borders between the two separate German states were not closed. People freely traveled from one country to the other. In Berlin, thousands of people crossed the border every day. However, the differences between West and East Germany grew over time. The American Marshall Plan and Western investment allowed West Germany to experience an economic miracle, and citizens enjoyed political freedom. In contrast, immediately after the war the Soviets dismantled factories in its occupation zone and shipped them back to the Soviet Union as war reparations, stunting economic growth in East Germany; citizens were also denied their political freedoms. Soon, more and more people from the East moved to the West. In order to stop this exodus, the East German government, backed by the Soviet Union, erected a wall between East and West Germany and around West Berlin in 1961 (see "Berlin" on website). Transportation routes were cut, the windows of houses

FIGURE 2.19 Western Europe: Paris, France. The Arc de Triomphe de l'Etoile stands on the hill of Chaillot and is the center of radiating avenues such as the Champs Elysées. In 1806 Napoleon conceived of a triumphal arch in the spirit of ancient imperial Rome that he could dedicate to the glory of his armies. Designed by Jean Francois Thérè Chalgrin (1739-1811), it was completed in 1836. Now a symbol of French patriotism, a huge French flag is hung from the ceiling of the arch on national holidays.

(a)

(b)

FIGURE 2.20   Western Europe: Germany. (a) Divided among the Allied powers. (b) Berlin: The historic Reichstag building is the location of reunited Germany's parliament. The new glass dome symbolizes the transparency of German democracy. *Source: (a) Data from Jones 1994; data from Heffernan 1998.*

facing the wall were bricked up, and families were separated. The wall disrupted the lives of Berliners, both East and West.

In the late 1980s, the weakening economies of Communist Europe and the Soviet Union began to undermine the Communist governments. Soviet leader Mikhail Gorbachev urged reform throughout the Communist countries, but East German leader Erich Honecker refused. Rising discontent among citizens led to mass demonstrations and people trying to flee East Germany. Erich Honecker was replaced by the more moderate Egon Krenz with promises of reforms. However, growing crowds in East Berlin pushed their way past the border crossings to enter West Berlin on November 9, 1989, marking the fall of the Berlin wall. This event began the end of Germany's Cold War division. Less than a year later, the two Germanys once again became one country, and in 1991 Berlin was declared the capital of reunited Germany (Figure 2.20b).

Subregion: Northern Europe

Northern Europe is sometimes called "Norden." This subregion of cold climates is sparsely populated but rich in natural resources. From the 800s through 1200s, its Viking inhabitants were very powerful and extended their control far beyond their subregion. Denmark and Sweden later became powerful empires. At the height its power, which lasted from 1610 to 1718 and was known as "the Great Power period," Sweden controlled the areas now known as Finland and the Baltics, and northern areas of Poland and Germany. The Swedish army was able to defeat Danish, German, and Russian forces, often simultaneously, before it was permanently weakened in 1721 and Sweden lost most of its possessions. Sweden and Denmark gave up imperial ambitions in the early 1800s and have since frequently taken positions of neutrality in international relations. Modern Norway was controlled either by Denmark or Sweden through the years and did not achieve self-governance until 1814. Finland was within the Swedish kingdom from the 1100s to 1809, when it became a Russian possession. Finnish independence was declared in 1917. Today, the four largest countries in the subregion have some of the world's highest GDP per capita figures and standards of living with high levels of education and health care (Figure 2.21) (see "Geography at Work: Business Development," p. 71). These countries are also great supporters of human rights. Their high standards make them reluctant to join any international organizations out of concern that membership will lower their standards. In the organizations that they join, they often do not fully participate for the same reason.

Fishing and wood products are generally important industries in Northern Europe. The Swedish sawmill industry is Europe's largest and accounts for about 10 percent of the world's exports. Sweden is also a major world exporter of pulp and paper. Denmark is specifically known for Tuborg and Carlsberg beers and the toy company Lego, though Denmark has many other industries that manufacture furniture, handicrafts, and high tech medical goods, automatic cooling and heating devices, stereo equipment (Bang & Olufsen), and sensitive measuring instruments. Sweden excels in engineering, iron and steel, chemicals, and pharmaceuticals. Swedish inventions such as the ball bearing are indispensable to modern industry. Global companies include SKF, ABB, Ericsson, Volvo, Saab, Astra, and Pharmacia & Upjohn. The Finnish company Nokia is world renowned for its mobile phone, Finland's most important export product.

FIGURE 2.21 **Northern Europe: Copenhagen, Denmark.** Copenhagen's urban landscape combines modernity with Scandinavian flavor.

Subregion: Mediterranean Europe

Mediterranean Europe played a major role in directing the early course of Western civilization, from ancient Greek and Roman ideas to those of the Italian Renaissance. In the 1400s, Portuguese and Spanish exploration launched the Age of Discovery and soon marked the beginning of European colonization of other lands and peoples. With colonization, the Portuguese and Spanish transplanted their languages and Roman Catholicism around the world. By the Industrial Revolution of the 1800s, however, Mediterranean Europe had lost a lot of its political power and Greece was part of the Ottoman Empire.

In the 1900s, differences among the Mediterranean countries grew. Portugal and Spain were ruled by two dictators: Antonio Salazar, who came to power in Portugal in 1926, and Francisco Franco, who took control of Spain in 1938. Following the deaths of these dictators in the 1970s, Portugal and Spain became more democratic and outward looking, joining European organizations. Greece emerged as a nation-state in 1832 but did not take its current shape until the early 1900s when Ottoman control ended. Italy is the youngest of the major Mediterranean nation-states, coming into existence in 1861 and experiencing some boundary changes as late as the 1940s. Regionalism within Italy is very strong, with most Italians considering themselves Sicilians, Tuscans, Venetians, and so on first and Italians second. Italy became the most industrialized of Mediterranean countries, followed by Spain (Figure 2.22). Portugal and Greece still

FIGURE 2.22 **Mediterranean Europe: Barcelona, Spain.** Barcelona is world renowned for its Temple de la Sagrada Familia. Begun in 1882 and designed by the Catalan architect Antonio Gaudi (1852-1926), this church is far from complete.

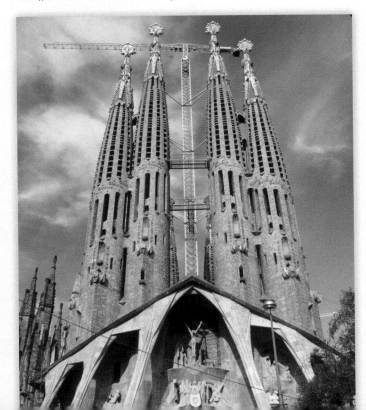

rely on agriculture, fishing, merchant marines, and tourism for much of their overseas income.

Industrialization began in Mediterranean Europe in the late 1800s in northern Italy and in the Catalonian region around Barcelona in northeastern Spain. Most modernization occurred after World War II and the incorporation of the Mediterranean countries in the EU. Industrialization in Greece and Portugal has been very modest. The two countries remained among the poorest economically of the EU countries and received large sums of regional development funds until countries of East Central Europe joined the EU in 2004.

Italy's GDP is almost twice the total of the other Mediterranean Europe countries and close to that of the United Kingdom, making it one of the world's largest economies. The country's economic power comes mainly from northern-based and high tech industries. The Po River valley between the Alps and the Apennines is the largest center of manufacturing in Mediterranean Europe. For example, Fiat manufactures automobiles in Turin. Venice, at the mouth of the Po River, is famous for its glass manufacturing. Milan, the largest city in northern Italy, is a major center of financial and other service industries as well as a producer of a diverse group of manufactured goods including tractors, domestic electronic goods, china, fashion, and pharmaceuticals. Milan and its surrounding towns produce nearly one-third of Italy's GDP and form one of the major growth areas of Europe.

Mediterranean Sea

Population growth in the 1900s, combined with crowding into coastal locations, industrialization, and the great increase in tourism, led to excessive pollution of the Mediterranean Sea—an important issue for many prospective tourists (Figure 2.23). The Mediterranean Sea is also one of the world's major shipping lanes (carrying 28 percent of the world's seaborne oil traffic) and has the world's highest level of oil pollution. It is an almost closed sea with a single narrow connection to the Atlantic Ocean at Gibraltar; water mixing dilutes the pollution, but it takes 80 to 150 years

for the Mediterranean to be completely renewed by the Atlantic Ocean. Confined within the Mediterranean, the chemicals and other nutrients from pollution decay and thereby deplete the sea's oxygen, killing sea life and creating large algae blooms that thrive in the anaerobic conditions. The Adriatic Sea has some particularly large algae blooms. The most polluted European areas of the sea are those near urban-industrial areas, such as Barcelona, Marseilles, Genoa, Naples, and Athens.

(a)

(b)

FIGURE 2.23 Mediterranean Europe: (a) Marbella, Spain. Crowded beach on the Mediterranean coast welcomes millions of tourists, particularly from Northern and Western Europe, during the summer months. Many buy permanent accommodations or time shares here. (b) The growth of tourism and oil spills have contributed greatly to the pollution of the Mediterranean Sea and the emergence of "hot spots," environmentally endangered areas. *Source: (b) Data from United Nations Environment Programme.*

Governments around the Mediterranean Sea met in 1975 and developed the Mediterranean Action Plan (MAP) to tackle pollution problems. Through MAP, the Barcelona Convention was adopted and required countries to reduce their pollution emissions. Sewage treatment improved in the wealthier countries such as France, but the poorer countries of North Africa, which also pollute the Mediterranean, have not been able to afford the necessary investment. Growing populations, tourist centers, and industrial projects have increased pollution, prompting six additional protocols and needed amendments to the Barcelona Convention in 1995. Even if pollution is reduced as countries get wealthier, the coastline and its delicate ecosystem are changed irrevocably when coastal wetlands are reclaimed and built over.

Subregion: East Central Europe

The countries of East Central Europe emerged as nation-states in the late 1900s and early 2000s. From the beginning of the nationalist idea in late 1700s to the end of World War I in 1918, most of East Central Europe was dominated by four great empires—the Russian, German, Austro-Hungarian, and Ottoman empires (Figure 2.24). The subregion enjoyed a brief period of independence between World War I and World War II, but then most of its countries had Communist forms of government and economies imposed upon them (see "Europe after 1945,"

p. 50). Most of these countries were directly controlled by the Soviet Union and were called Soviet satellite states. The Baltic countries were incorporated into the Soviet Union from 1945. Though Soviet domination ended by 1991, these countries share common experiences in moving from communism to more democratic forms of government and capitalist economies.

Prior to Soviet domination, East Central European countries also shared the common experience of being dominated by outside empires that treated their domains as colonies, preferring to extract resources and agricultural products rather than invest in industrialization. Thus the Industrial Revolution came slowly to East Central Europe. The Czech and Polish lands were somewhat exceptional in that a number of cities had factories. Hungary and Slovenia also developed industry. Nevertheless, a large number of people in the subregion still work in agriculture. Albania, for example, ranks as one of the most agricultural countries with 60 percent of its labor force engaged in farming.

After World War II, Communist economic policies were imposed on most countries in East Central Europe. The Communists improved the standard of living by encouraging industrialization, particularly in areas with coalfields, and by providing jobs for everyone, whether it made good economic sense or not. During Communist times, farming efficiency improved as industrialization provided tractors to replace horse-drawn plows. Great strides also were made in technology.

In terms of trade, Soviet Communists preferred not to be ensnared by capitalist practices, which they considered to be

FIGURE 2.24 East Central Europe: country boundaries in the 1900s. Compare these with the contemporary boundaries shown in Figure 2.1. Note the sizes, shapes, and locations of Poland, Latvia, and Yugoslavia at each time. *Source: Data from Demko & Wood.*

corrupting. Therefore, they kept trade among fellow Communist countries by setting up the Council for Mutual Economic Assistance (CMEA or COMECON), which also competed with the European Economic Community (EEC). COMECON linked the subregions' economies with one another and with the Soviet Union, forcing greater dependency on the Soviet system and preventing any country from realigning itself with the West. Cheap oil and natural gas from the Soviet Union also increased dependency, but left a legacy of contaminated soil and water. After the end of Soviet control in 1991, the countries of East Central Europe experienced economic crises as they reoriented themselves to the world economic system (see Chapter 3). Countries were challenged with breaking up business monopolies owned by the state, providing productive jobs, and cleaning up the environment.

Personal income, which was low during Communist times, resulted in very low consumer goods ownership compared to the rest of Europe. After 1991, the difficulties in moving to capitalism initially led to an immediate drop in GDP and personal income. Eventually, Poland, the Czech Republic, Hungary, Slovenia, and the Baltic countries had the greatest successes in adopting capitalist practices with their new economic policies raising GDP again. These countries were also the most industrialized before and during communism, and have had stronger traditions of democracy. Significantly, they also were closest to countries of the European Union (EU), making them easy beneficiaries of trade and investment from countries in the other subregions of Europe, especially from Germany.

In 2004, East Central European countries comprised most of the 10 new countries that joined the European Union (EU). These 10 countries have had such successful economic policies that their economies have grown faster than those of the first 15 members. If current trends for both groups continue, these East Central European countries will have standards of livings as high as France, the United Kingdom, and Germany within twenty years.

Before Yugoslavia was engulfed in war beginning in 1991, it had one of the strongest economies and highest standards of living of Communist Europe. Not a Soviet satellite, Yugoslavia pursued its own course. It employed the Communist idea of centralized planning, but compared to the Communist countries in the Soviet sphere, it also allowed a more genuine practice of the Communist belief that workers should manage their companies. Consequently, productivity was high in Yugoslavia. Travel to non-Communist countries was not as restricted as in the Soviet sphere. As a result, thousands of Yugoslavs, especially Slovenes and Croats, sought work as guest workers in countries such as Germany. These workers sent millions of dollars back to family members in Yugoslavia.

Croatia may have been just as prosperous as previously mentioned Slovenia, but independence in 1991 was followed by the devastation of war and little foreign investment. Now that the wars are over, Croatia is rebuilding and the economy is growing. Having suffered great devastation, Bosnia-Herzegovina's economy is still very weak. Serbia's economy faces great difficulties following economic boycotts during the wars and NATO bombings. Though the war is over, the country is receiving little foreign investment. Macedonia escaped most of the ravages of war, but its landlocked position among unfriendly neighbors and

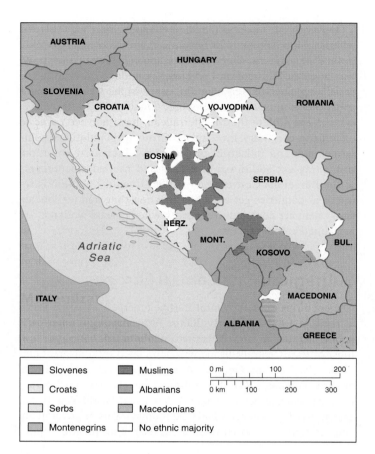

	Slovenes		Muslims
	Croats		Albanians
	Serbs		Macedonians
	Montenegrins		No ethnic majority

FIGURE 2.25 **East Central Europe: ethnic differences in former Yugoslavia.** Areas where a particular ethnic group comprises over 50 percent of the population. After 1991, Yugoslavia broke into five independent countries, although Serbia and Montenegro continued to call themselves "Yugoslavia" until 2002; Montenegro declared independence in 2006. Independence movements led to war in Croatia and Bosnia-Herzegovina. Violence in Kosovo (southern Serbia) in the late 1990s led to UN and NATO involvement in 1999.

its distant location from the wealthier countries of Europe attract little foreign investment to this largely agricultural country.

"Yugoslavia"

Most of East Central Europe transitioned peacefully from communism to democracy and market economies. The breakup of Yugoslavia in the 1990s was a major exception. In 1991 and 1992, all of Yugoslavia's republics except Serbia and Montenegro moved toward independence (Figure 2.25). The Serbs, who were the most numerous of Yugoslavia's ethnic groups, tried to keep the other republics from seceding from Yugoslavia, mostly because numerous Serbs lived in many of the seceding republics. Generally, the greater the number of Serbs that lived in a republic, the more problematic secession became. Thus Slovenia, which had very few Serbs, achieved independence relatively quickly and without much bloodshed. Secession was difficult for Croatia, where Serbs made up 12 percent of the population. After some warfare, Croatia

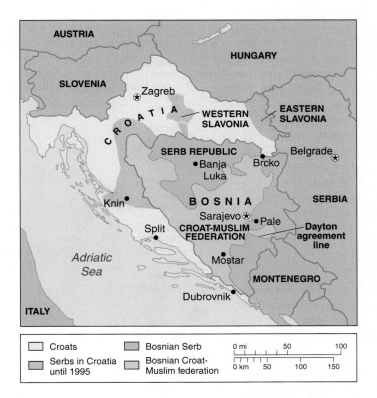

Croats
Serbs in Croatia until 1995
Bosnian Serb
Bosnian Croat-Muslim federation

0 mi   50   100
0 km   50   100   150

FIGURE 2.26 **East Central Europe: Croatia and Bosnia.**

achieved independence but lost control over its Serb-inhabited areas that resisted with the aid of the Serbian-dominated Yugoslav army. No ethnic group formed the majority in Bosnia-Herzegovina. The move toward independence was fiercely resisted by the large Serb population, which also received help from the Yugoslav army. International intervention in 1995 led to the division of the republic into two entities: a Muslim-Croat Federation and a Republika Srpska (Serb Republic) (Figure 2.26). While the war raged in Bosnia-Herzegovina, Macedonia avoided bloodshed, but to gain international recognition it had to officially change its name to the Former Yugoslav Republic of Macedonia (FYROM) to convince Greece that it would not attempt to annex the northern part of Greece, which is also called Macedonia. In 1999, UN forces took control of Serbia's southern province of Kosovo, which is inhabited by Albanians who desire independence. In 2002, Yugoslavia ceased to exist as the country's last two republics abandoned the name and called themselves Serbia and Montenegro. In 2006, Montenegro declared its independence from Serbia.

Contemporary Geographic Issues

Political Changes: European Union (EU)

After Soviet communism was firmly established in East Central Europe in the early 1950s, many Europeans felt that their nationalist notions and capitalist practices were threatened. Many European countries about to lose their colonies knew that they were too small to compete individually with the Soviet Union and United States. Moreover, the two world wars revealed the ugly sides of nationalist political competition coupled with capitalist competition and economic protectionism. Competition could lead to failure as well as success. On the other hand, cooperation in a non-Communist form was thought to result in everyone's success. Thus, to compete successfully again in the world economy over the long run, many Europeans in the non-Communist countries began to work together.

In 1949, not long after the end of World War II, Belgium, the Netherlands, and Luxembourg joined together in the Benelux customs union; the term **"Benelux"** is derived from **Be**lgium, the **Ne**therlands, and **Lux**embourg. In 1952, the Benelux countries joined together with France and West Germany to form the European Coal and Steel Community (ECSC). In 1957, the five ECSC countries plus Italy signed the Treaty of Rome to establish the European Economic Community (EEC), which was to evolve into the **European Union (EU)** known today (Figure 2.27). At the time, the EEC was expected to create a common market in which goods, capital, people, and services moved freely among countries. The European Commission became the executive arm of the community and was based in Brussels, Belgium. In 1967, the EEC changed its name to the European Community (EC) to emphasize the move from economic toward political goals. Other countries were motivated to join. Denmark, the Republic of Ireland, and the United Kingdom became members in 1973, Greece in 1981, and Portugal and Spain in 1986. By then, the European Parliament, with elected members from all these countries, was created and located in Strasbourg, France. It forms the legislative branch of the EU along with the Council of the European Union (often known just as the Council of Ministers). The president of the Council rotates every six months among representatives from each of the EU's member countries. Other EU organs are headquartered in Brussels and Luxembourg City (Figure 2.28).

The Single European Act, signed in 1986 when Spain and Portugal joined the EC, set out the steps to complete a single market. The Treaty of Maastricht (1991) attempted to set a timetable for monetary and political union and changed the name of the organization to European Union in 1993. Austria, Sweden, and Finland joined in 1995. The Swiss and Norwegian referendum votes on joining were close but rejected membership at the time. The Swiss most likely did not want to give up their tradition of neutrality, and it is believed that the Norwegians' desire to protect their fishing grounds and their practice of whaling played a major role in their rejection of EU membership. Both countries were probably concerned that membership would lower their standards. In 2002, 12 of the 15 EU members adopted a common currency known as the euro. It replaced the traditional national currencies such as the German mark, the French franc, and the Italian lira. The common currency facilitates the free movement of labor, capital, and goods to strengthen businesses and allow greater opportunities for consumers to buy the best products at the lowest prices.

In 2004, 10 countries, many formerly part of the Soviet bloc in East Central Europe, joined the EU in the organization's

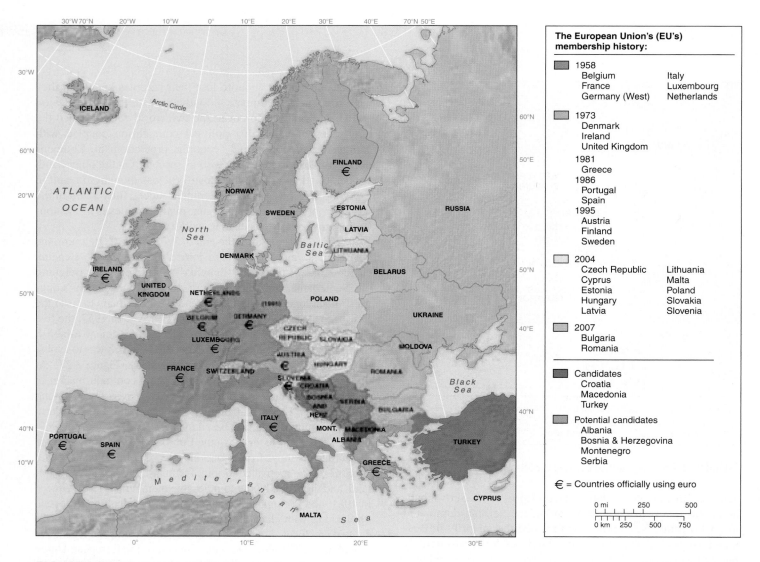

FIGURE 2.27 The European Union: its growth to date.

The European Union's (EU's) membership history:

1958
Belgium   Italy
France   Luxembourg
Germany (West)   Netherlands

1973
Denmark
Ireland
United Kingdom

1981
Greece

1986
Portugal
Spain

1995
Austria
Finland
Sweden

2004
Czech Republic   Lithuania
Cyprus   Malta
Estonia   Poland
Hungary   Slovakia
Latvia   Slovenia

2007
Bulgaria
Romania

Candidates
Croatia
Macedonia
Turkey

Potential candidates
Albania
Bosnia & Herzegovina
Montenegro
Serbia

€ = Countries officially using euro

FIGURE 2.28 European Union: Luxembourg. The EU flag flies alongside the national flag in front of the Luxembourg City walls.

largest expansion to date. Bulgaria and Romania joined the EU in 2007. Turkey's application has long been under consideration, and Croatia applied for membership in 2003.

European cooperation has proceeded for almost 50 years. Some see it as successful and others do not. In either case, the idea of individual European countries competing alone in the global economy or attempting to exert political clout is still a daunting prospect. Though the Soviet Union no longer exists, countries like the United States, Japan, China, and Russia are still much larger than any one European country. For example, the two largest economies in the world, the United States and Japan, have 300 and 129 million people respectively in 2006. In comparison, the most populous EU country is Germany with 82 million inhabitants. However, the combined population of the 27 EU members is almost 500 million people and has considerably more economic and political clout.

Though the EU is now very large with 27 members, the 12 new countries are mostly much poorer economically than the 15 older members and may be a financial drain. The East Central European countries have 53 regions similar to provinces. In 1998, before they became members, 41 out of the 53 regions were below 50 percent of the EU's GDP. Only the regions around Prague and Bratislava had per capita GDPs close to the EU average. The recent inclusion of the 12 new countries and possibly more places extra pressures on the European Regional Development Fund, established in 1975, which primarily sent funds to the western and southern margins of the EU, namely Ireland, Portugal, Greece, and southern Italy (the Mezzogiorno), and older industrial areas (Figure 2.29). The loans are used mainly for infrastructure projects, especially roads, telecommunications, water supplies, and waste disposal. Now EU regional development funds clearly need to be redirected to the new member countries of East Central Europe. Though the 2004 inclusion of 10 new countries and possibly more in coming years could make the EU a more powerful force than it is today, it could also delay or thwart further integration and actually undermine the goal of achieving greater economic and political clout through cooperation.

Cooperation may also lead to the loss of national sovereignty and the erosion of national identity. For integration to succeed, the member countries have to adopt similar—and in many cases, common—laws and economic policies. For the euro to work, for example, all member countries must limit their spending and keep their annual budget deficits within 3 percent of their GDPs. To do so, many EU countries may be unable, for example, to pay for their social programs and stimulate their economies as they see fit. They would have to give up a lot of what they value, and suffer

through economic slumps of high unemployment for the benefit of other member countries. If they choose to break the rules and spend, then they devalue the euro and damage the economies of the other member countries. In short, many Europeans are opposed to the euro because of the restrictions that come with it. The United Kingdom, Denmark, and Sweden have not adopted the euro and continue to use their own national currencies.

Integration also requires the removal of barriers, including border controls between member countries. States will not be able to stop the entry of foreigners, whether from other EU countries or from abroad as they enter through other member countries. This means that these foreigners may take local jobs, demand cultural rights, and generally be a visible foreign presence. It also means that countries will not be able to stop the surge of cheap, foreign goods and possibly contaminated food, driving businesses into bankruptcy and creating unemployment.

Two recent plans illustrate how many Europeans are concerned that European integration will lead to the loss of national sovereignty and the erosion of national identity (Figure 2.30).

(a)

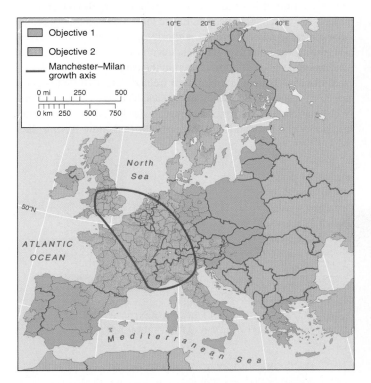

FIGURE 2.29 **European Union: structural funds 2004-2006.**
Areas eligible were the most disadvantaged regions (Objective 1) and those needing conversion following structural difficulties (Objective 2). All of Ireland and Portugal along with southern Italy and former East Germany received most available Objective 1 funds until recently; these funds are increasingly going to former Communist countries. Old industrial areas and some rural areas receive Objective 2 funds. *Source: © European Communities, 1995-2006.*

(b)

FIGURE 2.30 **European Union: reactions.** (a) Campaign posters in Ireland "for" and "against" the Nice Treaty in 2002. (b) Campaign posters in France "for" and "against" an EU Constitution in 2005.

The first example was expressed toward the Nice Treaty, which was drafted by the leaders of the EU's first 15 member countries in December 2000 at a meeting in Nice, France. The proposed Nice Treaty allowed for the admission of 10 new members, mostly from East Central Europe. However, it had to be ratified by all 15 member countries. Surprisingly, Irish voters rejected it in 2001, forcing Ireland's leaders to lobby for it strongly and put to a second referendum in 2002, when it was finally accepted by Irish citizens. The second example illustrating the resistance to European integration came with the drafting of a Constitution for the European Union (EU). In 2004, a Constitution was signed by the leaders of all the member countries but it had to be ratified by each member's government, some of which had the voters decide on the issue in referenda. Initially, nine parliaments ratified the constitution, and Spain's citizens voted for it in a referendum. The situation changed when French and Dutch citizens voted no in referenda on May 29 and June 1, 2005 respectively. Though Luxembourg's citizens and seven other Parliaments later voted in favor of the Constitution, unanimity was required. Thus, votes were postponed indefinitely in the remaining countries because the two no votes prevented the Constitution from ever being approved as it was written. The rejection of the Constitution by French and Dutch voters likely stemmed from the fear of loss of national sovereignty, especially following the EU's enlargement in 2004. Others argue that "all politics are local," and thus it was also an opportunity for voters to express their discontentment with their national leaders, especially those who were not protecting their nation's interests.

Though the European Union has its setbacks, nevertheless, it represents **supranationalism,** the idea that differing nations can cooperate so closely for their shared mutual benefit that they can share the same government, economy (including currency), social policies, and even military. During the Cold War, communism tried to offer a form of supranationalism, but after 1990 most Europeans abandoned the Communist experiment, leaving EU countries as the primary advocates of supranationalism. Members of the EU are still working out the details of their cooperation, but what they have accomplished is remarkable, considering that the more predominant nationalist idea, subscribed to by most of the world, holds that such cooperation is impossible between nations. Nationalism may still preclude ultimate political union in Europe.

It remains to be seen if supranationalism will work or not (Table 2.3). In many ways, the European Union is a grand experiment. Yet the combination of NATO security in the Cold War and linked economic policies enabled Europe to regain a major place within the global economic and political system by the end of the twentieth century. For example, European countries have many of the highest GDPs per capita in the world (Figures 2.31 and 2.32).

Devolution within European Countries

As Europe moves toward greater economic and possibly political integration, it is simultaneously experiencing **devolution,** the process by which local peoples desire less rule from their national governments and seek greater authority in governing themselves. The desire for complete independence is known as **separatism.** One example of devolution is seen in the case of the United Kingdom of Great Britain and Northern Ireland. Great Britain is comprised of England, Scotland, and Wales. The rise of Scottish and Welsh nationalism over the last few decades has resulted in greater autonomy for Scotland and Wales, with both now having their

TABLE 2.3	DEBATE: THE EU'S FUTURE
Reasons for the EU's Success	Reasons for the EU's Failure
Economic union pools together the resources of member countries and thereby strengthens the economies of every member country.	Economic union undermines the ability of member governments to make economic decisions that are in their national best interests.
Economic union allows for the free flow of capital and labor, permitting capitalist tendencies to strengthen members' economies.	The free flow of capital and labor undermines attempts by member governments to protect their national economies.
The euro, or common currency, further facilitates the movement of capital and labor by doing away with the costs of converting currencies.	The euro forces member governments to have monetary and budgetary policies that may harm their national economies.
Political union increases influence in regional and global politics because it combines the political and military strength of member countries.	Political union forces member countries to adopt foreign policies that go against their national interests (e.g., forcing some, like Ireland, to give up their neutrality).
Political union results in common laws and standards for individuals and the environment in member countries. It makes for better social and natural environments.	Common laws and standards among member countries undermine national needs and traditions, often watering down social and environmental laws.

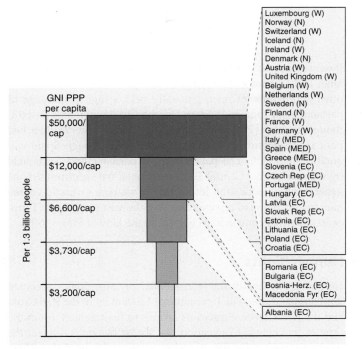

FIGURE 2.31 Europe: national incomes compared. The countries are listed in the order of their GNI PPP per capita incomes. Compare the relative material wealth of the countries in Western (W) and Northern (N) Europe with those in Mediterranean (MED) and East Central Europe (EC). Compare with Figure 1.20. *Source: Data (for 2005) from* World Development Indicators, *World Bank, and Population Reference Bureau.*

own parliaments with limited powers. Northern Ireland's situation within the United Kingdom is more complex. All of Ireland was in the United Kingdom when it was created in 1801. In 1922, independent-minded Irish succeeded in separating all but six counties of Ireland and creating the Irish Free State. Still part of the British Commonwealth, the Irish Free State had to swear allegiance to the British monarch until 1949, when it achieved its independence and adopted the name "Republic of Ireland." The six counties of the north remained part of the United Kingdom after 1922 as Northern Ireland. The violence in Northern Ireland often is depicted as a religious conflict between Roman Catholics and Protestants, but it is really a conflict between Irish nationalists and Unionists, also called Loyalists. Irish nationalists, feeling oppressed by the UK government, seek to unite the six counties of Northern Ireland with the Irish republic to the south. Because most Irish nationalists are Catholic and have the goal of uniting Northern Ireland with a Catholic country, and because Unionists/Loyalists tend to be Protestant and want to maintain a union with a Protestant country, the issue is often described as a religious conflict. As the conflict became progressively violent, even ambivalent Catholics sought protection of other Catholics, and Protestants sought refuge with other Protestants. The conflict has led to religiously segregated

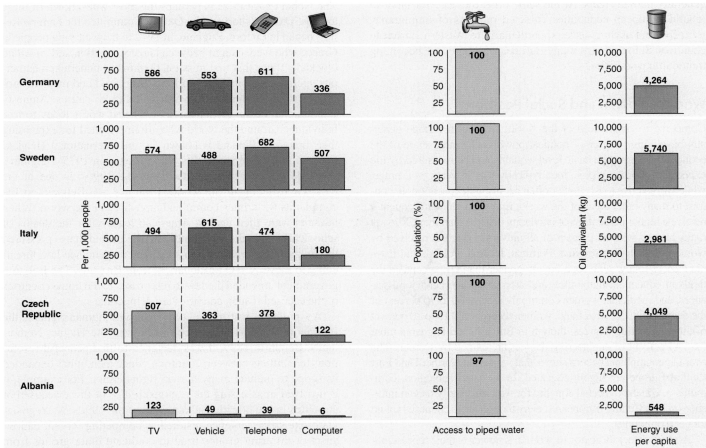

FIGURE 2.32 Europe: ownership of consumer goods and access to piped water and energy. Notice the diversity and summarize the differences among countries by subregion. Compare with Figure 1.19. *Source: Data (for 2002) from* World Development Indicators, *World Bank, 2004.*

communities, though almost as many violent acts are committed by Catholics against Catholics and Protestants against Protestants as across religious lines. Violence declined in the late 1990s as the UK and Irish governments have worked to create a shared government for Northern Ireland (like the devolved assemblies in Scotland and Wales). However, longstanding attitudes and distrust between groups make progress slow.

Europe has many other examples of devolution. Like the Northern Ireland case, the campaigns fought by the Basque peoples straddling the Spain-France border and the previously discussed disintegration of Yugoslavia are examples where violence has played a major role in the process. Other cases of devolution progressed relatively peacefully such as Estonia's, Latvia's, and Lithuania's independence from the Soviet Union and the division of Czechoslovakia into separate Czech and Slovak republics. The Czechoslovak situation was referred to as the "Velvet Divorce," coined from the "Velvet Revolution" of 1989 that peacefully ended communism in that country.

Multicultural Societies

Following the end of European imperialism after World War II, Europe led the way in promoting women's rights and more generally human rights. At the same time, the need for labor to rebuild European economies attracted millions of immigrants, refugees, and asylum seekers, particularly to Western European countries. Subsequently, many European countries are becoming multicultural societies.

Women, Power, and Social Position

Compared to other areas of the world, women in Europe generally have a high degree of political power and a large share of the economic wealth. At a basic level, equality can be measured by life expectancy. For the region, men typically live 75 years on average, while women live to 81. Power often begins with education. In contrast to many other areas of the world, European women frequently receive a higher percentage of university degrees than men. Norway ranks highest with 67 percent of all university degrees received by women. Bosnia-Herzegovina, Portugal, Iceland, Poland, and Estonia closely follow with each having greater than 60 percent. In the fields of education, humanities and arts, social and behavioral sciences, and medicine, women commonly receive 60 to 70 percent of the degrees. On the other hand, women receive only 15 to 20 percent of the engineering degrees, though in Bulgaria and Romania more than 43 percent of engineering degrees are earned by women. The overall geographical pattern shows that women in Northern and East Central Europe are highly educated. In the latter subregion, communism, which advocated equality for women, clearly had an influence on East Central European society though those countries are no longer Communist.

A more direct indicator of women's power is their representation in government. Northern European countries ranked among the highest in the world in this respect; Sweden placed second in 2006 with women holding over 40 percent of the seats in the lowest house in parliament, followed closely by Norway, Finland, and Denmark. The geographical pattern was varied in the other subregions. Women generally held a high percentage of parliamentary seats in Western Europe (the Netherlands is sixth) though Ireland (77th) and France (85th) were low. Women had poor representation in Mediterranean Europe though Spain was notably high (7th). The percentages of parliamentary seats held by women in East Central Europe spanned the spectrum despite generally high education levels among women. Overall, women had greater political representation in the vast majority of European countries compared to women in the United States (69th).

Human Rights

Europe is the hearth for international organizations concerned with human rights. The International Movement of the Red Cross and the Red Crescent traces its origins to Switzerland. From the time of its Geneva Convention in 1864, it has helped to set the modern standard for the ethical treatment of wounded soldiers and civilians during wartime. Sweden and Norway are known for the Nobel Foundation and the Nobel Institute, founded in 1900 and 1904 respectively. Alfred Nobel, the inventor of dynamite, placed his fortune in a fund upon his death. The interest from the fund awards people who have worked to benefit humankind. The Nobel Peace Prize is perhaps the most well-known of these awards. Oxfam (short for the Oxford Committee for Famine Relief) began in Oxford, England, in 1942 to help starving people in Greece who were caught between Nazi occupation and an Allied blockade. Since that first mission, Oxfam has undertaken numerous operations around the world to bring food and medical supplies to those suffering from war and similar causes. Amnesty International began in London in 1961 and works today to free individuals around the world who are imprisoned for expressing their opinions. Finland is known for the International Helsinki Federation for Human Rights, which began in 1975. Belgium's genocide law, which was enacted in the 1990s, is one of the most far reaching in the world. It allows non-Belgians to file complaints for crimes committed anywhere in the world. When lawsuits were filed against American leaders for the deaths of schoolchildren in the 1991 Gulf War, the United States pressured the Belgium government into amending its genocide law, threatening to move NATO headquarters from Brussels. The Belgium government amended the law to refer cases to the home countries of the accused if such countries are democratic.

A significant institution concerned with human rights is the International Court of Justice, located in The Hague, Netherlands. Begun in 1899, it was designed to find peaceful resolutions to conflicts between countries. Since then, it has expanded its scope to include many human rights issues. For example, it is involved in assessing the responsibilities of those accused of atrocities in the horrific Balkan wars of the 1990s. As Yugoslavia broke apart, violence resulted as competing fervent nationalists of differing groups tried to eradicate other groups from territories they desired (see Figure 2.26). Eradicating people from a territory because they are ethnically different is known as **ethnic cleansing.** Serb nationalists coined the term to refer to their own actions during the war in Bosnia-Herzegovina in the 1990s. They attempted to legitimize their claim to the republic by making the territory purely Serbian. Ethnic cleansing is now a term used to refer to similar acts perpetuated in other places around the world, not only currently but also in history. The Holocaust, for example, which refers to the Nazis' attempt to exterminate Jews (see "Patterns of Religion," p. 43) and other groups during World War II, is now referred to as ethnic cleansing, although the term did not exist at the time.

Widespread atrocities, including rape, in Bosnia-Herzegovina led to changes in international law. To deal with the atrocities, the International Court of Justice set up a special war crimes tribunal, known as the International Criminal Tribunal for the former Yugoslavia (ICTY). The tribunal tried numerous cases and continues to do so. Prior to the Yugoslav conflict, international law viewed rape as an unfortunate by-product of war, but the tribunal set the precedent that rape is used as a weapon in war and deserves greater punishment than before. Rape is now officially recognized as a war crime. In one case, the tribunal convicted a man for the crime of not preventing a rape.

Refugees and Asylum Seekers

The relatively high level of economic health and political freedom found in Europe, particularly Western Europe, has been attractive to political refugees and asylum seekers from other areas of Europe and the world. Germany, the United Kingdom, and France attract and receive the largest numbers of refugees. The wars in the Balkans greatly increased the number of refugees, with people from Bosnia-Herzegovina comprising the largest group in recent years. Germany has accepted by far the largest share, with over 300,000. Large numbers of refugees in Western Europe come from countries such as Sri Lanka, Turkey, Ghana, Iran, Iraq, Afghanistan, Somalia, Ethiopia, Democratic Republic of the Congo, Cambodia, and Vietnam. They contribute greatly to Western Europe's ethnic diversity.

Immigration and Evolving Identities

In addition to refugees and asylum seekers, Europe has received large numbers of immigrants looking for work. Immigration waves were created by World War II, for two reasons. First, the devastation of the war required massive rebuilding, but many young men of prime working age had been killed during the war. Second, the war weakened European countries with colonies, eventually forcing these countries to relinquish their colonies—a process known as **decolonization.** In short, European countries needed workers, and many were encouraged to come from the former colonies (Figure 2.33). The United Kingdom, France, and Germany have the largest numbers of immigrants. Yet all three have experienced and dealt with immigration differently over the decades and now face distinctly different problems.

The United Kingdom was the first to accept large numbers of migrants. In 1948, the British Nationality Act gave unrestricted entry to all subjects of the British Empire. The act also gave full political rights, which included the right to vote. Subsequently,

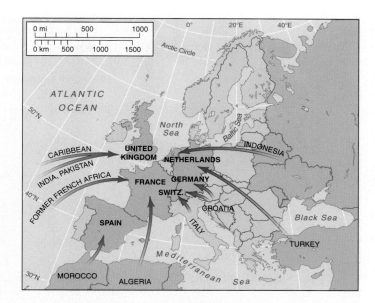

FIGURE 2.33   Europe: sources of guest workers, 1970 to the 1990s. Sources include people from the former colonies of the United Kingdom, France, and the Netherlands and countries around the Mediterranean Sea. What contributions do such workers make to the receiving and sending countries?

waves of migrants began arriving in the United Kingdom in the late 1940s from colonies in the Caribbean and the Indian subcontinent (India and Pakistan). Racial Relations Acts were enacted in 1965, 1968, and 1976 to outlaw racial discrimination. Race relations, however, primarily were seen as both a local and an urban issue. Thus emphasis was placed on local governments and urban redevelopment programs, particularly housing. As a result, racial minorities sought positions in local governments but success in achieving these positions varied from locality to locality across the country. Eventually minorities rose to higher levels of government and now have representatives in the British parliament. The 1999 Immigration and Asylum Act shifted emphasis from the British Commonwealth to all countries. In 2000, the Immigration Minister stated that migration was a "central feature" of globalization "with potentially huge economic benefits for Britain if it is able to adapt to this new environment." She also stated that the country is "in competition for the brightest and best talents," and emphasized that "Britain has always been a nation of immigrants." The United Kingdom has not been without its ethnic and racial tensions, but its open attempt to build a multicultural society has resulted in far fewer tensions than in France.

France also received huge numbers of immigrants from its colonies: Algeria and other African countries, and those in Southeast Asia, especially Vietnam (Figure 2.34). However, the migration waves began about a decade after they did in the United Kingdom. *Liberté, égalité, fraternité* (liberty, equality, fraternity), a concept stemming from the French Revolution and reaffirmed in the 1946 and 1958 constitutions, summarizes the French view of migrants. To the French, all citizens

FIGURE 2.34 Europe: the immigrant community. Immigrants are changing European societies. However, as indicated by these immigrants in France wearing the colors of France's flag, they also embrace French culture.

of France are French regardless of gender, language, religion, race, and ethnicity. Thus the French government generally does not recognize ethnicity and does not publish any related statistics. In the name of equality, the idea of multiculturalism and affirmative action programs (which ensure minority access to education, employment, housing, and any other sector) are shunned because they are seen as favoritism and thus also forms of discrimination. In 2004, the French government banned the wearing of headscarves in schools in order to reinforce the separation between church and state. The ban angered many French Muslims, though many young Muslim women supported the ban as a way of helping them break from the traditions of their families so that they could more easily enter into French society. Ironically though, the French belief in equality has prevented legislation that would stop discrimination and in turn help minorities integrate into society. Despite French ideals, minorities feel highly discriminated against and high unemployment has led to intense frustration and riots across the country that have resulted in the burning of thousands of automobiles. The French are holding firm to their view of what it is to be French, but that view is increasingly being challenged by France's minorities.

Germany has experienced and dealt with immigrants differently. Germany did not receive waves of immigrants from colonies because it did not have colonies. It initially recruited workers from other European countries like Portugal, Spain, Italy, Yugoslavia, and Greece. When not enough workers came from these places, workers were recruited from Turkey. Now more than 2 million Turks live in Germany. All these workers obtained the legal right to work but not citizenship and all the legal rights enjoyed by full citizens. The Germans coined a term for these foreign workers, *gastarbeiter,* or **guest worker.** Though allowed to stay as long as work was available, decades if necessary, the fundamental principle was that they would always be foreigners and would eventually go back to their respective homelands. The guest worker concept allowed Germans to hold to the idea that one was German according to ancestry and allowed Germany to preserve its "blood laws" concerning citizenship. By 1973 the labor shortage was over, and Germany stopped recruiting foreign labor. Soon, guest workers were encouraged to go back to their homelands, but many had been in the country for decades. They considered their original homelands "foreign" and would not leave. By the 1990s, the German population was aging and German's economy then needed younger workers for business and to support the social security system. Yet without the same rights as Germans, guest workers felt like second class citizens subject to discrimination. Germany's minorities became increasingly frustrated and tensions grew. Germans realized that the new global economy was a multicultural one and that Germany would have to open its arms to immigrants in order to survive. In 1999, the German government adopted the concept of "nation of immigrants" and changed its citizenship laws to allow immigrants to become Germans. Rules are strict. For example, a person is required to have irreproachably good conduct, show loyalty to the German constitution, sustain oneself and one's family without governmental support, and learn the German language. Tensions still exist within Germany because many immigrants have not been able to meet the standards. However, immigrants nevertheless have the ability to finally assimilate and show that Germany is becoming multicultural.

GEOGRAPHY AT WORK

Business Development

Geography is very useful to individuals working in business development. For example, Anne Hunderi (Figure 2.35), who received her master's degree in geography at the University of Oregon, works for the Hordaland County council in Norway. Hordaland County includes Bergen, Norway's second largest city. Since 1997, Anne has worked in various departments within the county. One of her first positions was that of project coordinator for the county. She oversaw a 6-million-dollar European Union project that identified the best ways to extend microcredit (small business loans) in six Western European countries. Later, she promoted Hordaland tourism in Scotland, coordinated study trips abroad, and arranged trade missions for local businesses. In her current position, Anne works primarily on developing the tourism industry and helping small businesses get started. Much of the Anne's help goes to specific groups like immigrants, young people, and women. For example, Anne has helped catering businesses and tailors to establish themselves. One interesting endeavour that Anne has helped to get its start is "Fresh," a small business that makes soaps and beauty products. Anne even used the resources of her position to open up her own bed and breakfast.

Anne believes that geography has great breadth and allows individuals with geographic knowledge to engage in many different jobs. For example, economic geography has helped Anne identify the location of resources, areas of supply and demand, and flows of trade. Political geography has helped Anne work with local land-use laws, international trade regulations, and immigration law. Understanding cultural differences and similarities has helped Anne work with many different people in her country and abroad. Geography's emphasis on the interconnections between these particular phenomena has been very helpful in Anne's career. In also applying the geographic concept of scale, Anne has helped individuals and local businesses become successful in the global economy.

FIGURE 2.35 Anne Hunderi, Using geography to promote local and international business development.

50°N 55°N 60°N 65°N 70°N 75°N 80°N

5°W
UNITED
KINGDOM

0°
BELGIUM

North
Sea

Norwegian
Sea

Orkney Islands

Shetland Islands

Arctic Circle

Svalbard
(Norway)

Franz Josef Land

5°E
NETHERLANDS
DENMARK

NORWAY
SCANDINAVIA
SWEDEN

Barents
Sea

LUX.
GERMANY

10°E
Baltic Sea

FINLAND

KOLA
PENINSULA

Novaya Zemlya

Kara Sea

Belyy

CZECH
REP.

15°E
RUSSIA
LITHUANIA
ESTONIA
POLAND
LATVIA

SLOVAKIA

Lake
Ladoga

Lake
Onega

Pechora

PECHORA
BASIN

GYDA
PENINSULA

20°E
Minsk
BELARUS

Dnieper

Northern Dvina

ROM.

25°E
MOLDOVA
Kiev
Chisinau

Moscow

Volga

Kama

URAL MOUNTAINS

Ob

WEST
SIBERIAN
PLAIN

UKRAINE

Don

RUSSIA

CRIMEAN
PEN.

30°E
Black
Sea

Volga

Ural

Tobol

Irtysh

Ob

40°N
CAUCASUS MTNS.

35°E
CASPIAN
DEPRESSION

TURKEY

Tigris

GEORGIA
T'bilisi
Yerevan

KIRGIZ STEPPE

Astana

SYR.
35°N
ARMENIA

Caspian Sea

Baku

USTYURT
PLATEAU

Aral
Sea

KAZAKHSTAN

KAZAKH
UPLANDS

AZERBAIJAN

IRAQ

Euphrates

UZBEKISTAN

Lake
Balkhash

30°N
TURKMENISTAN

Amu Darya

Tashkent

Bishkek

KUWAIT

Ashgabat

IRAN

KYRGYZSTAN

TIEN SHAN

BAHRAIN

25°N

TAJIKISTAN
Dushanbe

▲Ismoili Somoni
Peak
24,590 ft.
(7,495 m)

AFGHANISTAN

50°E 55°E 60°E 65°E 70°E 75°E 80°E 85°E

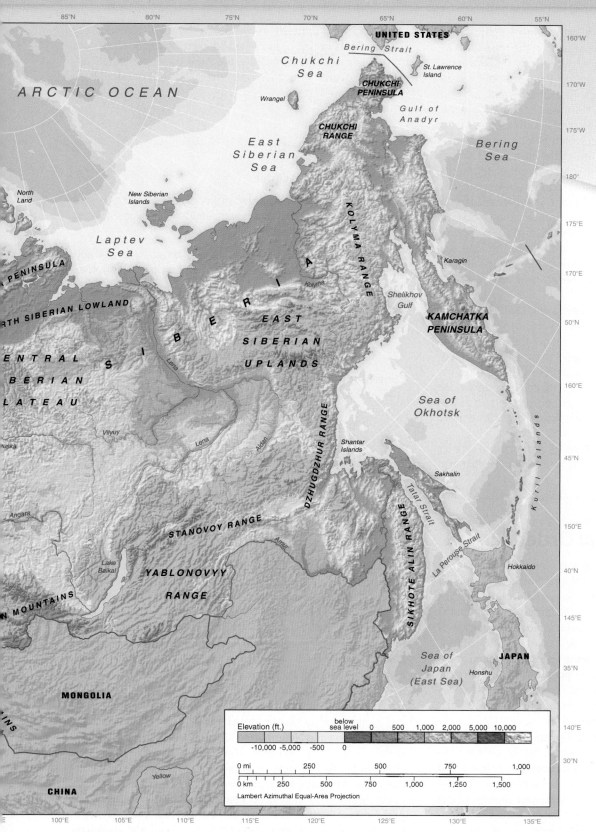

FIGURE 3.1 Russia and Neighboring Countries. Physical features, country boundaries, and capital cities. The region has mountains along its southern margin, with the Arctic Ocean to the north and the Pacific Ocean to the east. The land boundary with Europe in the west is marked mainly by lowland access. The interior lands are known for their climatic extremes: parched deserts in the south and harsh winters elsewhere.

Students voiced their discontent in both Georgia and Ukraine and helped to topple corrupt governments in the Rose and Orange Revolutions in 2003 and 2004 respectively. In Georgia, a student movement protesting corruption at universities evolved into *Kmara* ("Enough!"), a broad political movement seeking change through nonviolent civil disobedience. Through the use of large stickers and posters plastered in public places with slogans calling for change, together with marketing, media relations, public debates, and mass protests, the students of *Kmara* gained national attention and were joined by other segments of society. Though the country's 2003 general election was rigged in the favor of incumbent Eduard Shevardnadze, who was initially declared the winner, the continued and persistent use of civil disobedience forced Shevardnadze out of power in what became known as the Rose Revolution. The rose is a symbol of nonviolence and the symbol of democratically oriented opposition candidate Mikheil Saakashvili, who became Georgia's president.

A student organization by the name of *Pora!* ("It's Time") employed the same tactics and played a similar role in changing Ukraine's government in 2004. Though the government announced that Viktor Yanukovych won the presidential election against Viktor Yushchenko, many believed that the results were fixed. Massive protests and sit-ins ensued. Protesters wore Yushchenko's color orange in what later became known as the Orange Revolution following Yushchenko's victory in a new election a month later.

Communism ended in 1991, but many of the countries that gained independence from the Soviet Union, like Georgia and Ukraine, did not earnestly move down the road to democracy until students began organizing and exerting political pressure in the early 2000s. Significantly, the new, more democratically oriented governments have been more Western oriented than Russian oriented.

Chapter Themes

New World Regional Order

Distinctive Physical Geography: plains, plateaus, and major river valleys; southern mountain wall; continental interior climates; desert, grassland, forest, and tundra.

Distinctive Human Geography: cultural diversity; population; urban patterns and linkages; evolving politics; closed economies open up.

Geographic Diversity
- The Slavic Countries.
- The Southern Caucasus.
- Central Asia.

Contemporary Geographic Issues
- Is Russia still a world power?
- Human rights.
- Natural resources: oil and natural gas.
- Environmental problems.

Geography at Work: Wine Industry.

New World Regional Order

Until 1991, this entire world region was Communist and comprised of one country: the Union of Soviet Socialist Republics (USSR), more commonly called the Soviet Union. Of the Soviet Union's 15 republics, Russia was the largest and dominated politically and economically. The Soviet Union had many successes in its early years. However, in its last decades, its inflexible economic and political systems seriously weakened the country and allowed its discontented peoples to gain independence. The 15 republics of the Soviet Union became 15 independent countries. The three Baltic republics (Estonia, Latvia, and Lithuania), which had been part of the Soviet Union only since 1945, decided to seek greater cooperation with European organizations. The others, which comprise all the countries in this world regional chapter, share a more common political and economic legacy of the Soviet system. Recognizing the necessity of maintaining the trade flows established during Soviet times, the 12 former republics quickly formed the **Commonwealth of Independent States (CIS).** Merely an organization that allows member countries to discuss and work out economic needs, CIS members express no desire to pursue any form of economic and political integration like that of the European Union (see Chapter 2). Indeed, though heavily tied to Russia (formally called the Russian Federation), most are trying to break their dependency on the Russian Federation by establishing connections to other world regions. However, the Russian Federation is larger than all the other countries in the region put together (Figure 3.1), nearly twice the size of the United States; the smallest country, Armenia, is only the size of Maryland and Delaware combined. Not surprisingly, Russia is still the dominant political and economic force of this diverse region (Figure 3.2). Thus, we find it appropriate to call the region "Russia and Neighboring Countries."

Extending from the eastern edges of Europe to the Pacific Ocean, Russia and Neighboring Countries is the world's largest region in terms of land area. However, with much of its land either in the frozen arctic environments of the north or the hot deserts of the south, the region is the least populated on Earth. Great distances, river systems that tend to flow toward the edges of the region such as to the Arctic Sea, and extreme climates of hot and cold that make it difficult to build rail and road have prevented governments from integrating the region with highly developed transportation and communication networks.

FIGURE 3.2 Russia and Neighboring Countries: diverse landscapes and people. (a) New church, Kharbarovsk. (b) Georgian boys dance group "Preserve Our Culture." (c) Russian village, south Siberia.

Distinctive Physical Geography

The huge land area of Russia and Neighboring Countries results in a variety of natural environments. Vast plains and plateaus are bordered by mountains on the south. Winters are often long and harsh in some areas and summers hot and dry in other areas. Massive areas covered by a variety of natural vegetation types and soils provide a distinctive stage on which the development of human geographies occurred.

Plains, Plateaus, and Major River Valleys

Plains and low plateaus dominate most of the landscapes of this region. The North European Plain widens eastward from Poland into Belarus, Ukraine, and European Russia until it ends against the Ural Mountains (300–1,500 m; 1,000–5,000 ft.), which mark the line between Europe and Asia. Plains around the northern shores of the Black Sea and along the rivers leading to the Caspian and Aral seas extend into the nearly level, vast West Siberian Plains. Farther east, the relief again becomes hillier in the Central Siberian Plateau on ancient mineral-bearing rocks. The extensive areas of plains, interrupted by low hills, was a major factor in facilitating the invasions from eastern Asia and central Europe into Russia in medieval times and the later Russian imperial eastward expansion.

Across these low-lying landscapes flow some of the world's longest rivers. In the west, the Don River system flows into the Black Sea, which provides a strategic outlet to the Mediterranean; the Volga River flows into the Caspian Sea but is connected to the Black Sea via the Volga-Don canal. In Central Asia, the Amu Darya and Syr Darya rivers flow into the Aral Sea. The Ob, Yenisey, and Lena are the longest rivers of all and flow from the southern mountains of Central Asia and eastern Siberia northward to the Arctic Ocean; with their lower sections frozen for much of the year, meltwater frequently backs up and floods vast areas of wetland on both sides of their banks.

Southern Mountain Wall

The southern boundary of the region is marked by mountain systems. The Caucasus Mountains, between the Black and Caspian seas, are part of this line of mountain ranges that extends through the Elburz Mountains of northern Iran to the Tien Shan and the Pamir mountains along the southern borders of the Central Asian countries. These mountain ranges rise to over 7,400 m (24,000 ft.). They are snowcapped and in spring provide considerable meltwater for streams flowing through the dry areas of southern Russia and Central Asia. In the far east, the East Siberian Uplands and the volcanic peaks on the Kamchatka Peninsula parallel the Pacific coast. Among these high mountains are deep basins such as those filled by the Black Sea, Caspian Sea, and Lake Baikal.

Continental Interior Climates

Nearly all of Russia and Neighboring Countries lies north of 40°N and most of their area is north of 50°N. As such, these countries lie as far north as Alaska and Canada. The region contains places on Earth that are farthest from oceans, 500–2,000 km (320–1,250 mi.), and their moderating effects. They become extremely cold during winter and hot during summer, a situation described by the term **continentality** (see foldout world climate map inside back cover). The greatest temperature differences are in eastern Siberia, where large areas have January temperatures below −30°C (−22°F) and July averages of 12–16°C (56–60°F) (annual differences of 45°C, or 80°F). In the far east of Siberia, proximity to the Pacific Ocean results in more humid conditions, although winters remain long and very cold as Arctic winds sweep out from the continent's interior.

Western Russia also experiences continentality but not to such an extreme. Winter temperatures average −5 to −10°C (15–25°F), and summer temperatures average 15–20°C (59–68°F). With greater distance northward, the climate gets colder and the winters longer; northern Russia lies north of the Arctic Circle. The region's warmest climates are found around the Black and Caspian seas and in the Central Asian countries.

Precipitation declines with greater distances inland. Few parts of the region receive over 80 cm (32 in.) a year. Much of the region has moderate precipitation (40–80 cm, 16–22 in.) falling mainly in summer but producing a long-lasting snow and ice cover during the winter. Parts of southern Central Asia are arid because of high temperatures with high evaporation rates and distance from rain-bearing air masses. These areas rely on snowmelt from the mountains along their southern borders. Parts of far eastern Siberia have low precipitation and low temperatures.

Desert, Grassland, Forest, and Tundra

Latitude and climate (namely temperature and water availability) greatly influence the natural vegetation, which has also been greatly altered by human occupation. Hot deserts in the south change northward to steppe grassland, to deciduous and coniferous forest, and to tundra on the shores of the Arctic Ocean. These latitudinal distributions are interrupted only by mountain ranges (Figure 3.3a,b).

The deserts of the area east of the Caspian Sea contain few patches of grass and oasis vegetation and were once occupied by nomadic peoples. Trade routes via the region's oases provided long-established links between Europe and China.

North of the desert, the **steppe grasslands** grow on fertile **black earth soils (chernozems),** which are similar to the soils of the North American prairies. The steppes extend from Ukraine into Kazakhstan, one of the world's major arable regions. In the Middle Ages, the steppe grasslands provided the grazing lands through which wave after wave of invaders moved westward into Europe.

Farther north, trees can grow where evaporation rates decrease, and a zone of wooded steppe gives way to deciduous forests, linked to fertile **brown earth soils.** The trees have been largely cut to extend farmland.

Northward and eastward—where temperatures and precipitation decline—the deciduous forest gives way to forests dominated by hardy birch trees and evergreen pine, fir, and spruce. This **northern coniferous forest,** or boreal forests (taiga), dominates vast areas of land from Moscow northward and across most of Siberia. Pine and fir trees in the west give way eastward to larch. The taiga forms the world's largest forest area, covering 7.5 million km² (2.9 million mi.²; compare to 5.5 million km², or 2.12 million mi.², of rain forest in Brazil and 4.5 million km², or 1.7 million mi.², of related coniferous forest in Canada). It provides a huge reserve of biodiversity and is important in

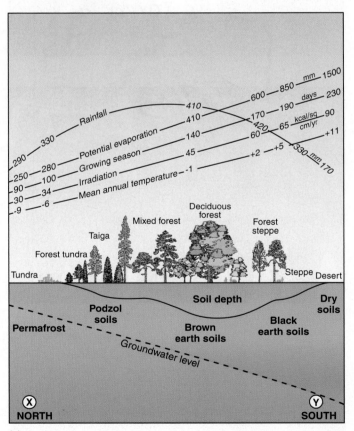

(a)

FIGURE 3.3 Russia and Neighboring Countries: natural vegetation and soils. (a) The section X-Y across the map illustrates the types of vegetation and soils in a north-south succession. (b) Map of natural vegetation and its relationship to soil types. *Source: Data from Brown etal. 1994.*

world climate change, absorbing one-third as much carbon dioxide as the tropical rain forest over a similar period. However, it is a very slow growing forest system compared to other forests. Underlying the taiga are poor **podzol soils** which have a gray sandy layer under a surface of slowly decaying leaves and above a layer where iron minerals accumulate, often becoming very hard and impeding drainage.

Around the shores of the Arctic Ocean and extending southward into the plateaus farther east is an area where trees will not grow. It is covered by grasses and low-growing shrubs. Such tundra vegetation is underlain by permanently frozen ground, or **permafrost,** that also extends southward beneath the taiga forest. The continuous and broken permafrost areas cover most of central and eastern Siberia and are up to 3,000 m (10,000 ft.) thick in parts of eastern Siberia. Farming is difficult if not impossible.

Distinctive Human Geography

Russia and Neighboring Countries remains ethnically diverse, reflecting three major influences: Christianity, penetrating the region from the southwest; Islam, entering from the south;

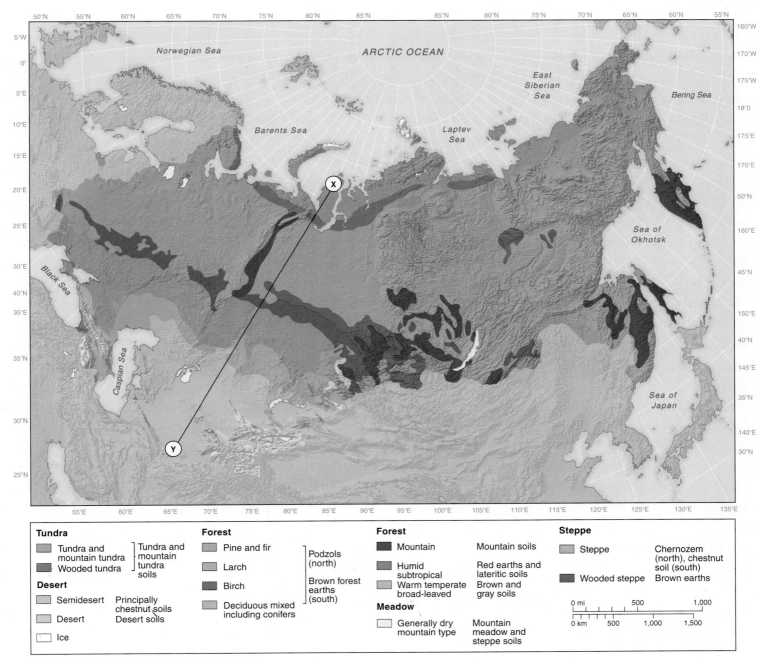

Tundra		Forest		Forest		Steppe	
Tundra and mountain tundra	Tundra and mountain tundra soils	Pine and fir	Podzols (north)	Mountain	Mountain soils	Steppe	Chernozem (north), chestnut soil (south)
Wooded tundra		Larch		Humid subtropical	Red earths and lateritic soils	Wooded steppe	Brown earths
Desert		Birch	Brown forest earths (south)	Warm temperate broad-leaved	Brown and gray soils		
Semidesert	Principally chestnut soils	Deciduous mixed including conifers		**Meadow**			
Desert	Desert soils			Generally dry mountain type	Mountain meadow and steppe soils		
Ice							

(b)

and Mongol culture, sweeping in from the east. At the same time, this world region is currently experiencing some dramatic changes. The Slavic countries and the Southern Caucasus are experiencing dramatic population decline, and Central Asia is showing great growth. The decline of economic subsidies given to people during Communist times to work in harsher physical environments has given people the incentive to migrate to more amenable climates within the region. Ethnic minorities, particularly Russians, are migrating back to their home countries such as Russia. Capitalism is changing urban patterns, causing some cities to decline while others grow with modern shopping facilities and global connections. The peoples of Russia and Neighboring Countries have had to cope with a dramatic transformation from rigid and inflexible Communist political and economic systems to more dynamic and uncertain democratic and capitalist systems.

Cultural Diversity

Russia and Neighboring Countries is inhabited by over 100 ethnic groups (Figure 3.4), but many of its peoples have as much in common with people of neighboring regions as they do with one another.

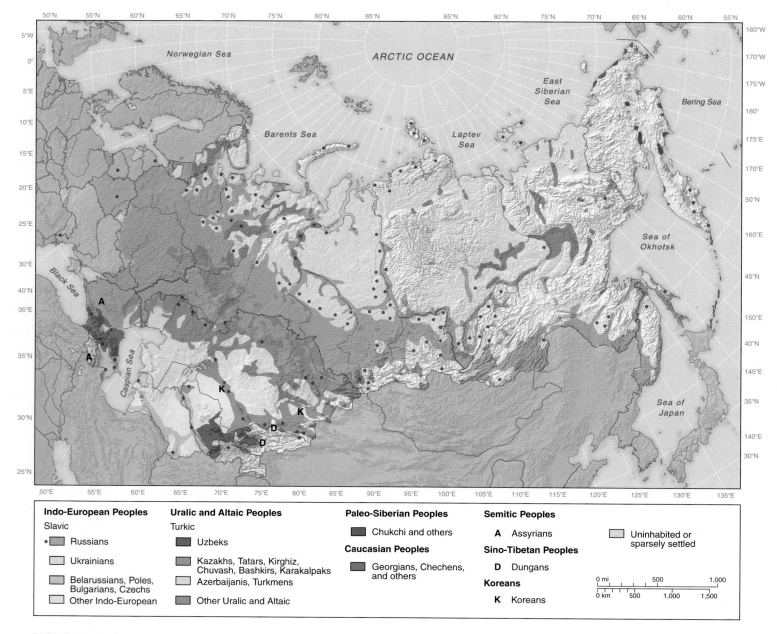

FIGURE 3.4 **The national groups in the former Soviet Union.** Note how the Russians spread across southern Siberia.

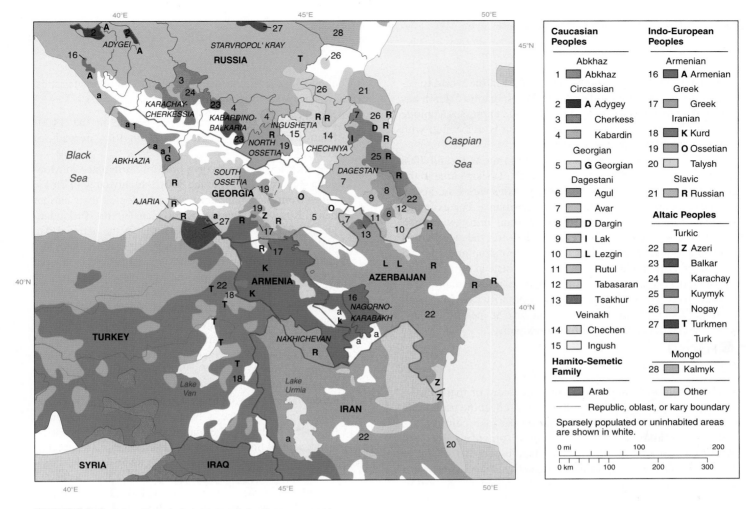

FIGURE 3.5 Ethnolinguistic groups of the Caucasus region.
Compare the distribution of groups to the location of political boundaries. Do these comparisons help to explain conflict in the subregion?

Eastern Slavs

Russians, Ukrainians, and Belarussians are Eastern Slavs, off-shoots of the broader Rus people and culture that emerged in the AD 800s and 900s. In 988, the Rus adopted Eastern Orthodox Christianity, which tied them to Constantinople. It also led to the adoption of the Cyrillic alphabet, which was derived from the Greek alphabet. All three Slavic groups are very closely related, as seen for example in the name *Belarussian*, which translates as "White Russian." Differences between them developed after Poles, Lithuanians, and Austrians ruled the western lands of the Rus from the 1300s to the 1800s. During this time, Ukrainians and Belarussians developed separate identities from Russians. Polish, Lithuanian, and Austrian influence is seen in the fact that nearly 20 percent of Belarus's population is Roman Catholic. A number of Poles and Lithuanians live in Belarus today, the result of Poland's boundary being relocated in 1945. Though many eastern Slavs live in Moldova, native Moldovans are very closely akin to Romanians and are not related to the Slavs.

The name *Ukraine*, meaning "borderland," illustrates history's shifting boundaries and influences. To the Russians, Ukraine is Russia's borderland, but to many Ukrainians, who have stronger ties with Europe than the Russians, Ukraine is Europe's borderland. Russian ties with Ukrainians and Belarussians were rees-tablished in the early 1800s, when the Russian Empire extended west. **Russification** began soon afterward and was particularly strong during Soviet times. Today, 81 percent of Belarus's population is Belarussian and 11 percent is Russian, but 63 percent of the population regularly speaks Russian and not Belarussian. Not surprisingly, Belarus's leaders have expressed interest in uniting Belarus with the Russian Federation. In Ukraine, Russians, living primarily in the industrial areas of eastern Ukraine, account for 22 percent of Ukraine's population. Many Ukrainians, however, fear Russian domination and cultivate their ties with the West.

Peoples of the Southern Caucasus

The Southern Caucasus is a small area of the world but it is culturally very diverse (Figure 3.5), partly because it lies at the historic contact zone between the Turkish, Persian, and Russian empires. The many languages spoken in the subregion are of different language families, so most have very little in common. Georgian is in the Caucasian language family. Armenian is Indo-European but stands alone on its own branch. Azerbaijani is a Ural-Altaic language (see Figure 1.13).

The Georgians (see Figure 3.2b) and Armenians both accepted Christianity early in history, in the AD 300s. The

Armenians claim their country was the first in the world to officially adopt Christianity. The Armenian Apostolic Church has been independent since the Middle Ages and expresses a unique view of Christianity. The Georgian Church is associated with Eastern Orthodox Christian churches.

Arabs introduced Islam to Azerbaijan in the 600s and 700s (Figure 3.6). In the 1500s, the Shia branch of Islam came to the country and now dominates. Despite close ties with Iran, Soviet secular policies greatly influenced Azerbaijanis. For example, Azerbaijani Muslims, unlike those in Iran, drink wine, and women are not veiled or segregated. In 1991, the Azerbaijani government also went against the wishes of Iran and adopted a modified Latin alphabet for Azerbaijani instead of the Arabic alphabet, the original alphabet of the Qu'ran used in many Muslim countries. The Latin alphabet is customarily used in Roman Catholic and Protestant countries but also in nearby Turkey, where the language is similar to Azerbaijani.

Central Asians

Modern Central Asians are settled, but many of their ancestors were nomadic, and their cultures still reflect the traditional ways of life. For example, in traditional Kazakh culture, it is customary to ask about the well-being of someone's livestock before inquiring about the person's health and that of his or her family. The dwelling of these nomadic peoples is the yurt, a circular tent consisting of a willow wood frame covered in wool felt. An opening at the top allows smoke to exit from the fire used for cooking and heating. Modern Central Asians no longer live in yurts, but they use them as decorative motifs for buildings or erect them in their yards and sleep in them during the summer. The yurt is an important symbol of national identity and appears on the national flag of Kyrgyzstan.

FIGURE 3.6 **Southern Caucasus.** Azerbaijani children.

Though Central Asians share some cultural characteristics, they also have some distinct differences:

- Kazakhs emerged as a distinct people in the 1400s from a mixture of Turkic and Mongolian nomads of Central Asia.
- The ancestors of the Kyrgyz probably originated in Mongolia (Figure 3.7a). After the 800s, they mixed with Turkic tribes from the south and west and adopted their current name, Kyrgyz, which means "40 clans" in the Turkic languages. The 40 clans are represented on the national flag with a sun that has 40 rays. During czarist times, both Kazakhs and Kyrgyz were called Kyrgyz, illustrating that the languages of the two peoples are very closely related.
- Turkmens trace their ancestors back to Oghuz tribes that inhabited Mongolia and southern Siberia around Lake Baikal. In the 700s, these tribes migrated into Central Asia and assimilated Turkic and Persian tribes, giving rise to the Turkmen. "Turkmen" probably means "pure Turk" or "most Turklike of the Turks." Carpet making and horse breeding are important national traditions. Five traditional carpet designs are incorporated into Turkmenistan's national flag. Akhalteke, a breed of horse well adapted to the desert, is the breed of national significance, appearing as the central figure in Turkmenistan's national emblem.
- Uzbeks are a Turkic people who moved into the lands now known as Uzbekistan in the 1500s. They are closely related to Turkmens but also have close ties with Tajiks. The Uzbeks, who prospered greatly from the Great Silk Road, wear clothing made with fine fabrics, color, and ornamentation (Figure 3.7b). All were expensive in earlier times, and the ability to incorporate as much as possible in one's dress showed one's wealth. This earlier culture practice is evident today. On any given day, it is still common to see individuals elegantly dressed, though they have no special function to attend.
- Tajiks probably acquired their name from an old Arab tribe. The Tajik language is a Persian language and was not distinguished from Persian (Farsi) until the Soviets designated Tajik as a unique language. At the same time, Tajiks had not differentiated themselves from Uzbeks, though the Uzbeks are a Turkic people. Both groups commonly spoke each other's languages.

Most Central Asians are Muslims. Those living closest to the Islamic heartland, such as the Uzbeks, converted to Islam as early as the 600s, soon after the rise of Islam. Those farther north, such as the Kazaks and Kyrgyz, converted to the religion only in the last 200 years. Some Central Asians seek ties with Iran (e.g., Azerbaijan and its Shia Muslims) or with other Sunni Muslim countries such as Turkey, Kuwait, and Saudi Arabia. These countries finance the building of mosques and other Islamic institutions in Central Asia. However, many Central Asians, like the Azerbaijanis, were heavily influenced by the secular ideas of Soviet communism and thus shun Islamic fundamentalists.

(a)

(b)

FIGURE 3.7 Central Asia. (a) Kyrgyz man drinking kumya—an alcoholic drink made from fermented horse milk. (b) Uzbek woman in front of Suzanne embroidery.

Women's Roles

Communist ideology professed that women were equal to men, and women began to achieve this equality following the revolution in 1917, before they were allowed to vote in the United States. During Soviet times, women in the region moved into positions traditionally held mostly by men in other societies (e.g., the United States), such as jobs in government, economics, medicine, or engineering. In some professions, women formed the majority. Most physicians, for example, were women. During World War II, some of the most decorated fighter pilots and infantry were women. Though communism gave women great freedom in their career choices, it also idealized the factory worker, not the physician or engineer. Consequently, women were given the freedom to move into professions that lost much of their prestige compared to the same professions in the capitalist world. At the same time, however, women also were encouraged to do industrial and construction work, holding more than 60 percent of the construction jobs.

Women achieved equality in their careers but usually not at home. After women completed a day's work at their jobs, many arrived home to undertake the traditional responsibilities of housework and child care. Men's attitudes and behaviors in regard to their wives did not change much during communism. As a result, between home and work, women ended up working long hours. As the region is transforming from communism to capitalism, the roles of women are changing. The idea that women are equal to men still exists, although more women than men have lost their jobs during difficult economic periods. Full and meaningful equality has yet to be realized.

Population

Decline and Growth

Not all of the countries of this world region are experiencing the same population dynamics. The Slavic countries and the Southern Caucasus are experiencing dramatic decline, and Central Asia is showing great growth.

Slavic populations may decline by several million by 2025 (Table 3.1). Growth rates typically ranged from −0.02 to −0.8 percent in 2006. Fertility rates were around 1.2, far below the fertility rate of 2.1 needed to maintain a population at its current number. The age-sex diagram for Russia (Figure 3.8) shows an uneven pattern of age groups as a consequence of World War II and a baby boom from 1950 to 1970, when large families were encouraged by the state. Women dominate the over-60 age groups to a greater extent than in other parts of the world as a result of male deaths in World War II and Stalin's persecutions. In 2006, birth rates were 9–10 per 1,000 of the population and falling, while death rates were 15–17 per 1,000

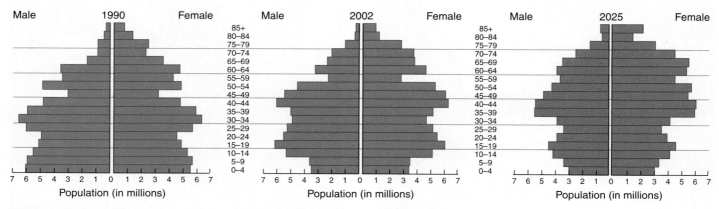

FIGURE 3.8 Russia: age-sex diagram. Account for the irregular shapes of Russia's population pyramids. How do they correspond to events of the last 50 years? *Source: U.S. Census Bureau; International Data Bank.*

TABLE 3.1

RUSSIA AND NEIGHBORING COUNTRIES: Data by country, area, population, urbanization, income (Gross National Income Purchasing Power Parity), ethnic groups

Country	Land Area (km²)	Population (millions)		% Urban	GNI PPP 2005	2005	Ethnic Groups (%)
		mid 2006 Total	2025 est. Total	2006	Total (US$ billions)	Per Capita (US$)	
SLAVIC COUNTRIES							
Belarus	207,600	9.7	9.4	72.0	76.7	7,890	Belarussian 78%, Russian 13%, Polish 4%
Moldova	33,700	4.0	3.8	45.4	8.6	2,150	Moldovan/Romanian 65%, Ukrainian 14%, Russian 13%
Russian Federation	17,075,400	142.3	130.0	73.0	1,514.5	10,640	Russian 82%, Tatar 4%, Ukrainian 3%
Ukraine	603,700	46.8	41.7	67.9	314.2	6,720	Ukrainian 73%, Russian 22%
Totals/averages	**17,920,400**	**203**	**185**	**65**	**1,914**	**6,850**	
SOUTHERN CAUCASUS							
Armenia	29,800	3.0	3.4	64.1	15.2	5,060	Armenian 93%, Azeri 3%, Russian 2%
Azerbaijan	86,600	8.5	9.7	51.6	41.5	4,890	Azeri 90%, Daghestani 3%, Russian 2%
Georgia	69,700	4.4	3.9	52.2	14.5	3,270	Georgian 70%, Armenian 8%, Russian 6%, Azeri 6%
Totals/averages	**186,100**	**16**	**17**	**56**	**71**	**4,407**	
CENTRAL ASIA							
Kazakhstan	2,717,300	15.3	16.0	57.5	118.2	7,730	Kazakh 46%, Russian 35%, Ukraine 5%, German 3%
Kyrgyzstan	198,500	5.2	6.6	35.0	9.7	1,870	Kyrgyz 57%, Russian 18%, Uzbek 14%
Tajikistan	143,100	7.0	9.3	26.4	8.8	1,260	Tajik 65%, Uzbek 25%, Russian 3%
Turkmenistan	488,100	5.3	6.6	46.8	28	5,840	Turkmen 77%, Uzbek 9%, Russian 7%
Uzbekistan	447,400	26.2	33.0	36.3	52.9	2,020	Uzbek 80%, Russian 6%, Tajik 5%
Totals/Averages	**3,994,400**	**59**	**72**	**40**	**218**	**3,744**	
Region Totals/ Averages	**22,100,900**	**278**	**273**	**54**	**2,203**	**5,000**	

Source: World Population Data Sheet 2006, Population Reference Bureau. Microsoft Encarta 2005.

and rising, yielding one of the fastest rates of natural decrease in the world that year.

For the Russian Federation, such figures hide variations within this huge country. In some parts, deaths outnumber births by two to one. The most significant geographic varia-

tion is between the western regions of the Russian Federation and the more recently settled regions of the north and Siberia. In western Russia, birth rates are low, death rates high, and the population is aging quickly. Yet the total numbers show less of a decline because other Russians are immigrating to western Russia from the north and Siberia and from the former Soviet republics.

Economic decline has been the major cause of population decline in the Slavic countries. Lack of government subsidies for industries in the north and Siberia is one factor. Elsewhere, many industrial cities have high unemployment because they have not been able to adapt to market economies. Many who are still working have not been paid for months or years. Abortion rates are high, a legacy of Soviet times when abortion became a common means of birth control. Health care has also suffered greatly with economic decline. Hospitals have frequently run out of basic supplies such as drugs and anesthetics. Economic hardship also has increased the incidence of alcoholism, particularly among men. For many hospitals, 90 percent of emergency cases have been alcohol-related. Environmental contamination (e.g., nuclear and water pollution) is a major problem too. The entire situation has given women little incentive to have children.

Some countries of the Southern Caucasus subregion experienced some decline after 1991. By 2006, growth rates ranged from 0.1 in Georgia to 1.1 in Azerbaijan. However, fertility rates ranged from 1.6 to 2.0, signaling future population decline if the situations do not change.

In contrast to the rest of the world region, the five Muslim countries of Central Asia anticipate growth from the 2006 total of 59 million people to 71.5 million by 2025. In 2006, Kazakhstan, which is more than 30 percent Russian, had the lowest growth rate (0.8 percent per year) and Kyrgyzstan the highest (1.4 percent). Death rates are very low—a formula for increasing population. Fertility rates fell from around 6 to between 3 and 4 from 1970 to 2006.

Distribution and Patterns

In the Russian Federation, the greatest concentration of people is in western Russia (Figure 3.9). Higher population densities continue eastward along the Trans-Siberian Railway to the southern end of Lake Baikal and Vladivostok on the Pacific coast. The extensive areas of mountains and desert in the south and of

FIGURE 3.9 Russia and Neighboring Countries: distribution of population. For Russia, comment on the location of the highest and lowest densities of population. How do these reflect climatic conditions, initial historic superiority, the line of the Trans-Siberian Railway, and other factors?

permafrost-ridden lands in the north and east contain few people. The more fertile lands of western Russia, Ukraine, Belarus, and Moldova have higher population densities and are punctuated with greater concentrations of people around industrial cities.

The Caucasus Mountains create uneven distributions of people in Georgia, Armenia, and Azerbaijan, with few people in the mountains and most in the plains. The distribution of population in Central Asia is marked by differences between areas of few people in the arid and mountainous zones and areas of higher densities in the irrigated lowlands.

During Russian imperial and Soviet times, all the peoples of this world region lived together in one country. To maintain control of this vast country, the government in Moscow encouraged Russians to move to other republics and the minority peoples, sometimes against their will, to move great distances from their territories. Consequently, when the fifteen republics of the Soviet Union became independent countries in 1991, many people found themselves outside their designated countries. Twenty-five million Russians, or 17 percent of the Russian population of the Soviet Union, live in the "Near Abroad" (Figure 3.10).

High concentrations live in Estonia, Latvia, northern and eastern Ukraine, and northern Kazakhstan. In many locales in the Crimean Peninsula (part of Ukraine), Russians constitute more than 90 percent of the population. Many of the new countries see their long-established Russian minorities as threats and have insisted that these Russians assimilate (e.g., learn the native language and way of life) or go back to Russia. By 2001, it was estimated that as many as 5 million Russians emigrated from the independent republics to the Russian Federation (Figure 3.11). Time has not decreased hostility toward Russians in many countries. After making life miserable for Russians in Turkmenistan, the leader of that country ended dual-citizenship for Russians in June 2003. Russians choosing Russian citizenship desperately tried to sell their property quickly because non-Turkmen were not allowed to own property. However, with only a few weeks' notice, many Russians had to sell at very low prices.

In the Southern Caucasus, most Armenians and Azerbaijanis do not live in their respective countries. Of the 6.3 million Armenians in 2001, less than 3.0 million lived in Armenia. Many lived in Azerbaijan and Georgia. Of the 19 million Azerbaijanis, less

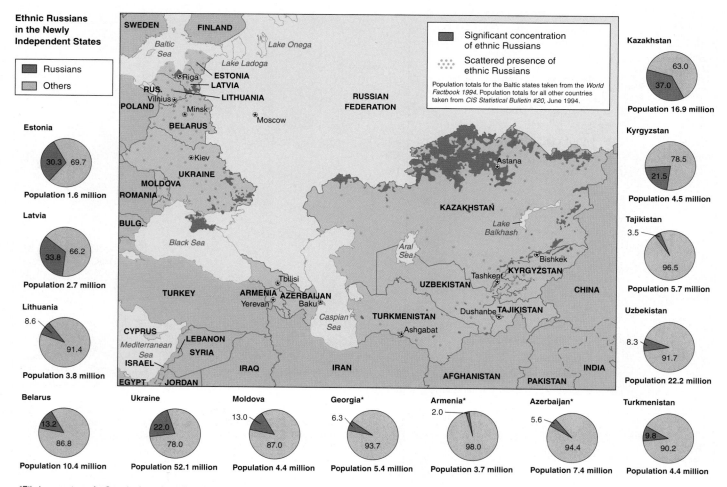

*Ethnic percentages for Georgia, Armenia, and Azerbaijan taken from the 1989 Soviet census; they may not accurately reflect present conditions.

FIGURE 3.10 Ethnic Russians as minorities in neighboring countries. In which countries are Russians the largest minorities? Do Russians outside of Russia enhance or hinder the political power of Russia in the region?

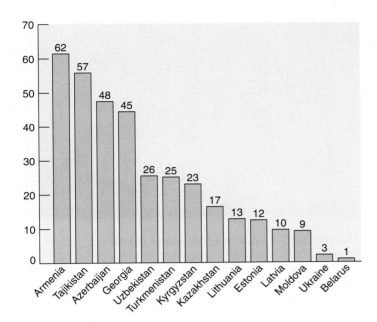

FIGURE 3.11 Migration of ethnic Russians from the former Soviet republics, 1989 to 1998. Each number represents the percentage of ethnic Russians who have emigrated from that country's total population of ethnic Russians.

TABLE 3.2	Populations of Major Urban Centers in Russia and Neighboring Countries (in millions)	
City, Country	**2003 Population**	**2015* Projection**
SLAVIC COUNTRIES		
Moscow, Russia	10.5	10.9
St. Petersburg, Russia	5.3	5.2
Kiev, Ukraine	2.6	2.6
Minsk, Belarus	1.7	1.7
Kharkov, Ukraine	1.5	1.4
Novosibirsk, Russia	1.4	1.3
Nizhniy Novgorod, Russia	1.3	1.2
Yekaterinburg (Ekaterinburg), Russia	1.3	1.2
Samara, Russia	1.2	1.1
Chelyabinsk, Russia	1.1	1.0
Kazan, Russia	1.1	1.1
Omsk, Russia	1.1	1.0
Rostov-na-Donu, Russia	1.1	1.1
Dnepropetrovsk, Ukraine	1.1	1.0
Ufa, Russia	1.0	1.0
Volgagrad, Russia	1.0	1.0
Donetsk, Ukraine	1.0	0.9
Odessa, Ukraine	1.0	1.0
THE SOUTHERN CAUCASUS		
Baku, Azerbaijan	1.8	2.0
Tbilisi, Georgia	1.1	1.8
Yerevan, Armenia	1.1	1.0
CENTRAL ASIA		
Tashkent, Uzbekistan	2.2	2.3
Almaty, Kazakhstan	1.1	1.1
*estimated		

Source: United Nations Urban Agglomerations 2003, with estimates for 2015 (2003).

than 8.1 million lived in Azerbaijan. Most lived in neighboring Iran. Azerbaijan is 90 percent Azerbaijani, but 90 other groups lived there as well. In contrast to Armenians and Azerbaijanis, most Georgians live in Georgia, which had 4.4 million inhabitants. However, 30 percent of Georgia's population is composed of other groups such as Armenians, Russians, Azerbaijanis, Ossetians, and Abkhaz. The peoples of the Central Asian countries also live in one another's countries. For example, the boundaries of Kyrgyzstan, Tajikistan, and Uzbekistan weave through the fertile and densely settled Fergana Valley, leaving many Kyrgyz, Tajiks, and Uzbeks in their neighbors' countries.

Urban Patterns and Linkages

Urbanization

This world region's highest rates of urbanization are in areas that emphasized urban-industrial development. The Russian Federation, Ukraine, and Belarus are the most urbanized, while Moldova and most of the Central Asian countries with their traditional ways of life are among the least urbanized. The Russian Federation has the greatest number of cities and the largest cities of the region (Table 3.2).

Urban Landscapes

In 1917, at the time of the Bolshevik Revolution, 17 percent of Russians lived in mainly small, provincial towns and cities. Moscow and St. Petersburg had grand designs for buildings and

roads ordered by the czars (Figure 3.12). In the Southern Caucasus and Central Asia, the cities, which lie on ancient and medieval trade routes, are very old with architecturally beautiful structures that attract tourists (Figure 3.13).

Joseph Stalin's five-year plans emphasized intensive centralization and rapid industrialization. Industrial centers expanded and new specialist resources (e.g., oil, mining) centers emerged, often in new towns in remote regions. Urbanization increased to 48 percent of the region's population in cities by 1959 and to more than 70 percent in the 1990s. Stalinist urbanization produced distinctive, standardized cities throughout the Soviet Union. Housing for the workers consisted of numerous poor quality, high-rise, concrete buildings with few shopping opportunities in their vicinity because Soviet planners did not foresee the need for much shopping.

One of the major trends to emerge since the breakup of the Soviet Union in 1991 is the building of suburbs, particularly around Moscow. The congestion and pollution of inner city

FIGURE 3.12 Russian urban landscape. The cathedral of the Intercession, commonly known as St Basil's Cathedral, is located on Red Square, the center of Moscow. Ivan IV (The Terrible) had the cathedral built to commemorate Russia's victory over the Mongols. Originally intended as eight smaller, clustered cathedrals to represent Russia's eight victorious battles over the Mongols, the result was a single cathedral with eight domes and chapels signifying the victories.

FIGURE 3.13 Central Asia: Samarqand, Uzbekistan. Historic Muslim buildings highlight the importance of Islam in Central Asia. Madrasah (theological seminary) buildings of the 1500s and 1600s border Registan Square. Typical features of Islamic art include arches, tilework, domes, and minarets.

FIGURE 3.14 Moscow: IKEA. In recent years, European and American style super shopping centers have been appearing in the region. Notice how this IKEA has its name written in Cyrillic as well as Latin text. As elsewhere, the shopping center, which is along the northwestern beltway (MKAD), depends upon and promotes greater automobile usage.

streets due to growing use of cars and trucks, together with increased urban crime, caused richer families to move out. The mayor of Moscow contributed to the movement by ejecting families from apartment blocks designated for renovation and commercial sale. Financial institutions, including banks, fund the new suburban residential developments and assess the customers' ability to pay.

A new focus of the market economy was not only the construction of new private housing but also the conspicuous appearance of suburban shopping malls with supermarkets, clothing chains, electronics stores, movie theaters, and foreign-owned hotels (Figure 3.14). In the 1990s, most were built in Moscow, followed by St. Petersburg. By 2003, 28 percent of Russia's retail sales were in Moscow though only 2 to 4 percent of Russians shopped in modern facilities. As a result of the fierce competition in Moscow, investors began searching for new cities, usually regional capitals with more than 1 million inhabitants. Kazan, Samara, Yekaterinburg, Krasnodar, and Kaluga have emerged as new cities of investment. For example, Marriot Renaissance opened a hotel in Samara in late 2003, and IKEA chose Kazan to locate its first store in Russia.

Global City-Regions

Many urban areas are found in Russia and Neighboring Countries, but only Moscow ranks as a global city as identified in Figure 1.23. Almaty (Kazakhstan), Kiev (Ukraine), St. Petersburg (Russia), and Tashkent (Uzbekistan) offer some global city services. The lack of global connectedness is the direct result of the Communist legacy. Communism precluded the development of the types of accounting, advertising, banking, and law that are so characteristic of capitalism and globalization. Refusing to engage in global capitalist competition, the Soviet government likewise focused inward, on self-sufficiency, and only engaged in trade with fellow Communist countries. Consequently, earlier global cities of this world region lost many of their global connections. Since the end of communism in 1991, the capital cities, especially Moscow but also cities such as St. Petersburg, are developing accounting, advertising, banking, and law firms and have received the greatest amounts of international investment. In fact, Moscow is developing more rapidly than other cities, creating a large gap between Moscow and the rest of the country.

Evolving Politics

Russia has been the dominant country of this world region for almost a millennium. Prior to Soviet times, the Russian Empire was one of Europe's great empires. Unlike the European empires such as France and the United Kingdom, which conquered lands overseas, the Russian Empire conquered peoples and land adjacent to it, extending its borders as it annexed new territories with very diverse populations. The Soviet Union, founded in 1922, had many successes but internal problems caused it to

disintegrate in 1991. Though the 15 newly independent countries contain many peoples of ancient ancestry, many of the new countries had scarcely been independent before.

The Russian Empire

The Russian Empire traces its roots back to the principality of Muscovy, which began to expand in the 1400s. Muscovy's leader, Ivan III (the Great) married Sophie Paleologue, the niece of the last Byzantine emperor. After Constantinople fell to the Ottomans in 1453, Ivan was seen as the true inheritor of the Christian realm. He adopted the title "czar," derived from the Latin *caesar* for emperor. In 1613, Mikhail Romanov ascended the throne. The Romanov dynasty lasted until the last czar was deposed in 1917.

From 1480, Muscovy's territory continued to expand, especially to the east into Siberia and Central Asia but also to the west, bringing the Russians into contact with Central and Western European culture. The greatest territorial expansions were made under Peter the Great (1682–1725), Catherine the Great (1762–1796), and Alexander I (1801–1825). By the early 1900s, the Russian Empire achieved its greatest extent, stretching from Europe to East Asia and the Pacific Ocean (Figure 3.15).

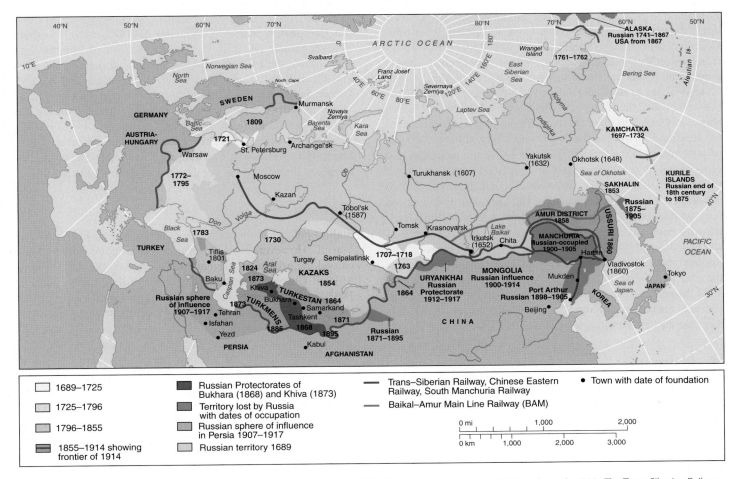

FIGURE 3.15 Russian Federation: history of growth. The expansion of the Russian Empire from the 1600s to the early 1900s. The Trans-Siberian Railway was built through territories that were regarded as safely Russian. It was rerouted after territory was lost. The Baikal-Amur Main Line Railway (BA M) was begun in 1974 and completed in 1988. It opened up more of the vast mineral and forest resources of eastern Siberia and the Russian far east to greater exploration.

Russia was a world power, but its strength was derived from its sheer size. Internally, it had many problems. Politically, the czars remained in total control and continued to claim to speak in the name of God, but absolute power inhibited the creation of an efficient government. Technologically, Russia was decades behind Western Europe. Most Russians remained enserfed peasants in a feudal system, farming with horses or oxen. At the same time, Russia had expanded its boundaries to include more than 100 different peoples. After the rise of nationalism in Europe in the mid-1800s, Russia's nationalities began clamoring for independence. This situation worsened as Russians themselves became nationalistic and began suppressing non-Russians. They developed Russification policies to force Russia's minorities to become more Russian ("Russified").

The Founding of the Soviet Union

World War I (1914–1918) exerted great stress on the Russian Empire. By 1917, starvation and a huge death toll worked together with long-standing opposition to czarist rule. Revolutionaries deposed Czar Nicholas II and set up a provisional government that proved to be weak. A number of national groups staked their claim to territory and declared independence. In late 1917, the Bolsheviks, a group of Communists also known as "Reds," overthrew the provisional government. Civil war ensued when anti-Communists, called "Whites," tried to dislodge the Bolsheviks from power. By 1922, the Reds gained the upper hand, expelled their enemies, and reclaimed many of the lost territories.

In 1922, under the leadership of Vladimir I. Lenin, the Bolsheviks established the Union of Soviet Socialist Republics (USSR), commonly called the Soviet Union. This new country succeeded the Russian Empire, but the Bolsheviks changed the old system of government and economy. They abolished the monarchy and government by a privileged few and replaced them with "soviets." Soviets were workers' and soldiers' councils that drew their members from the common citizens. The Communists denounced religion because the clergy sanctioned the rule of political leaders who oppressed the common people. Consequently, the Bolsheviks attacked churches and mosques, dynamiting many and turning others into scientific centers such as planetariums to show the error of religious belief. After Joseph Stalin came to power in 1924, he purged millions of his enemies, suspected enemies, and potential enemies, including about half of the officers in the military (see the "Human Rights" section in this chapter, p. 101).

World War II

Stalin entered into a nonaggression pact with Hitler in 1939, allowing the Soviet Union to take control of territories that formerly belonged to the Russian Empire: the Baltic states, eastern Poland, and Bessarabia (now Moldova). Finland was also a target, but the Finnish forces successfully defended their country from the Red Army. The victory of the small Finnish military over the huge Red Army underscored the weakness of the Soviet Union and its military.

On June 22, 1941, Hitler launched Operation Barbarossa, the invasion of the Soviet Union. The campaign was initially devastating. The Soviet Union survived largely because of its sheer size. A vast amount of land was lost and millions of people were killed, but even larger areas were left unconquered. The Soviets moved war production farther east, out of the range of the Nazi military, including its air force. With help from the United States and the United Kingdom, production increased. Finally, the harsh Russian winter brought the German Nazi armies to a standstill. In the meantime, Stalin moderated many of his harsh policies, including the persecution of the Russian Orthodox Church. After a fierce battle at Stalingrad in 1943, Soviet forces began rolling back the Nazi invaders. The people of the Soviet Union had turned the tide of war and began winning what they call the "Great Patriotic War." By May 1945, the Red Army had swept through East Central Europe and occupied much of Germany, including the capital, Berlin.

Its victory in World War II allowed the Soviet Union to annex the Baltic countries and Moldova, former territories of the Russian Empire the Soviets invaded in 1940, and new territory in East Central Europe never before governed by the Russians (e.g., East Prussia, the northern half of which is now Kaliningrad, and areas of Poland). Victory also allowed Stalin to establish and support Communist governments in East Central European countries. To counter NATO and the Marshall Plan (economic aid from the United States) and later the European Economic Community, Stalin created the Warsaw Pact and Council for Mutual Economic Assistance (CMEA, also known as COMECON). Stalin also began persecutions again, accusing entire ethnic groups and nations of treason during the war. He moved millions from their homes, mostly to Siberia. After Stalin died in 1953, his successors were deliberately less harsh. However, Stalin's system of government and economy remained intact until the demise of the Soviet Union in 1991.

Glasnost

Stalin's policies allowed the Soviet Union to rebuild after World War II. However, rigid governmental structure and policies caused the economy to eventually decline by the 1970s. After Mikhail Gorbachev became the leader of the Soviet Union in 1985, he immediately set out to reform his country's political and economic systems. In doing so, he highlighted two concepts: *glasnost* (informational openness) and *perestroika* (economic restructuring) (see "perestroika" in the next section). *Glasnost* referred to the policy of providing government information to citizens. It was intended to have a positive effect on the country by empowering citizens with knowledge. However, freedom of information also allowed citizens to learn about corruption, government abuse, forced labor, and many of the other problems of the Soviet government. Non-Russians vented their anger at the Soviet government for its Russification policies. In contrast, Russians complained that the Soviet government suppressed Russian culture, especially the Russian Orthodox Church, and distributed Russia's resources to the country's minorities. To avoid the wrath of citizens, many politicians echoed their anger, championed local and regional causes, and turned against the central government in Moscow. Even Boris Yeltsin, the elected leader of the Russian Republic in 1990, called for Russia's independence from the Soviet Union.

Mikhail Gorbachev believed in communism and tried to preserve the Soviet Union by implementing reforms. Ironically, he became the Soviet Union's last leader as his policies led to the unraveling of the country in 1991. Led by Lithuania, the Soviet republics all declared independence by December. Economically tied to one another, all the republics, except for Lithuania and the other two Baltic republics of Latvia and Estonia, formed the Commonwealth of Independent States, though Georgia waited a couple of years before joining. However, while Gorbachev's policies led to the rapid demise of his country, his reforms paved the way for further changes; without these reforms, the country would have probably disintegrated anyway, perhaps more violently.

Closed Economies Open Up

Prior to 1917, Imperial Russia was largely agricultural with most people living as peasants on the land. Industrialization was only beginning in such cities as Moscow and St. Petersburg. When the Communists came to power, they were suspicious of capitalists who they saw as exploiting the people. To prevent this from happening, they took complete control of the economy through a concept of a "planned economy" (see Chapter 2, p. 50). The distrust of capitalism also led the Communists to close the Soviet Union's economy as much as possible from the global economy because it was primarily capitalist in nature. Communist economic policies were successful in the early decades of the Soviet Union but caused the country to disintegrate in its latter decades. This world region is now opening up to the global economy.

Five-Year Plans

With the Soviet economy initially largely agricultural, farmers could not relate to the urban-industrial ideology of communism. When Joseph Stalin came to power in 1924, he believed it necessary to forcefully industrialize the Soviet Union's economy to achieve the Communist ideal of a "workers' paradise." It was also necessary to overcome the technological advantages of the West.

To transform the economy, Stalin developed the idea of the **five-year plan.** The first one was launched in 1928 and called for collectivization and industrialization. Collectivization was a way of making farmers into factory-like workers and thus more sympathetic to the Communist way. Under collectivization, small family farms were merged together to create large farms, thousands of acres in size. The large farms were better designed to use modern farm machinery, then under production in the new factories. The government became the owner of the collectives and farmers became employees. With collectivization, farmers became more like factory workers, even living in tightly packed housing like urban factory workers.

Government owned all industries. In what is known as the **command economy,** the government set quotas favoring heavy industry over production of consumer goods. The five-year plans also established **central planning.** In contrast to capitalist economies, supply, demand, or profit making did not dictate what would be produced. With central planning, the government decided how many goods and services were needed by society, almost without cost considerations. Rapid and forced industrialization prevented the Soviet Union from experiencing the world economic depression of the 1930s, as the government kept investing in the economy and providing jobs. Significant amounts of the production came from the slave labor of the millions of people that Stalin had purged.

Communism at an Economic Standstill

When Stalin died in 1953, his series of five-year plans had industrialized the Soviet Union's economy, despite the serious setbacks of World War II. The West itself, however, continued to develop economically so that the Soviet Union was still behind, despite spectacular scientific breakthroughs such as the space program. By the end of the 1950s, the Soviet Union had caught up only to the West of the 1920s, the time when Stalin began the five-year plans. In the meantime, industry in the West evolved to adapt to new materials such as plastics and other synthetics, and used new fuels such as petroleum and natural gas. Western industry also adapted to constantly changing consumer demands that required continual retooling and new locations for production facilities. The Stalinist Soviet economy, however, was so rigid that it could not adapt to change.

The weaknesses of the Soviet economic system compounded over time. For example, the military accounted for a large portion of the country's economy and saw little need to be efficient. With the government owning all businesses and an absence of competition, managers saw no need to use fuel wisely or search for fuel alternatives in a country rich in natural resources. The communist guarantee of a job for everyone meant that labor costs were high though wages were low and no attempts were made to increase productivity by updating machinery and computers. Finally, the practice of central planning meant that government bureaucrats, not supply and demand, determined what was produced. The bureaucrats proved to be highly inefficient.

Resources such as oil, building materials, and equipment became scarcer because they could not be adequately exploited or delivered. The various ministries of the government hoarded them, attempting to become self-sufficient because other agencies could not supply them. This practice created further redundancies, shortages, and squandering of resources. Bureaucrats played it safe by locating new enterprises where they could best obtain supplies. In most cases, the locations were big cities. This led to excessive migration to cities, creating congestion, high living costs, and environmental problems. Ironically, though government bureaucrats contributed to the ruination of the Soviet economy, they became the unhappiest segment of society and greatly wanted change. As the privileged within the Soviet Union, they were displeased to see their standard of living drop below that of the poorer groups in the materially wealthy capitalist countries.

Perestroika

For *perestroika*, Gorbachev believed that it was necessary to divorce economics from politics, allow more local control, and introduce free market practices. Such policies went against Communist ideals and established interests. The bureaucrats fought Gorbachev and his policies, creating great political turmoil. As political battles raged, Gorbachev's economic policies ran opposite of what had been practiced since the 1920s. A crisis ensued as the economy attempted to reverse direction. Companies, having existed in a noncompetitive environment, now had to meet their own costs and find their own customers. No longer having the ability to waste resources, they had to cut costs. To generate income, they cut production and raised the prices of their goods, but they soon found that they could not afford to buy supplies from one another. Furthermore, the lack of a capitalist banking and financial system meant that cash could not flow easily. Poor transportation and communication systems only exacerbated the situation. Production declined. To cut costs, companies laid off workers, increasing unemployment. The Soviet economy continued to spiral downward with frequent labor strikes and rampant crime.

Today's Economic Geographies

After the Soviet Union's disintegration, the new countries of this world region faced two economic problems. First, the heavy industry developed during Soviet times was far out-of-date (Figure 3.16). Since 1991, the independent republics have worked to update their factories, attract foreign investment, engage in the global economy, and develop more service industries. Second, Soviet economic policies tied the republics' economies very closely together. Many of the countries now prefer greater independence from the Russian Federation, but old connections are hard to break and new relationships difficult to form. For example, during Soviet times, eastern Ukraine was the most important iron- and steel-making region of the Soviet Union. Eastern Ukraine can provide for Ukraine's economic independence today, but the area's industries depend on imports of oil and natural gas, primarily from the Russian Federation, to meet 85 percent of their energy needs. As another example, during Soviet times, Moscow limited industrial development in the Southern Caucasus and encouraged agricultural production. Georgia supplied over 90 percent of the Soviet Union's tea and citrus fruits. Central Asia is a major supplier of cotton as well as mineral resources such as coal, iron, chromium, oil, and natural gas that are important to Russian factories. Many of the countries of the world region can find markets other than the Russian Federation for their products, but the transportation network leads to the Russian Federation.

Following the demise of the Soviet Union, many people around the world wondered if Russia would remain a world power because it did not have a world power's economic strength. By 2005, Russia produced only 2.5 percent of the world GNI PPP. The GNI PPP per capita within Russia and the other CIS countries were far below those of capitalist countries such as EU countries and the United States (Figure 3.17). Almost 12 percent of Russia's people officially live in poverty, earning less than US$2 per day. The ownership of consumer goods reflects this modest income (Figure 3.18). TV set ownership is high, as television provided mass communication from the government

FIGURE 3.16   Kerch, Crimea. Following the end of communism in 1991, outdated, inefficient factories like this one were abandoned. In the 2000s they have been seen as sources of raw material. Thus they, like this one, have been stripped of their bricks, steel, and other materials which were then sold.

FIGURE 3.17   Russia and Neighboring Countries: national incomes compared. The countries are listed in the order of their GNI PPP per capita for 2005. CA = Central Asia, SC = Southern Caucasus. Compare with Figure 1.20. *Source: Data (for 2005) from World Development Indicators, World Bank; and Population Reference Bureau.*

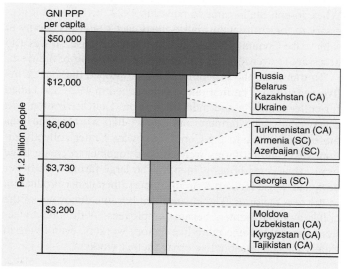

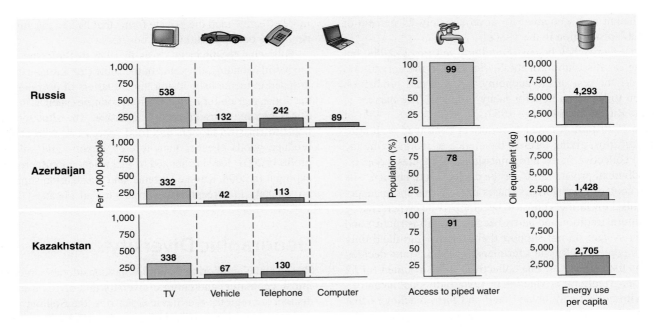

FIGURE 3.18 **Russia and Neighboring Countries: ownership of consumer goods, access to piped water, and energy usage.** What is the significance of relatively high television ownership? Compare with Figure 1.19. *Source: Data (for 2002) from* World Development Indicators, *World Bank, 2004.*

Agriculture

The Slavic countries' northerly latitudes and distant locations from moderating maritime influences result in dry conditions. Low temperature climates in the north and high temperature ones in the south make farming difficult almost everywhere. Even in the best locations, the late arrival of spring or the early appearance of winter can ruin a year's crops. The most productive farming areas are in the west of Russia, Ukraine, and Moldova in the former steppe and deciduous forest areas (see "Geography at Work: Wine Industry," p. 105). Ukraine's vast plains of black soil produce abundant crops of wheat and make the country agriculturally self-sufficient. Moldova also has low relief and good soils like western Ukraine. Fruits, vegetables, wine, and tobacco are the main farm products. Belarus, however, is a country with low relief but mostly poor soils. Having been shaped by continental glaciation, the southern part of the country is covered by the Pripyat Marshes and the northern part by glacial moraines. The sandy glacial soils are good for potatoes. Toward the north in the Russian Federation, around Moscow, arable land gives way to livestock farming. East of the Urals, farming is restricted to areas that have sufficient water and length of summer growing season. The warm climates of the Southern Caucasus make it possible to grow citrus fruits, tea, tobacco, cotton, and rice.

In the 1950s, Soviet leader Nikita Khrushchev sought to increase agricultural production with his **Virgin Lands Campaign**

to make the Soviet Union self-sufficient and independent of the global economy. This campaign promoted farming in lands where it had never been done before, but many of these lands had poor soil quality because they either did not have enough water or heat to grow crops. For example, many were in the semidesert and desert areas of Central Asia, especially in the Kazakh Republic and the adjacent dry areas of the Russian Republic, much of it stretching across southern Siberia. It was believed that these dry areas would be productive by using new irrigation systems. From 1917 to 1987, irrigated land in the Soviet Union increased from 3.5 to 20 million hectares. At the same time, large drainage networks were built in the lands north of Moscow to remove water from the waterlogged soils. From 1956 to 1987, drained land increased from 8.6 to 19.4 million hectares.

The Virgin Lands Campaign proved to be a huge failure over the long run. Production increased, but many problems resulted. Massive amounts of water were diverted from rivers feeding the Caspian and Aral seas, wreaking havoc with their ecosystems. Pouring water on desert soils only led to soil salinization. Drainage projects were no more successful. Farm production fell on many of the newly established collectives. Eventually, the Soviet government stopped funding the Virgin Lands Campaign. Much of the land was slowly abandoned, though Kazakhstan is still one of the biggest producers of wheat and cotton.

During Communist times, most farmland was allocated to either state farms or collective farms. Typical state or collective farms consisted of thousands of acres and thousands of livestock with several hundred workers, each receiving salaries and benefits like other kinds of workers. Though state-owned farms provided job security, they were extremely inefficient. Private plots, making up only 3 percent of agricultural land, were much

to the people during the Soviet era. Rising oil prices, which began in 2004, have helped Russia and the other oil-producing republics to greatly improve their economies.

more efficient and accounted for approximately 25 percent of agricultural production in the 1980s.

Mikhail Gorbachev began agricultural reforms in 1988, but following the dissolution of the Soviet Union, the agricultural sector, like the rest of the economy, suffered and production fell, often to less than half for many products. The harvest in 1995 was Russia's worst since 1963.

In the 1990s, the Russian government continued with agricultural reforms, giving farmers the choice of reorganizing the state and collective farms into joint-stock companies, cooperatives, individual private farms, or the choice not to change. The first two choices gave workers shares in the farms and management rights. Few farmers chose to became private owners, partly out of cultural tradition and partly because the infrastructure and economic system do not support them. Private farmland only grew to 5 percent of Russia's farmland by 1995. Many decided to stay on the collectives. The collectives still accounted for 52 percent of agricultural production in 1996. Individual incentive, mixed with collectively owned land and farm machinery that was better integrated into the economic system, proved to be

more effective than the private farms that lacked machinery and needed to be run as individual businesses.

Following the financial crisis in 1998 that affected much of the world economy, the value of the ruble (the Russian currency) dropped tremendously, losing three-quarters of its value against the dollar. Agricultural imports from countries such as the United States became extraordinarily expensive. The situation gave a boost to Russia's farmers and food processors, who could then produce goods cheaper than foreign imports and still earn a profit. In 2001, Russia changed from a wheat importer to a wheat exporter. In 2004, Russia, Ukraine, and Kazakhstan supplied approximately 11 percent of the world's traded wheat.

Geographic Diversity

Russia and Neighboring Countries has considerable internal political, economic, and cultural diversity that divide it into three distinct subregions: the Slavic Countries, the Southern Caucasus, and Central Asia (Figure 3.19 and see Table 3.1).

FIGURE 3.19 Russia and Neighboring Countries: subregions (the Slavic countries, Southern Caucasus, and Central Asia). Note the distribution of the major cities and their concentration in the western part of the region.

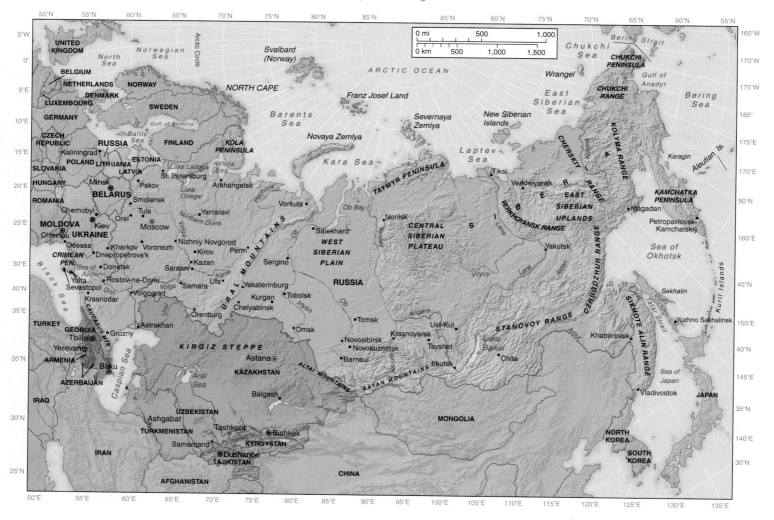

Subregion: The Slavic Countries

The Slavic countries of this subregion are the Russian Federation, Ukraine, and Belarus. Moldova also is included because many Slavs live there, and it is closely tied to the Slavic countries. The Russian Federation is by far the largest in land area of any country in the subregion and in the world. It is nearly twice as large as Canada, the United States, or China. In 2005, Russia had 77 percent of the CIS area and 50 percent of the CIS population.

Though the Russian Federation and the other countries of the subregion are experiencing economic hardship as their economies move from communism to capitalism, the Russian Federation still exerts considerable power. The Russian Federation remains a nuclear power and continues to hold one of the five permanent seats of the UN Security Council, the most powerful organ of the United Nations. Because it contains substantial portions of the world's natural resources, the Russian Federation has considerable economic potential. In the mid-1990s, the Russian Federation joined the Group of Seven (G7), an informal organization representing the world's most wealthy countries, changing the organization to the Group of Eight (G8). It held the presidency of the organization in 2006.

The Slavic countries seen on the map today became independent only in 1991 with the boundaries they had as Soviet republics. This situation is also true for the Russian Federation, though the Russians controlled the Soviet Union and its predecessor, the Russian Empire. Prior to 1991, Ukraine was only independent for a brief period after World War I until it became part of the Soviet Union in 1922. Before that, it was part of the Russian Empire. Belarus was never independent before 1991 and was usually either part of the Russian Empire or the Polish-Lithuanian Kingdom. Of the former Soviet republics, Belarus is the most closely tied to the Russian Federation. Belarussians have considered creating a Russian-Belarussian Federation since 1991.

Moldova, too, was never independent and only has been a distinct territory for fewer than 200 years. It was part of the Romanian province of Moldavia until the Russian Empire annexed the part east of the Dnieper River in the 1800s and named it Bessarabia. Romania annexed the territory after World War I, but the Soviet Union annexed it again after World War II. Many Romanians and Moldovans hoped to unite their two countries after the dissolution of the Soviet Union in 1991, but the Russian military stationed in the country prevented this. Worried about a union of Moldova and Romania, Russians and Ukrainians living in Moldova declared their own republic in the Transnistria region. Other ethnic minorities have made similar proclamations. The government has not been able to suppress these independence movements completely.

The Russian Federation

The Russian Federation is the modern political state representing the land known as Russia. To many Westerners, Russia is a mysterious land, hidden in cold, dark forests on the eastern and northern fringes of Europe. Europeans have regularly included the Russian heartland within Europe but at the same time have considered Russians too "Asiatic" to be European. For centuries, Europeans struggled to understand Russia, exemplified by Winston Churchill's remark that "Russia is a riddle wrapped in a mystery inside an enigma." Depending on their relationship with Europe and their desire to be within Europe, Russians themselves frequently alternate between emphasizing their European qualities and their wider role spanning eastern Europe and northern Asia. After the fall of communism in 1991, Russia's internal political geography and its economic and social relationships all dramatically altered.

Political Divisions

The internal political geography of the Russian Federation includes a mixture of political units. To a large extent, the country's political geography was inherited from the Soviet system, though some changes have been made since 1991. The units fall into two categories: administrative and autonomous. Administrative units consist of 6 federal territories (*krays*), 49 regions (*oblasts*), and 2 federal cities (Moscow and St. Petersburg). Much like the states, counties, and municipalities of the United States, they were created to administer the large country. In contrast, the autonomous units, consisting of 21 republics, 1 autonomous region (*oblast*), and 10 autonomous districts (*okrugs*) (Figure 3.20), are able to craft many of their own laws and govern themselves somewhat differently than the rest of the Russian Federation.

Together, the 89 political units represent different levels of size, resources, and political power. Unlike the United States, where the federal government has the same relationships with lower levels of government (e.g., states, counties, etc.), Moscow has an asymmetrical relationship with its political units, particularly the autonomous territories, in which each political unit negotiates its own relationship with Moscow. The resource-rich republics tend to exercise the greatest authority over their own governance.

The autonomous territories were established by the Soviet Union to reflect the presence of ethnic minorities, such as the Tatars, Sakha (Yakuts), and Buryats (Figure 3.21). Soviet law protected minority languages, religions, and cultures. However, only 52 percent of the Russian Federation's approximately 30 million non-Russians live in the autonomous territories today (Figure 3.22). As a way of controlling their vast country, the Soviets drew boundaries for the republics that deliberately left many members of ethnic groups outside their intended territories, and included many Russians within them. Also, many recognized nationalities did not receive republic status. Of the 90 numerically significant groups recognized in the 1989 census, only 35 had a homeland. In addition to the earlier mentioned policies of Russification, boundary drawing was clearly a means that the Soviets used to divide and dilute non-Russian groups. The situation is particularly true in the North Caucasus, where the native peoples have most fiercely resisted Russian rule through history. For example, the Karbardians were grouped together with the Balkars, though they had more in common with their neighbors the Cherkessians.

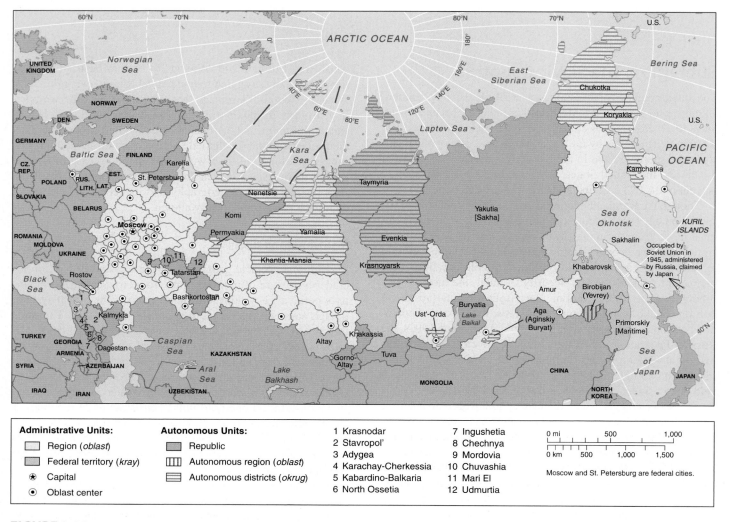

FIGURE 3.20 Russian Federation: administrative divisions. Notice the uneven distribution of power.

Administrative Units:
- ☐ Region (*oblast*)
- ▨ Federal territory (*kray*)
- ✱ Capital
- ⊙ Oblast center

Autonomous Units:
- ▨ Republic
- ▥ Autonomous region (*oblast*)
- ▤ Autonomous districts (*okrug*)

1 Krasnodar
2 Stavropol'
3 Adygea
4 Karachay-Cherkessia
5 Kabardino-Balkaria
6 North Ossetia

7 Ingushetia
8 Chechnya
9 Mordovia
10 Chuvashia
11 Mari El
12 Udmurtia

Moscow and St. Petersburg are federal cities.

Heartland and Hinterland in Russia

Another basis of regional differences in Russia is **heartland** and **hinterland.** The heartland lies west of the Urals and includes many of the original eastern Slav territories. It contains the greatest concentration of Russian people and accounts for much of the country's economic and political activity. It is also known as the Russian homeland. The Moscow and St. Petersburg urban regions, the Volga River valley, and the Urals contribute to the heartland's prominence. The Moscow region, approximately 400 km (250 mi.) square, is home to 50 million people and was the focus of Soviet central planning and transportation routes linking the entire country. Local manufacturing includes vehicle, textile, and metallurgical industries. St. Petersburg, a major Baltic port north of Moscow, is a smaller manufacturing center but still produces around 10 percent of total Russian output, including shipbuilding, metal goods, and textiles.

Southeast of Moscow, the Volga River is lined by a series of industrial cities that use the river transport, connected since the 1950s by a canal outlet to the Black Sea. Around 25 million people live in the Volga River region, which was developed for manufacturing during and after World War II—at a distance from advancing German armies and helped by the discovery of major local oil and natural gas fields. Manufactures include specialized engineering and the Togliatti car plant built by Fiat of Italy.

East of the Volga River basin, the Urals contain metal ores and are only a minor barrier to communications. Like the Volga region, the southern Urals were developed during and after World War II, principally as a metals center. Oil and natural gas fields are located to the south and east.

The Russian Federation's hinterlands include Kaliningrad *oblast* along the Baltic Sea, the Arctic regions around Murmansk, the mining areas east of the Urals, resource-rich Siberia, and the Pacific region inland from Vladivostok. Large expanses of the hinterland are virtually empty of people and economic activity and remain inaccessible to development. Siberia forms a huge area that can be divided into the more de-

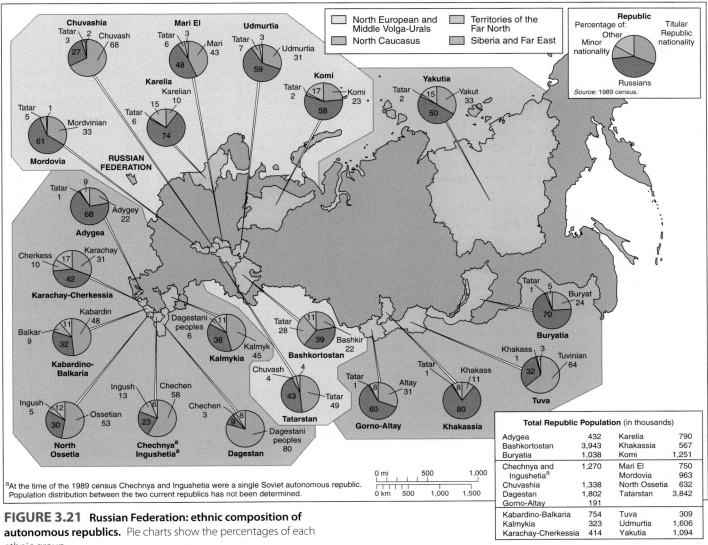

Chuvashia
Tatar 3 — Chuvash 68 — 2 — 27

Mari El
Tatar 6 — Mari 43 — 3 — 48

Udmurtia
Tatar 7 — Udmurtia 31 — 3 — 59

Komi
Tatar 2 — Komi 23 — 17 — 58

Yakutia
Tatar 2 — Yakut 33 — 15 — 50

Legend:
- North European and Middle Volga-Urals
- North Caucasus
- Territories of the Far North
- Siberia and Far East

Republic
Percentage of:
Other — Minor nationality — Titular Republic nationality — Russians
Source: 1989 census.

Karelia
Karelian 10 — Tatar 6 — 15 — 74

Mordovia
Tatar 5 — 1 — Mordvinian 33 — 61

RUSSIAN FEDERATION

Adygea
Tatar 1 — 9 — Adygey 22 — 68

Karachay-Cherkessia
Cherkess 10 — Karachay 31 — 17 — 42

Kabardino-Balkaria
Balkar 9 — Kabardin 48 — 11 — 32

North Ossetia
Ingush 5 — Ingush 12 — Ossetian 53 — 30

Chechnya[a] Ingushetia[a]
Ingush 13 — Chechen 58 — 6 — 23

Dagestan
Chechen 3 — 8 — 9 — Dagestani peoples 80

Kalmykia
Dagestani peoples 6 — 11 — Kalmyk 45 — 38

Tatarstan
Chuvash 4 — 4 — Tatar 49 — 43

Bashkortostan
Tatar 28 — 11 — Bashkir 22 — 39

Gorno-Altay
Tatar 1 — 8 — Altay 31 — 60

Khakassia
Tatar 1 — Khakass 11 — 8 — 80

Tuva
Khakass 1 — 3 — Tuvinian 64 — 32

Buryatia
Tatar 1 — 5 — Buryat 24 — 70

Total Republic Population (in thousands)			
Adygea	432	Karelia	790
Bashkortostan	3,943	Khakassia	567
Buryatia	1,038	Komi	1,251
Chechnya and Ingushetia[a]	1,270	Mari El	750
		Mordovia	963
Chuvashia	1,338	North Ossetia	632
Dagestan	1,802	Tatarstan	3,842
Gorno-Altay	191		
Kabardino-Balkaria	754	Tuva	309
Kalmykia	323	Udmurtia	1,606
Karachay-Cherkessia	414	Yakutia	1,094

[a]At the time of the 1989 census Chechnya and Ingushetia were a single Soviet autonomous republic. Population distribution between the two current republics has not been determined.

0 mi — 500 — 1,000
0 km — 500 — 1,000 — 1,500

FIGURE 3.21 Russian Federation: ethnic composition of autonomous republics. Pie charts show the percentages of each ethnic group.

FIGURE 3.22 Russia. Buryat singers at Datsan Temple, southeast of Irkutsk.

veloped southern margins along the line of the Trans-Siberian Railway and the northern lands.

Siberia is an essential part of Russia, making up three-fourths of its land and providing a large proportion of its raw materials—a common feature of hinterland regions. In 1990, Siberia produced 73 percent of Russia's oil, 90 percent of its natural gas, 61 percent of its coal, all of its diamonds, and 30 percent of its timber and electricity. It also has a growing fishing industry on the Pacific coast.

Along the southern strip of Siberian Russia are a number of centers that were industrialized according to the Soviet planning processes and are separated by large tracts of sparsely settled land. In the Kuzbas region around Kuznetz, 2,000 km (1,200 mi.) east of the Urals, a major coalfield was developed initially to supply coal to the steel manufacturing centers of the Urals. This led to local industrialization, supported by iron ore and the return transport of bauxite from the Urals. Aluminum, steel, and products using them are major outputs. The world's largest aluminum plants at

Krasnoyarsk, Bratsk, and Sayansk provided metal for the Soviet aircraft and missile industries.

Farther east, centers close to Lake Baikal, including Irkutsk, form a narrow belt of industrialization using hydroelectricity generated in the headwaters of streams flowing to the Yenisey and Lena river systems. Mining and lumbering are also important in isolated localities, while larger cities provide services to extensive areas in northern Siberia where population and mining activities are scattered.

The far east, flanking the southern Pacific coast of Russia, has ports such as Vladivostok and Nakhodk, and metal industries along the Amur River. In the 2000s, Japanese corporations began investing in the region's mines. The presence of a large population in neighboring China, coupled with a new Russian law in 2003 that allows foreigners to lease land in Russia for 49 years, has attracted many Chinese farmers.

Soviet centralized planning often chose locations for production facilities for political, rather than economic, reasons. Special defense-related factories, for example, were built in remote locations where it was expensive to maintain them. Workers were paid more to work in these faraway facilities, and the cities that grew up around them received many subsidies. After 1991, the Russian Federation stopped giving financial support to these poorly located factories and cities. Without subsidies, many factories were not able to compete economically and had to shut down. In turn, unemployed people emigrated to the Russian heartland. Subsequently, regional differences and local characteristic lifestyles are now more pronounced within the Russian Federation.

The realities of such factors as proximity to consumer markets or ports with world trading connections are causing geographic shifts in regional production patterns. Moscow, for example, has had the most success in the transition to capitalism. As much as 20 to 25 percent of Moscow's population is middle class, the highest percentage in the Russian Federation. In other areas, less than 10 percent of the population has attained middle-class income.

Subregion: The Southern Caucasus

Georgia, Armenia, and Azerbaijan straddle the Caucasus Mountains and are frequently called the Transcaucasus, meaning "across the Caucasus" (Figure 3.23 and see Table 3.1). This term reflects a Russian ethnocentric view of the region as these countries are on the other side of the Caucasus Mountains from Russia and were once Russian colonies. The more neutral term "Southern Caucasus" is used to refer to these countries, and "Northern Caucasus" is applied to the part of the Caucasus in the Russian Federation.

Armenia and Georgia are both very mountainous with many peaks rising above 5,000 m (15,000 ft.). Azerbaijan is moun-

tainous in the west along the borders of Georgia and Armenia, but the eastern areas of the country, formed by the Caspian Sea coast, are flat with areas below sea level. The origins of the word *Azerbaijan* are not clear, but one version derives from Persian words that mean "land of fire." Azerbaijan certainly contains great quantities of petroleum. "Land of fire" could refer to the surface deposits that burned naturally in the past or to the oil fires in Zoroastrian temples that once dominated the region. Zoroastrianism is less prominent in this subregion today (although its modern-day adherents are in India and known as Parsis), but was a religious forerunner to Christianity and Islam, tracing its origins back to Azerbaijan. Zoroastrians believed that the Earth would be consumed in fire following judgment day. Interestingly, this belief developed in an area of the world where the land is oil-soaked and easily burns.

As former Soviet republics and current members of the CIS, Armenia, Georgia, and Azerbaijan struggle with establishing a viable political and economic existence. Ethnic conflicts ensued after 1991 and damaged these countries' economies. Toward the end of the 1990s, their economies steadily improved. All three countries have climates warmer than those of the other former Soviet republics and can produce agricultural products such as fruits, especially grapes, tobacco, cotton, and rice. They have also worked to reduce their dependency on Russia by greatly altering and increas-

FIGURE 3.23   The Southern Caucasus: countries, cities, and major physical features. Darkened areas within countries represent areas that have acted autonomously, often without the consent of their respective central governments.

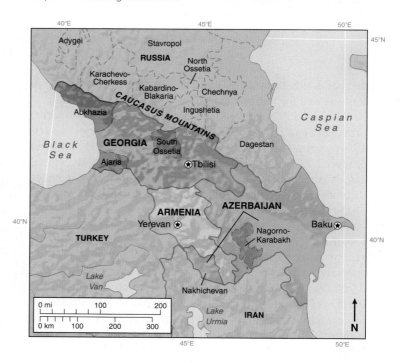

ing their industrial and service sectors. Georgia, located on the sunny, warm, eastern shores of the Black Sea, has great tourist potential. Azerbaijan, located on the Caspian Sea, will become one of the world's leading oil producers if it is ever able to fully exploit its oilfields. Until recently, because most pipelines ran through the Russian Federation, the Russian government was able to control Azerbaijan's oil trading and limit Azerbaijan's desire to establish relationships with other countries, most notably with Iran, Turkey, the United States, and those in Europe. The recently opened pipeline to the Turkish port of Ceyhan is helping to break the dependence on the Russian Federation (Figure 3.24).

Armenia-Azerbaijan

The persecution of Armenians has had a lasting effect on this part of the world. In 1895, the Ottoman government massacred 300,000 Armenians within its realm. Again in 1915, during World War I, the Ottoman government tortured, exterminated, and deported its Armenian population, claiming that the Armenians were a threat. Somewhere between 600,000 and 2 million Armenians were exterminated out of a prewar population of about 3 million in what can be referred to as the "Armenian genocide." Many Armenians became refugees, migrating across their traditional homeland

or leaving it all together. By 1917, fewer than 200,000 Armenians remained in Turkey.

It was not just this one period that inflicted a toll on Armenians. Over 1,000 years, foreign invaders wreaked havoc numerous times and scattered the population. Today, over half of the Armenian population lives outside of Armenia in a diaspora. About half of the diaspora community (e.g., those outside Armenia) lives in other CIS countries. The other half lives in communities from India across to Southwestern Asia, Europe, and North America, with a sizable number in the United States.

One of the largest conflicts among the former Soviet republics involved Armenia and Azerbaijan. In 1924, the Soviet government created an autonomous territory within Azerbaijan known as Nagorno-Karabakh. It was 94.4 percent Armenian. By 1979, Armenians represented only 76 percent of the region's population. Armenians began to fear their loss of numbers and objected to Azerbaijani laws that restricted the development of the Armenian language and culture. Clashes between the Armenians of Nagorno-Karabakh and Azerbaijanis began in the 1960s and developed into war by 1992. Armenian forces of Nagorno-Karabakh seized most of the territory and advanced westward to link their territory with Armenia. Afterward, they moved into Azerbaijan proper, but the Turkish and Iranian governments warned the Armenians

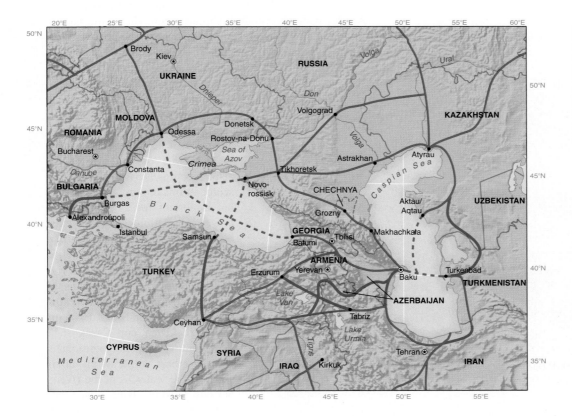

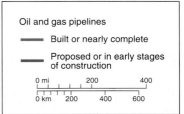

FIGURE 3.24 Russia and Neighboring Countries: pipeline outlets. Russia would like oil and gas exports from Central Asia and the South Caucasus to be piped to its Black Sea ports. The other countries and their international oil company partners are examining and building routes through Turkey to Ceyhan or through northern Iran. How does this situation illustrate the interdependence of countries in the world economic system and the importance of individual countries and their actions?

to cease hostilities. Peace talks sponsored by the UN, Russia, Iran, and other countries ended the shooting war in 1994. In addition to Nagorno-Karabakh, Armenian forces continue to control approximately 20 percent of Azerbaijan. An official agreement on the governance and the political status of territories has not emerged.

Subregion: Central Asia

Central Asia's five countries were once part of a land called Turkestan. Turkestan was not a single, consolidated country but a loose confederation of tribes. Beginning in the 1800s, the Russian Empire expanded forcibly into Turkestan and took firm control of most of it by the end of the century. In the 1920s, Soviet authorities drew boundaries for new republics which emerged as the countries of Central Asia after the Soviet Union broke up (Figure 3.25 and see Table 3.1). With more than 40 percent of the population of the five Central Asian countries and the only one that borders all the others, Uzbekistan is the dominant country.

The Central Asian countries share similar landlocked situations, arid or semiarid climates, and Muslim faith of many of their peoples. The subregion's fertile river valleys of the Amu Darya and Syr Darya are one of the cradles of human civiliza-

tion. They played important roles in the trading of goods and ideas from the time of the earliest civilizations in Mesopotamia, China, and the Indus River valley. Straddling old trade routes between east and west, most notably the Great Silk Road, great cities such as Bukhoro (Bukhara) and Samarqand (Samarkand) emerged. During the 700s and 800s, Bukhoro became one of the leading centers of learning, culture, and art in the Muslim world. Its grandness rivaled the other Muslim cultural centers of Córdoba (Spain), Baghdad (Iraq), and Cairo (Egypt). Some of Islam's greatest historians, geographers, astronomers, and other scientists came from the area. In the late 1300s, Timur (Tamerlane) emerged as the dominant leader in Central Asia and conquered lands far to the west, south, and east. A new flowering of culture began as numerous scholars and artisans came to reside in his imperial capital, Samarqand. Timur's grandson was one of the world's first great astronomers. Literature flourished and great religious structures and palaces were built (see Figure 3.13).

Following independence in 1991, GDPs dropped considerably and are generally below their Soviet era levels though they are improving. Soviet infrastructure still keeps Central Asia dependent on the Russian Federation. Pipelines and transportation lines are of Soviet specifications and run primarily to the Russian Federation. Though Central Asians are concerned with Russian domination, and seek to diminish Russian influence within their countries, they willingly joined the CIS. Political instability or poor political relations also make trade with or through neighbors like China, Afghanistan, and Iran risky.

All five countries produce similar commodities: oil, natural gas, and cotton. Thus, rivalry rather than cooperation is most common, especially in attempts to attract foreign trade. Kyrgyzstan and Tajikistan, however, do not have large fuel reserves like the other three. Consequently, the other three countries have shut off fuel supplies as a political lever against their neighbors.

The issue of water resources also encourages rivalry. Water is scarce in this dry area of the world but important for agriculture and power generation. Disputes exist over the allocation of water that flows in rivers passing through a number of the countries. Major rivers originate in the mountains of Kyrgyzstan and Tajikistan and flow down to the lowlands of Turkmenistan, Uzbekistan, and Kazakhstan. Each country frequently complains that the others withdraw an unfair share of the water from the rivers. Serious tensions have arisen between Kyrgyzstan and Uzbekistan over water in the Fergana Valley,

FIGURE 3.25 Central Asia: countries, cities, and major physical features. These countries have a mountainous southern border. Under the Russian Empire, they were carved out of the former Turkestan area. Only the cities of Almaty and Tashkent have populations of over 1 million people.

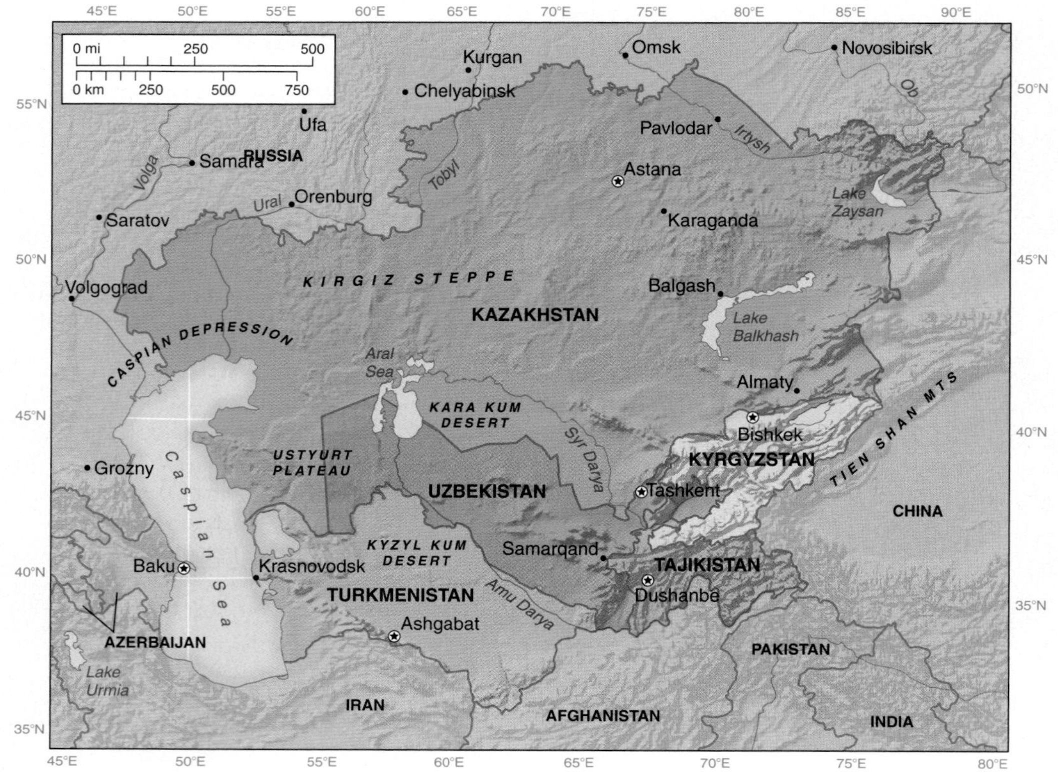

where agricultural reform and land privatization programs are endangered by the water disputes.

Contemporary Geographic Issues

Is Russia Still a World Power?

The fall of communism in East Central Europe and the Soviet Union in 1991 brought about profound change for Russians. On the positive side, it promised political and economic reform, with greater freedom of expression and a better standard of living. On the negative side, it coincided with the loss of much of their country, the open expression of great anti-Russian feelings from people within their world region, and the questioning and potential end of Russia as a world power. These experiences were a tremendous blow to the prestige of the Russian people.

When the 14 non-Russian republics declared their independence from the Soviet Union in 1991, it seemed that the peoples of these republics simply were expressing their right of self-determination. The Russians, however, believed that the Soviet republics were Russian lands regardless of who lived in them. They saw the Soviet Union as rightfully theirs because it was created from the Russian Empire, lands that they struggled for and acquired over the course of centuries. Ukrainian and Belarussian independence particularly is difficult for Russians to accept because much of the land and peoples of these republics were part of *Rus*—the original Russian heartland and people. Though these Soviet republics declared independence, Russians do not see the 14 new countries as foreign. Instead, they see them as part of their "Near Abroad" and feel that they have an exclusive voice in both the internal and international relations of these countries.

As noted in the "Population Distribution and Patterns" section (p. 83), approximately 25 million Russians live in the "Near Abroad." The adjustment from majority to minority status has been difficult for Russians, both in and out of Russia. The policies of these independent republics have caused over 5 million Russians to migrate to Russia.

Countries of the "Near Abroad" also are crucial to Russia's status as a world power. The Baltic republics housed key military installations. The Crimean Peninsula had been a major Soviet naval facility, but this became part of the Ukraine when the Soviet Union broke up in 1991 and the Ukraine Republic became independent. The Russian Republic tried unsuccessfully to reclaim the facility as it would take decades and billions of dollars of investment for Russia to re-create a comparable naval facility along the Russian coast of the Black Sea. Kazakhstan was the center of the Soviet space program. Russians see the Soviet space program as their accomplishment, one that only Americans have matched. Control over the oil in the Caucasus and Central Asia is also key to Russia's role as a world power.

Coping with the loss of the "Near Abroad" is compounded by struggles by non-Russians within the Russian Federation for in-

dependence. Chechnya is the most vexing of them all. As the Soviet Union disintegrated in 1991, the Chechens moved toward independence from the Russian Federation and declared it in 1994. The Russian military responded by supporting a Chechen rebel group that sought to overthrow the Chechen government and keep Chechnya within Russia. The military campaign was not as quick as the Russian government had hoped. Military operations have continued, though intervals of negotiations bring relative subsidence of hostilities. The campaign, however, has been brutal. By early 1996, an estimated 40,000 to 100,000 people, mostly civilians, had been killed in a republic of just 1 million. Street-to-street fighting and bombings by the Russian air force have made many of Chechnya's cities uninhabitable. Accused of gross human rights violations, Russia's international reputation has suffered from the war in Chechnya. However, Russians fear that if they grant Chechen independence, then Russia's other minorities will also move toward independence. In addition, Chechnya, especially Grozny, is a major transit point for oil leaving the Caspian Sea region (see Figure 3.24). Keeping the oil flowing through this pipeline is crucial for Russia. If proposed pipelines through Georgia and Iran succeed in taking business away from the Russian pipeline, Russia will lose billions in income, along with ability to influence international affairs.

Russian influence in world affairs diminished in other ways in the 1990s (Figure 3.26). Soldiers in Russia's military are vastly underpaid, unprepared, and demoralized from their experience in Chechnya, which followed the earlier defeat in Afghanistan in the 1980s. Though Russia is a nuclear power, the United States, viewing itself as the sole victor of the Cold War, acts unilaterally in international affairs. At the same time, the North Atlantic Treaty Organization (NATO) expanded into East Central Europe to include countries formerly dominated by the Soviet Union. For old and new members of NATO alike, the expansion was a defensive move. For the Russians, NATO's moves were provocative, particularly since they were taken without Russian consent. NATO was a Cold War institution created to defend Western Europe from Soviet communism. Because Soviet communism no longer existed and Russia was very weak, NATO expansion suggested that the organization was not defensive but offensive. It played on Russian fears that NATO was nothing more than another force, not unlike Napoléon or the German armies of the two World Wars, to menace Russia.

By the late 1990s, Russians felt ostracized by the United States and Western European countries. Closer relations with China led to the signing of a friendship treaty in 2001. The treaty is a significant agreement because it sets aside decades of tension caused mostly by border disputes. The treaty allows for cross-border trade. For example, Russia will send oil and military supplies to China. Both China and Russia agree to crack down on Islamic fundamentalists who straddle their borders and threaten the territorial integrity of their countries. In short, a Russo-Chinese alliance is a signal to the rest of the world that Russia is still a world power.

Vladimir Putin, the leader of Russia, was able to change his country's relationship with much of the world after the events of

(a)

(b)

FIGURE 3.26 **The Russian military presence.** (a) Russian paratroopers march through Red Square, Moscow, during Victory Day celebrations, May 9th, 2003. (b) Russian soldier and the wreckage of a Mi-8 helicopter south of Grozny, Chechnya, March 23rd, 2005. The Mi-8 helicopter belonged to Russian interior Ministry troops and crashed while preparing to land after suffering a technical malfunction.

9/11. Putin complained less about American international diplomacy, though it offended long-held Russian interests. Instead, he sought to be more conciliatory to the United States in hopes of gaining U.S. cooperation and increasing Russia's international influence. Certainly, the U.S. attempt to suppress Islamic fundamentalism in Afghanistan paralleled a Russian concern that led to the Soviet invasion of Afghanistan in the 1980s. At the same time, U.S. criticism of the Russian military campaigns in Chechnya turned to sympathy when al-Qaeda fighters were taking refuge in Chechnya and aiding the Chechen cause. Russian leaders were able to describe the war in Chechnya as another front on the war on terrorism. Russia was also able to press the Georgian government to do something about Chechen rebels hiding in the Pankisi Gorge. Russians are sensitive about foreign activity in a "Near Abroad" country such as Georgia, but with Russian-American cooperation in the war on terrorism, Russians voiced little concern about the United States dispatching elite military forces to Georgia to take care of a problem that vexed Russia. Relations began to cool with the United States again in 2006. As Putin has tried to improve government effectiveness and the economy with an old-style heavy-handedness that led to the Russian government retaking control of many businesses and curtailing freedom of speech, the U.S. government has been critical, and European governments leery, of Putin.

Internationally, Russia exerted its power by determining the fate of the Kyoto Protocol. It solidly ranks as the world's second greatest CO_2 emitter with 17.4 percent of such emissions, be-

hind the United States at 36.1 percent but ahead of third-ranked Japan at 8.5 percent. Therefore, the Russian Federation was able to play a key role in determining if the Kyoto Protocol would receive the required ratification of Annex I countries representing 55 percent of CO_2 emissions. The United States opposed the Protocol but needed Russian opposition to keep the Protocol from going into effect. However, European countries and Japan supported the Protocol and needed Russian support for it to go into force. Finding itself the crucial decision maker, the Russian government delayed its decision, allowing each side opportunities to sway it with offers. By the end of 2004, Russia decided to join with European countries and Japan and ratify Kyoto.

Since 2000, Russia has become a more full-fledged member of the informal organization representing the world's most wealthy countries, once known as the Group of Seven (G7) but, with Russia, increasingly known as the G8. In 2002, NATO agreed to give Russia a voice in NATO affairs, though Russia is not a full-fledged member. Russia also has offered to supply the United States and its allies with oil. Showing a preference to buy oil from Russia rather than from many of the countries of Southwestern Asia, such as Iraq and Iran, the United States has become more supportive of the Russian oil pipeline from the Caspian Sea (see "Natural Resources: Oil and Natural Gas," p. 102) though it encourages the countries of the Southern Caucasus and Central Asia to find alternative routes around Russia for their oil and other exports. With an improving economy, more effective political leadership, and a huge nuclear arsenal built during Communist times, the Russian Federation has the opportunity to exert itself as a world power in the changing climate of international affairs, though it can harm its own reputation too (Table 3.3).

Human Rights

In Imperial Russia, the czars sent their political opponents to Siberia, far from the political center of the empire. After the Bolsheviks came to power in 1917, they continued the practice of banishing their opponents. Joseph Stalin raised the practice to a new level when he created the "Main Directorate for Corrective Labor Camps," known in Russian as *Glavnoe upravlenie ispravitel no-trudovykh lagerei* or *gulag* for short. Besides places to send political opponents, Stalin also saw the *gulag* as a useful tool for economic development. Slave labor was essentially free, costing only a few bowls of thin soup and a couple slices of bread a day for each laborer. Slave labor could be used to complete dangerous work that no free worker would agree to do. After Stalin's death in 1953, the number of people sent to the *gulag* dramatically decreased but did not stop. It is difficult to know exactly how many perished in the *gulag*, although estimates are in the tens of millions.

Since the end of communism, human rights abuses have decreased but not ended. The system of police practices and prisons that violated human rights did not completely disappear, particularly not in many Central Asian countries. Economic hardship has prevented many prison reforms. Court systems need to be reformed. Defendants sit in jail before their cases are heard, and low pay makes many judges prone to accepting bribes. Juries are rare. Judges frequently serve as prosecuting attorney as well as judge. Not surprisingly, during Boris Yeltsin's administration of the 1990s, the conviction rate was 99.6 percent. The war in Chechnya caused a new series of human rights abuses.

Democratic reforms since 1991 allowed groups working for human rights to blossom, especially in the Slavic Countries. For example, Amnesty International now has as many as 40 chapters across the region. Some human rights groups are outgrowths of international organizations, though Putin has since limited their activities and made some illegal. Others are unique to the concerns of the region's people. The Committee of Soldiers' Mothers of Russia, "Mother's Right" Foundation, and the Moscow Center for Prison Reform are a few examples.

Human rights, however, have worsened in Central Asian countries, where leaders rule like Stalin once did. For example, Saparmurad Niyazov has ruled Turkmenistan since independence in

---

TABLE 3.3   DEBATE: IS RUSSIA STILL A WORLD POWER?

No Longer a World Power	Still a World Power
Russia lost possession of 14 republics (now independent countries) and with them sizable populations and resources.	Russia's heartland remains together, and Russia is still the largest country in the world, with a sizable population and considerable resources.
Russia's international power is undermined by the presence of 25 million Russians who now find themselves as ethnic minorities in other countries. Russia must be careful not to offend these other countries and thereby endanger the Russians who live in them. Russia may have to make economic concessions to these countries to protect Russian minorities in them.	Russians living as ethnic minorities in other countries increase Russia's political influence in those countries because they can vote and hold political office. Russia can also pressure those countries on behalf of Russians living in them. The Russian minorities also strengthen Russia's trading links with these countries.
The Russian government has been ineffective in controlling independence movements in its own republics. It has been condemned for human rights violations in Chechnya. Its military is underpaid, unprepared, and demoralized. Russia was unable to control the conflict in the former Yugoslavia and the expansion of NATO.	Russia gained sympathy from the United States for its "war on terrorism" in Chechnya. Russian soldiers are very patriotic, as in the past, and Russia is a nuclear power. With peacekeeping troops in the former Yugoslavia, Russia influences events in that area of the world. Russia also now has a voice in NATO's affairs.
With rampant corruption, Russia's economy is weak and has floundered since the end of communism in 1991. Russia is unable to capitalize on selling its vast resources on the world market. Its control of Caspian Sea oil is questionable. Russia has little influence in global economic organizations.	Russia's economy is steadily improving, and the government is fighting corruption. Russia has signed agreements to supply oil to other countries and is able to control the flow of oil from the Caspian Sea. Russia has joined the Group of Seven (G7), making the organization the Group of Eight (G8).

1991. In 1999, he had himself elected president for life and built a huge statue of himself cast from pure gold. The statue revolves so that the sun is always behind it. Niyazov calls himself Turkmenbashi ("father of the Turkmens"). He has renamed the days of the week and the months of year, naming January after himself and April after his mother. Though Turkmenistan may have the most repressive regime, freedom of the press is almost non-existent across Central Asia. Political opposition groups are weak, and any of their leaders who do not operate outside their respective countries are most likely in jail, where torture is common.

Natural Resources: Oil and Natural Gas

Russia and Neighboring Countries has plentiful natural resources. The rocks contain huge quantities of a wide variety of mineral resources, including iron and gold as well as many minerals that are in short supply elsewhere in the world. Diamonds and other valuable products are common. Many of these resources occur in the ancient rocks of the Siberian plateaus or

have been washed by erosion into the intervening areas of sedimentary rocks. Fuels include coal, oil, natural gas, and uranium (Figure 3.27—compare with Figure 3.24). The Caspian basin and the Siberian lowland contain substantial oil reserves, perhaps as much oil as the Persian Gulf area (see Chapter 7).

The Russian Federation can use its huge oil and natural gas reserves to generate income for industrial modernization, raise the Russian standard of living, rebuild the Russian military, and exert itself as a world power. If current trends continue, by 2030 Russia will supply Europe with 94 percent of the oil and 81 percent of the natural gas that world region consumes. Poland, Slovakia, and Hungary already receive 100 percent of their oil from Russia, and the Baltic states, Slovakia, and Romania currently obtain 100 percent of their natural gas from Russia. Russian companies also have purchased significant shares of East Central European oil and gas companies. In 2006, Gazprom (Russia's state-controlled gas company) signed a deal with the German firm BASF to increase its share of gas distribution in Germany, and explored the possibility of making a bid to buy Centrica, a company that controls the United Kingdom's gas network. In the east, the Russian govern-

FIGURE 3.27 Russia and Neighboring Countries: oil and gas fields. Assess the availability of oil and natural gas to Russia, the other neighboring countries, Europe, China, Japan, and the potential for developments in Siberia.

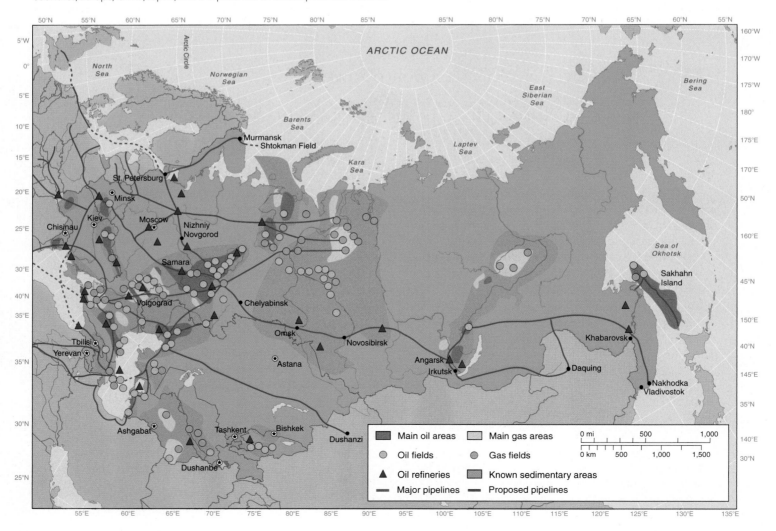

ment has been negotiating deals to supply both China and Japan with oil and natural gas. It seems that Russia has been moving slowly to encourage both China and Japan to compete with each other with offers to pay for the costs of new pipelines and other infrastructure. Still another plan to drill in the Shtokman gas field in the Barents Sea could lead to the sale of liquefied natural gas to the United States and open the door for Russian companies to gain shares in American gas distribution companies.

A number of incidents in the 2000s illustrate the economic and political importance of oil and natural gas. In 2005, Gazprom acquired the private oil firm Sibneft. Also, the state-controlled oil firm Rosneft obtained the main production operation of the privately owned oil company Yukos after the Russian government put Yukos' leader in jail and ordered Yukos' dismantling. In 2005, the German and Russian governments announced a plan to build under the Baltic Sea to Germany a pipeline which can supply natural gas to Western Europe. Gas supplies were temporarily shut off to Belarus in 2004 during a dispute between Belarus' leader and the Russian government. In January 2006, during a very cold winter, gas supplies were also temporarily shut off to Ukraine

following the Ukrainian government's resistance to pay higher prices for Russian natural gas. It is also believed that the Russian government was retaliating against Ukrainians for defeating a pro-Russian government in Ukraine and putting a pro-Western one in its place ("the Orange Revolution"). The fuel stoppages affected and angered Western Europeans. If the Baltic pipeline is built as planned, then Russia will be able to stop the oil flow to its neighbors and East Central Europe without disrupting the flow to Western Europe and vice versa. The willingness of the Russian government to use oil and natural gas as a political weapon has led the United States and other countries to encourage the countries of the Southern Caucasus and Central Asia to find alternative routes around Russia for their oil and other exports.

Environmental Problems

Environmental damage of the tundra, forest, and desert areas are a serious issue for Russian and its Neighboring Countries today (Figure 3.28). For centuries, inhabitants of the region

FIGURE 3.28 Russia: areas of environmental degradation.

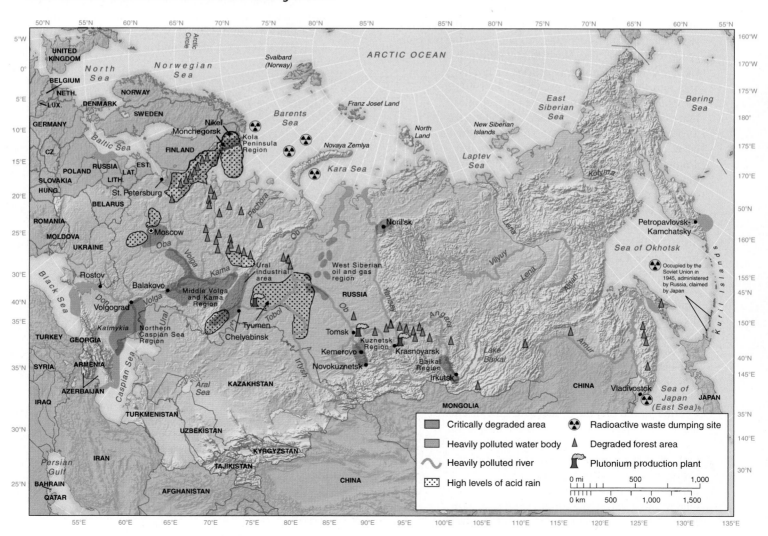

thought that the vast quantities of resources in their lands were inexhaustible. Later, communism preached that nature should be transformed to serve human needs. Rapid industrialization led to the massive exploitation and extraction of minerals and the construction of enormous factories, often built quickly with little thought to environmental protection. Emphasis on chemicals, steel, and the nuclear industry coupled with faulty storage of toxic rocket fuel and oil, the testing of weapons, and the mining of metallic minerals laid waste to huge areas, many of which remain unreported and unreclaimed. A few examples illustrate the range of environmental problems facing Russia and Neighboring Countries.

- One of the major legacies of the Soviet Union is frequent oil pipeline breaks and leakages. Literally, hundreds occur every year. Some of the resulting oil spills are small but others cover hundreds of square kilometers. Inefficient construction practices are the major cause. However, one incident in 2004, which resulted in a 540 km² (209 mi.²) oil spill in the Samara region, was caused by an incision carelessly made to illegally siphon oil.
- The city of Norilsk in western Siberia has the most polluted environment in Russia. The region contains some of the world's richest deposits of mineral ore: 35 percent of the world's nickel, 10 percent of the world's copper, significant cobalt reserves, and 40 percent of the platinum group of metals. The Norilsk Metallurgical Combine releases millions of tons of pollutants into the atmosphere each year. In 1991, for example, the amount was nearly 2.4 million tons, with the sulfur dioxide content at 40 to 50 times legal limits. The stinking acidic haze blackens snow, kills trees for miles around, and poisons the river. Many locals, most of whom work in the factories, believe that the noxious fumes inoculate them against disease, but local physicians report high incidents of respiratory illnesses and shortened life expectancy, as low as 50 years.
- In addition to industrial pollution, nuclear contamination caused major environmental damage. The 1986 Chernobyl nuclear reactor explosion, which became a symbol for such disasters, affected most of the area immediately around it north of Kiev, but the fallout was worst to the north in Belarus (Figure 3.29). Nuclear dumps on islands, peninsulas, and in seas, especially along the Artic Ocean, have also released radioactive pollution. Well before Chernobyl, the "Kyshtym incident" in 1957 in the Urals exposed almost 500,000 people to harmful doses of nuclear radiation. Coupled with other industrial activities, nuclear contamination puts the Urals high on the list of the most polluted areas of the CIS.
- One of the greatest environmental disasters in the world occurred in the lands east of the Aral

Sea (Figure 3.30). Although Central Asia is arid, snow on the high mountains to the east melts in spring, providing water to such rivers as the Amu Darya and Syr Darya that flow into the Aral Sea. This water provided the basis of local irrigation farming and small urban settlements through history, but the Soviet Union adopted ambitious plans to use the water on irrigated cotton farms inside and outside the main river basin. Supported by growth-oriented, unbending bureaucrats who often acted against the advice of government scientists, the project was taken too far and so much water was extracted that the rivers stopped flowing into the Aral Sea. The sea is now less than half its previous size, all transportation on it has ceased, and the supply of moisture for the mountain snows has dried up. Glaciers in the mountains are in retreat, groundwater levels have fallen accompanied by ground subsidence, and dust storms now affect areas that had seldom experienced them. In 2005, an $86 million grant from the World Bank provided for a 13 km (8 mi.) long dam to regulate the flow of the Syr Darya. The dam is allowing the northern Aral to rise 42 m (138 ft.), bringing the sea within 13 km (8 mi.) of the town of Aralsk.

- The Caspian and Black seas are threatened by oil, toxic waste, ships' ballast water, and nutrients from fertilized fields. Only five of the 26 fish species caught in the 1960s in the Black Sea were still found in 1992.

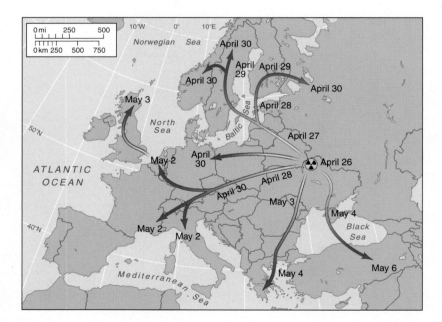

FIGURE 3.29 Russia and Neighboring Countries: impact of a nuclear disaster. The areas affected by the Chernobyl nuclear power plant explosion that released clouds of radioactive gases in 1986. The winds at the time carried the radioactive gases in numerous directions in the following days. In Ukraine, over 1.5 million people were affected, 10 percent of whom were severely radiated. All 600,000 members of the cleanup team suffered minor illnesses. Local death rates increased as a result of delays in evacuating inhabitants after the explosion. Twenty years later, people still could not return to their homes. Farther afield, the radiation was detected over Scandinavia, and some sheep rearing areas in the United Kingdom still face restrictions on pasture area in 2006.

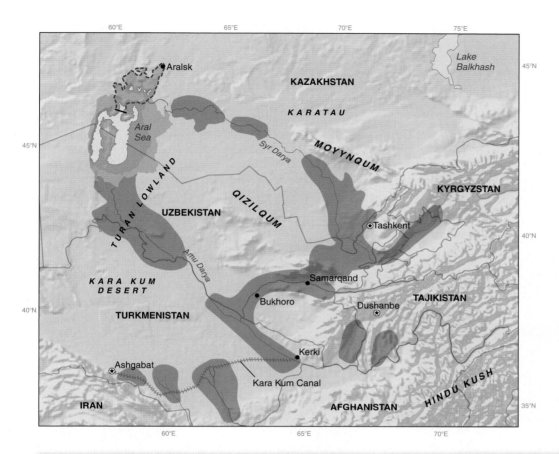

FIGURE 3.30 Central Asia: Aral Sea environmental disaster. The Aral Sea, Kazakhstan, and its surroundings. The sea is a basin of inland drainage in an arid region. It is supplied by meltwater from the mountains to the southeast. Irrigation projects such as those supplied by the Kara Kum canal have been steadily diverting water from the Aral Sea, causing it to dry up.

GEOGRAPHY AT WORK

Wine Industry

Robert W. Hutton, a retired subject cataloguer for the U.S. Library of Congress, finds geography important to understanding the wine industry. In his cataloguing activities, Robert found a lot of books on Russian wine but the wine experts he worked with knew little of winemaking in Russia. Having a bachelor's degree in Russian from Haverford College and a master's in geography from Columbia University, Robert found his passion with so many books on Russian wine. Many of the books in Russian contained a lot of information on the former Soviet republics, especially in the warmer southern republics stretching from Moldova to Kazakhstan. In 1989, Robert was able to travel to Georgia, his first trip to the region (Figure 3.31). A second trip was to Moldova in 1993, and then one to Ukraine and Crimea in 1998. In each case, Robert was able to talk to the local winemakers in technical Russian winemaking language. Such cultural understanding gave a good impression of Americans.

Robert finds that each wine is unique to the area where it is grown, and geography's focus on both the physical and the human world takes into account all the factors explaining where and how wine is made. The term *terroir* refers to the physical geography of

winemaking, that is, such factors as climate and soil. However, economics and cultural attitudes are just as important in understanding the wine industry. Whether during czarist or communist times, Robert found that the concern seemed to be with quantity over quality. Since the end of communism in 1991, winemakers have become more interested in producing better quality wines, but they find that their customers from the region give a high priority to high alcohol content over flavor.

Currently, Robert is working with a winemaker from the region who is writing a comprehensive book on winemaking in Ukraine. It will be the first book written on wines in Ukrainian since the end of Soviet times, and the English edition which Robert is helping to produce will be the first such book in English.

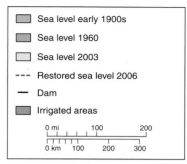

FIGURE 3.31 Robert Hutton, retired subject cataloguer for the Library of Congress, Washington, D.C.

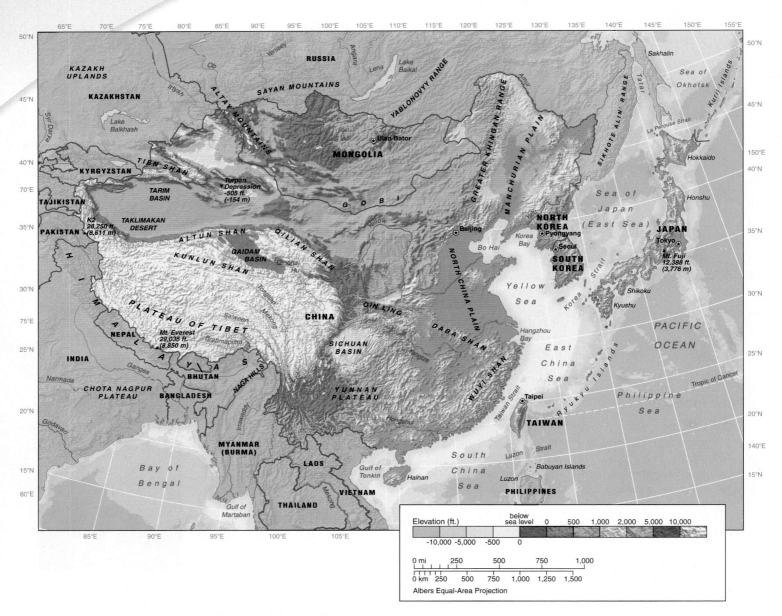

FIGURE 4.1 **East Asia: mountains, lowland plains, major rivers, countries and capital cities.** The region includes China and Japan, contrasting North and South Korea, Mongolia, and the island of Taiwan. The land boundaries are mainly mountains, but with significant breaks in the northwest, northeast, and southwest. Island chains separate the China and Japan (or East) seas from the Pacific Ocean.

Angela is the English name of a tour guide who works in China's growing tourist industry. In 2005, China became the world's fourth largest destination for international tourists (after France, the United States, and Spain) with receipts increasing eightfold since 1990. Today, tourism accounts for about one-fifth of the country's gross domestic product.

Angela was born in northeast China but moved with her parents to Beijing when she was five. There they were allotted a one-family apartment but had to share a single water tap and a toilet with other tenants. It was not until the 1990s that the family could move to better accommodation. After high school, Angela studied for four years at the Beijing Tourism Institute. After gaining her degree, she spent six months as an intern in the state travel industry, guiding one to four people at a time around Beijing and gaining confidence by 1998 to embark on the nationwide tours. Further reform changes from the late 1990s affected Angela's way of life. She now works on contract to a tour company and is paid when working. Tips have become an important part of the tour guides' incomes. She has her own pension and health care plans. In 2002, Angela bought her own apartment with modern facilities, and married a lighting engineer in 2003.

Many young people like Angela prefer the freedoms of reform to the old centrally planned system. Some aspects, such as land ownership and certain industries, are still centrally controlled. In 2008 Beijing will host the Olympic Games, and preparations are underway to improve airports, major roads and infrastructure. Angela looks forward to being a part of this special event that is expected to draw tourists from all around the globe. In early 2007, over a million tickets to the games had been sold.

Chapter Themes

Defining the Region

Distinctive Physical Geography: mountains and major rivers; subtropical and temperate climates; forests; grasslands, and desert; natural resources.

Distinctive Human Geography: population distribution and density; population growth; ethnic groups; urbanization; economic development.

Geographic Diversity
- Japan.
- The Koreas.
- China, Mongolia, and Taiwan.

Contemporary Geographic Issues
- China: A new world power.
- The rise of Japanese and Korean multinationals.
- Population policies in China.
- Globally connected cities of East Asia.
- Human rights in China and Japan.

Geography at Work: China's landscapes and global change

Defining the Region

East Asia (Figure 4.1) includes the countries of China, Japan, North and South Korea, Taiwan, and Mongolia: some of the world's materially wealthiest and some of its poorest. The region's present character builds on millennia of cultural and technological development and interactions with the natural environment. From the early 1800s, European colonial powers used force to draw East Asian countries in, but all countries in this region resisted European colonization.

After 1945, U.S. postwar reconstruction in Japan, South Korea, and Taiwan included the installation of democratic governments and connections to the U.S. economy. After its establishment in 1949, the People's Republic of China was governed by the Communist Party with central direction. Led by Japan, China, South Korea, and Taiwan, the region reemerged as a world political, economic, and cultural force in the 1990s. These East Asian countries not only interact with external globalization trends, but also contribute to those trends (Figure 4.2). By contrast, Mongolia is a much smaller country between Russia and China that follows poli-

cies it believes will placate its neighbors, although the end of the Soviet regime in 1991 led to greater political freedom. North Korea remains in the thrall of an inward-looking ruler who continues to lead the country into poverty as an international outcast.

Chinese culture dominated the region's history. From ancient times, the Chinese instigated many of the most significant human advances from technology to art. Until well into the 1900s, the Chinese saw themselves as the "Middle Kingdom"—in the center of the world. Chinese, Koreans, and Japanese share many cultural characteristics including Buddhism, the inclusion of Chinese characters in their scripts, close family life and kinship links, and a focus on communal organization (Figure 4.3). These common elements contributed much to modern attitudes toward the global economy. Today's business phenomenon of the "Asian Way" built on this common culture and is marked by a hierarchy of relationships that define how people work and live with each other. While the similarities are important, differences between the Chinese, Japanese, Mongols, and Koreans give character to each of the modern countries.

FIGURE 4.2 Pudong: China's new commercial hub. The Pudong area across the Huang Pu from central Shanghai. Some believe the TV tower completed in 1998 is too *yang* (male), and that later buildings around it are more rounded and modest *yin* (female) to give it balance.

FIGURE 4.3 East Asia: cultural features. A temple beside Dongting Lake, lower Chang Jiang, China. People come to pay homage to the Buddha and /or local deities, and it is a tourist stop.

Distinctive Physical Geography

The natural environments of East Asia have affected the cultural history and have added distinctive local features to the region. Large numbers of people crowd into relatively small areas of well-watered lowlands. They are much affected by natural events, especially in rural areas, and they themselves have major impacts on the physical environment by modifying natural features and processes.

Mountains and Major Rivers

Mountain systems and relatively small areas of lowland (see Figure 4.1) form the diverse relief of East Asia. Although the high proportion of rugged terrain suggests a difficult environment for human occupation, the region supports some of the highest densities of population in the world. High mountains extend eastward from the Himalayan Mountain ranges and the Tibetan Plateau, occupying over one-third of China. On the border with Nepal and India, the Tibetan Himalayas are the world's highest mountains.

The volcanic and earthquake-prone islands of Japan form active landscapes with major shocks and eruptions every decade or so as tectonic plates clash. In southern China, some of the most distinctive hilly landscapes are in areas of limestone rock, where the almost vertically sided hills are riddled with caves in what is known as a karst landscape.

In summer, meltwater from the high mountain snowfields to the west combines with intense rains toward the coasts, supplying high flows in some of the world's most active rivers. The three major rivers of China are, from north to south, the Huang He (Yellow River), the Chang Jiang (Yangtze or Long River), and the Xi Jiang (West River)—with its wide lower section, the Zhu Jiang (Pearl River). Inland, these major rivers carve deep valleys with steep slopes and boulder-strewn streambeds. The rivers carry large quantities of water toward the sea, together with eroded mud, sand, and rock fragments. As they near the coast, they drop their load of sand, silt, and clay to form wide fertile plains and deltas. The tidal range and wave activity are both low around these coasts, allowing the large loads of silt and clay to build deltas at river mouths.

Subtropical and Temperate Climates

East Asia has a variety of climatic environments centered on the continental temperate regime (see foldout world climate map inside back cover) that resembles the eastern parts of North America, from Cuba and Florida in the south to New England and Newfoundland in the north.

Southern China has a **subtropical rainy** climatic environment. In summer, southeasterly winds are drawn into central Asia as the continent heats up and rising air leads to low pressure. These winds bring moisture from the tropical oceans and heavy rains. In winter, the winds blow outward from the high pressure over the cold continent and are cooler and drier. In mid-to late summer, the heating of the Pacific Ocean makes the region subject to typhoons (the Asian equivalent of hurricanes) with strong winds and rain. Temperatures in Guangdong Province average 13°C (55°F) in January and 28°C (81°F) in July, with annual rainfall of 1500 mm (60 inches), most of it in the summer.

Farther north, coastal China, the Koreas, and Japan have continental temperate climatic environments with summer rains and drier winters. Although summers are as warm as those farther south, icy winds blowing from Siberia make winters much colder. Japan receives more winter precipitation than the continental countries; winds blowing from the continental interior pick up moisture on crossing the Sea of Japan and precipitate snow on the west-facing mountains.

The western parts of China and the whole country of Mongolia have arid or semiarid climatic environments because of their distance from the ocean: humid ocean air loses its moisture by raining before reaching the interior. Winters are extremely cold. In the westernmost parts of China, high altitudes in the mountains and Tibetan Plateau cause even greater summer-winter and daily ranges of temperature. Lhasa in Tibet has an average temperature of 0°C (32°F) in December and 17°C (60°F) in summer.

Forests, Grasslands, and Desert

The range of climatic and relief environments in East Asia produces a variety of natural vegetation types, although human land uses later replaced or modified these ecosystems. The subtropical broadleaf forests in southern China covered extensive areas dominated by a great variety of species, including teak forests. Farther north, temperate deciduous and evergreen forests were largely cleared from the lower, more cultivable areas but remain on steep slopes in Northeast China, the Koreas, Taiwan, and Japan. Inland, the increasing aridity causes the forests to give way to grassland and desert in northwest China and Mongolia. The Gobi Desert occupies much of Mongolia and large areas of northern China.

Natural Resources

The major natural resources of East Asia are surface water flows, fertile soils, and minerals. The mineral resources of East Asia (Figure 4.4) include coal, oil and gas, iron, gold, and

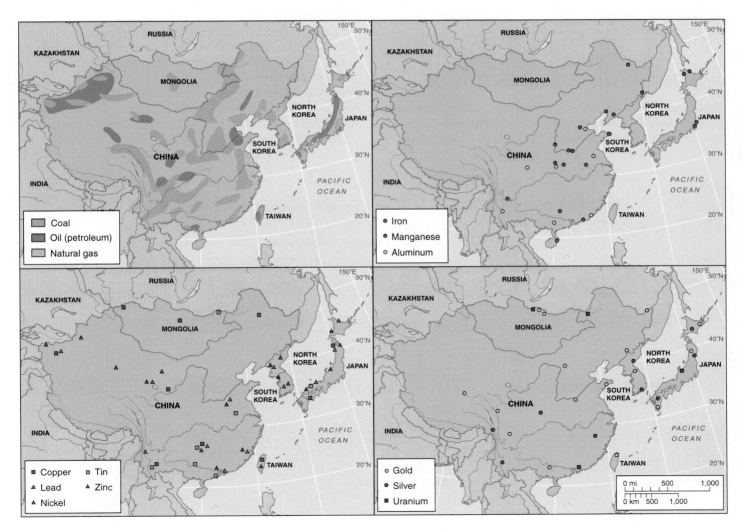

FIGURE 4.4 East Asia: mineral resources. Compare the resources of East Asian countries. (a) Fuels: coal, oil (petroleum), natural gas. (b) Metals: iron, manganese, bauxite (aluminum ore). (c) Metals: copper, lead, nickel, tin, zinc. (d) Precious metals: gold, silver, uranium.

precious stones. Coal deposits are widespread in China, while major deposits of oil and natural gas occur in western China and in many offshore locations that are still being explored. Japan has few mineral resources and relies on imported raw materials for its industries.

Although the initial Chinese civilization near Xi'an was based on millet, the growing of wheat in the north and wet rice in the south made it possible to feed much larger numbers of people and maintain high population densities. The water and the annually renewed fertile **alluvial soils** of the lower valley floors enabled abundant crops to be harvested.

Distinctive Human Geography

In 2006, East Asia was home to 1.5 billion people, 23 percent of the world's population. Despite an emphasis on population control measures, this could rise to 1.7 billion by 2035 (Table 4.1). Since the population lives on only 11 percent of the world's land, East Asia is marked by some the world's highest population densities. The combination of large populations, difficult environments, and involvement in the world economy leads to the growth of massive cities (Table 4.2).

Population Distribution and Density

East of a line drawn across China from northeast to southwest, just west of Harbin, Beijing, Lanzhou, and Chengdu, and extending across the Koreas, Taiwan, and Japan, population densities are over 25 per km^2 and exceed 100 in many areas. The densest areas are along the eastern and southern coasts of Japan, the western parts of the Koreas and Taiwan, north-central China, and the coasts and major river valleys of southern China. The western half of China and all of Mongolia are desert or mountain environments with densities of less than one person per km^2, except along major transportation routes (Figure 4.5).

Population distribution in China is marked not only by the contrasts between environmentally different areas, but also by the contrast in the proportions of growing urban populations and declining rural populations. Today, China is experiencing rapid urbanization of its people.

Japan is densely but unevenly populated. The main concentrations of people are on the small areas of lowland, and 70 percent of the total population lives in the Pacific coastlands from Tokyo to Osaka. This urbanized zone is often known as the "Tokaido **megalopolis**"—a series of almost continuous metropolitan centers with urban functions that exchange flows of people and goods with the surrounding areas and the rest of the world. As urban populations grew, the more remote and environmentally difficult areas in the mountains and along parts of the western coasts of Japan suffered depopulation. Such regions often have up to 25 percent of their population over 65 years old, and government grants are available to help people stay or settle there.

Mongolia has a dispersed population with falling natural increase and life expectancies now in the 60s. Just over half of Mongolia's population lives in urban places as the result of the pull of growing industrialization and the push of rural poverty.

TABLE 4.1 **EAST ASIA: Data by subregion, country, area, population, urbanization, income (Gross National Income Purchasing Power Parity), ethnic groups**

Country	Land Area (km²)	Population (millions)		% Urban	GNI PPP 2005	2005	Ethnic Groups (%)
		mid 2006 Total	2025 est. Total	2006	Total (US$ billions)	Per Capita (US$)	
JAPAN							
Japan	377,800	127.8	121.1	78.7	4,014.1	31,410	Japanese 99%, some Koreans, indigenous
THE KOREAS							
Korea, Republic of (South)	99,020	48.5	49.8	81.5	1,059.7	21,850	Korean 99%
Korea, People's Dem. Republic (North)	120,540	23.1	25.8	60.0	—	—	Korean 99%
CHINA, MONGOLIA, AND TAIWAN							
China, People's Republic of	9,596,960	1,311.4	1,476.0	36.9	8,655.3	6,600	Han Chinese 92%, Zhuang, Mongolian Tibetan, Uygur
China, Macao SAR	8	0.5	0.6	98.9	—	—	Chinese 95%, other Asian, Europeans
China, Hong Kong	1,040	7.0	8.1	100.0	242.3	34,670	Chinese 96%, Europeans, South Asians
Mongolia	1,566,500	2.6	3.1	57.4	5.6	2,190	Mongolian 90%, Kazak, Chinese, Russian
Taiwan	35,760	22.8	23.6	78.0	—	—	Taiwanese 84%, mainland Chinese 14%
East Asia Totals/Averages	**11,797,628**	**1,544**	**1,708**	**74**	**13,977**	**19,344**	

Source: World Population Data Sheet 2006, Population Reference Bureau, Microsoft Encarta 2005.

TABLE 4.2	Populations of Major Urban Centers in East Asia (in millions)	
City, Country	**2003 Population**	**2015* Projection**
JAPAN		
Tokyo, Japan	35.0	36.2
Osaka-Kobe, Japan	11.2	11.4
Nagoya, Japan	3.2	3.3
THE KOREAS		
Seoul, South Korea	9.7	9.2
Pusan, South Korea	3.6	3.4
Pyongyang, North Korea	3.2	3.5
Inchon, South Korea	2.6	2.8
Taegu, South Korea	2.5	2.5
CHINA, MONGOLIA, AND TAIWAN		
Shanghai, China	12.8	12.7
Beijing, China	10.8	11.1
Tianjin, China	9.3	9.9
Hong Kong, China	7.0	7.9
Wuhan, China	5.7	8.0
Shenyang, China	4.9	5.2
Chongqing, China	4.8	5.8
Guangzhou, China	3.9	3.9
Chengdu, China	3.4	3.9
Xi'an, China	3.2	3.6
Changchun, China	3.0	3.6
Harbin, China	2.9	2.9
Nanjing, China	2.8	3.0
Dalian, China	2.7	2.9
Zibo, China	2.7	3.0
Jinan, China	2.6	2.9
Taiyuan, China	2.5	2.8
Qingdao, China	2.4	2.7
Taipei, Taiwan	2.5	2.4

*estimated

Source: United Nations Urban Agglomerations 2003, with estimates for 2015 (2003).

Mongolia's capital, Ulan Bator, houses nearly one-fourth of the country's population, including those living in large tented (yurt) areas.

Population Growth

Changes in East Asia's population structure and distribution have important economic and social implications. China supports about a fifth of the world's population on only 7 percent of the world's arable land. As population policies in the 1950s and 1960s fluctuated between pro- and anti-growth, the country's population continued to rise, despite the soaring death rates and plummeting birth rates that characterized the period known as the "Great Leap Forward" when China experienced severe famine. Fertility reduction became a national priority in the 1970s. Later marriage, longer intervals between births, and fewer children were advocated, while modern contraception was made widely available.

The one-child policy was instituted to further decrease population growth. Growth rates fell from 2 percent per year in 1965 to 1.4 percent in the 1980s and to 0.6 percent in 2006. From 1965 to 2006, total fertility fell from 6.4 to 1.6. Even this major achievement merely slowed the continued growth of China's huge population. More children survived early childhood despite fewer babies being born, as infant mortality fell from 200 per 1,000 live births in 1945 to 27 in 2006. The numbers of people living to old age also increased as life expectancy more than doubled from 35 to 72 years. Investments in education cut adult illiteracy from 80 to 19 percent. (See discussion in the Contemperary Geographical Issue, Population Policies in China, page 127.)

Japan's total population increased from around 90 million in 1960 to 128 million in 2006. The combination of current low fertility (1.3) and population growth rates (natural increase of 0.0 percent) leads to estimates that the total population will peak at around 128 million in 2010 and fall to 121 million by 2025. Infant mortality at 2.8 per 1,000 live births is the world's lowest, and life expectancy at 82 years is the highest. As a result, Japan's population is aging. In 1970, 7 percent of the population was over age 65; by 2001, the proportion rose to 17 percent and is currently 20 percent. Meanwhile, smaller nuclear families averaging three persons per household in the 1990s became common. The aging of the population increases the needed pension provision, medical costs, and leisure provision for the retired, but it also adds to firms' wage costs as more employees reach senior positions. The Japanese are now worried that there will not be enough labor for their industries or funds to support the increasing numbers of elderly people. Encouragements for couples to have more children are being considered.

South Korea's population of 48.5 million people in 2006 is likely to rise slowly to around 50 million by 2025 because of a low rate of population increase. Life expectancy rose from 47 years in 1955 to over 77 years today. As in Japan, the population in older age groups will increase markedly in the next 20 years. North Korea had 23 million people in 2006, and this is predicted to rise to nearly 26 million in 2025. These two countries differ in that South Korea's modest rise is related to growing affluence, while North Korea's has a context of increasing poverty and lack of opportunity.

Ethnic Groups

The **Han Chinese** make up 92 percent of the Chinese population. The Han culture diffused by conquest and movements of people to administer new areas since the AD 200s and 300s.

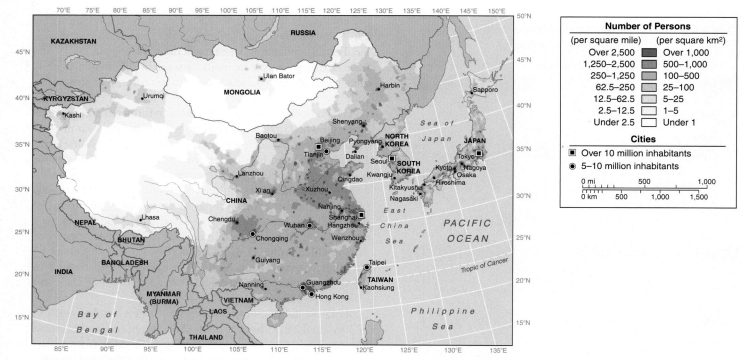

FIGURE 4.5 **East Asia: distribution of population.** Relate the main areas of high and low population density to physical features. Is this a satisfactory basis for explaining the differences?

Most minority groups live along the northern borders with Russia and Mongolia and in western China. Central Asian ethnic groups dominate the north and northwest. Minorities such as the Mongols in the north, the Tibetans and Uighur in the west, and the more dispersed Manchu and Muslim Hui, are allowed limited jurisdiction over their affairs, particularly the expression of their culture in music and drama. Still, local resentment of the Chinese takeover of their lands causes the eruption from time to time of anti-Chinese demonstrations in the minority strongholds of Tibet and Xinjiang, but these are suppressed by the central government.

Most Japanese people stem from an ancient mixture of Asian continental and Pacific island stocks. However, three small minority groups remain important. On the northern island of Hokkaido the Ainu retain ethnic and language differences. On the largest island of Honshu, the Burakumin, numbering around 1.2 million people, were outcasts and allowed to work only in the lowest occupations until their emancipation in 1871. The third group consists of foreign nationals, three-fourths of whom are Koreans, moved forcibly to Japan in the early 1900s. Until the 1980s, they found it difficult to gain Japanese citizenship and are still subject to discrimination. Migration into Japan is restricted, and only 1 percent of the population is of foreign origin.

North Korea and South Korea—officially the People's Democratic Republic of Korea and the Republic of Korea, respectively—have a common ethnicity and similar history until the mid-1900s. After the Korean War (1950–1953) the two countries moved in different directions ideologically, politically, and economically. Ethnically, the Mongol people dominate Mongolia, although it also has 10 percent who are Chinese, Russians, or Kazakhs.

Urbanization

The degree of urbanization in East Asian countries varies considerably from a low of 37 (36.9%) percent in China to approximately 80 (78.7% & 81.5%) percent in both Japan and South Korea (see Table 4.1). The region has some of the largest cities in the world, led by the massive Tokyo region in Japan with a population of 35 million. The growth of global and intraregional trade in East Asia led to the emergence of global city-regions (see Figure 1.23) as essential nodes linking this region to the rest of the world through business and political exchanges. See the Contemparary Geographic Issue, Globally Connected Cities of East Asia, page 129.

Japan and the Koreas

While the region has always had large and important cities, rapid urbanization took place in the 1900s. In the short period from 1920 to the early 2000s, Japan changed from a largely rural country, in which industrial towns housed 25 percent of the population, to one in which almost 80 percent now live in cities. At the end of World War II, half of the Japanese population was still rural, but further industrialization from 1945 to the mid-1970s multiplied jobs in manufacturing and service sectors, encouraging migration to expanding urban areas.

In equally urbanized South Korea, the high proportion living in towns is a sign of modernization. The South Korean urban population increased from 25 percent in 1950 to 50 percent in 1980 and 80 percent in 2001. The main population and industrial facilities are in the northwest, centered on the capital Seoul, Inch'on, and Taejon and along the southern coast, including Pusan, Taegu, and Ulsan. Almost totally destroyed in the Korean War of the 1950s, Seoul was rapidly rebuilt.

North Korea's population is also concentrated around its western coasts, in the capital city, Pyongyang, and Namp'o. Country-wide, only 60 percent live in towns. Although possessing greater mineral resources than South Korea, particularly coal and metallic ores, North Korea has not developed a major manufacturing capability that draws more people to the towns.

Urban vs. Rural Population in China

Until the mid-1900s, most Chinese cities were market and administrative centers, often with walls that restricted expansion. The exceptions were some port cities, of which Shanghai became the largest, growing because of external trade contacts dating back to the 1800s. Today the largest cities in China (see Table 4.2) include the capital Beijing, Shanghai, Tianjin, and Chongqing, all of which are treated as political provinces. Hong Kong, the next largest, is now a special area within China after the end of its British colonial government in 1997. Shenyang, Wuhan, Guangzhou, Changchun, Chengdu, Xi'an, and Harbin each grew from around 1 million or fewer people in 1950 to multi-million cities. It is likely that further growth of the major cities will take Beijing, Shanghai, and Tianjin to over 10 million people each by AD 2015, when China will have over 100 cities with populations of more than 1 million people.

In China, policies toward urban growth fluctuated from the mid-1900s. Communist rule from 1949 alternately favored urbanization and movements back to the countryside. In the Mao Zedong era, the wish to encourage modernization through industrialization led to rapid urbanization from 1949 to 1960. In this period the urban population more than doubled, from 49 million to 109 million, representing an increase from 9 to 16 percent of the total population. Part of this urbanization resulted from industrialization around previously administrative cities such as Xi'an and Nanjing. Another part came from the development of new inland centers such as Baotou, the iron and steel center in Inner Mongolia. Public rhetoric, however, exhorted the Chinese to avoid the environmental dangers of urbanization.

From the 1960s to the mid-1970s, the rural-urban basis of the registration system (*hukou*), established in the 1950s, was used to prevent rural-born Chinese from becoming legal residents of cities. The Cultural Revolution was launched by Chairman Mao in 1966 to ensure that China would remain true to Maoist ideology and not slip back into traditional "bourgeois" ways. As part of the movement that involved mobilizing the country's youths, over 20 million young urbanites, as well as many others, were moved to rural areas to be "reeducated" in the virtues of rural life. By 1978, the urban population accounted for around 13 percent of the total, but living conditions in urban areas had declined.

Recent Urban Migration in China

From 1978 to the early 2000s, economic reforms placed an emphasis on urban jobs, better incomes, and better access to services and led to a further rapid rise of urban population num-

bers. During this period the official urban population rose from 13 to 36 percent of the Chinese population, although the *hukou* system still operates in theory.

From the late 1970s, new urban-based industries recruited labor from the countryside on temporary contracts; such workers now make up to 10 percent of urban inhabitants. Many new urbanites do not bother to acquire temporary residence certificates. The growing middle-income class in China is almost exclusively urban, working in administration or as managers and technicians. Their numbers more than tripled from 1980 to the early 2000s. Office, business, and factory workers rose from 23 million to 40 million.

Over the same period, increased investment in housing and urban infrastructure tripled the housing space built each year, mainly in large apartment blocks. Better building standards, sewage services, transportation, and domestic water and gas supplies improved the lot of urban dwellers. The huge new apartment and condominium blocks, however, encroach on farmland that is particularly valuable in a country with so many people to feed and insufficient farmland to grow the necessary food. Although a debate continues over the relative merits of expanding the largest industrial cities or placing more resources in widespread small and medium-sized towns that are integrated with the surrounding rural areas, Chinese people are increasingly making their own decisions and moving to large urban areas.

Economic Development

Most countries of East Asia have a major involvement in the global economic system. They cluster toward the top of the world range of gross national income (Figure 4.6) and ownership of

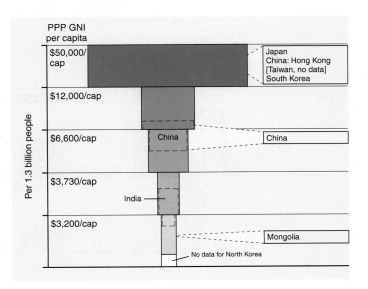

FIGURE 4.6 East Asia: country average incomes compared. The countries are listed in order of their GNI PPP per capita. Note the relative positions of Japan, China and Mongolia. *Source: Data (for 2005) from World Development Indicators, World Bank, and Population Reference Bureau.*

consumer goods (Figure 4.7) For U.S. traders, Japan is an established major competitor, while the People's Republic of China, South Korea, and Taiwan are the main emerging big markets. North Korea and Mongolia, however, remain largely outside the world economy.

The recent economic growth of these countries was at first based on exporting products to American and European markets, but by the 1990s was also bolstered by trade among countries within this region and Southeast Asia. Asian internal trade is now as great as external trade, questioning whether the region's economic future lies with intra-regional trade or a global scope. However, the gradual evolution of the Asia-Pacific Economic Cooperation (APEC) forum reflects the global trading links from East Asia to Southeast Asia, the South Pacific, and the Americas.

From Poverty and Defeat to Renewed Eminence

After World War II, poverty, political disruption, and cultural confusion were the rule in East Asia. In 1945:

- Defeated Japan was economically devastated and hated in the region for its wartime acts.
- The Koreas struggled with independence following decades of Japanese occupation and exploitation. Splitting Korea into North and South led to the 1950s Korean War, which destroyed much of both countries.
- In China, the warring Nationalist and Communist factions resumed their prewar conflict. When the Communists prevailed in 1949, the remnant of the Nationalist army fled to Taiwan.

Manufacturing provided the basis of modern Japanese growth. Modernization in Japan built on economic stability and government encouragement. New factories built in the coastal zone from Tokyo to Osaka made iron and steel, ships, automobiles, constructional engineering products, and textiles.

The newly industrializing South Korea, Hong Kong (before and after it became part of China in 1997), and Taiwan increased their incomes six- to sevenfold from 1980 to 2000. South Korea and Taiwan built economies on exporting to the rest of the world. Over nearly 30 years from 1949 to 1976, the People's Republic of China under Mao Zedong built national cohesion, albeit at tremendous cost in terms of lives and policy changes. China experienced its most rapid economic growth from 1980, increasing its total GNI PPP from US$202 billion in 1980 to US $1,065 billion in 2000 and US$2,264 billion in 2005. However, neither North Korea with its declining economy nor Mongolia with its small economy experienced such economic expansion.

Will Miracle Growth Continue?

The East Asian economic growth in the later twentieth century was termed the "East Asian Miracle" in a 1993 World Bank report, which argued that distinctive Asian characteristics motivated and fueled economic growth:

- Successful countries had strong governments that managed stable business environments by such measures as keeping inflation low. Their tax structures distributed some of the growth rewards throughout communities. Domestic savings increased.

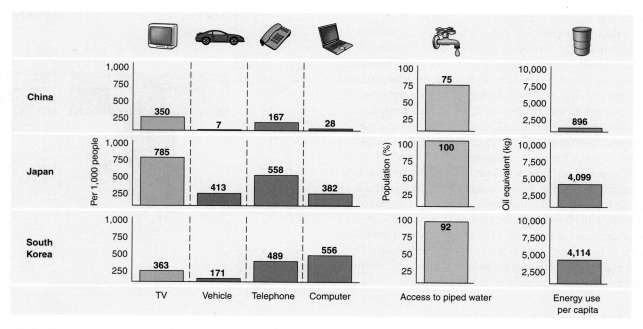

FIGURE 4.7 **East Asia: ownership of consumer goods, access to piped water, and use of energy.** How do these figures relate to economic growth? *Source: Data (for 2002) from World Development Indicators, World Bank, 2004.*

- Governments encouraged export competitiveness that led to partial global integration.
- Education programs focused on primary and secondary education for both boys and girls as a priority over prestigious higher education for a few.

By the mid-1990s, the rapid economic growth in some countries began to slow. It was geographically uneven and subject to interruptions. For example, in China, the rising incomes of people along the coast, who have access to global trade, contrasted with the continuing extreme poverty of many inland areas.

In 2001, a World Bank report, "Re-Thinking the East Asian Miracle," reanalyzed the "East Asian Miracle." Its alternative explanations centered around a combination of global trends and local attitudes:

- Economic growth up to the mid-1990s resulted mainly from increased inputs of capital, labor, machinery, and infrastructure (transportation, power, etc.), rather than from rising productivity (e.g., increasing output per employee).
- The advantages of activist government industrial policies and government-supported liaisons between bankers and industry appeared less convincing when large corporations used government funds wastefully or corruptly.
- The policy of encouraging export sales and protecting home markets was a poorer motivator of economic growth than openness to total trade in the interdependent world of globalization.
- Increasing global links showed the weakness of such social institutions as strong government, family control of businesses from the smallest to the largest corporations, and regulatory agencies without sufficient powers.
- Many of the profits from export earnings went into real estate speculation rather than further investment in productive facilities. Money bound up in such speculation was not available to meet a crisis.

The late 1990s economic crisis dealt a blow to assumptions that the "Asian Way" would produce continuing economic expansion. Geographic factors, beginning with the development of distinctive cultures in specific places through history and interactions of people with the natural environment, bring a greater degree of diversity to this region than one simplistic prescription could change in a few years. The International Monetary Fund made loans in exchange for an end to protectionist policies that kept Western foreign investments out of East Asia. South Korea recovered its economic impetus within a couple of years, but in that time it also suffered unemployment, corporation failures, and an increase in long-term poverty.

Geographic Diversity

East Asia (see Table 4.1) is dominated by two powerful countries, Japan and China. At present, Japan's strength is economic. China's strength is political-military, cultural, and increasingly

economic. Mongolia is linked to China because of its situation between China and Russia. Taiwan is also linked to China, which claims possession of the island. Although recent trends bring the two closer to reunification, many Taiwanese bitterly oppose such a move. As separate countries, North and South Korea have very different levels of global involvement but were united until 1950 and may reunite in the future.

Japan

Small and densely populated, Japan became the world's second-wealthiest country following a dramatic economic recovery after 1945. Japan's physical environments—from snowy Hokkaido in the north to subtropical Okinawa in the south and from the volcanic mountain spine to the coastal lowlands—combine with distinctive contrasts in urbanization, manufacturing emphases, and rural ways of life. Japan consists of four main islands: Hokkaido, Honshu, Shikoku, and Kyushu (Figure 4.8).

Ancient traditions place Japan's foundation in 660 BC. The principle of **Shinto** encapsulated Japanese traditional values in a national religion built on animism, ancient myths, and customs.

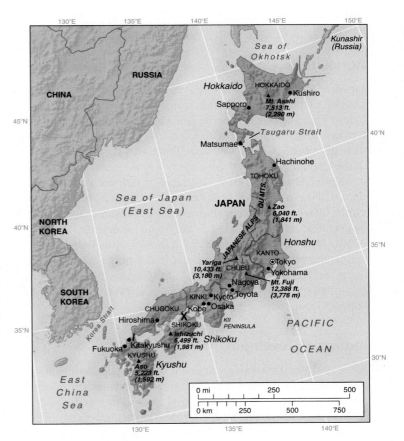

FIGURE 4.8 Japan: Major islands, districts and cities. Assess from the map the significance of the mountainous spine and the long coastline for Japan's geography. The 'X' on the map shows where the Seto Great Bridge links the islands of Shikoku and Honshu.

It is a pantheistic religion of many gods and involves emperor worship and the concepts of Japanese superiority. In Shintoism, both living and nonliving objects possess spirits, with Mount Fuji and other mountains being given special reverence.

Japanese emperors resided at Kyoto, retiring from public life and delegating administration of the country to leading families of court nobles. From the AD 1100s, these feudal lords took over the imperial administration, and armed samurai warriors under a *shogun* (military leader) acted over the heads of the powerless court. When the last shogun resigned in 1867, the Meiji emperor regained his position as titular head of government. The royal capital moved to the shogun center of Edo, later called Tokyo ("eastern capital"). However, in the 1889 constitution, the emperor became a figurehead and political power was taken by leading figures in the parliament.

At the end of the 1800s, Japan's imperialist ambitions, coupled with its new military power, enabled the country to take over Taiwan and Korea, and to win a war against Russia that gave it Manchuria in northern China. During World War I, Japan advanced its economy and generated a trade surplus. However, its involvement with Axis powers during World War II and the dropping of atomic bombs on Hiroshima and Nagasaki in 1945 wiped out earlier economic gains. Even in 1960, Japan, despite 100 years of modernization, still had a gross national income per capita that was one-eighth that of the United States.

Agriculture

Japanese farms occupy small areas of arable land within the hilly and mountainous country, often intermingled with housing and industry which compete for the land and pay higher prices than farmers could afford without subsidies. Government support through subsidies maintains agriculture as a significant sector of the economy, although the country's agricultural population is waning. After 1970, only 12 percent of farmers worked full-time without off-farm jobs. The size of most farms remained stable, however, because land increased in value and many owners held on to their land as a family investment. In the early 2000s, Japanese farmers faced new shocks. Although Japan's agriculture remains highly protected from foreign competition, improved transportation links and refrigeration brought large quantities of Chinese tomatoes, eggplants, onions, and garlic bulbs to Japanese supermarkets.

Industrial Development

After 1911, the Japanese government became more democratic, and political parties and labor unions were established. In the period up to World War II, industrial expansion was based on government-business links, cheap labor, large numbers of small businesses, and improved infrastructure, aided by cheap raw materials on the world market. Overseas, Japan became known for its cheap goods, often of mediocre quality. The 1930s economic depression—when other countries raised tariff charges against cheap Japanese imports—led to an internal pact be-

tween Japanese military leaders and the elite families, the *zaibatsu*, who owned the biggest industries. The military leaders took over political power and shifted industrial production to naval ships, air force planes, and other military equipment. This phase culminated in World War II.

In the following decades, Japan moved rapidly to the forefront of the global economy. Japan's export-led growth raised its gross national income one-hundredfold between 1960 and 2000 at current dollar values. From 1960 to 1990, Japan's total GDP grew from 9 to 53 percent of the U.S. total and from 3.5 to 14 percent of the world total. From the 1970s, major Japanese industries had sufficient resources to invest in the United States, Europe, and Southeast Asian countries to manufacture goods for their domestic markets. Japan then invested in China. Japanese companies produced goods in China and Southeast Asia for their brands that sold worldwide. In the 1990s, however, as the U.S. economy surged further ahead, the Japanese economy slowed and fell back to one-third of the U.S. total GDP in 2000 (Figure 4.9).

Japanese industrial development is strongly localized, with heavy industry in coastal locations on reclaimed land because of the need for space and proximity to ocean transportation links to raw material suppliers and markets. Light industry is more widely distributed and intermixed with housing, with older firms retaining inner-city sites. Location in the inner city takes advantage of the availability of labor, linkages with other firms, and market outlets. Local governments encourage the establishment of light industrial estates on city margins, including science parks for high tech industries.

The Japanese government assisted and advised industry, with the **Ministry of Economy, Trade and Industry** (METI;

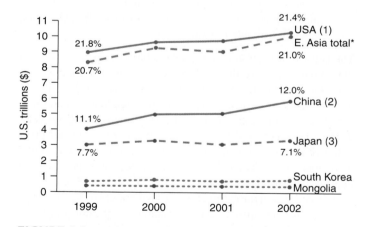

FIGURE 4.9 **East Asia: an increasing proportion of world GDP PPP.** The percentages given are of the world total GDP PPP for each year. The figures in parentheses are the country's world position. First, this graph shows the East Asia total closing economically on the United States. No data are available for North Korea or Taiwan; their contribution would bring equality or a slight advantage to East Asia. Second, it shows how China's economy is gaining ground while Japan's falls back. Third, the East Asian countries showed more of a dip in 2001 than the United States. *Source: Data from Human Development Reports (United Nations) 2001–2004.*

formerly the Ministry of International Trade and Industry) encouraging export sales through a worldwide network of market intelligence-gathering offices. Diversification required greater investment but raised the output of light industries such as those producing cameras and household appliances. After copying others' technologies and designs, Japanese firms became initiators. The increased application of technology brought gains in productivity and international competitiveness to industries including electronics, robotics, and new materials. Japan's successful exports became known for their quality, reliability, and market-leading technology. In the 1980s, other countries built huge deficits of payments to Japan, causing the Japanese yen to double in value compared to other international currencies.

In the later 1900s, Japan's industrial economy expanded from dependence on importing raw materials and producing goods for export markets. It moved from iron and steel manufacturing to making autos and high tech goods. A slowing of Japan's economy in the 1990s resulted in its making investments in other countries. The increase in service occupations, from education and health care to retailing and tourism, widened the range of employment and contributed to a diversified economy that is not so dependent on the fluctuations of world markets as a manufacturing-dominated economy.

Japan's popular culture ranks high among the country's latest exports. Revenue from royalties and sales of Japanese music, video games, anime, comics (*manga*), films and fashion rose 300 percent over the last decade, garnering the country US$12.5 billion in 2002.

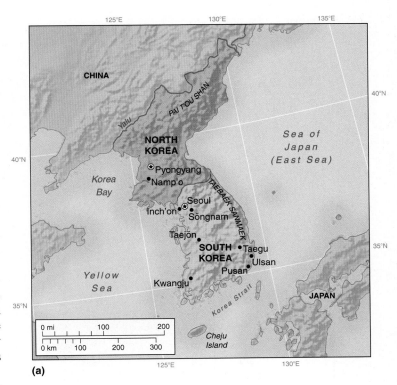

(a)

FIGURE 4.10a The Koreas, North and South. After Japanese occupation in the early 1900s, the Koreas prefer the term "East Sea" to "Sea of Japan", but have not convinced others to change the name.

Renewed Economic Challenges and New Directions

By the early 2000s, Japan's economy again needed to change. It suffered from falling prices and business collapses, its established systems of government-business linkages were undemocratic and increasingly inefficient, and its banking systems came under pressure because of huge debts, particularly resulting from investments in Southeast Asia. While Western nations imitated Japanese production processes and competed against Japanese goods in world markets, the Japanese looked more closely again at Western ways, particularly in relation to financial control.

The Koreas

Sharing a common ethnicity and culture, North Korea and South Korea (Figure 4.10a) also had a common history to the mid-1900s but after that moved in different directions. The countries occupy the hilly Korean peninsula that extends southward from the Chinese border along the Yalu River.

The area that is now North Korea was the center of the first Korean kingdom, around which the peninsula was unified in the AD 600s. After the Mongol invasion and retreat, Confucian principles of a hierarchical system of government and society

were adopted. The Yi dynasty, founded in Seoul in AD 1392, ruled the kingdom of Korea until 1910, although it became subject to the Chinese in 1644. Close proximity to Japan attracted aggressive interest from that growing military power. In 1894–1895, as rebels opposed the Korean government, the newly strengthened Japanese forces overran Korea, defeating both Chinese and Russian attempts to aid the Koreans. Japan annexed Korea in 1910.

At the end of World War II, the U.S. and Soviet armies defeated Japan and divided the "freed" North and South Korea at the 38th parallel. North Korean forces invaded South Korea in 1950. The United Nations sent forces under U.S. commanders to support South Korea, and, following destruction over much of the territory as warfare moved back and forth, the two sides agreed on an armistice in 1953. The agreement resulted in the formation of a Demilitarized Zone between the two countries (Figure 4.10b).

After World War II and the Korean War, South Korea established import substitution industries that developed manufacturing skills, substituted imports with locally produced goods and hence reduced the need to pay hard currency for foreign goods. Its dictators forced the people to work for low wages while the economy improved. The transportation infrastructure built by the Japanese when they ruled Korea in the early 1900s and rebuilt with external funds after the Korean war helped develop the economy in its early stages of growth. South Korea

(b)

FIGURE 4.10b **The Koreas, North and South.** The Demilitarized Zone. A bridge crossing the frontier river has been destroyed and trees cut to maintain an open vista.

built a series of major highways, and its public transport system is one of the world's best.

The devastating Korean War took place when three-fourths of the South Korean people still earned a living from farming. In postwar land reforms, the government took over large, inefficiently run estates and encouraged a new group of small landowners to apply imported fertilizers. Farm productivity rose, and South Korea became largely self-sufficient in food. The farming and rural emphasis gave way to manufacturing, services, and urban living. From the early 1960s to the early 1990s, few other countries' economies grew so fast. Exports rose from US$33 million in 1960 to US$150 billion in 2000. A country of subsistence farmers under largely feudal control was transformed in a generation to the world's main maker of large ships and memory chips, the fifth-largest automaker, and eleventh-largest world economy—with greater output than the whole of Africa South of the Sahara.

Contrasting Koreas

Although mostly noted for its economic growth, South Korea is increasingly making a regional contribution in popular music that is both cultural and economic. Young pop stars such as Boa develop alongside other youth culture inputs in movies, TV shows, computer games, and fashion. This trend relates partly to South Korean consumers preferring local products

to the pervasive Western music and movies. It is also part of South Korea's development of knowledge-based service industries. South Korean TV dramas are highly rated in Vietnam. China, with its huge and growing market for Korean pop music (K-pop), is a prime target for record companies, which envisage setting up an academy for discovering young Chinese pop stars. The main problems for a South Korean venture in China are the rogue bands copying songs and the rampant piracy and copying of CDs.

While South Korea became one of the newly emergent urban-industrial countries with support from the United States after the Korean War, North Korea remains poorly developed and isolated from the rest of the world under a Communist regime. After the loss of Soviet Union subsidies in 1991, North Korea depended increasingly on international aid to make up for internal economic collapse and famine. By the early 2000s, the GDP per capita in North Korea was one-eighth that of South Korea's. In 2000, the death rate rose 50 percent higher than in 1994, indicating widespread famine and disease. At this time, North Korea was the world's largest recipient of famine relief. The best food produced in North Korea feeds the army and party officials, while other people depend on donated food, often of lower quality. The country remains under suspicion by other countries of having nuclear and biochemical weapons. In October 2006, the North Korean government announced that it had successfully completed its first nuclear weapons test, although some scientists remain skeptical of the success of the test.

By the late 1990s, there were signs of closer ties developing between the two Koreas. Presidential visits led to reunions of Korean families separated for 50 years. Talk of reunification of the two countries gathered some momentum, but both countries had reservations about such a move. North Korea did not wish to lose its independence, while South Korea counted the potential cost. In 2000 South Korea president Kim Dae Jung created the Sunshine Policy, aimed at improving trust and economic linkages between the two countries. South Korea has provided the North with food, fertilizers, medical supplies, and clothing, while North Korea has responded by selectively allowing divided families to meet and be reunited. The South Korean government favors diplomatic and conciliatory measures in engaging with its northern neighbor.

China, Mongolia, and Taiwan

The People's Republic of China (PRC) is the world's most populous country, with 1.31 billion people in 2006, and the third-largest by area (Figure 4.11). Hong Kong (now also commonly called Xianggang, the pinyin version of Mandarin Chinese) and Macau, previous colonies of the United Kingdom and Portugal respectively, were returned to PRC control in 1997 (Hong Kong) and at the end of 1999 (Macau). Mongolia (2.6 million) and Taiwan (22.8 million) are tiny by comparison but have distinctive roles in relation to China and the rest of the world. These two countries are discussed together after the section on China.

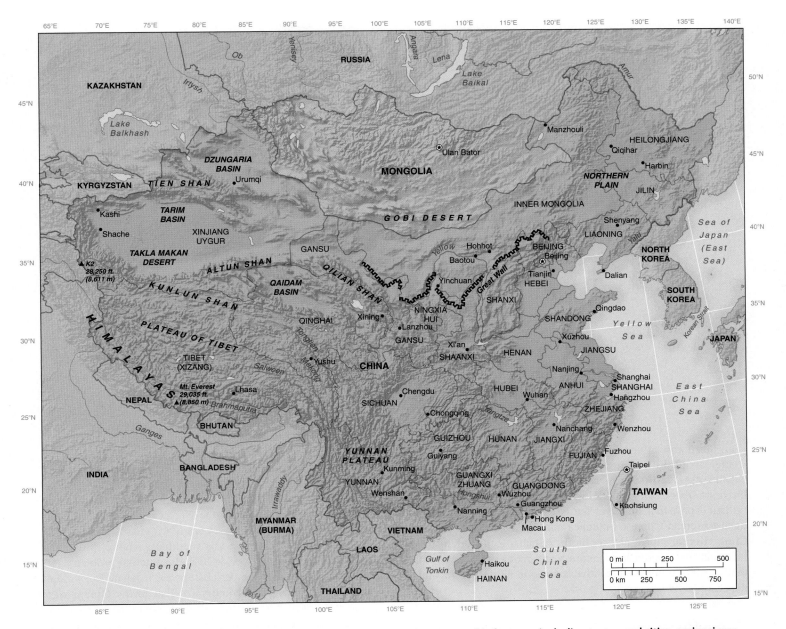

FIGURE 4.11 **The People's Republic of China, Mongolia and Taiwan: major geographic features, including towns and cities, major rivers, and provincial boundaries.** China has 23 provinces (capital letters), four cities with provincial status (Beijing, Shanghai, Tianjin, and Chongqing), and five autonomous regions that have large minority populations (Tibet, Inner Mongolia, Guangxi, Ningxia, and Xinjiang).

China: Early Civilizations and Kingdoms

By 2000 BC the first Chinese civilization developed in the Huang He (Yellow River) valley (Figure 4.12), based on wealth from agricultural surpluses and craft skills and controlling much of the lower Huang He and lower Chang Jiang (Yangtze River) basins.

The Zhou dynasty (1122–256 BC) used Iron Age technology, irrigation, and deeper plowing of the soil to support economic growth. During the political and social upheavals of this period, **Confucius** (Kong Fuzi), an administrator, called for better-trained and more able managers to focus on orderly conduct and proper relations in all social contexts. **Daoism,** the teaching of Laozi, disdained the Confucian system, preferring a return to local, village-based communities with little government interference. Daoism is based on the apparently opposing concepts of *yin* and *yang*: complementary, interdependent principles operating in space and time, emblems of the harmonious interplay or balance of all pairs of opposites in the universe (see Figure 4.2).

Around 220 BC, the Qin dynasty (spelled "Ch'in" in older forms of romanization, giving China its Western name), imposed strict laws to weld the separate feudal states into a

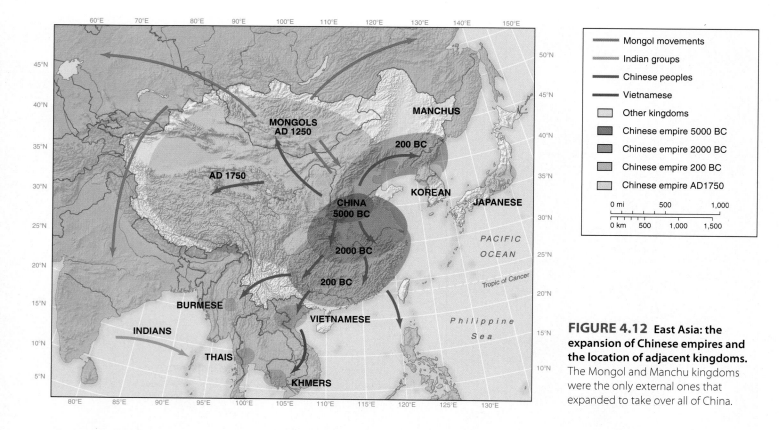

FIGURE 4.12 **East Asia: the expansion of Chinese empires and the location of adjacent kingdoms.** The Mongol and Manchu kingdoms were the only external ones that expanded to take over all of China.

centralized and culturally uniform empire with a standardized written script. Rulers sited the Qin capital near modern Xi'an. Southeast Asian Buddhism reached China in the early years of the first millennium AD, adding a spiritual dimension to materialistic Confucianism. Over the next three centuries, the Tang dynasty spread Chinese influence into the northeast of modern China, Korea, Japan, and northern Vietnam. Zen Buddhism, a form that is centered in meditation and moments of insight in the course of mundane activities, arrived in Korea and Japan in the early 1200s.

Mongol Invasions

In the 1200s, the Mongols invaded China. Kublai Khan, the Mongol leader, established his capital in northern China at Beijing. He and his descendants ruled as Chinese emperors, driving the center of native Chinese intellectual and economic development southward. In the 1300s, growing Chinese resentment of the new rulers, disruptions, and taxes imposed on them, combined with crop failures and famines, opened the way for rebellion and the proclamation of the Ming dynasty in 1368.

Later Empires

Renewed Chinese expansion occurred under the Ming dynasty (1368–1644), during which the economy, literature, and education made advances. Chinese naval power under Admiral Cheng-Ho (1405–1433) ventured beyond India to the Persian

Gulf, Red Sea, eastern Africa, and Madagascar. Although there was trade, the Chinese did not conquer new lands.

The succeeding Qing (Manchu) dynasty ruled from 1644 to 1911. By the mid-1700s, its rulers had expanded the Chinese realm to its farthest limits in the Northeast, Mongolia, Xinjiang (the Northwest), Tibet, Burma (Myanmar), and Taiwan. In 1820, China was the world's wealthiest empire, producing nearly one-third of the world GDP. But internal problems left China vulnerable to the imperialist forces of Japan, Russia, and the West. Following the country's defeat in the Opium wars of the mid-1800s, China ceded Hong Kong to the British, was forced to open its ports to foreign trade and allow foreigners to travel and live within its territory.

Disrupted Chinese Republic

In 1912 Sun Yat-sen and his Kuomintang party (Guomindang, or the Chinese Nationalist Party) claimed overall political power. Attempts to reunify China under a common government in the 1920s and 1930s, however, contended with strife among local warlords, a Communist rebellion, and Japanese attacks. The main Japanese invasion of China began with a 1933 expansion out of the northeast territory that it had occupied earlier in the 1900s, followed by a general declaration of war in 1937. The Kuomintang and Chinese Communists united in resistance.

The Chinese Communist Party came into power in 1949 and established the People's Republic of China. The new government's first aim was to turn a backward, feudally divided agricultural country into a united, advanced, and centralized socialist industrial country. Mao Zedong, the leader and dictator of China from 1949 to his death in 1976, wished to make rapid and massive changes.

Collectivization and Communes

Modeling changes on the Soviet Union experience, China's Communist government prioritized large-scale industrialization and the collectivization of agriculture. Although little investment went into improving agriculture, rural social structures were revolutionized. Before 1950, most Chinese were landless tenants, working tiny parcels of land. The landlords or rich peasants who owned over 70 percent of the cultivated land made up fewer than 10 percent of the population.

In 1952, 300 million landless peasants received their own plots of land, houses, implements, and animals, free from debt. By early 1956, over 90 percent of the rural population joined cooperatives. In "advanced" cooperatives, the land became the property of the organization, apart from tiny garden plots for each family. A typical advanced cooperative included approximately 150 households farming a total of around 400 acres of land. This process, known as **collectivization,** was completed by late 1956.

The government also established communes. A **commune** combined agriculture, industry, trade, education, and the formation of a local militia. The commune leadership planned the agricultural year, other economic activities, and social functions. They guaranteed food, clothing, education, housing, and arrangements for weddings and funerals to members.

The Great Leap Forward and Cultural Revolution

The **Great Leap Forward** of the late 1950s aimed to increase China's rate of industrialization by producing basic industrial products, such as iron and steel. It attempted to integrate rural and urban areas by setting up small-scale production units in both. Throughout much of China, community efforts devoted to iron smelting, often in small backyard furnaces, caused a neglect of farm work and made worse the drought-based famine of 1960, in which 25 to 30 million people died. The plan was a disaster. Both industrial and agricultural production dropped, and China was forced to import wheat from Western countries. Stung by failures and misunderstandings, the Chinese leaders abandoned most Great Leap Forward policies.

After a few years of higher production, Mao Zedong attempted to change the whole basis of Chinese society and its administration, and the country once again turned inward. The **Cultural Revolution** of 1966 to 1976 diverted attention from increasing economic productivity to protests against unchanging local authorities and leadership cliques. Mao's "Little Red Book" summarized his wish for revolutionary actions to purge society. The Red Guards, composed mainly of young adults, carried out his tenets, destroying management structures at all levels of society and trashing cultural relics. Such fear-based

tactics kept Mao in power until his death. His policies, however, swept aside many of the short-lived economic gains of the early 1960s in a massive persecution of the educated classes, technical experts, and former leaders.

China Joins the Wider World

After 1976, Mao's successor, Deng Xiaoping, brought in reforms that abolished most of the communes and engaged with the global economic system. His views, expressed by the saying, "It doesn't matter whether it's a black cat or a white cat, as long as it catches mice," enabled the adoption of more open policies in the economy. These policies attracted investment from outside China and generated the world's highest rate of economic growth from the mid-1980s to the early 2000s. Foreign investments led to the rapid economic growth of many coastal regions of China. The central Communist government, however, stayed in power, and its often oppressive measures restrained personal freedoms.

After Deng Xiaoping died in 1997, his successors developed the policies that brought economic growth and involvement in the global economic system. In 1998, further reforms enabled and encouraged families to buy their own housing, leading to a growth of speculative building. The acquisitions of Hong Kong and Macau in the late 1990s highlighted the process of reclaiming long-term Chinese territories and the end of foreign ownership of Chinese lands.

Through this period, many families' material expectations increased. In the 1970s, the desired items were a watch, a bike, and a sewing machine. Through the 1980s, they were TVs, washing machines, and refrigerators. In the 1990s, attention shifted to hi-fi equipment, microwave ovens, and air conditioning units, and by the 2000s focused on computers, cars, and apartments as people were encouraged to buy rather than save.

Farming and Rural Living in the 2000s

Chinese farming is both land and labor intensive as the country attempts to feed 21 percent of the world's population from 7 percent of the world's arable area. Output per unit of land is high. Most agricultural land is cropped two or three times per year, and competition is increasing between land-intensive crops—wheat, corn, soybeans, and cotton in the north and rice and sugarcane in the south—and labor-intensive crops that produce greater value per unit of land, such as vegetables and fruit.

The more profitable specialist crops, including industrial crops (cotton, soybeans), fruit, and vegetables, together with livestock products, threatened the output of some of the traditional staples, particularly wheat and rice. By encouraging grain growers to increase yields by using more imported potash fertilizer, China produced enough grain to put annual surpluses into storage.

In the 1980s, commune-based agriculture gave way to the **household responsibility system,** allowing families to decide which crops to grow. Small groups entered short-term contracts, leasing land to achieve production quotas set by the government.

Farmers could sell any surplus crops or animal products in local markets (Figure 4.13).

While successful in terms of increasing production and raising incomes in many rural areas, the individualistic policy of household responsibility led to the neglect of some aspects of rural land use that communal working had encouraged, including the long-term processes of planting woodland and maintaining road and canal infrastructure. In the early 2000s, other problems facing agriculture included long-term soil erosion and the expansion of urban-industrial areas that removed good land from production.

Growth in Chinese Manufacturing

After 1978, the drive to greater efficiency and an open-door policy to encourage foreign investment became more important than self-sufficiency, and led to what has been termed the "Great Leap Outward." Today, China's manufacturing sector comprises a complex and uncertain intermarriage of local enterprise, foreign multinational capital, and public ownership. The country's industries grow rapidly in southern and coastal China, but elsewhere need new plant and infrastructure. China's economy is now a mixture of state-controlled enterprises, military-industrial diversification, and foreign multinational and local investment in manufactured goods for export.

From the 1980s, reduced central government subsidies and the "management responsibility system" — the industrial equivalent to the household responsibility system—placed more decision-making in the hands of firms that were still government owned. By the early 2000s, 95 percent of industrial production

sold at market prices. The central government, however, now focuses its own investment on only 1 percent of existing publicly owned enterprises, allowing others to be merged, taken over by workers, or go bankrupt.

In the late 1990s, bank crises due to overlending affected much of the rest of East Asia, slowing the rapid economic growth rates in the region's countries and reducing the markets for Chinese-made goods. Part of the problem faced by other East Asian countries was the expansion of Chinese output, causing oversupply to common markets.

One of the projects initiated by the Chinese government to support industry and transportation is the massive Three Gorges project on the Yangtze River. China has the world's greatest hydroelectricity potential, but even if it were all developed, it would supply just 6 percent of the country's needs. The building of the Three Gorges Dam beginning in 1993, with much of the dam and its ship locks completed in 2003, and a final stage due in 2009, illustrates the environmental and social issues facing major hydroelectricity projects. The outcome should generate enough electricity to save the yearly burning of 45 million tons of polluting high-sulfur coal, and will help flood protection and navigation. The dam, however, created a lake 600 km (450 mi.) long below Chongqing, and 1.3 million people are being resettled. Farmers on good valley land have been moved to poorer uplands, and submerged industrial towns have been rebuilt on higher land. But this project is Chinese government policy and, although questioned by environmentalists and internal advisers, is on target for completion.

Mongolian Isolation

Mongolia is a landlocked country of mountain ranges, the Gobi Desert, and semiarid grassy steppes akin to the northern steppes region of China. It has a small population and struggles to maintain its independence as a buffer state between China and Russia. The capital, Ulan Bator, is connected by rail to both China and Russia.

After being a Chinese province, Mongolia became independent in 1921 with Soviet Union backing, and a Communist government was installed in 1924. The Soviet Union modified Mongolia's East Asian character, replacing the Mongolian alphabet with Cyrillic script and reorganizing the education system. Mongolia became dependent on Soviet aid and trade. Russia's military forces were withdrawn only in 1992. In 1996, a coalition of democratic groups won an election to introduce reforms. After they lost Russian support, the Mongolian people strove to earn a living while attempting modernization in a landlocked, semiarid environment. In 2000, the former Communist Party was overwhelmingly reelected on the basis of policies that combine reform, social welfare, and public order.

FIGURE 4.13 China: rice paddy near Yinchang, Chang Jiang Valley. A farmer uses a water buffalo to prepare land for planting rice. Nearby paddies already have rice plants. The older structure that housed both humans and animals has been augmented by a newer house for the farmer's family. Although Chinese farmers include the poorest people, many have experienced better incomes in the last 20 years.

Mongolia's Limited Resources

Mongolia has a strategic location in the heart of Asia but restricted economic potential. Many people still gain a living from herding livestock on the semiarid grasslands. Farmers responded to privatization and deregulation of meat prices in the early 1990s by massively increasing their herds. Cropland areas declined as agricultural cooperatives were split and land abandoned because there was no equipment to cultivate it.

For the future, Mongolia seeks to increase its output of minerals such as copper, gold, and molybdenum and encourages foreign investment to provide alternatives to the previous Russian links. Private enterprises became the main source of economic growth in the 1990s, with rising production of gold, although a major fall in world copper prices and weaker prices for cashmere woolens slowed growth. Mongolia suffers from a limited range of resources and the problems of making the transition to a market economy.

Taiwan

Taiwan has 10 times the population on one-fourth the area of Mongolia. It is an island country with a mountain "backbone" along its eastern coast. At the end of the 1895 war, victorious Japan took Taiwan, along with Korea. After World War II, Taiwan returned to China. In 1949, Kuomintang forces escaping from China took control of Taiwan. The Kuomintang leaders brought their families to join several thousand military personnel who had been stationed on Taiwan to quell anti-Kuomintang riots. Taiwan prospered with high tech export industries and became a major investor in mainland Chinese industries.

Taiwan is densely populated. Originally settled by aborigines of Malay-Polynesian origin, the island's population today is nearly all of Chinese origin. The Taiwanese have demographic characteristics similar to those of other materially wealthy countries, including life expectancies of 75 years. Taipei is the capital and largest city of Taiwan, which has nearly three-fourths of its population living in urban areas as a result of the growing industrial and service-centered economy.

While Taiwan maintains its independence, the People's Republic of China insists it remains a part of the mainland country and intends to reincorporate it. The PRC attitude places Taiwan outside the normal niceties experienced by independent countries, such as membership of the United Nations. For 50 years after 1949, the PRC threatened to invade Taiwan and backed off only under U.S. military threats.

China's leadership determines to apply to Taiwan the "one country, two systems" formula adopted for the incorporation of Hong Kong. Around half the Taiwanese population already supports reunification with China based on a Hong Kong type of provision that would preserve Taiwan's prosperity and its role as a major investor in Chinese economic growth. Many Taiwanese, however, watch events in Hong Kong before deciding to become fully part of China again and, in the meantime, continue to strengthen their military capabilities. Supporters of Taiwanese independence argue that it is different from Hong Kong: Taiwan is 170 km (100 mi.) offshore, has no built-in deadline date for return to China, and has a more mature democracy than Hong Kong and much of the rest of East Asia. The independence position has guarantees of Western support, especially from its strongest ally, the United States. But pres-

sures grow on Taiwan as many of its best workers and managers move to the mainland, unemployment increases, and World Trade Organization members ask questions about the Chinese being able to use the Taiwanese ports and airports freely.

In the early 2000s, Taiwan became more enmeshed in mainland China's economy. China is Taiwan's third largest trading partner, following the United States and Japan. Nearly half of the investment in Shanghai's Pudong development project (see Figure 4.2) is Taiwanese. Although Taiwan's government officially limited Taiwanese investments in China, over $60 billion followed indirect routes through Hong Kong, despite few return profits.

Taiwanese Economy

After 1949, the Kuomintang government encouraged rapid industrialization, rural change, and urbanization. The increasing rice surplus gave way to more diversified farm products, including sugar, tea, vegetables, and fruits. Expansion of urban-based industrial jobs and buildings, plus the mechanization of farming, reduced the farm population from 6 million in 1965 to fewer than 4 million in the 1990s, while an initial agricultural trade surplus gave way to increasing food imports.

Starting in the mid-1960s, export processing zones focused attention on manufacturing for export. By the 1980s, the emphasis switched to high technology developments. Private ownership of firms increased from 52 percent in 1960 to over 80 percent in the 1980s. The government retained control of banks, interest rates, and exchange rates to maintain a consistent financial environment. It cultivated a strong domestic manufacturing industry, at first based on transistor radios, followed in the 1970s by the assembly of Japanese electronic goods, and then promoted links with companies such as Philips (Netherlands) to build its own integrated circuit industry. Taiwan now leads the world in such technologies as integrated circuits, laptop computers, modems, and data communications.

By the 1990s, Taiwan had one of the largest world trade surpluses and was able to export capital, particularly to China and Thailand. As is the case with Japan and South Korea, direct investment in Chinese manufacturing contributed to many "Taiwanese" products. As Chinese costs undercut Taiwanese producers, Taiwan continues to invest in new technologies such as biotechnology and new integrated circuit designs. Such investment, however, demands so much capital that it poses major risks for the Taiwanese government such as overspending and eventually losing in competition with the Chinese. Taiwan's furniture and textile industries have already moved to mainland China, leaving increased unemployment on the island, and its electronics industries are under a major threat.

Large financial surpluses meant that major financial problems in the late 1990s affected Taiwan less than South Korea and Southeast Asia. Taiwan's other positive economic features include mainly small corporations and more flexible conditions of operating that enable failing companies to cease trading. Such conditions result in higher productivity compared to the South Korean *chaebol* and Japanese conglomerates (see section "The Rise of Japanese and Korean Multinationals," page 125).

Contemporary Geographic Issues

China: A New World Power

China has the potential to become a major world power alongside Japan. It is the world's most populous country, and between 1980 and the early 2000s, its "open door" policies were the basis for economic growth at the fastest sustained rate of any country. This economic growth multiplied the value of its output fivefold, and raised 600 million people out of absolute poverty. Chinese factories now make many goods bought in Western stores, from electronics to clothing.

In 1991, around 100,000 private businesses employed 1.8 million people; by 2001, 24 million worked in 1.5 million large private businesses and around 30 million in smaller, individual concerns. In that decade, private industry grew from a few percent to 40 percent of total Chinese industrial output.

Current foreign direct investment to China is fourth in the world, after the United States, Germany, and the United Kingdom, and outstanding among the world's less materially wealthy countries. Foreign direct investment in China totaled US$60.3 billion in 2005. Companies from the United States, Europe, Japan, South Korea, and Taiwan have moved much of their low-end assembly to China, although there is a rise in high-end manufacturing and some services as well.

To support such economic investment, China's huge supply of labor includes college graduates and computer engineers who are paid one-tenth as much as their Taiwanese or Japanese counterparts. However, Chinese entrepreneurs themselves face difficulties. Their government is slow to expand the list of companies allowed to go public with share offerings. Many entrepreneurs find it difficult to persuade Western financiers to support them in joint projects after intense scrutiny by external analysts and investors. Most investments in China come from Chinese in Hong Kong and Taiwan, and from other Chinese living abroad—but also increasingly from Japan.

Western and Asian multinational corporations trading in China include oil companies, electronics groups, auto manufacturers, and restaurant chains. For example, Kentucky Fried Chicken (Figure 4.14) opened a huge outlet in China next to Tiananmen Square in central Beijing in 1987. Despite the 1989 incident in this square, when Chinese army tanks attacked peaceful demonstrators, McDonald's followed in 1992 and by 2001 had nearly 400 outlets throughout China. Both KFC and McDonald's plan to expand further. The first Starbucks outlet in China—serving coffee in a tea-drinking country—opened in 1999, and by 2005, 120 shops held franchises in 13 Chinese cities, although concentrated mainly in Beijing and Shanghai. The most contested site for Starbucks was in a Beijing Forbidden City souvenir shop, with 70 percent of 60,000 people surveyed opposing it. But it continues

FIGURE 4.14 China: central Beijing. A KFC outlet alongside Chinese stores, together with a BMW and lots of bicycles: a mixture of global economy and local trends.

to serve coffee. Such symbols of an intruding global economy, however, generally excite little public antagonism. Even at the height of anti-American demonstrations over the bombing of the Chinese Embassy in Belgrade, Yugoslavia, in 1999, the Chinese who stoned the American diplomatic missions left KFC and McDonald's largely untouched. Many Chinese regard the expansion of multinational food franchises as a positive sign that they are involved in the global economy.

China joined the WTO in November 2001. The Chinese leaders expect WTO membership to boost exports and foreign investment, but it will also force its industries to become more competitive in world markets in both quality and price. WTO rules expect China to cut import tariffs and allow foreign businesses to compete in its highly protected areas such as telecommunications, automaking, insurance, and banking industries.

China's GDP growth has surpassed that of the Asian "miracle" economies. It is the world's top location for outsourced manufacturing. It has witnessed exponential increase in its exports and trade. By 2005, China was tied with Mexico as the United States' second-biggest trading partner (after Canada), and in 2004 was the European Union's second-biggest trading partner (after the United States). China was Japan's biggest trading partner in 2004.

But growth in sectors like construction and manufacturing are slowing down. Now China is looking to become a global power in software and services as well. It is making inroads into outsourced global services such as processing medical claim forms, applications for loans and credit cards, and even grading examinations from the developed countries of the world.

Following a strategy of more regionally balanced development, new centers for back-office work are located away from the coasts for example, in Dalian, a city in northeastern China and Xi'an, the capital of Shaanxi province. Many of Dalian's residents speak Japanese and Korean due to historical connections, giving it an advantage in performing back-office services for these two countries. The Xi'an High-Tech Industries Development Zone is one of China's largest technology parks. Currently housing 7,500 companies, it covers 35 km^2 and is expected to expand to 90 km^2. China's disciplined and trained workforce that includes a large number of university graduates well educated in basic computing and mathematics, and the lower wages paid to Chinese engineers and programmers make the country highly competitive in the global arena for this work. Well known Internet technology companies like IBM, Hewlett-Packard, Microsoft, Siemens, and Infineon all have been in China for several years now. However, it may take China years to catch up with India, as the latter has the advantage of a large English-speaking educated population and greater dominance in research and design.

The positive changes of greater economic and cultural exchanges have also facilitated the easy transmission of diseases. A case in point is the severe acute respiratory syndrome (SARS) epidemic of 2003. Believed to have emerged in southern China, the disease spread from there to nearby regions. According to World Health Organization (WHO) estimates, more than 8,000 people contracted SARS by July 2003. Regional and national economies were disrupted, while the tourism and hospitality industries were most affected as visitors stayed away from areas affected by the disease.

The surrounding countries remain wary of China's military and new economic strength. They fear a Chinese people who continue to perceive a sphere of influence extending to lands it held during its period of greatest expansion in the Ming dynasty.

The Rise of Japanese and Korean Multinationals

Japanese and South Korean exports of manufactures and foreign investments led to the growth of huge multinational corporations. From Japan, automakers such as Honda, Toyota, Nissan, and Mitsubishi Motors, together with electronics and electrical goods makers such as Sony, Sharp, Fujitsu, Panasonic, and Matsushita, built worldwide brands manufactured in many countries. From South Korea, multinationals such as Hyundai, Samsung, Daewoo, and LG (Lucky Goldstar) penetrated markets around the world. Many Japanese and South Korean employees from managers to technical staff worked abroad in subsidiary companies of the multinationals. In the 1980s and 1990s, around 50,000 Japanese were involved in such situations and the South Korean numbers increased from 150,000 to 250,000. Government employees in overseas embassies also doubled during this period. These figures are similar to those in Europe and North America, demonstrating the growing role of these two East Asian countries in the global economic system.

In the early 2000s, Japan was the world's third-largest economy after the United States and China (see Figure 4.9). Japanese firms, especially those involved in electronics or automobile production formed conglomerates known as *keiretsu*.

These industrial groups were built around financial nuclei that usually included a major bank, and produced giant trading and manufacturing companies.

The Example of Toyota

The Japanese company Toyota is the world's second (2006) largest automotive group in sales after General Motors (Figure 4.15). In the early 2000s, its capital resources were double those of the U.S. company. Toyota is a Japanese version of multinational corporations. Toyota's experiences illustrate some of the ways in which multinational corporations (MNCs) grew to worldwide prominence in the later 1900s and some of the challenges they face in the 2000s. Its policies remained successful as it expanded exports and manufacturing abroad. As globalization becomes more pervasive, Toyota may have to become more transnational by allowing control of some of its activities to occur outside Japan.

Kiichiro Toyota first researched gasoline engines in 1930, switching the family textile business into cars; family directors still sit on the Toyota board. Although Toyota's headquarters are in Tokyo, its Takaoka base is a sprawling collection of 35-year-old factory buildings amid the rice fields of Aichi Prefecture to the southwest. Forty years ago, the city of Koromo was so impressed with the company's local investment that it renamed itself Toyota City. The Toyota City factories produce 700,000 vehicles a year and provide the model "mother factory" for five overseas plants.

At first, Toyota copied—and improved—others' innovations, such as Ford's moving production assembly line. Costs were tightly controlled and profits plowed back into production. The company continued to be financially flexible and lean. At Takaoka, Toyota developed *jidoka*—automation with human intelligence, combining high quality, short delivery times, and low-cost production. Combined with *kaizen*—continuous improvement—and just-in-time delivery of components, *jidoka*

FIGURE 4.15 Japan: Toyota plant. The new Toyota Prius hybrid sedan comes off the production line at Tsutsumi plant, Toyota City in October 2003. Demonstrating the company's technological lead, the Prius was soon in great demand, at six times the monthly target of 3,000 cars.

led to strong demand for the reliable Toyota vehicles. Such methods were taken up by MNCs worldwide.

In the early 2000s, Toyota developed CCC21 (Construction of Cost Competitiveness for the 21st Century), a program aimed at cutting billions of dollars from its costs. The company is reviewing its design, manufacturing, and other costs with the objective of better use of equipment and human resources. It will not, however, fire any of its 36,000 workers at Toyota City but will reduce the 4,000 temporary workers. Toyota also negotiates with component suppliers to reduce costs by up to 30 percent. Denso, its chief supplier (and fourth-largest component maker in the world), employs 85,000 people and oversees more than 300 suppliers. Toyota holds an average 25 percent interest in most of its component makers. Such policies and links illustrate a typical Japanese conservative attitude that builds in linkages that are not quickly removed when inefficiencies occur, because of commitments to lifetime employment and other long-term agreements. However, Toyota's competitors such as Nissan and Mitsubishi Motors are breaking apart their supplier chains to achieve cost savings, and this will eventually cause Toyota to rethink its policies.

After 1980, its adaptability enabled Toyota to change from a Japanese auto exporter into a global operator with 41 manufacturing subsidiaries in 25 countries. Toyota's successful strategy began with making mid-priced cars for the growing numbers of middle-class Japanese, and it still has 42 percent of its domestic market. The large home sales are a major strength in light of international currency fluctuations. When the company expanded overseas, local production within the United States and Europe minimized problems of currency exchange rate shifts. Toyota also targeted local demands, producing light trucks for the U.S. market, which is the source of two-thirds of its current profits. Europeans bought smaller cars. The Lexus luxury brand sells in both regions.

To keep up with market trends, Toyota launches a new model every month. That requires huge investments that are questioned by the increasing numbers of foreign shareholders who replace more compliant Japanese bank shareholders. The realization that even such large single companies cannot satisfy all the world's market areas led to Toyota joining an alliance with Peugeot-Citroën of France to market small cars in eastern Europe. It is also entering joint research projects with other manufacturers.

Korean Manufacturing and the *Chaebol*

In South Korea, the iron and steel, shipbuilding (Figure 4.16), chemicals, automobile, and textiles industries established from the 1950s continue to be important, although in the 1990s there was a swing toward high tech industry.

The South Korean government supported the larger family-owned companies that developed into huge conglomerates—the *chaebol*, such as Hyundai, Daewoo, LG, and Samsung. The *chaebol* became diverse in their products, starting up subsidiaries that often had little chance of success, taking on massive debts, and providing major problems for the country.

South Korea's mode of industrial organization into these large conglomerate firms threatened to weaken much of the country's economy and political stability from the late 1990s. When Halla, the twelfth-largest *chaebol*, went bankrupt in 1997 with debts of 20 times the company's assets, Hyundai, the largest, was found to have guaranteed around 15 percent of the debts. This highlighted the fact that the founders of the two *chaebol* were brothers and had strong cross-company links. Family loyalty—part of Korean (and Confucian) culture—was exposed as making problems for running large corporate businesses in a free-market context of increasing transparency. The *chaebol* also controlled large parts of the South Korean financial markets, providing an environment for building up debts that were shared through many links.

South Korea was once called the "Republic of Hyundai" after one of its most powerful *chaebol*. After the 1997 crisis exposed its huge debts and damaging grip on the economy, Hyundai was restructured around several independent companies. Hyundai Motor, however, is the only one that makes profits, driven by its U.S. sales. Roh Moo-hyun, who became president of South Korea in 2003, pledged to make the *chaebol* more account-

FIGURE 4.16 South Korea: Ulsan city shipyard. The Hyundai *chaebol* is a major builder of large ships. Note the space required for assembling the materials and the range of equipment used in construction.

able. In this, he has the support of young reformers who wish to change the country's conservative, business-driven political system to one that is more liberal.

South Korea was a major investor in China from the 1990s, taking advantage of cheap labor costs in the nearby Chinese province of Shandong. The initial rush of small companies to invest in Chinese locations is giving way to investment by major South Korean corporations such as Daewoo, Samsung, and Hyundai, which link to hundreds of smaller companies producing their components and products.

In the crisis years of 1997 and 1998, economic problems spread from Southeast Asian countries to South Korea. In late 1997, indebtedness reached the point where the South Korean currency, the won, and its stock exchange collapsed. Devaluation of its currency increased South Korea's indebtedness to other countries. Bank closures, bankruptcies, and unemployment stirred militant trade unions into political action against the stringent conditions proposed by external loan providers. By the early 2000s, the South Korean economy had largely recovered, helped by the breakup of some family-owned *chaebol*, and the merging of several of the new firms with U.S. and European corporations.

Population Policies in China

When an influential nuclear scientist, Song Jian, attended a population symposium in 1979, he made some rough calculations of Chinese population growth on the basis of the poor 1964 census results and formulas used to predict missile trajectories. When asked to flesh out these predictions, he worked with four other scientists and gave results based on whether families would have one, two, or three children. It was the first computer-based forecast in China and demonstrated that a rigorously implemented one-child policy would keep China's population to 1 billion by 2000, falling to 700 million by 2050 with advances in medical procedures such as female sterilization, abortion, and other birth control methods. His projections for two- and three-children families suggested continuing increases in the future.

Rapid Population Growth

Chinese population change since 1950 was marked by major shifts in policies and attitudes. Although the Maoist government watchword of the early 1950s was "strength in numbers," the government soon realized that around 20 percent of its annual income increase would merely enable the country to keep up with the basic needs of the extra people. China made several attempts to restrict fertility.

The census of 1953 registered a total population of 583 million, and the first birth control campaign was launched in 1956 in urban areas. Its impact was minimal, however, in comparison with the 25 million to 30 million who died as a result of the "Great Leap Forward" beginning in 1958. Birth rates increased

until China launched another program of birth control in the early 1960s, based on delayed marriages and a wider distribution of birth control pills. It had some effect but was then overtaken by the disrupting Cultural Revolution, which swept aside many of the government-established groups that administered family planning. Once again, birth rates and population totals rapidly increased.

One-Child Policy

In 1979, Deng Xiaoping approved the one-child policy before further debate, and this key plank of Chinese reforms was implemented without any law being passed by the National People's Congress. Initially a "temporary measure," it has lasted for over 25 years with its complex regulations, harsh punishments, and 80,000 full-time family planning workers to enforce them (Figure 4.17). Despite the new economic freedoms granted to peasant farmers, the policy subordinated individual interests to those of the country; local officials could monitor the most private aspects of individuals' lives.

Previously, some one-child policies had been implemented by urban authorities in Shanghai and other cities in the 1970s. Urban dwellers living in state units could be tracked. But the countrywide one-child policy was directed at rural areas where the communes (and their controls) were being disbanded, food production was increasing, and extra labor was needed.

The policy included "sticks" and "carrots." Families complying with the policy received certificates that enabled them to claim free health care for the mother during pregnancy and delivery, free health care for one child, free education for one child, promotions and better pay for parents of one child, free family planning supplies and operations, and sometimes free vacations. However, families not complying lost free health care and free education for the first child, received lower pay or lost jobs, were excluded from social clubs and organizations, and came under pressure and were even forced to have abortions.

FIGURE 4.17 Chinese population policies. Family planning poster in Chengdu. The single child depicted is a girl; the one-child policy collides with a cultural preference for boys. What is the purpose of the English writing?

Uneven Implementation

The one-child policy was not executed evenly, however. Crowded urban areas often had stricter enforcements than western rural areas. The southern industrializing area in Guangzhou was not subjected to the policy in order to attract foreign investment to an area of plentiful cheap labor. In Xinjiang province in the far west, the minority people were allowed two and three children. In some areas, the houses of those resisting the set quotas were burned, and women were imprisoned and subjected to late abortions, with some babies even being killed as they were being born. Peasants rioted in some rural areas, killing local party officials and family planning workers until the policy was softened in 1984 and parents were allowed to have second children under certain circumstances. Elsewhere in rural areas in particular, baby girls were killed and the act glossed over before officials knew about it, creating imbalances between males and females in those areas.

One population projection showed that, with one child per family, China would reach a maximum population of 1.25 billion in 2000 and would then decline to under 500 million by 2070. With two children, a maximum of over 1.5 billion would be reached in 2050 and subsequent decline would be slow. With three children per family, the population could rise to over 4 billion by 2080.

The policy did not meet its target, for the 2000 population was 1.27 billion and continuing to increase instead of slowing markedly. Even that figure is debatable, since peasants mislead officials and officials misreport statistics. In Hubei province, when county officials had insufficient emergency relief food, they dis-covered that the total population was 10 percent higher than the census figure, with most families having three children, because local officials had reported numbers to match the official quota. Occasional events bring to light families with up to six children. In 2002, a pharmaceutical company revealed that it produced 25 million vaccines for infants, although the estimated baby population was only 20 million, indicating a higher birth rate.

Current Trends

So, would China have been better off without this draconian policy (Table 4.3)? Comparisons with India suggest it might have been. In India, which has only moderately successful policies on family planning, fertility dropped, and the country claims it has already avoided 230 million extra births and will stabilize its population in 2040, the same year as China hopes for this result. A 1986 study in two counties of Shanxi province allowed peasants to have two well-spaced children. By 1996, the growth rate had fallen, fewer third children were born, fewer abortions were carried out, and less female infanticide occurred. Less co-ercion went with better relations between the peasants and the authorities. This is an important factor in any country.

The policy continues and is justified by the China Population and Information Center as being effective in some parts of the country and subject to better application of the policy, more consultation, and better advice. In the areas of greatest compli-ance, such as Beijing, two-thirds of couples held single-child certificates. This is a marked change from Chinese traditions

TABLE 4.3   DEBATE: POPULATION POLICIES IN CHINA

Arguments for One-Child Policy	Arguments against One-Child Policy
The one-child policy results in fewer births that place claims on resources.	Family planning is most successful when parents act on their own initiative, perhaps guided by information.
It brings slower population increase and a more balanced population structure in the early stages.	It upsets the population balance between sexes (abortion and murder of baby girls) and age groups (in the later stages when there are small numbers of young people and increasing numbers of older people).
It allows better planning for urban expansion, housing, education, job availability, and transport and utility infrastructure.	It ignores other factors in population growth, including migration, the effect of natural disasters (as during the Great Leap Forward) and diseases, and the impact of political chaos (as in the Cultural Revolution).
It provides control by the central government and local officials in the interests of the whole population.	It often requires draconian methods of control that violate human rights.
It brings a rapid realization of the importance of family planning in the face of imminent overpopulation disaster. Other policies had failed.	It ignores cultural and economic factors that favor large families.
It is important to maintain this process.	The one-child family is at risk and fragile in placing all hopes on a single child: most families would consider two children ideal. Only children are often unduly pampered and obese.

that favored large families. The center claims that "family planning has become a social habit."

The future of China's population is linked to potentially political issues such as the problem of caring for the elderly as they grow in numbers. Traditionally, extended families looked after older people, but urbanization often breaks close family ties. In the face of small government pensions for old age and the closure of many state-owned employers that made better provisions, Chinese families strive to increase their savings for later years.

Globally Connected Cities of East Asia

The growth of global and intraregional trade in East Asia led to the emergence of global city-regions (see Figure 1.23) as essential nodes linking this region to the rest of the world through business and political exchanges. Tokyo, Japan, is one of the world's top four global city-regions—with New York, London, and Paris—while Hong Kong is in the second tier. Seoul (South Korea), Osaka (Japan), Taipei (Taiwan), and Beijing and Shanghai (China) are also major cities with significant employment growth in international business services.

Tokyo

Tokyo, the seat of the Japanese government, displaced Osaka as the country's commercial capital after World War II, when many Japanese companies moved their headquarters to the city. The growth of Tokyo paralleled the rise of Japan in the global economy. In the 1970s with the expansion of the Japanese economy and associated social change, Tokyo was transformed into the now familiar cityscape with skyscrapers, expensive shopping districts, hotels, and corporate headquarters. The Ginza area in Tokyo is Japan's most exclusive and expensive shopping and entertainment district. It has the most expensive real estate on earth; 1 square meter here costs US $10,000 (Figure 4.18). Today Tokyo is a world city that is a control point for international finance and transnational production and marketing. In 1960 only one multinational headquarters was located in Tokyo; by 1997 the city housed 18 of the world's top 100 corporate headquarters, including five of the largest.

In 1986, as well as being the center of government administration, Tokyo became the world's second financial center after New York and ahead of London. In the early 2000s, Japan was the world's third-largest economy with 8 of the 10 largest world banks, having built its world prominence on huge capital surpluses. Japanese banks have continued to increase their absolute and relative asset strength. Top ranked global banking groups such as the Mitsubishi Tokyo Financial Group and the Mizuho Financial Group have their headquarters in the city. The Mitsubishi Financial Group merged with UFJ Holdings, another large Japanese bank, to form the world's largest privately owned bank with about $1.6 trillion in assets. Headquartered in Tokyo, the megabank has rapidly increased its overseas presence.

Tokyo has the largest metropolitan economy in the world; its GDP of US$1.32 trillion would put it in seventh place among na-

FIGURE 4.18 Japan: Ginza area in Tokyo. The Sukiyabashi pedestrian crossing in Tokyo's ritzy shopping and entertainment district.

tional economies of the world. It is the center for Japan's transportation, publishing, and media industries. Nearly 70 percent of Japan's anime production companies are located in Tokyo, and recently the first global animation center was opened in the city. It is also an important center of international advertising, housing five of the world's top 20 advertising agencies. However, critics claim that Tokyo continues to be a predominantly Japanese rather than a global city, lacking a cosmopolitan culture and truly global linkages. The city does not attract as much international labor (less than 2 percent of Tokyo's labor force is foreign) or foreign capital.

Seoul

Seoul has been the capital of Korea for over 600 years. It evolved from a monocentric city during Japanese occupation to a growing metropolis with multiple centers during the latter half of the 1990s (Figure 4.19). Seoul houses most of the country's industry and is connected via "technobelts" (systems to link industry and research via different modes of communication) to new centers of research and development.

In 1996 South Korea abandoned its policy of state-led capitalism and switched to a neo-liberal market-oriented economic model. Following the deregulation of capital outflows and tremendous increases in foreign direct investment, Korean banks, investment and insurance firms have joined the global financial market. As Seoul's economy globalized, its connectedness with the rest of Asia and the world increased. Seoul-based multinationals such as Samsung, Hyundai, Daewoo, and LG have led the way in making overseas investments, reaching beyond investment in China and Southeast Asia. They now have operations in Russia and the countries of Eastern Europe and Africa. Numerous foreign banks and financial organizations operate in the city. Seoul hosts 71 percent of Korea's overseas-based

FIGURE 4.19 Seoul, capital of South Korea. The Chongno tower and other modern buildings dominate the center of the city, which was largely destroyed in the 1950s war against North Korea. The volume of road traffic reflects the country's economic growth.

service industries, half the nation's international hotels and trading companies, and nearly all of its communication services. Practically all of the country's stock brokerages, foreign bank offices, offices of foreign media and broadcasting networks are in the city. Seoul has also become an important international airline hub. Increases in the number of air linkages to cities around the world and the number of international flights to and from Seoul speak to the city's increasing global connectedness.

Employment growth in international business services in Seoul has been significant over the last 20 years. In addition to flows of global capital, Seoul has also witnessed an influx of foreign migrant workers. The Industrial Training Program instituted in 1993 permits recruited foreign labor to work for specified lengths of time in low and unskilled jobs. Workers from developing countries of Asia including China, Vietnam, Bangladesh, the Philippines, Nepal, and Indonesia came to South Korea through the program, although many continue to stay as illegal workers.

The Seoul Metropolitan Government identifies the four future images of Seoul as a "historical, humane, cultural, and world city." As stated previously, Seoul exports Korean popular culture to East and Southeast Asia. The 1988 Summer Olympic games held in the city and the 2002 World Cup soccer finals, co-hosted in Seoul and several Japanese cities, further enhanced Seoul's international profile and its global reach via television, webcasts, and other media.

Hong Kong

Hong Kong comprises an island (centered on Victoria) and mainland peninsula (Kowloon) that were ceded to the British in the mid-1800s. A more extensive area of mainland, the New Territories, was leased to the United Kingdom for 99 years in 1898. All returned to Chinese rule in 1997. The Basic Law, under which Hong Kong operates within China, guarantees retention of its free-market system for 50 years and treats the city as a complementary financial center to Shanghai. By 2007, rivalry continued between Hong Kong and Shanghai over becoming China's dominant financial sector. While Hong Kong has its own dollar currency, linked to the U.S. dollar, and gains in international markets, Shanghai's currency is the Chinese yuan, not easily exchanged with Hong Kong dollars, but giving it great advantages within China.

The world's busiest port, Hong Kong, has an excellent natural harbor with deepwater access and extensive water frontages (Figure 4.20). Goods received from inland China are sent to worldwide destinations, while others collected from the rest of the world are sent into China. These are the features of an **entrepôt.** As the Chinese economy opened up and expanded after 1976, particularly in the special economic zone around Shenzhen, the value of Hong Kong's trade multiplied sevenfold. At the time of political transfer, Hong Kong was already deeply enmeshed with interior China in trade, investment, and personal contacts. In 1996, China accounted for 37 percent of Hong Kong's trade and Hong Kong for 47 percent of China's trade. Over 20,000 trucks crossed the Hong Kong-mainland border each day. As part of China, Hong Kong began with advantages over its internal and external rivals for dealings inside the growing country. New ports opening along the Zhu Jiang delta in the early 2000s relieved the huge pressure on Hong Kong's port facilities.

The population of Hong Kong, 7 million in 2006, is almost totally Chinese. As the main urbanized area became overcrowded from the 1950s, new towns spread into the New Territories. They now accommodate around one-third of the total population. The high population density causes half of government expenditure to go for roads and public transportation. In the late 1970s, electrification upgraded the railroad route inland to Guangzhou (Canton) and its use increased. Two cross-harbor tunnels and an underground transit system opened in the 1980s. A new Hong Kong International Airport was built offshore north of Lantau Island and linked by bridges to the mainland. It replaced the older airport in Victoria Harbor that could not be expanded on its limited site.

Hong Kong's growth as a manufacturing center took off in the 1970s, and by 1990, manufacturing provided one-third of all employment. The production of electronic goods and scientific equipment increased, but the textiles and clothing industries declined because of foreign competition and rising labor costs in Hong Kong. In the 1980s, manufacturing moved out of Hong Kong and expanded just over the border in southern China, with

FIGURE 4.20 People's Republic of China: Hong Kong. View from Victoria Peak on Hong Kong Island to Kowloon across the harbor. Hong Kong has some of the highest population densities in the world.

its cheap land and labor costs. The higher wages in factories and service industries force farming labor costs up because farms have to match the wage levels.

By 2000, the shift out of manufacturing as factories moved inland led to service industries accounting for 85 percent of Hong Kong's GDP and 80 percent of employment (from around 50 percent in 1980). Hong Kong is one of the world's major finance centers and the third-largest gold market, accounting for up to 15 percent of the world total. Hong Kong provides expertise in finance, trade, and public administration and has an important stock market. It is the regional headquarters for 800 foreign companies. Tourism became Hong Kong's second major industry, with 11.7 million visitors in 1996. Many new hotels catered to this influx of people that added an extra 10 percent to the population at peak times. After Hong Kong's transfer to China in 1997, tourism continued at a high rate. The numbers of visitors topped 23 million in 2005. Today, Hong Kong is considered a city with a high degree of global connectivity. Its geographical position and history make it a prime location to service clients in the growing Chinese market.

Human Rights

Traditional Chinese concepts of rights emphasize the duty of the state to ensure stability and encourage economic prosperity rather than civil liberties for its citizens. Up to the late 1970s, rural life in China remained harsh. In 1978, the country estimated that 33 percent of its rural population (260 million people) lived below its defined poverty line. As the initial impacts of 1980s agricultural reforms doubled output and rural incomes, rural industries created over 100 million new jobs, and the incidence of rural poverty declined from 33 percent to 9 percent (from 260 million to 97 million people). Despite these socioeconomic improvements, Western countries have criticized China's record of human rights violations since the formation of the Communist state, pointing to the 25 to 30 million deaths during the Great Leap Forward when a combination of natural and man-made disasters caused a drastic reduction in food grain production and resulting famine. The Chinese government did not seek international aid at this time even as its people were dying. The lower numbers in the 40- to 44-year age group in the 2002 age-sex diagram (Figure 4.21) for China reflects the impact of this major population disaster.

Gender Equality

While most countries in East Asia have made significant strides in achieving gender equality, much remains to be achieved. The age-sex diagram for China shows that there are more women than men in the older age groups in the country but fewer in the younger groups. This gender imbalance is attributed to a clash between the one-child policy and a traditional preference for sons in Chinese society, resulting in the abortion or even the murder of girl babies in order to save the family "space" for a boy child. The resulting disproportion between the sexes could pose major social problems for China. Demographers predict that by 2025, tens of millions of marriageable men will not be able to find wives. The shortage of women is already reflected in the increased trafficking of women and girls from inland China

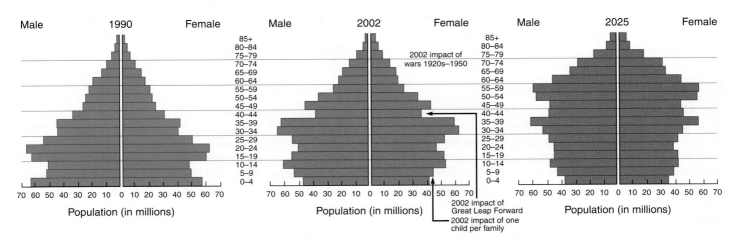

FIGURE 4.21 China: age-sex diagrams for 1990, 2002, and 2025. *Source: U.S. Census Bureau; International Data Bank.*

and the neighboring countries of Burma and North Korea for prostitution and marriage.

Mao Zedong in the 1950s said that "women hold up half the sky." Women's position in Communist China improved as their greater participation in economic and political arenas was encouraged. But as capitalism makes greater inroads into China, many of the gains are being lost. Women are increasingly working in manual, lower paid jobs in agriculture and industry. Women workers in village and township enterprises routinely hold the low-level positions and are the first to be laid off. They also have to retire at age 50, five years before the retirement age for men.

Throughout Japan's modernization, its people maintained a traditional culture that encouraged attitudes based on loyalty to the emperor, family, and workplace. The main Japanese traditions, from Shintoism to Buddhism and Confucianism, agree on the subordinate status of women. Although women's status is changing with their greater education and involvement in the labor force, they still work for lower wages, and few attain management positions or lifetime employment status.

Currently a woman cannot ascend to the Chrysanthemum Throne and become the Empress of Japan. A bill presented to the parliament by Prime Minister Junichiro Koizumi in early 2006 sought to amend the Imperial House Law and allow women to succeed to the throne. Although progressive Japanese and the members of the prime minister's advisory council support female succession, traditionalists and nationalists have so far strongly opposed the idea.

In South Korea as well, traditional Confucian mores have resulted in an overall lower status for women. The country has one of the world's highest rates of sex-selective abortions and homicide of females. Gender disparity is still prevalent in access to education, job opportunities and social status, although there are efforts toward greater equality by the government and NGOs. The country's female labor force participation has risen to over 42 percent, but many of these workers are in temporary and part-time positions, few reaching top level and prestigious jobs. An interesting reversal of traditional gender roles can, however, still be found in the coastal villages of Cheju Island, where women dive for marine products such as oysters and seaweed and are considered the main breadwinners of the family.

Civil Liberties and Political Rights

The Chinese government handles the tensions associated with globalization in various, often contradictory, ways. While educated Chinese now have increased opportunities to shape careers, go abroad, or pursue research interests without party interference, the army and security forces continue to respond firmly to dissent and unrest. The justice system is poor and subject to political pressures that prevent magistrates from acting fairly: for most people outside political and business elites, trials and imprisonment remain arbitrary and often long term.

Experiments of local democracy in rural areas often lose their impact because central ministries impose restrictions on powers of action and officials continue to demand illegal fees. The endemic corruption requires a new system of independent courts and accountability. Many mid-ranking officials, however, understand "political reform" to mean merely streamlining the bureaucracy. Even among Communist Party members, liberals vie with old-guard Maoists, and free-market capitalists with leftists. Each group has its own journals and Internet websites. Although more opinions are aired than was possible in 1989, multiparty democracy is not yet mentioned as even a remote possibility.

The government diverts internal pressures for greater democracy by a renewed emphasis on Chinese nationalism through a historical focus on the pre-Communist triumphs of past civilizations and greater publicity for Chinese cultural and athletic achievements. However, a rise in the number of registered NGOs in 2005 to about 280,000 of which 6,000 are foreign, points to a greater role played by such agencies in Chinese life. While acknowledging deficiencies, China asserts that its human rights situation is improving.

The PRC government argues that human rights should focus on "positive rights" such as adequate food, shelter, and clothing rather than on "negative rights" of freedom of speech, press, and assembly. "Positive" rights demand a more active role by the state while "negative" rights suggest non-interference by the state. Editors of newspapers and magazines that stray from government directives or are critical of the People's Republic are fired or their publications banned. Opposition and organized protest against the government are not tolerated, and are often met with brutal retaliation. In 1989, the Chinese army opened fire on unarmed protestors in Beijing's Tianamen Square, killing between 400 and 2,600 people, mostly students, and injuring several thousands. The brutal military operation to crush a peaceful protest by students who occupied Tianamen Square for seven weeks demanding democratic reform was condemned by the international community. Incidents of torture, forced confessions, and forced labor are also reported. According to Amnesty International, in 2004, 3,400 people were put to death in China for a variety of crimes, including those of a nonviolent nature, accounting for almost 80 percent of all such executions globally.

Minority Rights

Several ethnic minority groups in China, Japan, and Korea reportedly receive poor treatment. Minorities, such as the Tibetans and the Uygurs who live in rural western China, are politically and culturally suppressed for fear that they will push for independence (Figure 4.22). At the same time, significant numbers of Han Chinese are moved into these areas, arguably to help develop them, but with the result that local cultures and populations are diluted. Journalists and local activists report high levels of repression in the minority strongholds, including the confiscation and burning of local history texts, forced labor

FIGURE 4.22 Urumqi, Western China. Ethnic Uygur men leave a mosque after a religious service in Urumqi in the largely Muslim Xinjiang Province. Note the mix of traditional and Western attire.

on government projects without pay, and suppression of information about human rights abuse, including arbitrary detention and torture. In Japan, social outcasts such as the Burakumin, native ethnic minorities like the Ainu, and foreign ethnic minorities like the Koreans still suffer from discrimination. They are often not considered for positions in good companies and are trapped in poorly paid jobs with little hope for advancement.

While religious freedom is guaranteed in China's 1982 constitution, there are widespread claims that members of religious groups such as the Falun Gong are arrested and imprisoned, sometimes for long periods without trial. All religious organizations must register with the government. Contrary to the official position on cultural freedom for minority groups, young people in Xinjiang are not allowed to attend mosques or receive religious instruction until they have completed nine years of compulsory schooling. Similar restrictions occur in Tibet, where the Buddhist community is repressed.

The government of China restricts Internet access to what it considers sensitive subjects, such as websites with content that is religious, pertains to Tibetan and Taiwanese independence movements, freedom of speech, democracy, and police brutality as well as indigenous and minority rights. Censored sites include Voice of America, BBC, Yahoo!, Google, Geocities, and Wikipedia.

GEOGRAPHY AT WORK

China's Landscapes and Global Change

Erle Ellis and his team of collaborators study the impacts of population growth, economic development, and the use of modern industrial technologies and fossil fuels on environment and agriculture in China (Figure 4.23). The country stands as one of the few places where agricultural systems that had sustained high yields continuously for centuries might still be available for study. By using a multiscale approach integrating regional data on land cover, terrain, and climate with higher-resolution satellite imagery, household surveys, interviews with village elders, and field measurements of soils and vegetation, the team measured ecological changes with unique precision at five rural sites selected across environmentally distinct regions of China.

Erle's early hypothesis that the secret to sustaining long-term high agricultural productivity was through planting nitrogen-fixing legume "green manure" crops and by efficient nitrogen recycling, was disproved. However, he discovered that nitrogen was added to the soil via sediments harvested from canals and purchased oilcakes (residues from pressing oil from imported soybeans and rapeseed). Another important finding was that because of the ready availability of synthetic nitrogen, sediment was now left to fill in village canals and had increased soil nitrogen concentration over time, producing a dramatic net increase in nitrogen storage across the region's rural landscapes. Potentially, this nitrogen "sink effect" has global implications. As

FIGURE 4.23 Dr. Erle Ellis and his team carry out a humidification study in China.

carbon storage is tightly linked to nitrogen storage in soils, this nitrogen sink can help explain the "missing sink" for atmospheric carbon dioxide that continues to complicate our understanding of anthropogenic climate change. Erle's research shows that fine-scale local changes in landscape management could have major regional and global consequences that are observable only by detailed site-based measurements.

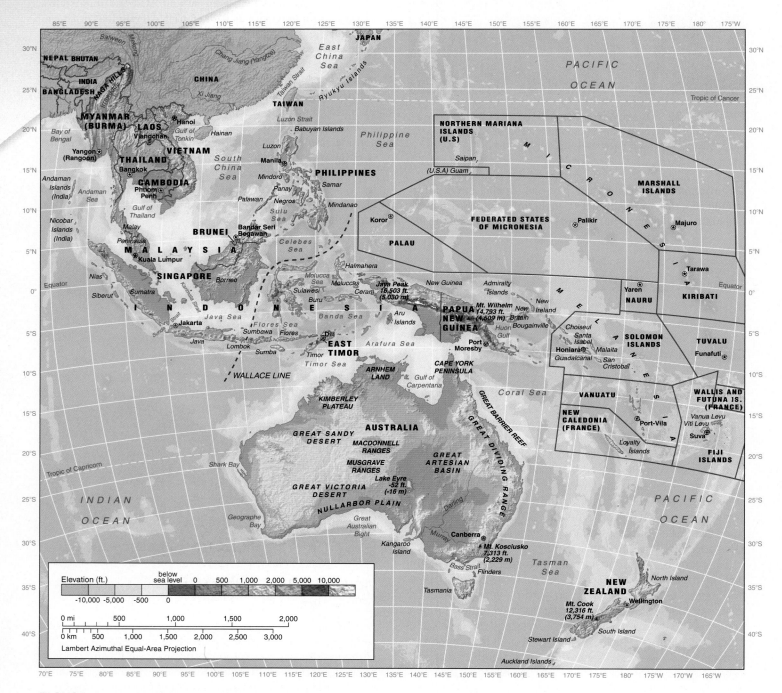

FIGURE 5.1 Southeast Asia, Australia, and Oceania, including Australia, New Zealand, many Pacific Ocean islands, and Antarctica. Physical features, country boundaries, and capital cities. The region has the Indian Ocean to the west, the Pacific Ocean to the east, and a land boundary with China to the north. Included in the region, but not shown on this map, are thousands of islands to the east, including Tonga, Samoa, French Polynesia, and Easter Island (see Figure 5.19), and Antarctica to the south (see Figure 5.22).

"I am Indra, a middle-aged mother living in Malaysia in a suburb of the capital, Kuala Lumpur. I have seen major changes in our way of life over the last 50 years. Kuala Lumpur has grown from a small town to a huge city with new buildings extending in one direction to the coast and inland to the new administrative center at Putrajaya. New supermarkets and chain stores arise almost daily. Our way of life changed in many ways as our government encouraged economic and population growth. We now have many electrical devices in our homes. Our family ties remain strong. My children have done well in school and live and work nearby. They see a great future for Malaysia. Although there are tensions in our multiethnic society, especially among Malays, Chinese, and Indians, and among Muslims, Christians, and Hindus, there are also many mixed marriages, such as a Thai father and Malay mother."

"New Zealand is a great country to live in. I, John, farm in North Island and enjoy a lifestyle as good as any in the world. It is relaxed and affluent for most. That was not always true, especially after the United Kingdom joined the European Union and many of our markets for meat and dairy products were reduced. Our new links with East and Southeast Asian countries have opened up new markets, often for new products, and we have to make changes to meet demands. Lots of visitors envy me living in a country with a small population and so many natural wonders: tourists from Europe, North America, and Asia are pouring in, along with movie makers. Our main problem is when our young people look for jobs: many leave for opportunities in Australia, the UK, or the United States, all a long way away."

Chapter Themes

Diverse Region, Closer Internal Ties

Defining the Region

Distinctive Physical Geography: continents and islands; varied climates; distinctive ecosystems; natural resources; natural hazards.

Distinctive Human Geography: cultural diversity; population distribution and dynamics; politics of independence; growing economies.

Geographic Diversity
- Southeast Asia.
- Australia and New Zealand.
- Oceania.
- Antarctica: a region?

Contemporary Geographic Issues:
- ASEAN and APEC.
- Conflicts: ocean space and piracy.
- Australian identity.
- Singapore: global magnet.

Geography at Work: the tsunami of December 2004

Diverse Region, Closer Internal Ties

Southeast Asia, Australia, and Oceania is a region increasingly united by internal and external flows of people, goods, and information. It is part of the Pacific Rim, in which new trade and political links join it with East Asia and the Americas. It also looks west to the Indian Ocean and even European countries. The increasingly global view that marks this region stimulates the growing internal associations among geographically diverse places.

Southeast Asia, Australia, and Oceania (Figure 5.1) incorporates all or part of three continents (Asia, Australia, and Antarctica) and thousands of islands. The variety of physical geographies includes tropical climates from all-year rainy to monsoon and arid, as well as temperate climates and the perpetual frosts of Antarctica; biomes include tropical rain forest, deciduous forest, grassland, desert, tundra, and ice sheet. The region has a history of people movements in and between the mainland and islands over millennia, followed by European trading and colonialism from the late AD 1400s. The diversity of environments and peoples (Figure 5.2) that resulted in many discrete local regional identities then became parts of colonies. The colonial

territories formed the basis of new independent countries from the late 1950s.

Defining the Region

This region has only one short land boundary (see Figure 5.1) with China (Chapter 4) and India (Chapter 6) in the north. The other boundaries are in the ocean, with wide stretches of water separating this region from others.

The composition of this region and its boundaries are not agreed on by all geographers. Some would not place the Southeast Asian countries, Australia, New Zealand, Oceania islands, and Antarctica in the same region because of the physical and human differences among them. Despite such differences, the factors that caused their isolation from each other are growing less significant. In a globalizing world, close proximity to each other results in common trade and security concerns.

Distinctive Physical Geography

Southeast Asia, Australia, and Oceania extends from north of the equator to the South Pole and from the eastern margin of the

(a)

(b)

(c)

FIGURE 5.2 Southeast Asia, Australia, and Oceania: a multicultural region. (a) "Long-necked" Karen children, whose families moved to western Thailand from Burma and who present a tourist photo opportunity. (b) Australian aborigines after shopping in a store in Nangalala, Northern Territory. Tensions arise between traditional ways and Western lifestyles. (c) Australian children with European backgrounds wave flags at an international cricket match.

Indian Ocean across the Pacific Ocean. It is one of the largest world regions, but over 80 percent of the area is ocean.

Continents and Islands

Tectonic plate movements over the last 200 million years led to the huge continent of **Gondwanaland** breaking up to form this region. Earthquake and volcanic activity along convergent plate margins accompanied the movements. Volcanic eruptions and earthquakes continue today, part of the Pacific Ocean "Ring of Fire." Large fragments became the separate continents of Australia and Antarctica. As Australia on the Indian plate (see Figure 1.5) drove northward toward Asia, it combined with the Philippines and Pacific plates pushing in from the east to raise the Southeast Asian islands. In the west the Indian plate dived beneath the Eurasian plate and created the high mountain ranges of Myanmar and the volcanic line of islands extending from Sumatra and through Java. To the east, the plate movements formed the Philippines, and brought New Guinea and other continental fragments into collision with the central group of islands that are mainly Indonesia today. From the south, the plate movements brought Australia closer to these islands. Rising ocean levels at the end of the Ice Age formed islands, and drowned land bridges, leaving most of the South China Sea and Java Sea as shallow continental shelf areas.

Despite being at a distance from plate margins, many Pacific islands, together with thousands of such features that do not reach the ocean surface, also have volcanic origins. Some were generated by eruptions at a divergent plate margin and carried westward on the plate; others formed over hotspots, erupting molten lava and ash to form the volcanoes.

Varied Climates

Tropical rainy climates affect a zone up to 10 degrees of latitude from the equator (see foldout map inside back cover), including Malaysia, Indonesia, Papua New Guinea, and many Pacific Ocean islands. Places receive 2,000–3,500 mm (80–140 in.) of rain per year and their temperatures vary little from 25°–30°C (80°F plus).

Local weather in the islands is dominated by daily wind reversals in convective circulations. During the day air heated over the land rises, drawing in cooler, humid air from the ocean. Such **sea breezes** bring moisture that rises and condenses to form clouds over the land in the afternoon, often leading to thunder and lightning and downpours of rain. Some islands have thunderstorms on over 300 days in the year. At night air cools over the land and flows back to the oceans; the skies clear.

Outside this zone, still within the tropics, the rains are seasonal and totals generally lower over the Asian continent, the Philippines, northern Australia, and many Pacific Islands. Temperatures range from around 30°C (80°F) in summer to 20°C (70°F) in winter. Parts of Vietnam and Thailand experience a more exaggerated,

monsoon climatic environment, variant with marked summer rains. Some Pacific Ocean islands have little relief, and those farther east receive little rainfall. Hilly islands often have rainy and arid sides depending on the direction of the trade winds: from the northeast in the Northern Hemisphere and from the southeast in the Southern Hemisphere. Windward slopes are wet; leeward slopes are drier. The tropical western Pacific Ocean is subject to destructive **typhoons** (hurricanes) that impact the Philippines and northern Australia.

Recent climate research shows that the weather in the tropics of this region is related to the **El Niño** phenomenon off western South America (see Chapter 9, p. 270). When the El Niño brings rain to Peru, tropical Southeast Asia, the nearby islands, and northern Australia are dry; for most of the time between these events, Peru is arid and rainfall is plentiful in the tropics of this region.

Australia lies across the **Tropic of Capricorn** and most of the continent is arid with high summer temperatures, but around 10°C (45°F) in winter. On the coasts outside the tropics and also in neighboring New Zealand, there are subtropical (in the north) and temperate (in the south) rainy climates with summer or winter maxima of rain with totals in the range of 500–1,500 mm (35–60 in.).

The Antarctic continent is almost all covered by ice that composes a large proportion of Earth's water in frozen form. Present rising temperatures and marginal melting of that ice is slowly causing ocean levels around the world to rise.

Distinctive Ecosystems

The diversity of climates, continents, and islands in this region supports a variety of natural vegetation and animal types. In the mid-1800s, Alfred Russell Wallace, a botanist, mapped the "Wallace Line" between Sulawesi and Borneo (see Figure 5.1) as a major division within the region's biogeography. The Asian lands to the west have tropical and monsoon forests typical of that continent. Australian biomes east of the line are dominated by plant species of eucalyptus (Figure 5.3) as well as animal **marsupials,** such as kangaroos, koalas, and opossums; bird life is colorful. Marsupials raise their young in pouches instead of

the womb. It was clear that Australia and some islands to its north and east were for long isolated from the Asian lands, although the reason was not known for 100 years after Wallace. It is now known that around 10–15 million years ago, plate tectonic movements pushed Australia and some islands north and west toward the rest of Southeast Asia, forcing the separate sets of plants and animals together.

On the Pacific islands palms are common, but few Pacific islands have many animals, apart from a diversity of bird species and a wealth of tropical fish varieties in the surrounding waters. Coral reefs are common around many islands (Figure 5.4). Reefs close to the coast are **fringing reefs,** those separated from the island by a lagoon are **barrier reefs,** and those surrounding a central lagoon without a volcanic island are **atolls.** The world's largest continuous chain of coral forms the Great Barrier Reef off the northeastern coast of Australia.

During European colonization, many species from this region were taken to adorn European gardens and external species were introduced. Domestic animals from Europe, such as dogs, deer, and rabbits found few local predators and require control programs. In New Zealand the unique forest and its fauna were cut and replaced by quicker-growing Douglas firs and pines; local animals and flightless birds, such as the kiwi, became extinct.

Natural Resources
Minerals

Southeast Asia, Australia, and Oceania possesses a wealth of mineral resources. Southeast Asia contains tin, precious stones, and iron. Much of the tin is mined from river deposits after the tin-bearing veins in the rocks were eroded and the heavy mineral dropped and concentrated in river transport. Australia is a major source of metals such as iron, copper, zinc, lead, silver, and gold. One of the world's largest copper reserves is on the island of Bougainville, part of Papua New Guinea. New Caledonia is the world's third largest producer of nickel. Other islands have deposits of guano (bird droppings) that are high in phosphate and extracted for fertilizers. The rocks of Antarctica

FIGURE 5.3 Australia: eucalyptus forest in humid Victoria state.

FIGURE 5.4 Oceania: coral island. This island has a hilly core of volcanic rocks, an outer coral reef, and a shallow lagoon between the reef and the land. A village is situated where a channel breaches the reef front.

contain numerous mineral resources, but at present are off-limits to development.

As well as metallic minerals, energy sources are common. Indonesia, Australia, and Brunei are important producers of oil and natural gas, while Malaysia has useful reserves. In 2004 Australia was the world's fourth largest producer of coal, 6.5 percent of the total after China (34 percent), the United States (18 percent), and India (7.4 percent). Most is exported. Australia also mines uranium for use in nuclear power plants.

Water

Water resources are plentiful in the tropical rainy areas. On the Asian continental area the major rivers are the Irrawaddy and Salween rivers in Myanmar, the Mekong River that rises in China and flows through Laos, Cambodia, Vietnam, and along the border of Thailand, and the Red River in northern Vietnam.

The Australian desert and many small islands with few hills to encourage rainfall are short of water. In Australia the Murray-Darling River system brings water to a semiarid area west of the Great Dividing Range, and to the north the **Great Artesian Basin** provides a source of water draining through rocks and emerging from wells under its own pressure (Figure 5.5).

Natural Hazards

The region shares vulnerability to natural hazards of different types, highlighted by the December 2004 **tsunami wave** that struck coastal Sumatra (see "Geography at Work," p. 163), Thailand, Malaysia, and places around the Indian Ocean. This was a recent example of the volcanic and earthquake events caused by the active plate tectonic margins around and through the region. Volcanic eruptions affect the Philippines, Indonesia, and North Island, New Zealand; frequent earthquakes also occur along the active plate margins.

Damaging weather systems include the typhoons that affect coastal areas of the Philippines and northern Australia. Intense tropical rainfalls cause the collapse of steep slopes and landslides of mud, especially where forest cover is removed. The alternating patterns of weather associated with El Niño lead to forest fires, smoke, and widespread breathing difficulties in parts of Indonesia, particularly Borneo and Sumatra. The rising ocean level as a result of global warming threatens the future of many of the low-lying coral atolls and other coastal settlements.

The tropical environments harbor many infectious diseases, such as malaria, cholera, typhoid, and rabies. Improvements in medical services, better sanitation, and the elimination of mosquito breeding sites have lowered the impact of such hazards and increased life expectation.

Human activities in natural environments create their own hazards. Loggers and farmers denude slopes and may be instrumental in soil erosion on steeper slopes. The overuse of fertilizers and pesticides pollute ground and stream waters. In the Vietnam War of the 1960s and 1970s, U.S. planes defoliated forests to expose Vietcong positions. The barren hill slopes created have still not recovered. Miners digging large quarries affect downstream water and air quality. Some islands in Oceania have been made uninhabitable by the extraction of metal ores and phosphate. The expanding cities create their own environments, and more controls are needed to reduce the pollution from vehicle exhausts and factories.

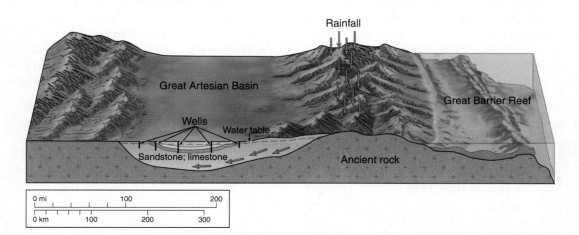

FIGURE 5.5 Australia: Great Artesian Basin and Barrier Reef. The geologic conditions of northeastern Australia produce a major groundwater source. Rain falling on the Great Dividing Range seeps into the rocks and flows downward through the rocks and into those beneath the basin. As the level of the water in the upper rocks is above those in the basin, water flows out under pressure without pumping. The Great Barrier Reef is the world's largest development of coral reefs, forming a feature off the coast of northeastern Australia that has become a major tourist draw.

Distinctive Human Geography

Cultural Diversity

The population of Southeast Asia, Australia, and Oceania is characterized by its multiethnic composition and by variations in distribution and population changes. Its present diversity resulted from thousands of years of peoples arriving from near and far and their interactions with the local people and natural environments.

Early People Movements

The earliest groups moved from the north, across the islands, and into Australia 40,000 and more years ago. By the early years AD local agricultural prosperity generated kingdoms. The Mon and Khmer peoples occupied present-day Cambodia to form the Khmer Empire. Later, Vietnamese, Lao, and Burmese peoples established other centers. Traders from India brought Hindu and Buddhist religions, leaving imprints of traditions and architecture such as the temples at Angkor Wat, Cambodia, and those on the island of Bali. The arrival of Islamic groups in the AD 1200s in-

troduced further cultural elements. Muslim merchants controlled trade in products such as spices. They established sultanates in Malacca (now in Malaysia), the islands of modern Indonesia, and Brunei. Such flows of people widened the range of languages spoken in the region.

Throughout this period, the region acted as a melting pot for people from China and India, living alongside and mixing with local people and interacting at various levels with those in Oceania and Australia. The Oceania islands had dark-skinned Melanesian, Micronesian ("small islands"), and lighter-skinned Polynesian ("many islands") people. In Australia the Aborigine people numbered between 200,000 and 500,000 in the late 1700s. They were nomadic hunters with many different languages, who produced rock paintings expressing their animistic nature worship. In New Zealand the Maoris came from other islands from the AD 800s, displacing the earlier peoples.

Colonization

From the late 1400s a series of European traders and colonists came into this multiethnic and multicultural complex (Figure 5.6).

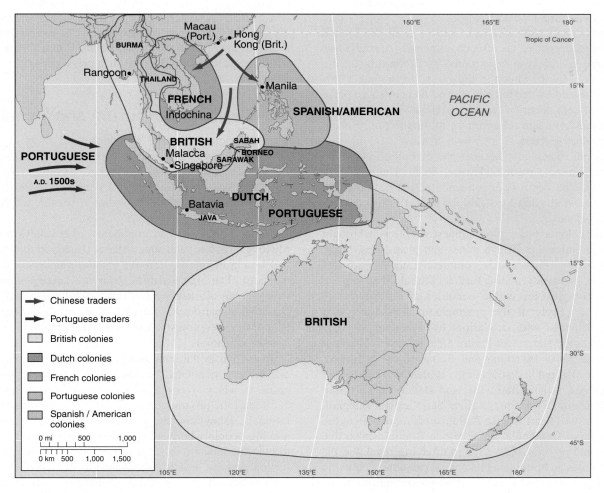

FIGURE 5.6 Southeast Asia, Australia, and Oceania: colonization. Portuguese, Dutch, British, French, and Americans took control of areas easily accessed from the ocean.

European trading demands reoriented local economies, political colonization determined modern country boundaries, and their ways of life increasingly impacted the region's cultures. Colonial languages joined the established creole varieties as common means of communicating.

The Portuguese first traded in the Philippines, and annexed Malacca in 1511. In the late 1500s Spain forced them out of the Philippines, building their capital at Manila on Luzon and bringing in priests to convert the local people to Roman Catholicism. The Dutch took control of trade in what became known as the Dutch East Indies or Spice Islands (modern Indonesia). The British then contested trade with the Dutch until an 1804 agreement resulted in a separation of interests: the United Kingdom took Malaya, building a new port at Singapore in 1819, and the Dutch took the East Indies.

During the early 1800s and 1900s the French occupied Indochina (modern Laos, Cambodia, and Vietnam), building roads and railways and encouraging manufacturing. In the late 1800s the British colonized Burma as part of the British Indian Empire, developing rice growing in the Irrawaddy River delta, building a new port at Rangoon (modern Yangon), and bringing in workers from India. Only the Kingdom of Siam (modern Thailand) remained independent with its strong monarchy, although it also imported Western ways.

In Australia and New Zealand, settlers from the United Kingdom largely replaced the indigenous people. Australia was the last major continent to be explored by Europeans, having been discovered by the Dutch in the 1600s. In the late 1700s the British claimed ownership and moved their offshore penal colony from North America (where the United States had just gained independence and refused to fulfill that role) to Botany Bay (now Sydney Harbor).

The Oceania islands also became part of Europe's colonial empires: the United Kingdom, France, Spain, United States, and Germany. The last lost its influence in 1919 after World War I, when Australia and New Zealand took trust responsibility for many islands.

During the colonial phase, more Chinese people were brought in to labor in Dutch and British mines and plantations. Many stayed and became important in commercial activities. Some converted to Christianity, adding further cultural diversity.

Cultural complexities within countries are illustrated by Thailand. The people of the mountainous north and west are mainly Buddhist with similarities to Myanmar; minority groups, such as the Karen (see Figure 5.2a) moved to Thailand after persecutions in Myanmar and China. In the northeast, long isolation resulted in distinct cultural identities and people share languages, artistic traditions, and Buddhism with Lao groups. The central Chao Phraya River basin around Bangkok is the Thai cultural heart, whose Buddhism is more akin to the Khmer people of Cambodia. Some Hindu Brahmins also live in this area and provide skills used in directing royal and official ceremonies and producing an annual calendar. Economic growth in this area attracted people from poorer regions of the country. To the south and along the Cambodian border Chinese settled from the 1800s as workers on sugarcane plantations and now work in timber mills and run small stores. Muslim Malay people increase in numbers toward the south of the narrow southwestern peninsula.

Human Rights

Ethnicity and political dominance are behind some major human rights abuses in the region. Myanmar's military dictatorship has persecuted both opposition parties and troublesome hill tribes, such as the Shan and Karen. Vietnam and the other countries of former Indochina have pursued vendettas against Chinese merchants and religious people. In Cambodia, the Communist Khmer Rouge ruled from 1975 to 1979, causing the death of 1.7 million people from disease, starvation, or execution. Many of the killers continued to live in Cambodia until 2001, when King Sihanouk put them on trial. Both Cambodia and Vietnam carried out genocidal eradication of hill tribes, linking political crusades to ethnic abuse.

Malaysia suffered ethnic riots in 1969 and implemented policies to correct an economic "imbalance" in which people of Chinese origin owned nearly all the businesses. By 1990, non-Chinese owned 30 percent of Malaysian businesses. However, reality intervened when wealthy Chinese businesspeople switched investment to other countries. Ethnic Chinese people in Indonesia are also subject to such pressures.

Population Distribution and Dynamics

The historic movements of people and their interactions with the natural environments resulted in variations in population numbers across this region (Table 5.1). In 2006 Southeast Asian continental and island countries had much higher densities of population than Australia and New Zealand (Figure 5.7).

Within Southeast Asia, continental river valleys (Irrawaddy, Mekong, and Red) and some islands (Luzon in the Philippines and Java in Indonesia) have the greatest numbers of people per unit area. The high densities today are based on the historic development of intensive wet-rice agriculture as well as modern urban expansion. Much of Borneo, northern Sumatra, and Irian Java, where forest has not been cleared until recently, have sparse populations.

In Indonesia the contrasts between highly populated Java and the less populated islands led to a **transmigration** program, in which over 6.5 million people were resettled from 1950 until the program ended in 2001. Despite its democratic rationale, many saw the program as political manipulation, moving people who were regarded as government supporters to outlying islands. Other problems stemmed from a poor selection process, which did not adequately consider age, skill levels, ethnicity, or appropriate farming systems. Furthermore,

TABLE 5.1 SOUTHEAST ASIA, AUSTRALIA, AND OCEANIA: Data by country, area, population, urbanization, income (Gross National Income Purchasing Power Parity), ethnic groups

Country	Land Area (km²)	Population (millions) mid 2006 Total	Population (millions) 2025 est. Total	%Urban* 2006	GNI PPP 2005 Total (US$ billions)	2005 Per Capita* (US$)	Ethnic Groups (%)
SOUTHEAST ASIA							
Cambodia, Kingdom of	181,040	14.1	19.6	15.0	35.1	2,490	Khmer 90%, Vietnamese 5%
Laos, People's Democratic Republic of	236,800	6.1	8.7	19.3	12.2	2,020	Lao 50%, Thai 14%, Meo and Yao 13%
Myanmar, Union of (Burma)	676,580	51.0	59.0	29.0	no data	no data	Burman 68%, Shan 9%, Karen 7%, Other tribes
Thailand, Kingdom of	513,120	65.2	70.2	32.7	550.6	8,440	Thai 75%, Chinese 14%, Malay 3%
Vietnam, Socialist Republic of	331,690	84.2	102.9	26.3	253.4	3,010	Vietnamese 90%, Chinese 2%, tribal groups
East Timor (Timor-Leste)	9,486	1.0	1.9	22.3	no data	no data	Complex Malayo-Polynesian, Melanesian-Papuan
Indonesia, Republic of	1,904,570	225.5	263.7	42.0	838.7	3,720	Javanese 45%, Sundanese 15%, Madurese, Malays
Malaysia, Federation of	329,750	26.9	34.6	62.0	277.5	10,320	Malay 59%, Chinese 26%, Indian 7%
Philippines, Republic of	300,000	86.3	115.7	48.1	457.2	5,300	Malay 95%, Chinese 2%
Singapore, Republic of	620	4.5	5.2	100.0	133.0	29,780	Chinese 76%, Malay 15%, Indian 7%
Totals/*Averages	**4,483,656**	**564.6**	**681.6**	**39.7**	**2,557.7**	**8,135**	
AUSTRALIA AND NEW ZEALAND							
Australia, Commonwealth of	7,713,360	20.6	24.6	90.7	629.8	30,610	White 95%, Asian 4%, Aborigine 1%
New Zealand	270,990	4.1	4.6	88.9	95.3	23,030	White 73%, Maori 12%, Asian 5%, Pacific Islander 4%
Totals/*Averages	**7,984,350**	**24.7**	**29.2**	**89.8**	**725.1**	**26,820**	
OCEANIA							
Fiji, Republic of	18,270	0.8	0.9	45.7	5.1	5,960	Fijian 50%, South Asian 45%
Kiribati, Republic of	730	0.1	0.1	43.5	no data	no data	Micronesian 99% with Tuvaluan minority.
Marshall islands, Republic of	179	0.1	0.1	68.0	no data	no data	Micronesian 99% German, Japanese, U.S., Filipino
Micronesia, Federated States of	702	0.1	0.1	22.3	no data	no data	Chuukese 41%, Pohnpeian 26%, other groups
Nauru, Republic of	20	0.01	0.02	100.0	no data	no data	Naurean 58%, other Pacific 26%, Chinese 8%
Palau, Republic of	458	0.02	0.02	77.4	no data	no data	Micronesian with Melanesian and Malay admixtures 70%, Filipino, 25%, Asians, Europeans
Papua new guinea, Independent State of	462,840	6.0	8.2	13.3	14.2	2,370	Melanesian 98%
Solomon islands	28,900	0.5	0.7	15.6	0.9	1,880	Melanesian 93%, Polynesian 4%
Tonga, Kingdom of	750	0.1	0.1	22.9	0.8	8,040	Tongan 98%
Tuvalu	30	0.01	0.01	47.0	no data	no data	Polynesian 96%
Vanuatu, Republic of	12,190	0.2	0.4	21.5	0.7	3,170	ni-Vanuatu (Melanesian) 94%, white 4%
Western Samoa, Independent State of	2,840	0.2	0.2	22.0	1.2	6,480	Samoan (Polynesian) 93%, Euronesian mixtures and others 7%
Totals/*Averages	**527,909**	**8.2**	**11.0**	**41.6**	**23.0**	**4,650**	
Region Totals/ *Averages	**12,995,915**	**597.5**	**721.8**	**57.0**	**3,305.8**	**13,202**	

Source: World Population Data Sheet 2006, Population Reference Bureau, Microsoft Encarta 2005.

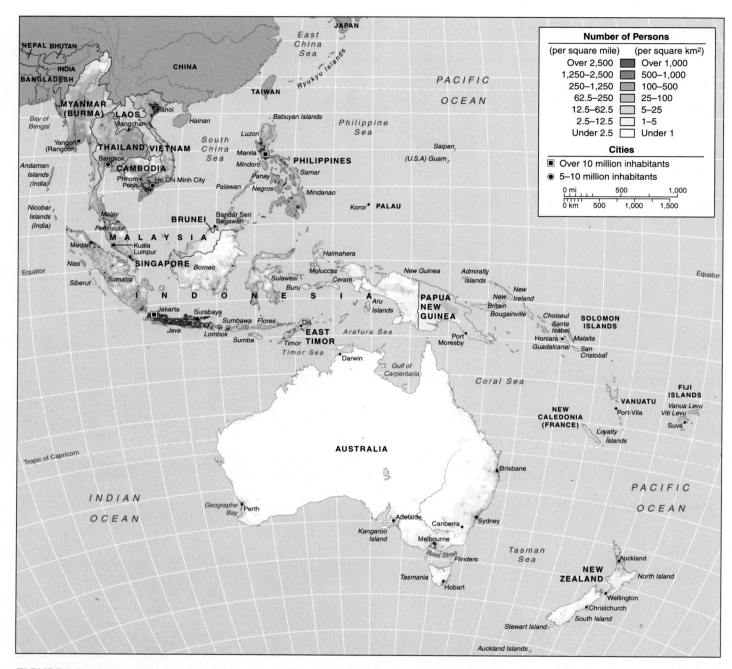

FIGURE 5.7 Southeast Asia, Australia, and Oceania: distribution of population. Compare and attempt to explain the areas of high and low density.

few unoccupied tracts of good land remained in the less settled islands. Nearly half the migrants failed to raise their own standard of living. Since 1997 local groups have murdered many settlers who cut the forest to plant oil palms or illegally produce timber.

The Oceania islands have small numbers of people, and the numbers decrease as distance eastward from Australia and Southeast Asia increases. In 2006 6 million of the 8.2 million people living on these islands inhabited Papua New Guinea.

Antarctica has no permanent population, but is the short-term home to groups of scientists.

Rapid Urbanization

Australia and New Zealand are highly urbanized countries with around 90 percent of their populations living in urban centers. Around half their total populations live in the metropolitan areas of over one million people (Table 5.2). The largest cities are all

TABLE 5.2	Populations of Major Urban Centers in Southeast Asia, Australia, and New Zealand in millions	
City, Country	2003 Population	2015 Projection
SOUTHEAST ASIA		
Jakarta, Indonesia	12.3	17.5
Bandung, Indonesia	3.8	5.3
Surabaja, Indonesia	2.6	3.5
Medan, Indonesia	2.0	2.7
Palembang, Indonesia	1.6	2.2
Bangkok, Thailand	6.5	7.5
Ho Chi Minh City, Vietnam	4.9	6.3
Hanoi, Vietnam	4.0	5.3
Haiphong, Vietnam	1.8	2.3
Phnom Penh, Cambodia	1.2	1.5
Singapore, Singapore	4.3	4.7
Kuala Lumpur, Malaysia	1.4	1.6
Yangon, Myanmar	3.9	5.3
AUSTRALIA AND NEW ZEALAND		
Sydney, Australia	4.3	4.8
Melbourne, Australia	3.6	4.0
Brisbane, Australia	1.7	2.0
Adelaide, Australia	1.1	1.2
Auckland, New Zealand	1.1	1.3

*estimated

Source: United Nations Urban Agglomerations 2003, with estimates for 2015 (2003).

FIGURE 5.8 Australia: Sydney. Sydney's harbor is made memorable by the opera house on the near promontory.

ports, which have high-rise central business districts and extensive suburban tracts with shopping malls (Figure 5.8). Sydney and Melbourne have sufficient business service activity to be classed as global city regions (see Figure 1.23).

Southeast Asia has the region's largest and fastest growing cities: Jakarta in Indonesia, Manila in the Philippines, and Bangkok in Thailand. Singapore is Southeast Asia's major global city-region, while Jakarta, Bangkok, Kuala Lumpur, and Manila are also increasingly involved in global transactions. However, the countries with the three largest cities still had less than 50 percent of their population living in urban centers. Malaysia had 62 percent. Vietnam, Cambodia, Laos, and Myanmar had under 30 percent.

Flows of people into the largest cities are responses to the availability of jobs, health care, education, transportation, and global connections there. However, the rate of people moving from rural to urban areas is often too rapid for the authorities and private firms to provide sufficient housing, jobs, and services. Incoming people build their own shantytowns and expand the informal economy. The increasing number of motorized vehicles adds to urban pollution.

The Oceania islands have low populations that do not support very large cities. On smaller islands, urban proportions are often high, but elsewhere the lower proportions of urbanization reflect a lack of economic diversification.

Population Growth

Like other aspects of this region, the contrasts of a few years ago are giving way to more similar conditions. The rapid population growth in Southeast Asia from 1965 (275 million) to 2006 (565 million) slowed as natural increase rates dropped from 2–3 percent to 1–2 percent. Average fertility rates in 1965 were around 6, but by 2005 were all under 4. In Thailand, family planning policies reduced fertility to under 2. The age-sex diagram for Indonesia (Figure 5.9) demonstrates the stability of birth numbers and greater life expectancies as a result of medical progress. However, the countries of former Indochina that had low rates of natural increase during the wars in the 1950s to 1970s, now have rising rates.

Australia and New Zealand doubled their populations from 1950, helped by immigration from Europe at a time of low natural increase. By the early 2000s Asian immigrants made up nearly half the total entrants. Australia is becoming a multiethnic and multicultural country.

Population growth rates vary among the Oceania islands. Places with high rates of natural increase, where the land and its resources cannot support people on a subsistence basis are on the point of overpopulation. Economic development to date seldom supports larger numbers of people. After World War II,

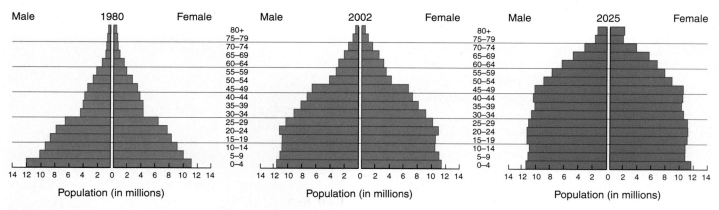

FIGURE 5.9 Indonesia: age-sex diagram. Suggest reasons for the changing population profiles shown by the three graphs and identify the social issues raised. *Source: U.S. Census Bureau, International Data Bank*

many islanders changed from their healthy diet of fruits, vegetables, and fish to imported canned meats and high-fat turkey tails. Obesity, diabetes, and heart disease became problems.

HIV/AIDS is an increasing problem affecting health and life expectancy, particularly in Thailand and Myanmar. Official denials of its expansion claim that the disease is confined to the communities of drug users, prostitutes, and their clients. However, the large sex markets partly aimed at foreign tourists may involve men who catch the disease and pass it on through sex with their wives. The ironic outcome affects Asian women who are unlikely to have extramarital sex.

In French Indochina the countries of Laos and Cambodia gained independence in the early 1950s and Communist fighters took the northern part of Vietnam from the French in 1954. They eventually added the southern part of the country when U.S. forces left in 1975 after 30 years of civil war. Thailand, which had never been colonized, stemmed the flow of the Communist advance from the 1960s, attracting much financial aid from the United States.

Most of the Oceania islands also gained independence after World War II, although some of the smaller ones remain dependent on larger countries. The French still govern French Polynesia as part of France.

Politics of Independence

New Countries

Australia and New Zealand were the first to become independent, establishing Dominion status within the British Commonwealth in 1901 and 1907, respectively. Groups of British people established new colonies in Australian coastal locations through the 1800s, and eventually became convinced that they should be part of a large federated country (Figure 5.10). New Zealand followed a similar pattern, but incorporated a respect for Maori land ownership in its agreement.

The end of colonialism for other countries in the region came within 20 years of the end of World War II. At first the new countries attempted self-sufficient nationalism. However, subsequent regional and global trends drew the countries into producing export goods, beginning in Southeast Asia and extending to the rest of this world region.

The republics of the Philippines and Indonesia gained independence in 1946 and 1950 following the growth of nationalist demands before World War II and the often horrific experience of Japanese occupation during the war. Burma gained independence in 1948 and was renamed Myanmar by its military leaders in 1989. In Malaya the British fought local Communist guerrillas, but made Malaya independent in 1957. The Federation of Malaysia was established in 1963, joining Malaya, Singapore, Sarawak, and Sabah. Brunei did not join, and Singapore was expelled in 1965.

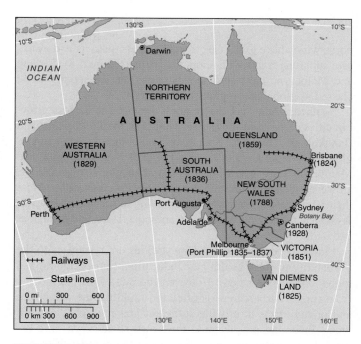

FIGURE 5.10 Australia: colonization patterns. From the late 1700s through the 1800s Australia's new ports became cities and centers of colonial, and later state, governments. The dates in parentheses indicate when the colonies were established. Many of the straight boundaries were drawn across the interior desert. Railroads link the main centers.

Growing Together

A combination of factors caused the very different new countries to move toward a greater regional cohesiveness. The trend began with the formation of the **Association of South East Asian Nations** (ASEAN) in the 1960s, when fears raised by the Communist advance through French Indochina forced the Philippines, Indonesia, Thailand, and Malaysia, backed by the United States, to form a united front (see Contemporary Issue, p. 158). Rapid industrialization in these countries, which was based on foreign investment, led to advances in economic well being. Brunei joined in 1984.

In the 1990s, as the perceived threat of attack receded, Vietnam, Cambodia, and Laos joined ASEAN. ASEAN switched from a political to an economic focus in a unique combination of democratic countries, Communist countries, and dictatorship-led Myanmar. The ASEAN countries then formed associations with China, Japan, and South Korea (ASEAN+3), Australia, and New Zealand.

This experience and the increasing amounts of trade with other parts of the world led to the **Asia-Pacific Economic Co-operation** (APEC) forum, which extended membership to a wider association of 21 countries by 2005. It includes countries in Southeast and East Asia, Australia and New Zealand, Canada, the United States, and some in Latin America.

Growing Economies

This region has a long history of regional and global associations. Before the European colonizations, Arab, Indian, and Chinese traders exchanged goods with coastal and island areas that had easy ocean access. Spices were traded back through the Arab lands to Europe. The trade attracted Europeans, first for high-value goods and later for raw materials. Colonization followed. Tin and rubber from Southeast Asian mines and plantations, and wool, dairy products, and minerals from Australia and New Zealand, were exported to Europe through much of the 1900s.

In the later 1900s these patterns changed. The countries of Southeast Asia added manufacturing for export, often funded by foreign investments, and developed major tourist industries. Australia and New Zealand, losing access to many of their former European markets with the development of the European Union, became more involved with supplying minerals, food products, and services to Southeast and East Asian countries. Intraregional trade increased. However, the emphasis on export goods makes most countries in the region competitors, and leads to tensions.

Differences of political policies add to these tensions. The logging of tropical rain forest in Malaysia and Indonesia has been criticized by Australia for insufficient environmental control and dispossession of local people from the land they have occupied. In response, Australians are told that this is none of their business. The clashes between Muslim and Christian groups on some Indonesian and Philippine islands, the tensions among the Malay, Chinese, and Indian groups in Malaysia, the difficult progress of the former Communist countries of Indochina, conflicts in East Timor, and the politi-

cal difficulties under the Myanmar military, raise questions of economic stability.

Countries in this world region have a range of material well being (Figure 5.11). This is reflected in their consumer goods ownership, access to piped water, and energy usage (Figure 5.12).

Geographic Diversity

Although there are signs that the countries in this region are moving closer to each other politically, culturally, and economically, there are still many differences at each geographic scale. There are contrasts among the subregions of Southeast Asia, Australia and New Zealand, and the Oceania islands; there is diversity among individual countries within the subregions; and there is diversity within the countries.

Subregion: Southeast Asia

Southeast Asia has an increasingly important position in the globalizing world. It lies between China to the north and Australia to the south, and between the Indian and Pacific oceans (Figure 5.13). It includes the continental countries of Vietnam, Laos, Cambodia, Thailand, and Myanmar, and essentially island countries of Malaysia, Singapore, Indonesia, East Timor, Brunei, and the Philippines.

Despite large distances between the countries, this subregion is where many human endeavors meet. The peoples come from Malay, Chinese, Oceania, Indian, and European stock. The "Eastern" religions of Hinduism and Buddhism mix with Islam and Christianity. The Philippines and East Timor are the only countries with dominant Roman Catholic allegiance. Political

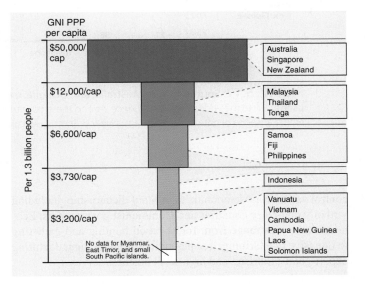

FIGURE 5.11 Southeast Asia, Australia, and Oceania: country per capita incomes compared for 2005 data. The data are GNI PPP.
Source: Data (for 2005) from World Development Indicators, World Bank; and Population Reference Bureau.

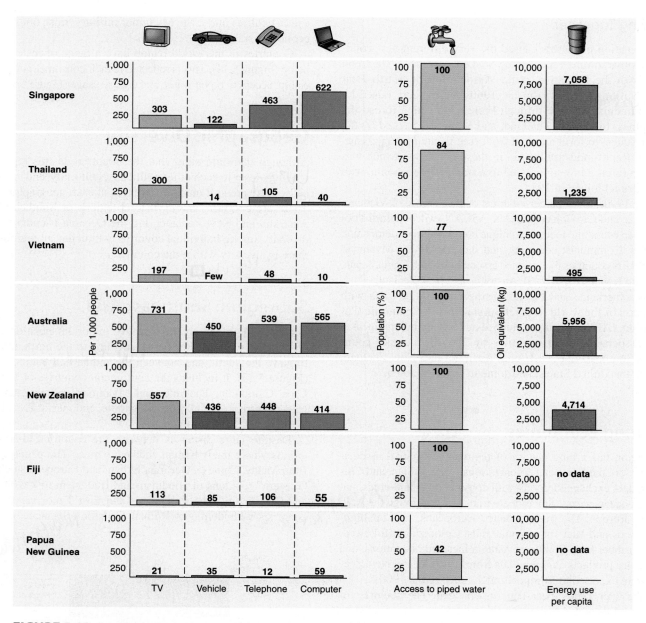

FIGURE 5.12 Southeast Asia, Australia, and Oceania: ownership of consumer goods, access to piped water, and energy usage. Note the wide range of affluence among these countries. See Figure 1.19 for details and compare with other regions. Gaps reflect absence of data for Fiji and Papua New Guinea. *Source: Data (for 2002) from World Development Indicators, World Bank, 2004.*

control varies from democratic to military dictatorship, including virtually one-party countries and Communist governments. Economic activities range from forest-based hunting and gathering to intensive rice farming, tree plantations, mining, manufacturing export goods, tourism, and high-level business services.

Mainland Countries

Thailand stands out as the one country that did not become a colony, retained a traditional monarchy, despite a number of military coups, and resisted the Communist advance after 1950. It also adopted an open attitude to the expanding global economy, developing export manufacturing based on foreign investment and a major tourist industry attracting 11 million foreign visitors in 2005. Thailand's GNI PPP is twice the total of the other mainland countries in this world region and per capita incomes are three times the best of the others. However, Thailand's economy reflects both sides of becoming global.

The income from major vehicle manufacturers and oil refineries leveled off in the 1990s, leading to investment switching

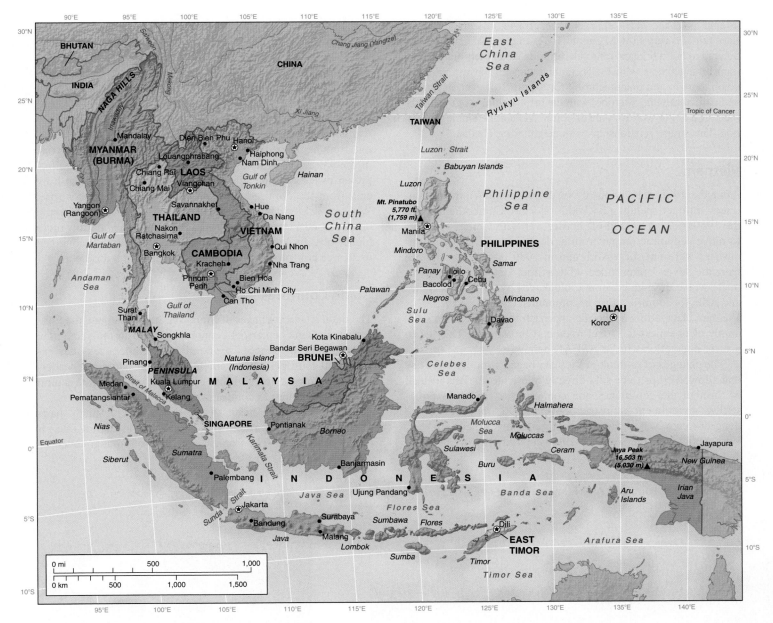

FIGURE 5.13 **Southeast Asia: countries and major cities.** How important are ocean routes in this region? In 1999 East Timor, which had been a Portuguese colony until 1974 and then occupied by Indonesia, gained its independence. Civil disturbances occur on other Indonesian islands.

to land and buildings. This asset "bubble" burst in 1997 with withdrawals from banks and pressure on the baht currency. The Thai economy recovered slowly with external help and further investment.

The tourist industry is subject to fluctuations in demand as Asian visitor numbers fell after such events as the late 1990s economic crisis in the area, the 9/11 attacks in the United States, and the Indian Ocean tsunami of late 2004. The illegal sex and drug trades generate crime and health crises. Imprisoned drug offenders rose by 10 times since 1990, swamping the courts and prisons. The drugs come across the border from southern Myanmar and merely take different routes when the Thai mili-

tary intervene. *Ya baa*, the methamphetamine "crazy pill," is of less concern than heroin to the world community, but is a major problem in Thailand. It was first used by workers needing to remain alert on long shifts, but became a recreational drug for students; children as young as 6 years become addicts with the prospect of years of uncontrolled violence.

Of the other mainland countries, Myanmar is a military dictatorship, while Laos and Vietnam have Communist governments coming to terms with a globalizing world. In 1989 Myanmar's military leadership adopted the new name to replace "Burma."

Since the 1990s, all these countries opened their economies to varied levels of external investment and associations, including joining ASEAN, although economic growth in Cambodia and Laos remains very slow. Myanmar, for example, paid Chinese contractors to build bridges, railroads, airports, and hotels and entered joint arrangements with European and Japanese investors to develop tourism and manufacturing for export. However, Myanmar and particularly the Wa tribe, with the tacit backing of the government, is a major world source of opium and amphetamine production. This strains relationships with neighboring Thailand.

Vietnam

After the 1975 unification of North and South Vietnam, continued military spending backed forays into Cambodia through the 1980s and an on-and-off border war with China. Attempts to move economic output from agriculture to heavy urban industries were not successful, and in 1989 the backup finance from the Soviet Union ended. At this time, Vietnam's plight was made worse through economic blockades by the United States and ASEAN countries. Still a Communist country, in the 1990s Vietnam changed its economic approach to follow China's example of openness to global trade and investment. This led to greater interaction with its ASEAN neighbors.

The changes of outlook in Vietnam began with the 1985 law that allowed crops to be sold directly in private markets instead of through the communes and cooperatives selling to price-setting government ministries. After 1990, the Vietnamese dong currency was devalued and foreign investment allowed. Within Vietnam, the southern area centered on Ho Chi Minh City found it easier to develop a marketplace economy than the northern area around Hanoi, which the Communist government controlled for a longer time. Vietnam's exports of rice now compete with India for third place in the world. Japanese and South Korean corporations were the main investors in manufacturing industries until the United States normalized relations with Vietnam in 1995, the year Vietnam joined ASEAN.

A further Vietnamese development in the 1990s was its progress from being a marginal coffee producer to rival Brazil as one of the world's largest exporters of the cheaper robusta coffee, grown in the central highlands. Farmers moved into the hilly area from the crowded lowlands as world coffee prices rose. When other world producers also boosted output, an oversupply crisis caused Vietnam to cut robusta production and grow more higher-quality Arabica coffee trees. Farmers were encouraged to change crops to rubber, peppers, cotton, and hybrid corn.

Island Countries

The main island countries of Malaysia, Indonesia, and the Philippines became centers of economic growth in the subregion. Although these countries are nominally democracies, they experienced long periods of single-party and autocratic leadership, with intervening periods of internal conflict for the central leadership and loss of control in outlying areas.

Farming remains basic to the economies of many parts of these countries, particularly the cultivation of wet rice, or padi (Figure 5.14). The tropical rainfall is often augmented by

FIGURE 5.14 Southeast Asia: farming. (a) Wet rice, or padi, cultivation in Bali. (b) A young farm worker in Thailand equipped to apply chemical sprays, including pesticides.

(a)

(b)

irrigation to obtain maximum crops. The new rice technology that developed at the research station of Los Banas in the Philippines in the 1960s led to the **Green Revolution.** Crop yields increased by combining high-yielding new varieties of rice with the use of fertilizers, pesticides, mechanization, and irrigation. Padi yields doubled from 1969 to 1989, allowing Indonesia and the Philippines to become almost self-sufficient at a time of rapid population increase. The main beneficiaries were owners of larger farms with good growing conditions, available labor, and capital to invest in the costly procedures. Smaller farms without access to capital gained little. The commercialization of farm production also had social impacts. Former communal land was sold to wealthy farmers or for urban expansion, creating many landless laborers who lost previous access to jobs at planting and harvest. Wage labor and mechanization destroyed traditional systems and resulted in large numbers of disaffected people who supported alternative Communist and Islamist pressure groups.

Under colonization, European companies established commercial tree crops, grown on large plantations and sometimes by local farmers on smaller plots, including rubber, palm oil, spices, and pineapples. Most plantations were taken over by country governments after independence. Good plantation management produces year-round harvests and employment. Malaysia dominated world production of rubber, but failure to recruit enough cheap labor led to Indonesia and Thailand competing in this market. In Malaysia, less labor-intensive oil palms provided an alternative crop in great demand for Chinese cooking, since palm oil produces higher temperatures than other vegetable oils. Indonesian plantation crops include coffee, spices (cloves, nutmeg, cinnamon), tea, and rubber. In the Philippines, coconut products, sugar cane, tropical fruits, and coffee make up one-fifth of exports.

The exports of tropical hard wood timber from Malaysia and Indonesia are greater than those from the Brazilian Amazon and Africa combined. From the 1990s, prices for such timber rose by 50 percent as crop and mineral prices fell or remained the same. Wood industries grew in importance in Kalimantan (Indonesian Borneo) and Sumatra, but rapid cutting caused social conflict as well as concerns for the future. In response to resented Western objections to unmanaged cutting, Malaysia claims to have responsible "sustainable logging" controls. However, it continues to export more tropical logs and sawn timber than any other country in this subregion, reducing forest cover by up to 2 percent annually.

Mining adds to commodity exports. Malaysia has a history of tin mining, dominating world production for many years. Bangka and Billiton, Indonesian islands, also produce tin. Oil has become the main mineral export of the region, with Indonesia being a major producer from its fields around Kalimantan and Sumatra. However, few new resources have been found and Indonesia is becoming an oil importer. Oil is also produced in the Malaysian territories of Sabah and Sarawak, together with neighboring Brunei. Indonesia is a major bauxite producer in the Riau islands and southwest Kalimantan. The bauxite is transported to the Kualatajay smelter near Sumatran coal mines to be processed into aluminum.

The main economic trend of the late 1900s was the shift to export manufactured goods and service industries. This shift intensified the income gaps between urban and rural areas. At independence, most countries had some factory-based industries and imposed tariffs and quotas to protect import-substitution industries producing for small local markets. Government licensing restrictions made local products expensive and **crony capitalism** arose from bureaucrats and entrepreneurs working together. The combination of government-owned processing facilities for mining and agricultural products and small-scale, often Chinese-owned, textile and consumer goods firms remains important in the economies of these countries. Multinational corporations based in the United States, Japan, and South Korea invested in new factories from the 1970s, especially in Malaysia and Thailand, shifting their emphasis from labor-intensive to high tech industries. Much of this export-oriented industrialization is labeled **ersatz capitalism,** since it relies on imported investment, management, machinery, and skills rather than local enterprise and savings.

The importance of government employment, and the growth of health, education, office, retail, and wholesale sectors, resulted in a huge expansion of service sectors and middle incomes in these countries. Tourist industries attracted increasing numbers of foreign visitors and foreign currencies. Thailand, Malaysia, Singapore, Indonesia, and the Philippines together went from 7 million tourist visitors in 1983 to 20 million in 1990, and 43 million in 2005. Their modern cosmopolitan cities, precolonial historic buildings, tropical weather, and spectacular beaches and upland areas with resort facilities are visited by people from Asia, Australia, North America, and Europe.

Of the three small countries, Singapore, with only 4 million people, is a world leader in business services (see Singapore: Global Magnet, p. 161) and Brunei enjoys prosperity based on oil wealth. East Timor at last gained its independence after decades of colonial and Indonesian domination, but continues to suffer internal conflict.

Indonesia

The Republic of Indonesia is the largest country in this world region in both area—when the land and ocean are combined—and population (see Table 5.1). The 5 million km^2 territory includes 1.9 million km^2 of land. The country extends over 5,000 km (over 3,000 mi.) from east to west and 1,760 km (1,100 mi.) from north to south. It incorporates around 15,000 islands, of which 6,000 are inhabited. Indonesia has an increasingly strategic position astride major sea lanes. It is marked by extreme internal diversity that is difficult for any government to control. The five main islands (see Figure 5.13) are Kalimantan, Sumatra, Irian Jaya, Sulawesi, and Java. Kalimantan (Borneo) is shared with Brunei and Malaysia; Irian Jaya shares New Guinea with Papua New Guinea, and Timor is shared with East Timor.

Indonesia is the world's fourth most populous country and the largest with a Muslim majority, with 88 percent declaring to be Muslim in the 2000 census. Archaeological remains and ancient buildings testify to a varied history. In the period from the AD 600s to 1300s, Hindu and Buddhist kingdoms formed on Sumatra and Java. Spice-trading Arabs brought Islam, which became the main religion, replacing collapsed Hindu and Buddhist kingdoms. In the 1500s Portuguese traders found many small vulnerable states in what they termed the East Indies. By the 1600s, the Dutch took over as the most powerful Europeans in this group of islands, relying on indirect rule to impose their will on, and extract income from, traditional local elites. The Dutch established large plantations and forced cultivation on Java, but brought in more liberal conditions in the 1900s. In 1942 Japanese armies captured Java to obtain local oil. They released General Sukarno, who acted as a puppet but also planned for independence, which Indonesia declared in 1945. The Dutch tried to reoccupy their colony, but international pressure caused them to acknowledge Indonesian independence in 1949.

Two generals governed the country for nearly 50 years. Sukarno (1949–1966) made a point of bringing the varied island groups under political control, but achieved little economic growth. This was the priority of General Suharto's government (1976–1997), although, like his predecessor, his long-term government ended with economic crisis and popular uprisings. In the Suharto years, annual per capita income rose from US$70 to US$1,142 and social services were the most extensive in Southeast Asia. However, in the mid-1990s the lavish lifestyles of Indonesia's political elite, centered on the Suharto family, drained the country's resources. The growing middle class found upward mobility capped by corrupt officials. As foreign debt doubled, the 1997 Asian economic crisis brought an end to Suharto's government. Subsequent elected governments strove to put matters right, but faced a reduction in oil income, conflict in several islands, the 2002 Bali terrorist bombing, and natural disasters of forest fires, atmospheric pollution, flooding, volcanic eruptions, earthquakes, and the 2004 tsunami. The island conflicts included independence movements in Aceh in north Sumatra and East Timor, fights between resettled and local groups on Kalimantan, and Muslim-against-Christian conflicts in Sulawesi.

The complexity of its island populations makes it difficult to assess the total Indonesian population. *Statistics Indonesia* quotes 221.9 million for 2005, but the *CIA Factbook* has 240 million. The island of Java has the world's highest population density, contrasting with parts of Kalimantan that have extremely low numbers of people. In the west most people are Malaysian, but in the east Melanesian characteristics prevail, and elsewhere mixtures and origins are complex. Many Javanese, Sundanese, and Batak peoples speak of themselves as nations rather than Indonesians. In Borneo the Dyak and Punan groups have totally different lifestyles. The Indonesian language, taught in schools, is closely related to Malay and began as a **lingua franca** to enable communication among more than 700 ethnic groups. The Pribumi people are considered natives of Indonesia and there are frequent conflicts between them and non-Pribumi people, especially those of Chinese origin.

Jakarta is the capital and largest city, with over 12 million people in the urban area. It was renamed from the Dutch Batavia by the Japanese to emphasize local wishes. It is a crowded and cosmopolitan city, with the local Betawi people and a large Chinese community dominating. Transportation by road and rail is overloaded in rush hours. Traditional bicycle rickshaws were banned from major roads in 1971 and finally dispatched in 1991. Jakarta is a center of universities, tourist attractions, and new shopping malls. Like many huge cities in developing countries, it is subject to overcrowding, especially during the working week because of poor transportation, and the wet season brings flooding by clogged sewage systems and waterways.

Indonesia is the only Asian member of OPEC outside Southwest Asia, although lack of investment in production equipment made it a net oil importer in the early 2000s. It also exports natural gas. In the 2000s the long-term importance of tin, bauxite, and silver production was largely replaced by copper, gold, and coal. External investment doubled coal production between 1999 and 2004 to 136 million tons, making it the world's ninth largest producer.

From the late 1980s, Indonesia encouraged foreign investment to produce clothing, machinery, transportation equipment, and footwear for export. However, the weak legal and financial systems made it difficult to collect debts or enforce contracts; banking practices allowed huge unsecured loans. Following the severe impacts caused by the late 1990s economic crisis and a drought that forced Indonesia to import rice, new investment stopped. By 2006 measures to improve these failings resulted in resumed economic growth, although rising world oil prices created further stress.

Logging in Borneo

Borneo is the largest of the many islands in this world region. It is mostly in Indonesia (where it is known as Kalimantan), but its northeastern coast is shared by Malaysia (Sabah and Sarawak) and tiny Brunei. A hilly-to-mountainous island, Borneo was once covered by tropical rain forest in which live a rich and distinctive fauna and flora, together with tribes such as the Dyak and Punan, many of whom continue to live on forest resources. From the 1950s, Christianity brought by missionaries led to the ending of slavery, head-hunting, and human sacrifice.

One of the world's last true wilderness areas is threatened by loggers and oil palm plantations. From 1985 to 2001, 56 percent of protected lowland forest was cut in Kalimantan. If the forest cover continues to be depleted at current rates, there will be hardly any left in 15 years' time (Figure 5.15). Cutting began outside the protected areas, but buffer areas and then the protected areas were also cut. Much of the logging was illegal.

In 2006 it was reported that a Chinese bank proposal backed by the Indonesian government would invest over US$7 billion to clear 1.8 million hectares along the Malaysian border. The forest would be cut over six years and the land planted with oil palms. Much of this land is part of the Betung Kerihun National

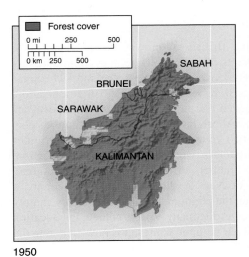
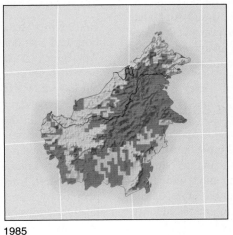

1950   1985   2020 (projected)

FIGURE 5.15 Logging in Borneo. Much of the tropical rain forest on this island is being removed without replanting. The Malaysian states of Sabah and Sarawak were first affected, but Indonesian Kalimantan is likely to be deforested by 2020. Logging, often against government regulations, and the planting of palm oil plantations are responsible. *Source: Data from The Times (London) 2006.*

Park (see Fig. 5.15), which at present can be reached only by four days in motorized canoe. It is the hilly core of Borneo, from which nearly all the main rivers radiate. Cutting the forest would not only destroy important ecosystems, but the smoke from burning forest "debris" would drift across Singapore and Malaysian cities. Upstream erosion and flooding and downstream sedimentation would follow. Moreover, oil palms grow much better on lowland areas than on such hilly terrain.

The Indonesian agriculture minister backs this project, which might bring 500,000 new jobs and reduce the prosperity gap between the Malaysian and Indonesian parts of Borneo. Palm oil is the world's most consumed edible oil, produced mainly in Malaysia (50 percent of world supply) and Indonesia (30 percent) with continued expansion expected. Palm oil is in demand for cooking and consumer products, and has a future in the production of biodiesel fuel. The latter could reduce Indonesian oil imports and produce major exports as world oil prices rise. The project is opposed by the World Wildlife Fund (WWF), which is lobbying for a Heart of Borneo protected area along the international border. The WWF quotes unheeded 1980s cautions over logging in Sumatra, where the rain forest has now nearly all vanished. Among the local people, some want to keep the forests and their ancestral way of life, but others look favorably to the prospect of palm oil jobs and the housing, education, health care, televisions, and karaoke machines they could buy. This is a common modern debate affecting local lives, ecologies, and sustainable economies, which are impacted by global investors and supported by government policy.

East Timor

East Timor (Timor-Leste in Portuguese, Timor Lerosa'e in Tetum), occupies around half of the island of Timor (see Figure 5.13) and is the newest of the countries in Southeast Asia. Portuguese colonial occupation of Timor in 1702 followed centuries of passage and occupation of the mountainous island successively by Australoid, Melanesian, and Malayan peoples, resulting in linguistic diversity. The largest group are the Tetum, concentrated around the capital city, Dili. Chinese and Indian trading networks from the 1300s valued the local sandalwood, slaves, honey, and wax. Pressure from the surrounding Dutch-controlled islands caused the western part of Timor to be ceded in 1859. The former Portuguese colony is 90 percent Roman Catholic.

After the World War II battles against Japanese forces, in which over 50,000 Timorese were killed, Portugal resumed control, but abandoned its colony in 1975, and independence was declared on November 28. Nine days later Indonesia invaded before international recognition by alleging that East Timor could have a Communist government. This overcame objections by Australia and the United States. Timorese guerrilla forces fought the Indonesian military, who punished the civilian population. It is estimated that up to 250,000 Timorese were killed out of the 1975 population of 600,000. In 1999 a UN-sponsored agreement among Portugal, Indonesia, and the United States led to a referendum and a vote for full independence. Violent clashes instigated by Indonesian forces and local militia were ended by an Australian-led peacekeeping force, although the militias continued to attack from their bases in West Timor.

The 1999 conflicts led to destruction of the poor economic structure and 260,000 people fled westward. UN efforts resulted in reconstruction of urban areas and rural infrastructure. By the time of full independence and UN membership in 2002, around 50,000 of those who fled had returned. Much remains to be done and East Timor has the lowest per capita income of any world country. Its economic future will be helped by the development of oil resources in the Timor Gap between it and Australia. In 2005, after some years of bickering over previously undefined boundaries between the two countries, an agreement was reached and joint development of the oil resources planned.

Subregion: Australia and New Zealand

Australia and New Zealand (Maori name Aotearoa) are dominions within the British Commonwealth (Figure 5.16). They have very different histories and populations than the countries of Southeast Asia, but are increasingly engaging with their neighbors as global pressures provide a common future. These trends affect people living in the two countries and the identity of their people (see "Australian Identity," p. 161).

Common History

Before they became British colonies from the late 1700s, both had low densities of indigenous peoples, dominated by Aborigines (Australia) and Maoris (New Zealand). The numbers of both groups were reduced after British colonization, but have now exceeded the numbers when the Europeans arrived. However, the Aborigines and Maoris continue to play small parts in the two countries. In Australia the Aborigines mainly form an underclass, with low income, poor education, and above-average numbers in prison. In 2006 the outgoing Governor-General of New Zealand spoke of her country's "dark secret," raising the plight of a similar underclass, largely peopled by Maoris. Despite many Maoris achieving education levels and jobs on a par with the whites, it is now clear that the majority lag behind.

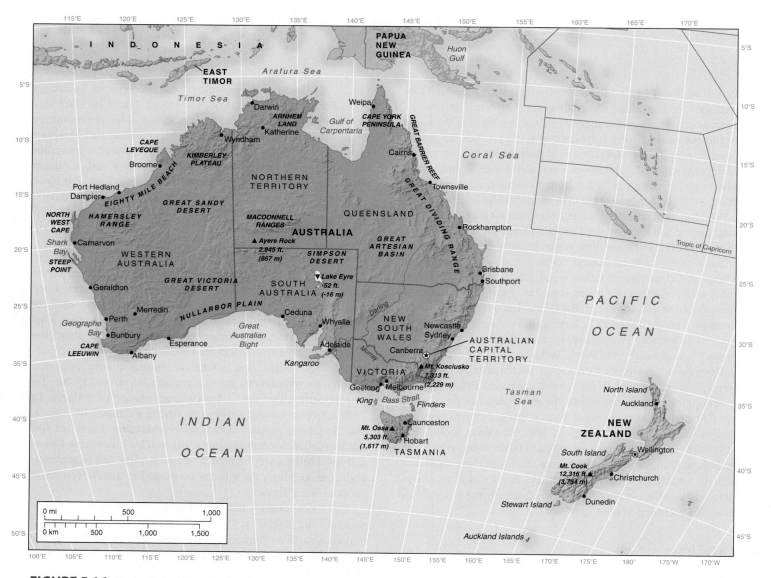

FIGURE 5.16 Australia and New Zealand: major geographic features. The political divisions, physical features, and major cities.

With a global position almost precisely opposite the home country, and initial difficulties in coming to terms with their Asian neighbors, the two countries became inward-looking and isolationist in the early 1900s, when their populations were largely white Britishers. Both countries depended on economic ties with Europe. When the ties were disrupted by World War II, their political and economic focus shifted to the United States. The ANZUS treaty between Australia, New Zealand, and the United States was signed in 1951. In the late 1900s, immigrants from southern Europe and later from Asia brought multiculturalism.

From 1973, the United Kingdom membership of the EU and the growth of Asian economies shifted foreign associations to the wider Pacific Rim. Today, both Australia and New Zealand have advanced economies, albeit based on primary products and services rather than export manufacturing. Their populations are predominantly urban: in Australia, over half the total population lives in the four major cities of Sydney, Melbourne, Brisbane, and Adelaide (see Table 5.2). Both countries are regarded by other Western countries as desirable places to live and are increasingly oriented to the global tourist industry (Figure 5.17). In 2005 cheaper flights brought 7 million visitors from Asia, North America, and Europe.

Both countries have cultures that make much of the outdoors and sporting activities: despite small populations, they have leading world positions in cricket, rugby, athletics, and swimming. Both have major movie industries, including a growing number of Australian movies and movie stars, and New Zealand locations of major productions (Figure 5.18).

And yet, the physical distances from similar materially wealthy countries in Europe and North America, together with their relatively small populations, result in social stresses. Greater ease of travel and global opportunities for educated and professional people generates wider prospects across the world. After decades of immigration dominating their population growth, both countries today experience emigration. Nearly 1 million Australians live abroad.

FIGURE 5.17 **Australia: tourism.** The Gold coast of Queensland attracts many visitors from Asia who wish to visit the Great Barrier Reef, which lies offshore.

FIGURE 5.18 **New Zealand:** *Lord of the Rings* **Premier Party, Wellington.** The producers and stars of the *Lord of the Rings* movies posed for photos at the film's world premier. The movie trilogy contributes to growth in New Zealand's motion picture and tourism industries.

Physical Contrasts

Australia is the only country that occupies a whole continent. New Zealand is composed of two major islands and is much smaller. The Australian continent is in the middle of a tectonic plate without significant earthquake or volcanic activity. Its climates are mainly tropical rainy in the north to arid over most of the country. Temperate rainy zones are limited to the southern coasts. New Zealand is crossed by a plate transform margin with volcanoes, earthquakes, and new mountain ranges. Its climate is temperate rainy.

Australia has a large empty arid zone in the center and west, and prospects for supporting a much larger population are small. In the early 1900s speculations envisaged 300 million people living there, although in the 1920s Griffith Taylor, a geographer, calculated that it would be difficult to support many more than 20 million—around the modern total. New Zealand's population never reached much more than 4 million.

Australia has one of the world's major stores of mineral resources in its ancient rocks. Similar to other areas of shield rocks in Africa, northern Canada, and Siberia, it has large deposits of iron ore and other metallic ores such as gold, platinum, uranium, and copper. The more recent rocks of the Great Dividing Range on the eastern side contain coal, silver, lead, zinc, and copper ores. Around its western and northern margins, Australia has huge deposits of oil and natural gas. New Zealand does not have such abundant metallic and energy resources, although the North Island has thermal volcanic zones.

Different Outlooks

New Zealanders disassociate themselves from Australian attitudes and policies. One expression of this is the New Zealand non-nuclear policy that resulted in the country refusing to allow

U.S. nuclear-powered or nuclear-armed ships access to its ports. The United States suspended its ANZUS treaty obligations. New Zealand also maintains close relations with many Pacific islands. Christchurch port is sometimes known as the "Gateway to Antarctica" as a result of several countries supporting their Antarctic bases from there.

Different Economies

Up to the mid-1900s, both countries relied on exports of farm products to Europe, especially the United Kingdom. New Zealand specialized in wool, lamb meat, dairy and fruit products and Australia in wool, beef and lamb, grain, and fruit. Australia also mined and refined metallic ores. Both countries developed import substitution industries, heavily protected by tariffs.

When the United Kingdom joined the European Union in 1973, further changes were forced on Australia and New Zealand at a time when Japan, the Koreas, the Southeast Asian countries, and later China were experiencing economic growth. In response to their increasing demands for industrial raw materials, energy sources, and changing food preferences, Australia in particular became Asia's "Farm and Quarry" while New Zealand's agriculture also marketed increasingly in the Pacific Rim area. However, neither country was able to develop major manufacturing industries that could compete in world markets and had to rely on the primary products as major sources of export income—unusual for materially wealthy countries. Service industries comprise 70 percent of GDP, including education, technology research, and tourism. While Australia maintains per capita incomes equivalent to many European countries, New Zealand has lower totals, reflecting the lower demand for its products and its smaller development of high-level services.

Subregion: Oceania
Many Islands, Few People

The southern parts of the Pacific Ocean contain thousands of small islands (Figure 5.19) covering an area where the ocean dominates: this is often known as Oceania. These islands face a potentially difficult future as a result of rising sea levels because of global warming.

Most islands cluster toward Southeast Asia, Australia, and New Zealand. The southwestern group of Papua New Guinea, the Solomon Islands, Vanuatu, Fiji, and New Caledonia, formerly known as Melanesia, includes the largest and most populous countries that are closest to the markets of Southeast Asia and Australia. A northwestern group (Micronesia) includes many small islands in the Marianas, the Marshalls, Micronesia, Guam, and Nauru.

To the east greater distances among islands bring greater isolation. Over the thousands of kilometers stretching to Latin

FIGURE 5.19 Oceania: the island countries. The country groupings, spread across a huge ocean area.

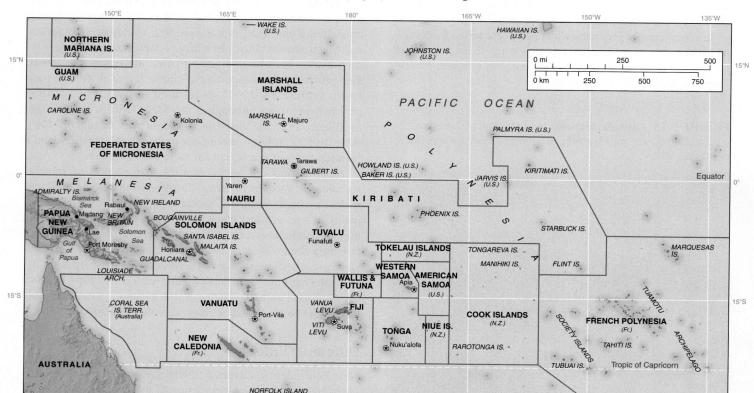

America, the Polynesian island countries include Fiji, Samoa, Tonga, and Tahiti, reaching to the most isolated Pitcairns and Easter Island.

These islands are grouped in small, mostly independent countries. Most gained independence in the 1970s, with Samoa in 1962 and the Marshall Islands in 1991. New Caledonia and French Polynesia are part of France for government purposes. Other islands have trust status associated with New Zealand, Australia, or the United States. Most countries remain financially dependent on outside aid. Few are invited to ASEAN or APEC meetings. The South Pacific Forum was formed in 1971, with Australia and New Zealand as major members, to develop closer economic cooperation, to promote independence for remaining colonies, and to try and stop nuclear testing in French Polynesia.

The island countries vary greatly in economic well-being. Some of the more prosperous are those with U.S. support or heavy French subsidies. Increasing air communications and the visits of growing numbers of tourists and cruise ships bring business to some islands, but the idyllic vision of the islands (Figure 5.20) does not always match reality and their people face economic difficulties.

In many of these countries the local products of copra—the white meat lining the insides of coconut shells that is a source of oil—sugar, bananas, and fish are insufficient to pay for imports of manufactured goods and modern social services. Stricter logging laws in Malaysia and Indonesia turned companies to the larger islands that had fewer restrictions. This led to the South Pacific Forum complaining that these islands were being "ripped off"

as local landowners received US$2.70 for timber that fetched US$350 on world markets and there were few attempts to re-plant. Fishing fleets from Japan, the United States, South Korea, and Taiwan take fish that make up around 40 percent of the total world catch, but not all pay access fees to the island countries.

The main natural resources are minerals on Papua New Guinea (copper, gold), New Caledonia (nickel), and Nauru (phosphate). The experience of Nauru illustrates some of the problems the islands face. From the mid-1900s surface mining of the phosphate rock brought initial prosperity, but left a jagged, barren surface over much of the small island of 21 km^2 (8 mi.^{2}). Opting for independence in 1968, the government bought out the mining company and sold phosphate to Japan and South Korea. The boom time income funded new schools, free medical facilities, and five new jets for Air Nauru. Few Naurians needed to work. Australian managers and Chinese shopkeepers moved in and money was invested in construction around the Pacific Rim. In the 1990s, however, world phosphate prices dropped, the mines were worked out, and property lost value. New ventures in the early 2000s included offshore banking, suing former colonial governments for mining-related degradation, and leasing land to Australia for a detention camp to house asylum seekers. Such income is temporary, and the once-affluent Nauru population faces poverty in a spectacular bust.

Papua New Guinea

Papua New Guinea (PNG) is the largest country in the Oceania subregion. It occupies the eastern half of the island of New Guinea and several offshore islands. "Papua" comes from the Malay word for frizzy Melanesian hair; "New Guinea" from a Spanish explorer who noted similarities to the people of Western Africa. Its population is extremely diverse, with over 700 local languages (10 percent of the world's languages) and around 1,000 culture groups among a population of 6 million: "For each village a separate culture." Most people live in small traditional societies that are acknowledged as desirable in the country's constitution and depend on subsistence agriculture. The "customary land title" given to 97 percent of the total land area provides a basis of inalienable tenure to these groups. Other lands belong to the government and can be leased privately. In 2005 only 13 percent of the population lived in towns, the largest of which was Port Moresby, the capital. Urbanization accelerated in the 1990s, leading to squatter settlements and social problems.

Papua New Guinea is a mountainous country, still largely forested. The highest point, Mount Wilhelm, rises to 4,509 m (14,793 ft.) and has glaciers on the equator. Such terrain makes transportation infrastructure difficult to develop, and airplanes are the only means of access to some places. Complex fault lines at the meeting of the Pacific, Philippines, and Indian plates give rise to frequent earthquakes and tsunami. Ecologically, PNG is related to Australasia, having been connected to Australia across what is now the Torres Strait before the postglacial rise in ocean level. Kangaroos, possums, and many birds demonstrate genetic links. Some of the PNG islands, however, were never connected to Australian land and lack many of the fauna.

FIGURE 5.20 French Polynesia: the idyllic view. However, this is where the French test nuclear armaments.

The earliest inhabitants came from Southeast Asia and developed agriculture around 9,000 years ago—one of the few world areas of original plant domestication. Later groups settled in coastal areas, introducing pottery making, fishing, and pigs. All of the major ethnic groups of Oceania live in PNG, together with immigrant Chinese, Europeans, and Australians.

European interest came late, toward the end of the 1800s, when Germans occupied the northern half of New Guinea and the United Kingdom the Papuan south. Australia took over both, the former as a League of Nations mandate after World War I, and the latter as a British possession and External Territory of the Australian Commonwealth. After World War II the two territories were combined and gained independence in 1975. There are three official languages, including little-spoken English, although it is becoming the main school language. Most people in the northern half, politicians, and local newspaper reporters use the creole New Guinea Pidgin (Tok Pisin) language. The constitutional rights of free speech, thought, and belief are upheld in the huge variety of Christian groups (96 percent of the total population in the 2000 census), although many combine that faith with traditional practices. Sea shells were abolished as currency in 1933, but still figure in some cultures' bride prices. However, cultural practices vary greatly. Some of the mountain groups have colorful gatherings (Figure 5.21). Kinship structures govern social relations and many groups place value on acquiring skills of hunting, farming, and fishing.

FIGURE 5.21 Papua New Guinea. Traditional dress—with sneakers.

Three-fourths of PNG inhabitants live by subsistence with occasional hunting, while there are also small farms with some commercial farm sales. PNG has rich natural resources, although the rugged terrain hampers exploitation. Minerals including oil, copper, and gold, make up three-fourths of export earnings. Coffee, timber, and shrimp also contribute. In 1990 the Australian-owned mining operation on Bougainville, which produced one-third of PNG's income, closed after years of local protest over environmental impacts and attempts to make the island independent. The PNG military prevented Bougainville independence, but the mining company could not operate under such conditions.

Antarctica: A Region?

Antarctica (Figure 5.22) is one of the world's seven continents and occupies 10 percent of Earth's land surface, but has no permanent settlement and is not divided into countries. International agreement prevents exploitation of mineral resources, and limits access to this frozen and largely ice-covered continent to scientists and a few tourists. Isolated from Latin America and Australia, Antarctica poses questions. Is it a major world region on its own? How important is it in a world regions course?

A Special Environment

For the rest of the world Antarctica's significance is its environmental context. Plate tectonic movements separated it from Latin America, Australia, and the Indian subcontinent. Left at the South Pole, the cooling of Earth's atmosphere some 20 million years ago led to thick ice accumulating, burying mountain ranges, and causing a large part of this continent to sink under the weight of ice. The frozen continent has its own climate with almost total darkness for several months in winter. In winter, the ice-covered area of the surrounding oceans doubles as the sea surface freezes, while in summer glaciers calve icebergs into the ocean and the sea ice breaks up.

The convergence of warm and cold water and air around 50° to 60°S generates frontal cyclone storms and high winds (known popularly as the "Roaring Forties") in a zone with few land areas. This ocean-atmosphere convergence effectively cuts Antarctica off from warming influences. However, small quantities of human-produced chlorine gases penetrated to produce an annual ozone hole above the continent at the end of the Antarctic winter in October (see Chapter 1, p. 13). The hole reduces the capability of the ozone layer in Earth's stratosphere to filter out solar energy elements that endanger surface living organisms; it is estimated that one-fourth of the oceanic plankton around Antarctica has died.

Antarctica is the world's largest accumulation of ice, which, if melted by global warming, would cause world ocean levels to rise with widespread disastrous consequences. During the late 1990s and early 2000s the breakup of its surrounding ice shelves was seen as a sign that this could happen, and in 2005 a report made it clear that temperatures across the continent were rising.

Antarctica is the center of a major ecosystem involving plankton, fish, seabirds including penguins, and marine mammals including seals and whales. In the late 1900s commercial

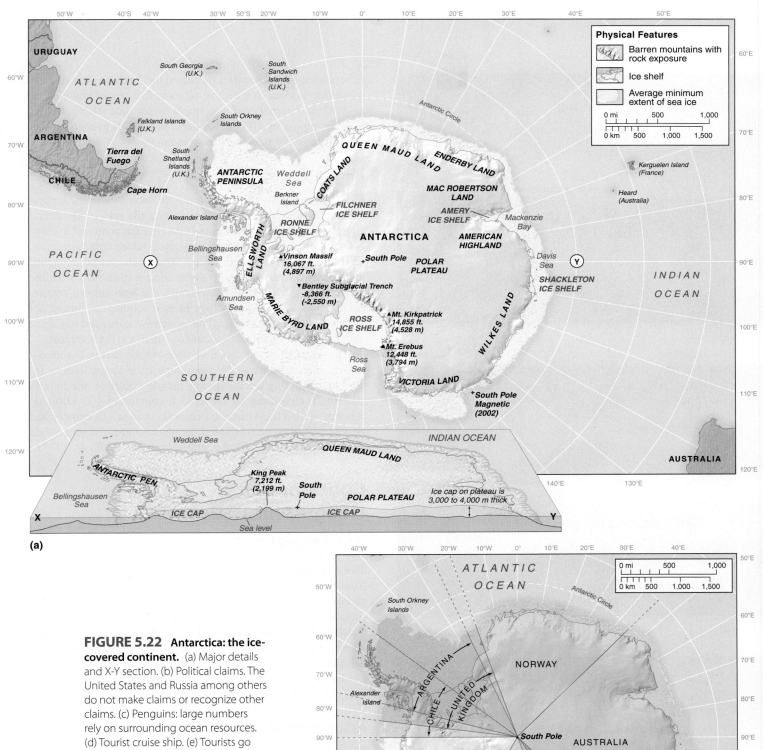

(a)

FIGURE 5.22 Antarctica: the ice-covered continent. (a) Major details and X-Y section. (b) Political claims. The United States and Russia among others do not make claims or recognize other claims. (c) Penguins: large numbers rely on surrounding ocean resources. (d) Tourist cruise ship. (e) Tourists go ashore at Danish scientific station.

(continued)

(b)

(c)

(d)

(e)

fishers from many countries increased their exploitation of these resources, leading to a decline of some fish stocks and whales. In 1982 an international agreement to regulate such activities was not signed by all countries and is difficult to monitor, but some species are now increasing in numbers.

Antarctic Treaty

From the late 1800s the first attraction of Antarctica was as a physical challenge to cross it on foot. During the mid-1900s several countries claimed jurisdiction over parts of the continent to carry out scientific studies that were necessary to understand the climate, ocean ecosystems, and glaciology. Thirty-nine countries signed the Antarctic Treaty in 1961, emphasizing non-military scientific cooperation, environmental safeguards, and international control. Additional clauses added in 1991 banned commercial mining activities. Countries not signing the agreement either did not agree with the restrictions or did not have an interest in the continent.

The treaty does not contain a code for the increasing tourist traffic, although informally tour ships keep to guidelines, including restricting the number of people ashore at one time to 100. In 2005, around 40 ships, each containing 100–600 passengers, visited the Antarctic Peninsula between December and February. Zodiac craft enabled people to land for short periods at specific sites, often near scientific research stations.

Contemporary Geographic Issues

ASEAN and APEC

Two organizations originating in this region reflect the shifts from inward-looking countries toward greater regional and global involvements (Figure 5.23). Formed in 1967, the Association of Southeast Asian Nations (ASEAN) claims to be a political, economic, and cultural organization of Southeast Asian countries, although it is widening its associations with neighboring countries. The Asia-Pacific Economic Cooperation (APEC) forum was formed on Australian initiative in 1989 and looks to wider international involvements.

Association of Southeast Asian Nations

The 1967 founding purpose of ASEAN and its members—Thailand, Indonesia, Malaysia, Singapore, and the Philippines—was to provide non-provocative solidarity against Communist expansion in Vietnam and internal insurgency. The Bali Summit in 1976 proposed economic cooperation, which did not gain much support until 1991, when Thailand proposed a regional free trade area. Brunei joined ASEAN in 1984 and Vietnam, Laos, Myanmar, and Cambodia in the late 1990s. East Timor may be the next member. In 2004, ASEAN countries had over 550 million people and total GDP PPP of US$2,172 billion, compared to the EU (460 million people and US$11,724 billion) and NAFTA (430 million people and US$12,890 billion).

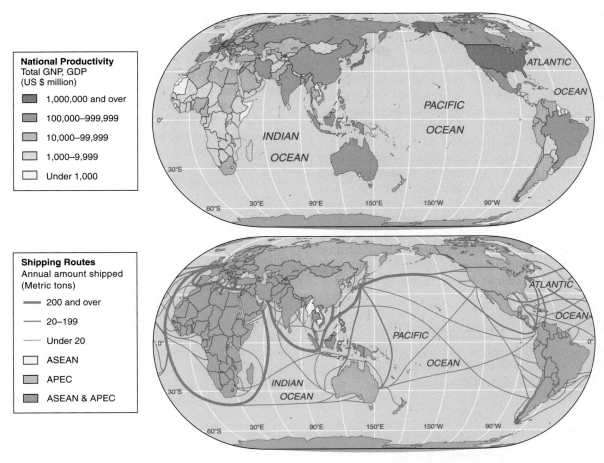

FIGURE 5.23 **ASEAN and APEC: the countries of this world region in a global context.** Compare the incomes of member countries with each other and those outside the membership, and comment on the location of major shipping routes.

ASEAN members are a politically, culturally, and economically diverse group of countries, including governments ranging from democracies to autocracies. ASEAN contains more Muslims (240 million, mostly in Indonesia, Malaysia, and Brunei) than any other group of countries. Buddhism is the religion of most people in Thailand, Myanmar, Laos, Cambodia, Vietnam, and Singapore. Roman Catholic Christianity predominates in the Philippines and East Timor.

The annual meetings of ASEAN leaders consider the economic and cultural development of member countries. From a pattern of meetings every three years from 1992, annual occasions took over from 2001, with a rotating list of summit hosts. Regular features of these three-day events are meetings with the three East Asian Dialogue Partners (ASEAN+3: China, Japan, South Korea) and two other regional Dialogue Partners (ASEAN-CER: Australia and New Zealand). In 2005 an East Asian Summit brought together ASEAN and six Dialogue Partners (as above plus India) and established an ASEAN-Russia Summit.

ASEAN has regular meetings with other countries through the ASEAN Regional Forum, an informal group of 25 countries concerned about Asia-Pacific security issues that first met in 1994. Other countries from this region—Australia, New Zealand, Papua New Guinea, and East Timor—join with India, Pakistan, China, Mongolia, Japan, North and South Korea, Russia, the United States, Canada, and the European Union.

Asia-Pacific Economic Cooperation

APEC is composed of a group of countries around the Pacific Rim, which meet to improve political and economic ties. It holds annual meetings by rotation in the member countries. Attending leaders dress in the national costume of the host country.

APEC arose partly from Australia's concerns that it would be isolated from relations with other countries after it had lost many of its European markets and received an initial lack of response by ASEAN. The first meeting in Canberra, Australia, attracted 12 countries. In 1993, U.S. President Clinton invited economic leaders to discuss promoting prosperity through cooperation in order to take forward global trade talks. APEC established headquarters in Singapore.

Early APEC goals focused on free trade and investment, including the reduction of tariffs below 5 percent. In the early 2000s security and terrorism became concerns after confrontations with demonstrators outside the summit venues and the 9/11 events.

APEC membership increased from the original 12 (Australia, Brunei, Canada, Indonesia, Japan, South Korea, Malaysia,

New Zealand, Philippines, Singapore, Thailand, United States) to include China, Hong Kong, and Chinese Taipei (Taiwan) in 1991, Mexico and Papua New Guinea in 1993, Chile in 1994, and Peru, Russia, and Vietnam in 1998.

In April 2006, at the Hanoi, Vietnam APEC Summit, Ambassador Tran Tong Tuan of Thailand, the Executive Director of the APEC Secretariat, laid out some personal thoughts about the changing nature and role of APEC. He considered that the rise of China in a cooperative and peaceful context necessitated a balance being achieved among the United States, China, Japan, and possibly India as a basis for prosperity in the Asia-Pacific. Human security is threatened by terrorism, epidemics, and natural disasters that are by their nature trans-boundary events that no single country can control. APEC has a major role in the Asia-Pacific region because of its wider-than-security basis and ability to pioneer needed developments. He concluded by claiming that APEC should consolidate its role as the primary means of international cooperation in the region and beyond.

Conflicts: Ocean Space and Piracy

Ocean space is very important in this region, illustrated by two different sources of conflict: territorial claims in the South China Sea and piracy in Southeast Asia (Figure 5.24).

South China Sea

The South China Sea is the world's largest sea body after the five oceans and is the world's second most used sea lane, being a major route for tankers delivering oil to China and Japan. It is almost surrounded by the countries of China, Vietnam, Cambodia, Thailand, Malaysia, Singapore, Indonesia, the Philippines, and Taiwan. There are 200 small islands, mostly uninhabited, and seamounts that do not break the surface. The main group is the Spratly Islands, the largest of which, Taiping Island, is only 1.7 km (less than 1 mile) long and 3.8 m (13 ft.) above sea level at its highest point. The whole area is subject to the surrounding countries' claims, which increased in intensity following discovery of oil and natural gas potential. The proven reserves of oil total 7.7 billion barrels and estimates predict 28 billion.

Names for the sea mirror the claims on it. The European name is followed by China, but many Vietnamese call it the Eastern Sea and Filipinos the Luzon Sea. The 1982 United Nations Law of the Sea allows each country to have an Exclusive Economic Zone of 200 nautical miles (371 km) beyond territorial waters. So all the surrounding countries claim large portions. The People's Republic of China claims almost all of the sea and is building an aircraft carrier battle group to secure oil and natural gas lines. That country seized the Paracel Islands from Vietnam in 1974, and a clash between these countries in the Spratly Islands in 1988 also led to deaths.

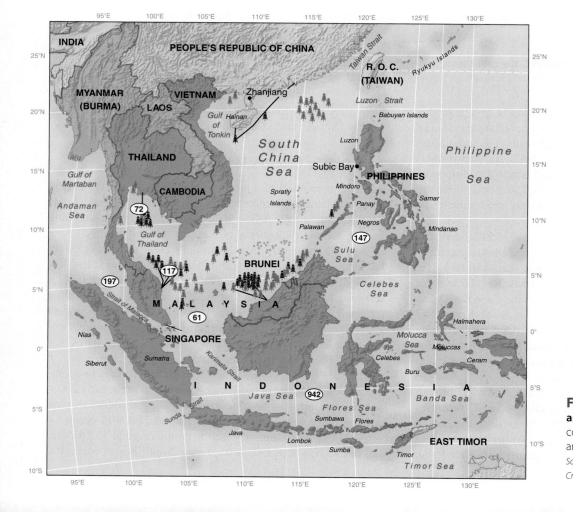

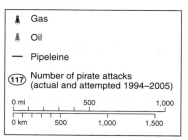

FIGURE 5.24 **South China Sea and Indonesia.** Notice the potential conflicts over the South China Sea area and the main occurrences of piracy. *Source: Pirate attack data from I.C.C. Commercial Crime Services, 2005.*

The ASEAN group, led by Malaysia does not want these territorial disputes to escalate. It established Joint Development Authorities where claims overlap, in which joint projects develop areas and divide profits without settling the territorial claims. This is working in the Gulf of Thailand.

Piracy at Sea

The waters around Southeast Asian lands are the most prone to piracy in the world. According to the International Maritime Bureau's Annual Piracy Report for 2005, almost 1,500 pirate attacks were recorded in the Indonesian area during the 12 years 1994 to 2005. The Malacca Straits between Malaysia and Sumatra recorded nearly 200. Malaysia, the Philippines, Singapore Straits, and Thailand recorded another 400 events over the same period. In each of these areas, numbers of piracies rose through the 1990s, leading to increased awareness, antipiracy watches, and law enforcement patrols. By 2005, the number of events decreased from the highest occurrences in 2003 in Indonesia and the Malacca Straits. Piracy in the Philippines peaked in the mid-1990s.

Tugs, general cargo ships, and tankers are all vulnerable to attacks by heavily armed groups of pirates in small, fast, or fishing boats. They do not often kill the crew members, but tie them up, take equipment and money, and even ransom the crews. Most attacks are short-lived, but some involve hijacking of the cargo ship, setting its crew adrift in a small boat. Cargoes hijacked in 2005 included tin, copper, and vegetable oil.

Australian Identity

The changes affecting Australians today highlight the issue of national identity in a globalizing world. In the later 1900s, Australian identity changed from isolation in this world region to incorporation.

In 1788 Britain established Port Jackson (modern Sydney), or Botany Bay, as a penal colony. It became part of New South Wales, in which free settlers soon outnumbered the convicts. The country of Australia increasingly represented an outpost of the British Empire in an alien Asian world. The 1800s saw colonies (later states) established around the continent's more hospitable coasts and the development of colonial commodity exports: Australia began its colonial role as farm and quarry with sheep and gold rushes. The indigenous Aborigines declined in numbers by disease and removal from productive lands as the settlers increased, their fate resembling that of native Americans.

The 1900s began with a federation of the colonies in the Commonwealth of Australia, largely independent within the British Empire. In the first part of that century Australians took on a distinctive identity as their politics, economy, and social life developed. It has been characterized as homogeneous and Anglophile, although that overstates the case. One feature, the informal White Australia policy, excluded nonwhite people from citizenship until it was abandoned in 1972. State and federal governments acted paternally, backed up by imperial support and guarantees for selling products back "home." Disputes between employers

and employees were kept quiet by wage arbitration. Industries, such as car manufacturing, were protected by import substitution with high tariffs. Many see Australia in this phase as a closed, sheltered, and insular country. However, economically it always had external ties to Europe and politically it contributed forces to the Boer War, World Wars I and II, and conflicts in Korea and Vietnam. Internally there were economic depressions in the 1890s and 1930s, and struggles over conscription and the rights of workers, women, and indigenous groups.

After World War II Australian identity changed, stimulated by the United Kingdom joining the EU in 1973, the rise of Asian and Southeast Asian economies, and intensifying globalization. The mainly Britisher population was bolstered by special programs attracting immigrants from southern Europe, and then from Asia. As multiculturalism extended further, the Aborigines experienced a resurgence of interest and growth in numbers. Industrial protection and centralized industrial relations gave way to market-based processes, leading to the closure of many firms and the foreign investment in others, and a different role for government. Australia became a service-oriented and urban country.

From the 1980s, Australian leaders emphasized the importance of the country's global role, but also supported a growing national awareness. The sporting achievements and the 2000 Olympics in Sydney highlighted this outlook. Australia played a seminal part in setting up APEC, is in close dialogue with ASEAN, and is a member of the East Asian Economic Forum. It looks for economic and military partners to encourage security and fair competition in its region. At the same time, the high-profile barring of asylum seekers from Vietnam and Indonesia landing in the country, and taking them to Nauru for "processing," emphasized continuing government control over immigration. Australians see national identity as important amid intensifying globalization. Perhaps it is more important than before. For many Australians it provides a buffer against an unruly world.

Singapore: Global Magnet

Singapore is envied by many other countries. It is the smallest country in Asia—an urbanized small island (Figure 5.25) without natural resources—but is by far the most materially wealthy in GNI PPP per capita in Southeast Asia (see Table 5.1). It is a global city-region equal to any but the world's three largest (see Figure 1.23), a multicultural and multinational center of international business services rivaling Hong Kong, Frankfurt, Sydney, and Chicago. However, not every Singaporean gains the benefits of its major world role (Table 5.3). Even in such a wealthy country, there are many poor—materially and socially.

Origins

Singapore ("Lion City") had a long history as a minor trading and fishing port before Sir Thomas Stanford Raffles of the British East India Company made a 1819 treaty with its owner, the Sultan of Johore (Malaya), to develop it as a trading point and settlement. Singapore's growth was rapid and it became a United Kingdom

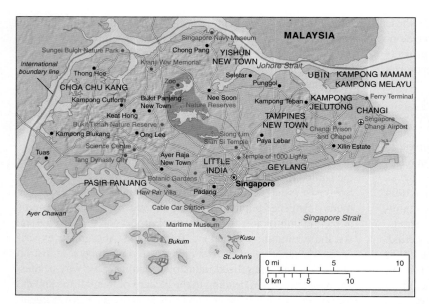

FIGURE 5.25 Singapore. The initial port and city center is where the name is printed. Changi Airport is one of the world's busiest. Most of the island is now built up with housing and commercial development, leaving a small area for nature reserves and gardens, but also keeping a number of historic buildings and the Science Center. What do the features shown on this map say about the tiny country of Singapore?

crown colony in 1867. Its strategic location on developing shipping routes between Europe and China made it an **entrepôt port**—where goods from surrounding areas were transhipped to larger ships and goods from farther away were distributed in smaller boats. During this period, Chinese and Indian people migrated to Singapore, with the former group now 75 percent of the total population.

Struggle for Growth

After World War II occupation by the Japanese from 1942 to 1945, Singapore became self-governing in 1959. In 1962 it entered the fully independent Federation of Malaysia with Malaya, Sabah, and

Sarawak. However, irreconcilable differences over ethnic policy led to the federation expelling Singapore in 1965, when it gained official sovereignty. At first, the newly independent Singapore struggled to deal with unemployment, housing shortages, and a lack of natural resources. The government took strong actions to reduce unemployment, improve living standards, build large-scale public housing, increase national defenses, and remove racial tensions. Most former rural land is now covered with housing projects and regional commercial centers to reduce overcrowding in the old central area on the south coast. Entry to the central area is controlled by electronic road pricing.

The Singapore government redirected its economy from port-centered activities to high tech industry. By the 1990s, Singapore firms manufactured 40 percent of world computer hard drives. As a transportation and communications hub, Singapore became the regional headquarters of multinational corporations. The government remains involved in the business and social life of the country. It is one of the cleanest, safest, and most orderly urban areas in the world. It spends 20 percent of its budget on education (compared to 4 percent in the United Sates). However, its actions attract criticism (Table 5.3), including the harsh punishment of seemingly minor crimes. It restricts length of stay for low-skilled and domestic workers, while encouraging well-educated businesspeople to settle. For some years, Singapore even encouraged intelligent and financially successful people to have more babies than the "less gifted."

Global and Multiethnic Country

Singapore has a multicultural population composed of people from many groups that speak different languages and have allegiance to different religions (Figure 5.26). The normally calm country occasionally boils over. In 2002, when U.S. actions in Afghanistan gained Singapore government support, some Mus-

TABLE 5.3	DEBATE: SINGAPORE
Advantages	Drawbacks
Tight government control provides structure and order for the citizens.	Citizens are not free to behave as individuals.
Government control produces extremely low crime rates: Singapore is one of the world's safest urban environments.	Singapore's punishments are extreme and inhumane: they are too harsh and do not fit the crime.
Singapore is one of the world's cleanest urban environments.	Singapore is sterile and contrived: more like a Disney theme park than a real urban environment or country.
There are no slums.	Materially impoverished people in Singapore lead segregated lives.
There is very low unemployment.	Street sweepers and "public toilet police" live restricted lives.
Citizens of Singapore enjoy high living standards, great socioeconomic indicators, and a high quality of life.	High quality of life comes with restrictions on civil liberties.

Southeast Asia's financial and high tech center. Singapore's main thrust is to promote investments as an international business center, and finance and manage developments in China and Vietnam. In the 2000s, in addition to its ASEAN membership, Singapore agreed to free trade arrangements with Australia, New Zealand, Japan, the United States, India, South Korea, Europe, and Chile. Its main trading partners are Malaysia, the United States, China, Japan, Taiwan, and Thailand.

GEOGRAPHY AT WORK

The Tsunami of December 2004

Geographic knowledge and geographic tools were utilized immediately to assist in recovery efforts following the devastating Indian Ocean **tsunami** on December 26, 2004 (Figure 5.27). Centered just off western Sumatra, the submarine movements of ocean floor set off by earthquakes generated huge waves that hit local shores and within a few hours caused damage and loss of life around the Indian Ocean, including western Thailand, Sri Lanka, and even eastern Africa.

FIGURE 5.27 **Indian Ocean tsunami: Koh Raya, Thailand.** People flee one of a devastating series of tsunami waves crashing on the shore near Phuket on Thailand's west coast.

The analysis and computer manipulation of satellite images and aerial photos from planes and helicopters using geospatial tools such as geographic information systems (GIS) began within hours of the tsunami's huge waves hitting the coasts around the ocean. Before-and-after imagery made it possible to assess losses in the human landscape (houses, stores, hotels, roads, bridges, and port facilities). They also highlighted changes to the physical environment, such as significant alterations to coastal landforms, losses of mangrove vegetation, and creation of new inlets and islands as others were eliminated.

An understanding of the physical and human geography of Southeast Asia and other affected areas was essential in coordinating the efforts of aid agencies, regional governments, and the international community. Recovery efforts and subsequent continuing support required an in-depth geographic appreciation of such aspects as social relations and burial practices, political relationships including local conflicts between government and insurgents, and transportation and communications networks available in the different areas affected.

(a)

(b)

FIGURE 5.26 **Singapore.** (a) Sri Mariamman Hindu Temple: figures of gods on external wall. (b) Waterfront at night.

lim Malays were arrested on suspicion of planning a bombing campaign. They complained about being an ignored minority, although Singapore provides Muslim schools.

Difficulties still exist in relationships with Malaysia. For example, Malaysia opened a new port, PTP, opposite Singapore and took some major port customers. When Singapore proposed a new bridge-and-causeway in 2006, Malaysia rejected the project. Singapore still depends on Malaysia to supply freshwater: catching the tropical downpours on Singapore Island supplies half of its needs but self-sufficiency is promised by extending reservoirs and building desalination plants.

In the 2000s, after successful recovery from the late-1990s economic crisis in the region, Singapore positions itself as

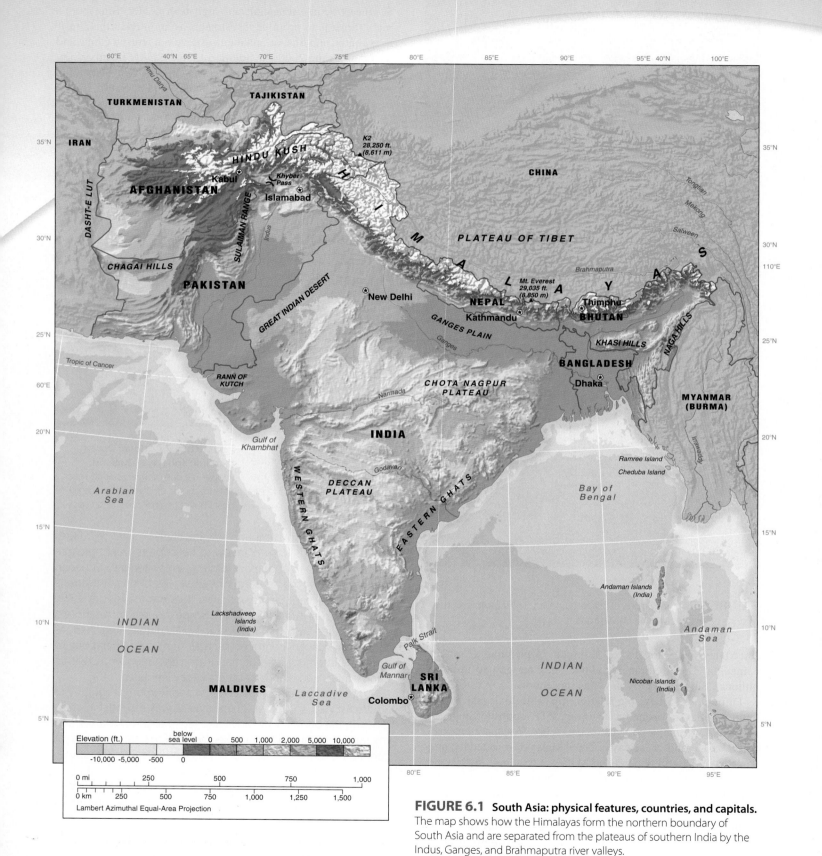

FIGURE 6.1 **South Asia: physical features, countries, and capitals.**
The map shows how the Himalayas form the northern boundary of
South Asia and are separated from the plateaus of southern India by the
Indus, Ganges, and Brahmaputra river valleys.

Priya is a 21-year-old college graduate who works at a 24/7 call center on the outskirts of the city of Bengaluru (Bangalore), India as a customer service representative. She arrives at work early in the evening in a company bus with fellow employees, ready to work the graveyard shift that begins at 6:30 p.m. and ends shortly before dawn. Her training included lessons to neutralize her Indian accent so that she would be understood by American customers. Priya is one of an estimated 350,000 people who are employed in back-office jobs in medical transcription, technical support, customer service, and animation in India, working for American and European companies. Although she had trouble adjusting to the odd work hours and the stress of dealing with irate customers, Priya is happy that she is earning her own living and has a well-paying job.

Meena, also 21 years of age, lives in a village 100 km from Bengaluru. A mother of two young children, she takes care of her household, tends the family's small vegetable garden, and helps her weaver husband. She has only been once to Bengaluru, which she found noisy, crowded, and overwhelming. She prefers village life, but worries that the future will not hold much for her children unless they move to the city. The contrasts between the lives of Priya and Meena are echoed in urban and rural areas in South Asia, a region that is largely rural but also has some of the world's largest cities.

Chapter Themes

Defining the Region

Distinctive Physical Geography: contrasting landscapes; monsoon climates, vegetation and soils; natural resources.

Distinctive Human Geography: cultural diversity; global linkages and the impact of colonialism; rural and urban contrasts.

Geographic Diversity
- India.
- Bangladesh and Pakistan.
- Mountain and island rim.

Contemporary Geographic Issues
- Population growth and patterns.
- Urbanization.
- Ethnic conflicts: intra-state and inter-state.
- Environmental problems in a developing region.

Geography at Work: battling infectious diseases

Defining the Region

South Asia, the smallest world region in terms of area, is the second most populous after East Asia. Most of the region consists of the Indian subcontinent enclosed by the Himalayan Mountains and reaching to the Indian Ocean (Figure 6.1). India, Pakistan, Bangladesh, Nepal, Bhutan, and Afghanistan are the countries located here. The region also includes the island countries of Sri Lanka and the Maldives.

South Asia bears the imprint of the rise and fall of civilizations and empires, the emergence and solidification of religions, and varied interactions between its people and the natural environment. Invading peoples, from the Aryans who entered the subcontinent in 1200 BC to the colonizing British who considered India the star in the crown of the British Empire, gave the region a shared history and perceived common culture (Figure 6.2) but uneven political, social, and economic geographies.

Dominating the region with an area of over 2.6 million km² (1 million mi²) and a population of over a billion, India is the world's seventh largest country in area and second largest in population. Pakistan, although second in area in this region, is only a third as large as India. This Islamic country, which was once a part of British India, now has closer ties to its Muslim neighbors in Southwest Asia. Bangladesh is also a Muslim country on India's eastern border and is almost completely surrounded by its larger neighbor. Comprising of mostly the floodplains and delta of the mighty Ganges and Brahmaputra river systems, it is one of the most densely populated countries in the world. Nepal and Bhutan are landlocked Himalayan countries. The former is open to outside influences, while Bhutan steadfastly limits its interac-

tions with the global system. Sri Lanka, whose relatively high level of social development sets it apart from other South Asian countries, continues to suffer from the effects of prolonged civil war. The Maldives consist of over a thousand islands and depend on fishing and tourism.

Distinctive Physical Geography

The natural environments of South Asia are impacted by the Himalayan ramparts, the great plains of its major rivers, coastal influences, and the long dry seasons and uncertain rainfall that are the hallmark of the dominating monsoon climate. The mountains, plains, and hilly areas influence the distributions of settlement and economic activities. Furthermore, South Asia's large population places stresses on its densely settled and fragile natural environments.

Contrasting Landscapes

World's Highest Mountains and Deep Valleys

The Himalayan Mountains, forming the northern boundary of the region (see Figure 6.1), include Mount Everest (8,850 m; 29,035 ft.), the world's highest mountain, and nearly 50 other peaks that rise over 7,500 m (25,000 ft.). This mountain wall is the southern margin of the high and broad plateau of Tibet (China). The combination forms a wide barrier to both climatic influences and historic incursions of invading peoples.

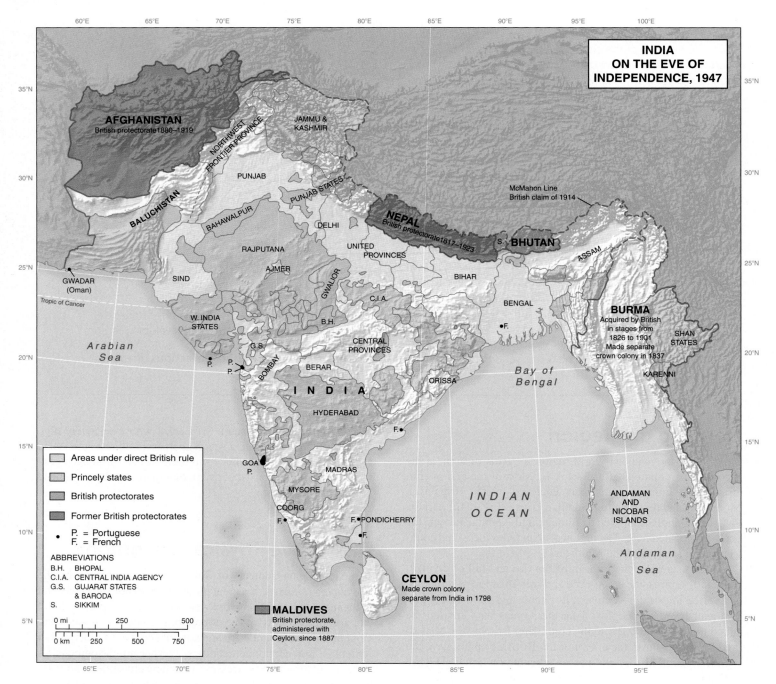

FIGURE 6.2 South Asia: the British Indian Empire just before independence in 1947. The princely states had working arrangements with Britain but were not ruled as part of the Indian Empire. At independence they agreed to be part of the new countries of India and Pakistan and subject to their governments. Kashmir is still disputed between India and Pakistan.

The Himalayas resulted from the continental plate fragment that is peninsular India crashing into the Eurasian continental plate (see Figure 1.5b). The resulting collision thickened the crustal rocks to produce the world's highest mountains. The continuing thrust of peninsular India beneath Asia, signaled by earthquakes, raises the mountains by 6 cm (2.4 in) per year. As they rise, however, glaciers and rivers cut into them and wear them

down. At present, an approximate balance exists between the rate of rise and the rate at which these mountains are being eroded.

The headwater streams of the Indus and Ganges river systems breach the Himalayan wall in the far northwest. Deep valleys, carved by glaciers and rivers, provided lower routes, such as the Khyber Pass through the Hindu Kush range, that give access to Afghanistan and Central Asia and formed historic points of entry for invading Aryan and Central Asian groups. At the northeastern end of the Himalayas, the ranges are lower as they turn southward into Myanmar, and include breaks eroded by the Brahmaputra River system that afford Indian access to and from China.

Peninsular Hills and Plateaus

Peninsular India is a tilted continental plate fragment of very ancient rocks. The highest points along the western coast—the Western Ghats—rise to just over 2,500 m (8,000 ft.). Much of the peninsula consists of plateaus and hills sloping eastward to the Bay of Bengal—a gradient followed by many rivers. Along the east coast, a broken line of hills, the Eastern Ghats, rises in places to 1,500 m (5,000 ft.). Coastal plains of varied widths border the peninsula. They are at their widest where rivers such as the Krishna and Godavari create deltas on the eastern coast.

As the continental plate fragment forming the peninsula of South Asia broke away from the other southern continents over 150 million years ago, layers of volcanic lava poured out through cracks in Earth's crust, covering large areas of older land. The lava flows form the Deccan Plateau in the northwestern peninsula.

Major River Basins

Between the Himalayas and the peninsular plateaus, three major river systems—the Indus, Ganges, and Brahmaputra—cross a wide lowland zone formed by their deposits. The melting Himalayan snows and the monsoon rains combine in powerful flows in these rivers. The strength of flow can be gauged from the fact that the Ganges and Brahmaputra sweep debris 3,000 km (nearly 2,000 mi.) out to sea in the Bay of Bengal.

Rock particles and fragments worn from the Himalayas and carried by these rivers built deposits of **alluvium** up to 3,000 m (10,000 ft.) thick beneath the plains. The surface of the deposits is generally flat but marked by small-scale relief of a few or tens of meters where the rivers cut new channels in older material. Along the northern margins of this plain, large **alluvial fans** of gravel mark the junction of the steep mountains with the lowlands. The sudden lowering of gradient as they enter the lowlands causes the rivers to drop the coarse gravel and sand they carry along their steeper-gradient mountain valleys. In the drier areas of the northwest are sections of **badlands topography**, where occasional rainfall runoff cuts dense networks of steep-sided gullies into unvegetated alluvial deposits. The badlands were traditionally home to *dacoits* (robbers) who preyed on travelers. At their mouths the Ganges and Brahmaputra join to form a huge delta of low-lying, flood-prone land that occupies most of Bangladesh. These plains are now densely populated and intensively farmed, using river water for irrigation. The Ganges is considered divine by Hindus for whom it has great religious significance. Hindus believe that bathing in the Ganges cleanses them of sins, while dying on its banks helps attain salvation (Figure 6.3).

FIGURE 6.3 South Asia: the sacred river Ganges. The river at Varanasi draws Hindu pilgrims to bathe in its waters. The river is regarded as a goddess. It is supplied from Himalayan glacier meltwater and is prone to flooding.

Monsoon Climates, Vegetation, and Soils

The monsoons dominate the climatic environment and life in much of South Asia (see foldout world climate map inside back cover). Monsoon winds bring heavy summer downpours of rain over much of the Indian subcontinent but little rain at other times of the year. Although the precise cause of the monsoons is still debated, these seasonal winds are affected by air pressure, the shifting position of high velocity jet stream winds in the upper atmosphere, and heating and cooling of the Indian subcontinent. (Figure 6.4). In the winter season of the dry monsoon, the Himalayan Mountains cut off South Asia from Central Asia and the freezing winds that blow from a high pressure area over the Tibetan Plateau. South Asia remains warm and very dry as winds flow outward from high atmospheric pressure over the northwest. Only northern Sri Lanka and southeastern India receive rain at this season—from winds that blow out from the continent, over the Bay of Bengal, become moist by evaporation from the ocean surface, and turn toward the land.

As the land warms in early summer, temperatures in South Asia become very high (over 30°C, 86°F) and air rises, lowering atmospheric pressure. When the wet monsoon breaks in June or July, winds are drawn into the low atmospheric pressure area over the continent. Southwesterly winds from the Indian Ocean bring moisture, causing heavy rainfall on the western coastal mountains of the peninsula. The lift forced by humid air flowing up and over mountains (the orographic effect) adds to the amount of condensation and precipitation. Some of the world's largest annual rainfall totals, over 10,000 mm (400 in.) per year, are recorded in the Assam hills of northeastern India.

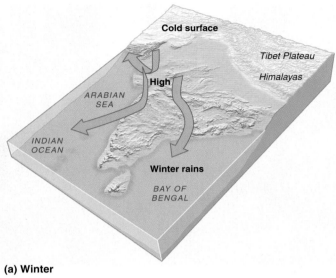

(a) Winter

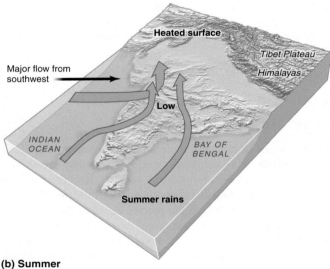

(b) Summer

FIGURE 6.4 South Asia: the monsoons of the Indian subcontinent. (a) In winter, high atmospheric pressure in the northwest of the region leads to dry winds that blow outward. (b) In summer, air rises over the heated continent, drawing in very humid air from the Indian Ocean—the source of water for the monsoon rains.

In the peninsula, most of the monsoon rains fall on the Western Ghat mountains, leaving the eastern lands in a rain shadow, which receive lower summer rainfalls that vary sharply from year to year. Precipitation falls as snow at high elevations on the Himalayas and the summer meltwaters add to river flows. Each year during the summer, a small number of tropical cyclones occur in the Bay of Bengal, bringing flooding and death to the Ganges-Brahmaputra Delta in Bangladesh. The concentrated rainfall and consequent flooding is also life giving, as it supplies the high water needs of crops like rice and jute and replenishes the rich alluvial soils of the river valleys and deltas.

The monsoon rains miss most of the northwestern parts of South Asia, including Pakistan and Afghanistan, which remain dry throughout the year, forming one of the world's major arid regions, the **Thar Desert.** The main source of water in this part of the region is the rivers, such as the Indus and its tributaries that are fed by the melting snows of the Himalayas.

Forests and Soils

Much of South Asia was originally forested, including the peninsula, northern river plains, and Himalayan foothills. After centuries of clearance by expanding populations for fuel and to increase the cultivated area, the teak forests of southern India and the forests on the Himalayan slopes are almost all that remain. These are being reduced further in size.

The most fertile soils occur in areas subject to annual flooding, on the lava plateaus or beneath the forests. Long use and the improper implementation of modern techniques have, however, reduced the soil quality. The drier area soils have been degraded by soil erosion, waterlogging, and salinization.

Natural Resources

The ancient rocks of the South Asian peninsula contain precious stones and mineral ores including iron and uranium, while the newer rocks on top contain one of the world's largest coal reserves. Deposits of oil and natural gas occur in the thick sediments beneath the major valley areas and just offshore.

Water is a crucial natural resource. Historically, the monsoon rains made it possible to support large populations at subsistence level. The Ganges, Brahmaputra, and Indus rivers brought economic life to the subcontinent's lowlands, especially after irrigation and water storage techniques were developed. The waters flowing northward from Sri Lanka's central hills also fostered a civilization based on irrigation. Water was instrumental in the Green Revolution. In the states of Punjab and Haryana, and parts of Uttar Pradesh, production rose and prosperity increased by irrigating more land, applying more fertilizer, and using more pesticides. Irrigated cropland across South Asia increased from 18 percent (132 million hectares; 326 million acres) in 1960 to 32 percent (174 million hectares; 430 million acres) in 1980 and topped 35 percent in 2000.

Natural Environmental Problems

People living in South Asia face a number of environmental hazards. Some are an integral part of the dynamic natural environment, but others are the outcomes of human clearance of land, population increase, and the context of the global economic system.

In the dynamic natural environment, the clashing plates that produced the Himalayas continue to set off earthquakes, such as those at Latur, Maharashtra, in 1993 and Bhuj, Gujarat, in 2001. More frequent earthquakes along the base of the Himalayas are usually less devastating, although very large ones occurred in 1905 and 1934. The October 2005 earthquake in Kashmir (Pakistan) measured 7.6 on the Richter scale. It killed at least

86,000 people and injured more than 69,000, as well as causing widespread destruction of buildings and roads.

Flooding is a major problem in the lower Ganges and Brahmaputra valleys and their combined delta. Snowmelt in the Himalayas combined with heavy rainfall, the funneling effect of the Bay of Bengal and storm surges from tropical cyclones, deforestation, low topography, and a very dense population make flooding a frequent disaster. Peak floods in 1998 and 2004 affected over 30 million people. It would be impossible, however, to either move so many people to unfloodable areas or construct effective flood prevention measures. But improved flood preparedness is needed for the vulnerable materially poor population. Increased flooding levels result from deforestation in the upper reaches of these rivers and their tributaries in India, Nepal, and Tibet, but there is little regional cooperation to reduce this effect.

In other parts of South Asia, drought is the main problem and the availability of water is critical. Water shortages raise major social issues in both urban and rural areas of India. While the urban middle classes can install storage tanks against times of shortage, the slum dwellers wait in line with buckets. While members of the village upper castes use good wells, the untouchables must find their water elsewhere.

In the future, the Maldive Islands and coastal areas such as much of the Ganges-Brahmaputra Delta that are only a few feet above sea level face drowning by a rising ocean level as global warming proceeds. The high costs of raising dikes and building sea defenses make such measures unlikely. Coastal areas are also at greatest risk during natural disasters such as the Asian tsunami in December 2004. The South Asian countries most affected were Sri Lanka, where over 31,000 people died, and India (primarily the state of Tamil Nadu), which reported over 10,000 deaths.

FIGURE 6.5 South Asia: pollution sources. The Union Carbide chemical plant and nearby poor housing in Bhopal, India.

Human-Induced Environmental Problems: Air and Water Pollution

The huge rise in South Asia's population and the necessary growth in national economies leave landscapes of exploitation, degraded resources, and pollution.

Making the herbicides and pesticides used in the Green Revolution resulted in toxic concentrations of chemical factories. In 1984, at the Union Carbide pesticide plant in Bhopal, India, water leaked into a methyl isocyanate storage tank, triggering chemical reactions and a cloud of toxic gases. At the time it killed 3,000 people immediately and another 12,000 subsequently in the adjacent slum areas (Figure 6.5). These official figures are thought to have been around one-third of the true numbers. The seriously injured numbered at least 50,000, with long-term effects on eyes, lungs, and immune systems. Over half a million people qualified for compensation, although payments were slow. Over 20 years after the event, the site, now owned by the state of Madhya Pradesh, remains highly contaminated. Both the state and Dow Chemicals (the U.S. parent company of Union Carbide) deny responsibility for cleaning up the site. This was a well-publicized, but not isolated, instance of the multinational corporation re-siting "dirty" manufacturing facilities in India.

Increasing damage of the Taj Mahal because of air pollution led to the closure of metal workshops in the surrounding area to reduce the sulfur gases that damage marble. However, it is only possible to take such expensive action on a local scale.

Nearly 40 percent of the Indian population lives in the middle and lower Ganges River basin. The land is intensively cultivated, and there are many factories and urban areas. Although considered sacred and health-giving by Hindus, the Ganges suffers from pollution from various sources and poses health risks to those who use its water (Figure 6.6). For example, heavily polluted effluent from a concentration of leather tanning works enters the Ganges River close to some of the main Hindu ritual bathing sites. In recent years, an upsurge of trash dumping in the Ganges, combined with funeral pyres along its banks, further worsened the water quality. The 1980s Ganga Action Plan aimed to clean up the river, but, although a few projects were implemented, lack of funding, continuing industrial malpractice, illegal dumping of wastes into the river, and illiteracy brought the program few successes.

FIGURE 6.6 **South Asia: water pollution.** Industrial waste and sewage flow into the Ganges River at Varanasi. Pollution is a continued problem despite the Ganga Action Plan's goal to clean the river.

Distinctive Human Geography

South Asia is a region of diverse peoples and cultures with its many languages and distinctive religious orientations. Home of one of the world's oldest civilizations (in Mohenjodaro, Harappa and Kalibangan in the Indus Valley) (Figure 6.7), the region has been impacted by the incursion and assimilation of different groups of people, the rise and fall of powerful local empires, and the effects of British colonial rule that lasted for over a century from the mid-1800s onwards. Waves of people, including the Aryans and Mongols from Central Asia, and Turks and other Muslims from Western Asia, entered through the northern mountain passes to access the rich plains. Later, traders including the Muslims (from AD 1100) came from the ocean, as did the colonizing Europeans (from 1500). Each group left its marks on the land, people, and culture. Among South Asian innovations that benefited the world are the decimal system and zero. Today, this region interacts with the rest of the world from a complex cultural base that fuses traditional values and modern ideas.

Cultural Diversity

Religion

The cultural diversity of South Asia includes numerous religions, languages, and dialects. Several ethnic religions (**Hinduism, Sikhism, Jainism**) and **Buddhism,** a global religion, originated in South Asia. Additionally, other global religions such as Islam and Christianity have a strong presence in the region. Hinduism, the religion of 80 percent of India's population (see Figure 1.14a), is believed to have crystallized in the Indus Valley in 1200 BC.

FIGURE 6.7 **South Asia: cultural mix and ancient origins.** Archeological remains of the ancient city of Mohenjodaro.

Most accurately described as "the religion of the people" and "a way of life," Hinduism is an ethnic religion into which people are born. Today, nearly all of the world's 800 million Hindus live in India. In the Himalayan kingdom of Nepal, almost 90 percent of the population is Hindu. In Sri Lanka, Hindus are a substantial minority group (about 15 percent of the population, mostly Tamil). About 35 percent of Bhutan's population is composed of Hindus of Indian and Nepalese origin.

Caste is a Hindu concept that was solidified by invading Aryans by at least 1500 BC. The **caste order** is similar to ethnic or class divisions elsewhere. Aryans grouped society into four categories: the priests (*Brahmins*—the top group); warriors and rulers (*Kshatriyas*); and commoner merchants and artisans (*Vaishayas*). Those outside these divisions, mostly non-Aryans, were the lowest caste of menials and servants (*Shudras*). Those who did not belong to one of these groups were outcasts and labeled "untouchables." Caste membership became hereditary, based on intermarriage within the group; birth determined status in society. Caste permeates all South Asian society and may still determine where people live, work, and who they marry, especially in rural communities.

Buddhism and Jainism were founded in the Ganges River valley in the 500s BC in reaction to aspects of Hinduism such as the rigid caste system and hundreds of gods. Asoka, the most celebrated ruler of the Mauryan Empire that held sway over most of the Indian subcontinent from 321 to 181 BC adopted Buddhism. He propagated it and sent Buddhist missionaries abroad to spread their faith. As a result much of Sri Lanka was converted to Buddhism, which also spread eastward and northward into Southeast and East Asia. After Asoka's death, Buddhism declined and Brahman Hinduism regained its importance in India.

Sri Lanka has a Buddhist majority (70 percent), composed of the dominant Sinhalese group. Bhutan is a Lamaistic Buddhist country like its neighbor, Tibet. Nepal has a small proportion of Buddhists who have had a major impact on the local character of Hinduism. Afghanistan was home to a fusion of Buddhist and Persian cultures evidenced in artifacts such as ancient giant Buddha statues in Bamiyan, which were destroyed by the Taliban government in 2001 despite international efforts to save them.

Jainism, which emerged in the 600s BC, is associated with its founder Mahaveera. Built around a code of nonviolence, it is designed to save followers from an endless cycle of birth and re-birth. This nonviolent code was later taken up by Mahatma Gandhi in his early-1900s campaign for India's independence from the British. Forbidden to farm (and kill worms and insects), Jains became an exclusive group. Many, especially in the Mumbai (Bombay) area, are wealthy traders. They make up a small proportion (less than 1 percent) of the Indian population today, but include over 3 million people.

Islam was brought to India by Arab traders in the AD 700s and spread by Sufi mystics and teachers in former Buddhist strongholds in northwestern and eastern India, particularly Punjab and Bengal. Muslim invasions from the 1200s onwards and associated conversions under Muslim sultans and emperors, especially the **Mughal** (**Mogul**) dynasty, solidified Islam's position in India. Under the Mughal dynasty, established in the 1700s by the Persian Turk Babar, India experienced developments in such fields as architecture: the Taj Mahal and many examples of Muslim buildings built in the 1600s (Figure 6.8a) survive throughout the region. Muslims form an important religious minority in India with about 120 million people who comprise 11 percent of the country's population, making India the country with the fourth largest Muslim population in the world. Pakistan and Bangladesh are Muslim states whose pre-independence history is closely tied to that of India. Arab (Muslim) control of medieval Indian Ocean trade routes included the Maldive Islands, which stretch southward into the Indian Ocean. The islands became solidly Muslim by the AD 1100s and remain so. Afghanistan is 99 percent Muslim, of which 84 percent are Sunni, and 15 percent Shiite.

Sikhism, which emerged in the early 1500s, combines aspects of Hinduism such as belief in reincarnation with strict monothe-

ism, borrowed from Islam. Sikhism preaches universal toleration and is marked by a strict code of conduct, many temples that have kitchens providing food for all. The Golden Temple at Amritsar is the holiest temple. Although small in number (approximately 20 million in India, composing 2 percent of the population) many Sikhs became wealthy from their bumper crops in Punjab. In 1980, extremist Sikhs in Indian Punjab occupied the Golden Temple in Amritsar and used it as a base for killing rival Sikhs until they were ousted in 1984 by a major military offensive. In revenge, they assassinated Prime Minister Indira Gandhi, after which Sikh areas of Delhi were burned by mobs.

Religion, with diverse expressions, remains important in South Asian life. Religious festivals, traditions, and rituals are celebrated with fervor; religious leaders, from the local Sai Baba of Shirdi to Sri Sri Ravi Shankar with his global Art of Living Foundation, have thousands of followers. However, religion is a tinder box within South Asia and minority religious groups remain vulnerable. Feelings of alienation have caused Christian groups in Nagaland State in northeastern India and the Sikhs in Punjab to lobby for a separate state or even an independent country, (Sikh

FIGURE 6.8b Afghanistan. The complexity of ethnic groups and their links across the political borders is a major factor in the difficulties of governing Afghanistan. The small number of border crossings in this mountainous country link some of the groups.

FIGURE 6.8a South Asia: historic buildings. The entrance to the mausoleum of Akbar, the greatest of the Mughal emperors, near Agra, India.

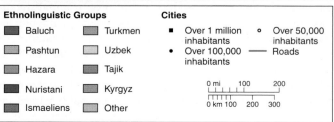

Ethnolinguistic Groups		Cities	
Baluch	Turkmen	■ Over 1 million inhabitants	○ Over 50,000 inhabitants
Pashtun	Uzbek	• Over 100,000 inhabitants	— Roads
Hazara	Tajik		
Nuristani	Kyrgyz	0 mi 100 200	
Ismaeliens	Other	0 km 100 200 300	

Khalistan, the land of the pure). In Sri Lanka, the dominant ethnic groups (Sinhalese, Tamil) have different religions (Buddhism, Hinduism) as well, adding to the tensions between them.

Languages and Dialects

The variety in languages in South Asia rivals its religious diversity. India alone has over 1,600 languages and dialects. Hindi, India's national language, is spoken and understood by about 40 percent of the population. It is merely one of 16 official languages recognized by Indian states and the one most used in movies and TV. Of India's official languages, the Indo-Aryan languages are prevalent in the north and Dravidian languages that originated in India predominate in the south. English remains the lingua franca, the common language used by the legal system, and is spoken by the educated elite. It is often used in modern Indian literature and university courses and is a common form of communication in commerce and national politics. Speakers of minority languages are at an increasing disadvantage as globalization affects more of India.

Bengali is the language of 98 percent of the people of Bangladesh, who originated from Indo-Aryan stock that mixed with local ethnic groups. The Biharis form a smaller group of non-Bengali Muslims who speak Urdu and migrated to East Pakistan after independence in 1947.

Urdu, a form of Hindi with Arabic script used by Muslim people before independence, is the official language of Pakistan. The Muhajirs, migrants from Hindu India to Pakistan in 1947, speak Urdu. Punjabi and Sindhi are also widely spoken in Pakistan. Many Pashtun groups live in the hills bordering Afghanistan. Rivalries among groups and the elitism of Urdu-speaking feudal landowners and military officers continue to dominate social and political life in Pakistan.

Fifty percent of Afghanistan's population speaks an Afghan form of Persian, which is the dominant language. Pashtuns dominate the south, while groups of Central Asian peoples, such as the Tajiks and Uzbeks, inhabit the north. Each of these ethnic groups has its own language and dialects (Figure 6.8b). In Sri Lanka, the Sinhalese speak Sinhala, a language that belongs to the Indo-European language family, while Tamil, a Dravidian language, is the mother tongue of the Tamils.

Global Linkages and the Impact of Colonialism

Historically, the material wealth of South Asia was built on internal gem and metal deposits and also on external trading links to African, Arab, and Southeast Asian lands. Although the region has had links with European countries for centuries, its seaborne trade was controlled by the Arabs since the AD 800s. The great wealth of this region became a magnet for European adventurers in the mid-1400s searching for an alternative route to get Indian products and riches to Europe that bypassed the Arab middlemen. In 1498, the Portuguese explorer Vasco da Gama reached India after sailing around southernmost Africa. The Europeans who arrived by sea found few defenses. During the 1500s, the Portuguese and Dutch established trading stations, such as Goa, around the coasts of South Asia. The first British trading post followed in 1612. During the 1600s, the Dutch forced the **(British) East India Company** out of the East Indies, and the British switched to India, ousting their Portuguese and Dutch rivals from most trading centers. The British East India Company took the Portuguese port of Bombay ("good bay," now Mumbai), built a new port at Madras (Chennai), and began trading with the most populous area of Bengal around modern Calcutta (Kolkata), which became the company's main center. At the beginning of the 1700s, India's GDP rivaled China's as the largest in the world.

The British Empire

By the early 1700s, political chaos caused by internal factions splintered South Asia into small and large kingdoms, ruled by Muslim or Hindu princes. Foreigners took control of the increasing overseas trade in cotton, cloth, rice, and opium. Over the next 100 years, the British East India Company, backed by the British Indian army that employed Indians as foot soldiers (sepoys), increased its hold on South Asia, taking a particularly strong position in Bengal and the peninsula. The company's area of influence was extended southward into Ceylon (now Sri Lanka) in 1798 and westward into Punjab in the mid-1800s.

A major mutiny of its Indian sepoy troops in 1857, which many Indians see as "The First Independence War," led the British government to take full political control of South Asia. It abolished the British East India Company and established the **British Indian Empire,** often referred to as the "British Raj" (*raj* means rule or government). It included almost all the subcontinent. However, 40 percent or almost one-fourth of the total population remained under 600 "independent" princely family governments, who governed their territories in harmony with British policies (see Figure 6.2).

The British saw their role as "civilizing" India through Western education, new technology, public works, and a new system of law. However, benefits went both ways. British interests redirected India's farms to produce raw materials for British industries. The region entered the expanding global economy of the later 1800s, when the export of Indian cotton compensated for negative British trade balances. Furthermore, the British Indian army became a tool of attempted imperial expansion into Afghanistan and Burma in the 1880s and into Tibet in 1903–1904. As British colonies, Ceylon and the Maldives were also incorporated into the world economic system that focused on the demands of the colonizing country and other wealthier markets rather than on local needs.

British imperial rule had massive effects on the geography of the peninsula. Its mercantile economy focused on primary products to be exported to Europe, defined the resources that could be developed, determined what was produced, altered patterns of land control, selected areas for development and cities for growth, and controlled external trade. British engineers irrigated land in the Indus and Upper Ganges river basins to produce cot-

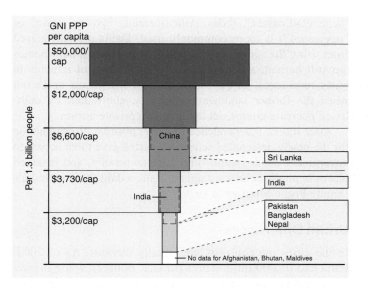

FIGURE 6.9 **South Asia: country average incomes compared.**
The countries are listed in order of their GNI PPP per capita for 2005.
Source: World Development Indicators, *World Bank; and Population Reference Bureau.*

ton for export to its textile mills while the Indian textile industry was suppressed in favor of British cotton goods. They built railroads from the main ports to move troops and exports. Former communal land was reallocated to larger and smaller landowners, forming interest groups that were expected, in return, to support the colonial administration. The cities of Calcutta, Bombay, Madras, and Karachi grew faster than others after the British East India Company designated them as foci where lines of overseas trade and internal communications met.

The British colonizers also set up schools to educate the native population and teach them English. As a result, English became the lingua franca and unifying language among the traders, lawyers, and elite families. Today, India has the second largest population of English speakers after the United States, which it is predicted to overtake by 2010. Fluency in this global language has helped South Asian participation in global arenas of trade, business, education, and a wide variety of professions. Within India, an English-speaking, college-educated work force is the basis of a low-cost service industry in "back-office" work such as call centers and other information technology-enabled services.

Although the countries of South Asia have made considerable progress in economic and social arenas and are increasingly involved in the global economy, they remain among the world's poorest (Figure 6.9). Low possession of consumer goods, access to piped water, and energy usage (Figure 6.10) show that even the most advanced countries here have a long way to go toward development.

Rural and Urban Contrasts

South Asia is a study in contrasts. Extremes of abject poverty coexist with tremendous wealth and modern luxuries; traditional village life in many rural areas contrasts with cities and their high-rise office blocks and prestige apartments, international hotels, modern conveniences, and global influences. Even though the region has some of the largest cities in the world, South Asia's population remains largely rural (Figure 6.11). Rural areas still have large proportions of the semi-subsistence and low-paid farming lifestyles that are linked to long-term poverty and lag behind urban areas in education and health care provision. By 2025, it is projected that India's rural population will be down to 55 percent of the total, but this will comprise a greater number of people than at present (762 million instead of 630 million) because of the overall population increase.

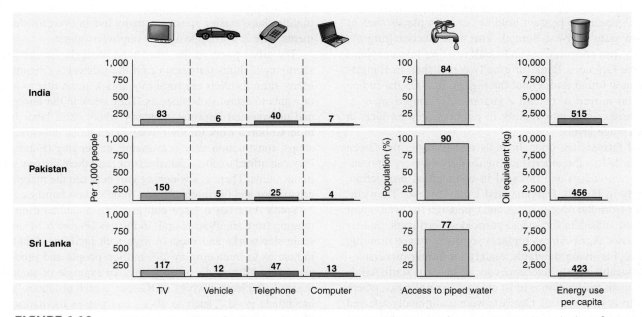

FIGURE 6.10 **South Asia: ownership of consumer goods, access to piped water, and energy usage.**   How do these figures compare with other Asian countries? *Source: Data (for 2002) from* World Development Indicators, *World Bank, 2004.*

FIGURE 6.11 **India: farming.** Farmers in the north India plains share a new tractor for fieldwork.

Rural Poverty

Laborers have little chance of escaping a lifetime of poverty, and debts are inherited by children. Cash incomes are small. Often unexpected obligations, such as sickness, the death of plowing oxen, or a demand for a dowry on the marriage of a daughter, deplete savings. In India, a manual laborer pays up to four times his annual wages for a daughter's dowry. Money lenders help with loans, but inability to repay leads to mortgaged and often lost land, while members of the family may be taken into bondage. Countermovements exist to combat rural poverty. Cooperative peasant holdings exist in places such as the Indian state of West Bengal. The most encouraging are the women-run cooperatives and small rotating loans projects such as the Grameen Bank, originally established in Bangladesh but now found throughout the region. In 2005 the Indian government moved to extend a guarantee of employment to all rural areas, with jobs mainly in public works projects at minimum wage levels.

Not all farmers are poor. In India and Pakistan the Green Revolution led to the emergence of wealthy farmers who enlarged their holdings and invested in commercial production, particularly in Punjab, Haryana, and Uttar Pradesh. They employ labor from the poorer states and campaign for even larger government subsidies for seeds, pesticides, fertilizers, and irrigation water. Agribusiness replaces semisubsistence farming, bringing higher living standards, mainly for farmer-owners.

While wealth largely determines social status in South Asia's cities, caste discrimination is most obvious in rural areas, where one's status is known to all. Outcasts were traditionally referred to as "untouchables" and later given the official designation of

"scheduled castes." Today, *dalit* (meaning "ground down" or "oppressed") is more commonly used. Dalits are still barred from using the same wells as upper caste Hindus, dalit homes are still burned, and caste-based political control remains in many rural areas. In the Indian state of Bihar, war broke out when the former landowning families, confronted by dalit-based guerrilla groups, set up their own private armies.

After Indian independence in 1947, positive discrimination for the newly designated scheduled castes gave them access to government employment, rural land ownership, and reserved places in public colleges. Jagjivan Ram, a dalit, became India's Deputy Prime Minister in 1979.

Urban Differences

In the cities the contrasts are equally obvious. As of 2004, India had 61,000 millionaires (in U.S. dollars), while the average Indian earns only $1.60 a day. One million extremely wealthy people (only 1 percent of India's total urban population) consume at the level of Western countries, and spend their money on lavish lifestyles that include luxury vacations and imported cars, appliances, clothing, cosmetics, and food. The very wealthy live in guarded colonies, bus their children to private schools, and send them abroad for higher education. They use air-conditioned cars, and have backup water and power supplies. They are cosmopolitan, even global, in education, livelihood, and social activities. They increasingly break with traditional culture.

The urban middle classes usually have secure jobs in the formal sector and buy consumer durables such as TVs and refrigerators. They have access to health care and educate their children in private English-language schools if funds permit. A growing young and fairly affluent set are expanding the middle class, especially in India. Upwardly mobile middle class households own or rent their own apartments. At the other end of the middle class housing spectrum, many live in overcrowded tenements. Life is a struggle against impoverishment.

The urban poor, on the other hand, struggle for existence in shantytown slums (*bustees*) or on the sidewalks (Figure 6.12). Many slum dwellers are rural migrants with no title to housing or rights to establish businesses. They work in the unorganized and unprotected informal economy where child labor is common. Children work for their living in tea stalls, industrial workshops, construction sites, scavenging, or begging (Figure 6.13). Few can afford to obtain an education, and there are few schools in the slums. There is a widening gulf between the marginalized urban poor and the more affluent, middle class families.

South Asia has a large and growing manufacturing sector ranging from small-scale craft industries (Figure 6.14) to large-scale steelworks and modern high tech facilities. Small-scale industries in India employ 140 million people and produce 35 percent of all manufactured goods. An example of such an industry is the leather works of Kanpur, which produces "quality handmade goods" such as shoes and purses for national and global markets in unsafe work environments, using poorly paid

FIGURE 6.12 **South Asia: shantytown.** Slum in Mumbai with dwellings made of sacking, plastic, wooden boards and metal sheeting. Compare these structures with the apartment blocks in the distance.

FIGURE 6.13 **India: informal sector.** A young girl carries bricks on her head at a construction site. Why is child labor still prevalent in South Asia?

FIGURE 6.14 **India: craft industries.** A merchant displays copper and bronze "mini temples" in a New Delhi suburban market.

and often, child labor. In contrast, quality specialty steel is produced using the most modern techniques in Jamshedpur, at prices undercut by only the best South Korean mills.

Geographic Diversity

A new diversity within South Asia emerged in separate countries following the end of the British Raj (Figure 6.15 and Table 6.1). The three groupings recognized here are:

- The Republic of India, the world's second-largest country in total population, has a dominant presence within the region.
- Bangladesh and Pakistan are two Muslim countries, each with over 100 million people, resulting from the breakup of the original Pakistan in 1971.
- The Mountain and Island Rim, comprising Afghanistan, Nepal, and Bhutan in the mountainous north and Sri Lanka and the Maldives in the island south, includes a group of smaller countries on the margins of South Asia.

India

India has the world's second-largest population and dominates South Asia in land area, resources, and economic activity. In 2006, the Republic of India's population of 1.122 (see Table 6.1) billion was

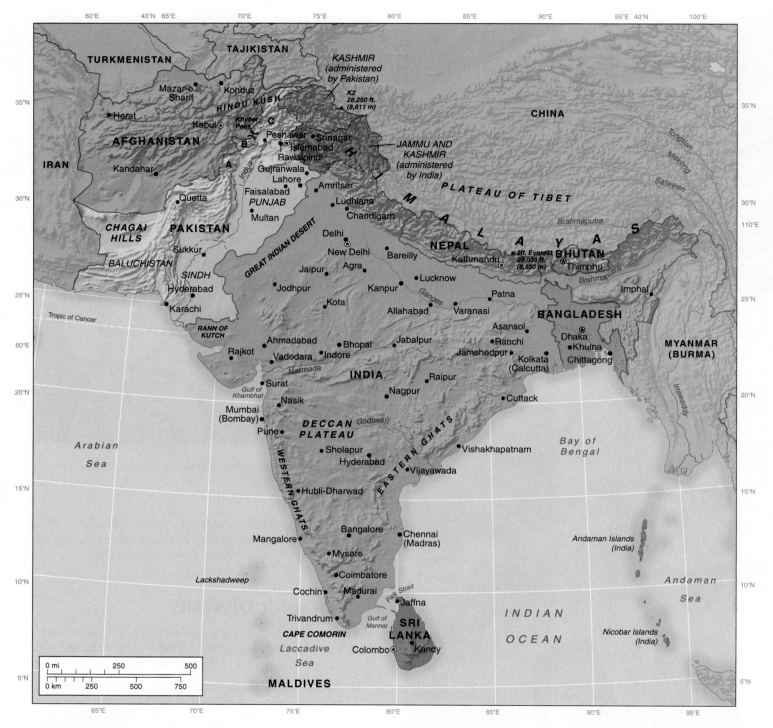

FIGURE 6.15 **South Asia: major geographic features, countries and cities.** India dominates the region by its size and centrality; Bangladesh and Pakistan are two Muslim countries created as one at independence but since broken apart. Afghanistan, Nepal, and Bhutan form the Northern Mountain Rim, while the Maldives and Sri Lanka are the Southern Island Rim. This is a region of huge and growing cities. In Pakistan A = Waziristan, B = Tribal Area, and C = Northwest Frontier.

second only to China. As an increasingly significant major power, India is a product of its history and despite limited resources to underwrite change, the country has an expansive vision of its future.

India became an independent country in 1947 as a republic with a federal union government. The former princely states and British Indian provinces were admitted as states within India's federation (see Figure 6.2). There are 29 states and six union territories, each with its own elected assembly. State boundaries relate to the distribution of languages, reflecting the huge range of languages spoken in the country. The number of states increased after independence, mainly as the result of the original states not coping with the interests of ethnic groups.

India is the world's largest democracy. Every adult has a vote. Women have a stipulated minimum number of seats in local government. The union (federal) government has responsibilities for defense, foreign affairs, and currency. It raises most of the taxes

TABLE 6.1	**SOUTH ASIA: Data by country, area, population, urbanization, income (Gross National Income Purchasing Power Parity), ethnic groups**						
		Population (millions)		**%Urban***	**GNI PPP 2005**	**2005**	
Country	**Land Area (km²)**	mid 2006 Total	2025 est. Total	2006	Total (US$ billions)	Per Capita* (US$)	**Ethnic Groups (%)**
INDIA							
India, Republic of	3,287,590	1,121.8	1,363.0	28.7	3,881.4	3,460	Indo-Aryan 72%, Dravidian 25%
BANGLADESH AND PAKISTAN							
Bangladesh, People's Republic of	144,000	146.6	190.0	23.4	306.4	2,090	Bengali 98% Punjabi, Sindhi, Pashtun, Muhajir (Indian origin)
Pakistan, Islamic Republic of	796,100	165.8	228.8	33.5	389.6	2,350	Pakistan: Punjabi (45%), Pashtuns (15%), Sindhis (14%), Seraikis (11%), Muhajirs, ex-India (8%), Balochis (4%) (Data from Wikipedia.org)
MOUNTAIN AND ISLAND RIM							
Afghanistan, Islamic State of	652,090	31.1	50.3	22.0	—	—	Pashtun 38%, Tajik 25%, Hazara 19%
Nepal, Kingdom of	140,800	26.0	36.2	14.2	39.7	1,530	Nawar, Indian, Tibetan, Gurungi, Sherpa, Bhhutia 50%, Nepalese 35%,
Bhutan, Kingdom of	47,000	0.9	1.3	30.9	—	—	Sharchops 10%
Sri Lanka, Dem. Socialist Republic of	65,610	19.9	22.2	20.0	89.8	4,520	Sinhalese 74%, Tamil 18%, Moor (Arab) 7%
Maldives, Republic of	600	0.3	0.4	26.8	—	—	Sinhalese, Dravidian, Arab, African %'s?
Total/Average	**5,133,790**	**1,512**	**1,892**	**24**	**4,707**	**2,790**	

Source: World Population Data Sheet 2006, Population Reference Bureau. Microsoft Encarta 2005.

and makes grants to the states. In the 1990s, states increased their functions, particularly in the delivery of social programs including health and education.

Agriculture

The country's economic profile is moving from a farm base to factories and offices. Between 1965 and 2002, farming decreased its contribution to GDP from 44 to 23 percent without causing famines. Due to the adoption of Green Revolution technologies that use high-yielding crop varieties, chemical fertilizers, pesticides, and irrigation, food grain production in India almost quadrupled from 50 million tons in the 1950s to over 190 million tons today. In 2002, India became the world's second-largest exporter of rice in good monsoon years (after Thailand) and second-largest producer of wheat (after China). However, the benefits of the new technologies are geographically and socially uneven. The farmers with larger land holdings in the states of Punjab and Haryana benefited the most. Among the drawbacks of the Green Revolution is the need to buy new hybrid seeds for each crop cycle and use increasing amounts of fertilizer and pesticides to maintain the high yields. Large areas of the hilly peninsula that have inadequate irrigation grow coarse grains such as millet and sorghum that withstand dry conditions.

The vagaries of the monsoon still determine the harvest: good years enable India to export food; bad years require imports. Despite adequate food to feed its population, hunger and malnutrition are still prevalent. In the ten years 1990–1992 to 2000–2002, the hungry in India decreased from 25 to 21 percent,

but the numbers increased from 216 to 221 million. Important commercial crops are cotton (India is the world's third-largest cotton producer in the world), tea grown on plantations in the hills of northeastern India and high ranges of Tamil Nadu and Kerala in southern India, and jute, a commercial fiber cultivated on the western margins of the Ganges-Brahmaputra Delta.

Manufacturing

India has a large and growing manufacturing sector ranging from small-scale craft industries to large-scale steelworks and modern high tech facilities. The distribution of manufacturing in India reflects historic craft specialties, raw material locations (mines, farms), the British legacy of transportation hubs, ports, and administrative centers, varied encouragements offered by the union and state governments since independence, and entrepreneurial skills.

The most mineral-rich part of India is the Chota Nagpur Plateau in the northeastern part of the peninsula (see Figure 6.1), where plentiful iron ore and coal deposits formed the basis of local steelmaking. India's rocks also contain other metal ores such as bauxite, copper, gold, and manganese. As part of its self-sufficiency policies after independence, the Indian government developed a large-scale, heavy industrial sector of steel, chemical, and aluminum works, many built in conjunction with Western corporations such as Union Carbide and ICI.

The Indian government followed economic policies of self sufficiency and import substitution in the decades soon after independence. In the 1990s, it adopted new policies that involved

India more in the global economy by raising production and exports and being more open to foreign investment. Major multinationals such as General Motors, Ford, Chrysler, Peugeot, Volvo, Hyundai, Mitsubishi, Daewoo, and Volkswagen have begun producing cars and trucks for the local market. Indian metal forging companies make car parts and the Indian pharmaceutical industry is challenging world markets.

Between 1965 and 2006 industry rose and then stablized at 20 percent, while services rose from 34 to over 60 percent of the GDP. Textiles and garments (mostly of cotton, but also blends with synthetic fibers and jute) comprise over one-fourth of India's exports, mostly to the United States and European Union countries. Further economic modernization, however, depends on India's ability to supply the infrastructure of power, telecommunications, and road transportation that will attract and keep productive businesses. Unfortunately, power shortages still affect many areas. India remains deficient in transportation infrastructure. Its railroad system has a major role, but needs updating to compete with the gradual improvement of roads and the efficient new private airlines.

Human Development

Over the years since independence, India achieved self-sufficiency in food, literacy rates doubled, life expectancy rose from 33 to 63 years, infant mortality fell (from 165 per 1,000 live births in 1960 to 58 in 2006), income poverty was reduced, and overall income substantially grew. Forecasts expect India's population to have a growing work-age group in contrast to most other countries.

Despite these improvements, in the early 2000s India's overall Human Poverty Index (HPI) was still high at over 30 percent, 60 million children under age 4 were undernourished, 50 percent of the population was illiterate, women remained at a disadvantage in society, and rural poverty affected about 40 percent of the population. Progress is slow. In 2005, India's total GNI PPP was fourth in the world (China was second).

On the political front, in the 1990s the world's largest democracy became increasingly caste-based and regional. The Congress Party, a dominant force since independence gave way to local groups in the states, and some countrywide political groupings began to have greater roles. In 1998, the Bharatiya Janata Party (BJP) became the new government, with a strong Hindu nationalist mandate. In 2004, the Congress Party regained control of the union government. India's problems of poverty, unemployment, weak demand, and low levels of investment continue to be major challenges.

Bangladesh and Pakistan

Bangladesh and Pakistan were partitioned from British India at independence in 1947 on the basis of their Muslim majorities, to form a single country (Pakistan) divided into East Pakistan and West Pakistan. Perceptions that resources from East Pakistan were being siphoned off to benefit West Pakistan resulted in the East Pakistanis revolting for a separate state. In 1971, after a brutal civil war, Bangladesh (formerly East Pakistan) was established as a separate country with India's military help. War-ravaged Bangladesh built itself anew with huge quantities of aid from the world's materially wealthier countries. Since independence, its internal politics have been marked by military takeovers, coups, and assassinations. In the 1990s, a settled democracy brought major positive changes. Although over 90 percent of the population of both countries is Muslim, Bangladesh is a secular republic, unlike Pakistan whose 1973 constitution is founded on Islamic law.

In Pakistan, General Pervez Musharraf seized power in 1999 and dismissed the democratically elected Pakistan parliament. Although the Pakistan Supreme Court backed this move, it insisted that an election be held within three years: the general was reelected in 2002, but in 2004 agreed to stand down as head of the military so that he could remain in office until 2007.

From 1980 to the early 2000s, Pakistan's literacy rate doubled, although it was still only 45 percent; electrification of rural households rose from 16 to 61 percent; and television reached 75 percent of city dwellers and 50 percent of the rural population. In late 2001, as a result of its support for the U.S. antiterrorist coalition, sanctions against Pakistan (imposed after its 1998 nuclear tests) were dropped, and it was given increased access to European markets as a reward for opposing terrorists and the drug trade.

Agriculture

Bangladesh and Pakistan are among the world's poorest countries. Agriculture is the mainstay of both countries' economies. Pakistan is a country of arid lowlands and high mountains, whereas Bangladesh is mostly low-lying apart from the small hilly eastern region inland of Chittagong, and is well watered by rain and rivers. Water is scarce in Pakistan, but flooding is a major problem in Bangladesh. Over half of the Bangladeshi and Pakistani labor force is engaged in farming, producing 23 percent of GDP in 2002. Bangladesh grows over half of the jute that enters world trade (Figure 6.16), as well as rice, tea, and sugarcane. The floodplains and deltaic lands that support most of Bangladesh's agriculture are low-lying and vulnerable

Figure 6.16 South Asia: jute harvest near Tangail, Bangladesh. The fiber is stripped from the plant stalk using the plentiful water of the Ganges River. What is jute used for?

to ocean surges during tropical cyclones. The slow expansion of agricultural production in Bangladesh at 2 to 3 percent per year cannot support a rapidly growing population.

Cotton is Pakistan's chief commercial crop. Wheat, rice, and sugar cane are grown as irrigated food crops. Some farmers are beginning to look to more profitable labor-intensive crops, such as vegetables and fruit. Much of Pakistan's best farmland is owned by a few wealthy landlords. The rich landowners' linkages with politicians gains them access to jobs in the government bureaucracy and the ability to veto measures that might decrease their dominant position in Pakistani society. Land is treated as a source of political power rather than a resource to enhance productivity.

Manufacturing

Bangladesh's limited manufacturing sector was traditionally based on the processing of agricultural products such as jute (see Figure 6.16), rice, tea, and sugar cane. The jute industry and other agro-industry units require restructuring, but the Bangladesh government faces the opposition of trade unions in trying to privatize the industry.

In the 1990s, following the adoption of structural adjustment policies that reduced trade restrictions and encouraged the external financing of garment factories, Bangladesh doubled its exports. The garment industry, with retailing customers in Europe, the United States, and East Asia, has become a major export earner. Enterprise zones at Chittagong and Dhaka (Dacca) are further attempts to expand industrial income and employment. Low wages for apparel workers in Bangladesh (9 to 20 U.S. cents per hour compared to 20 to 30 cents in Pakistan and over $8 in the United States) as well as the use of child labor has drawn criticism. During the 1996 elections, political strikes of half the 1 million poorly paid garment workers led to economic reforms that returned people to work and the economy to growth. Privatization continues, but severe infrastructure bottlenecks (roads, power stations, communications) slow the economy's expansion. Losses by state-owned enterprises continue to be a heavy burden for the limited national budget to bear.

Pakistan has a longer experience of manufacturing than Bangladesh. In exports, primary products decreased from 48 to 10 percent of the total between 1975 and the early 2000s, while manufactured goods rose from 52 to 89 percent. Textile manufacture, especially of cotton goods, dominates industries that include food processing, chemicals, and car assembly. During the 1990s, growth in output and productivity in manufacturing increased jobs, even in rural areas, at a rate that equaled the rate of new entrants to the labor force. However, such gains did not benefit the poorest groups of people. The new government elected in 1997 gave signs that it would help make producers more internationally competitive. It also wished to tackle problems that kept so many of its people poor and socially deprived. A military government replaced the elected government before these aims were met.

Both Bangladesh and Pakistan have small reserves of oil and natural gas, although neither exploited them fully until the 1990s. Pakistan's important chemical industry is built on deposits of gypsum, rock salt, and soda ash. Fishing in the coastal areas of both countries is poorly organized.

Service industries employ a growing proportion of the labor force in Pakistan (36 percent of women and 18 percent of men) and are expanding in Bangladesh (30 percent of men, 12 percent of women), and both countries produce around half of their GDP from such industries. Health, education, and financial services are growth areas. Neither country attracts many tourists.

Human Development

The HPI is just under 50 percent for both countries. Pakistan's GDP total is one-third greater than that of Bangladesh, but both have adverse trade balances and depend on aid donations. In Bangladesh, the problems stem from overpopulation (very high densities with a low resource and infrastructure base). Fewer than half the population has access to sanitation or legitimate electricity supplies and only 14 percent to garbage disposal.

In Pakistan, the inequitable distribution of wealth still leaves large numbers very poor, although the average per capita income is higher than that of Bangladesh. Pakistan lagged behind other low-income countries in social conditions into the mid-2000s, with low levels of literacy among women, high infant mortality (79 per 1,000 live births, compared to the average of 57 in all low-income countries), and continuing high total fertility. Following the failure of earlier efforts, plans in the 1990s included a social action program that claimed major advances in the first three years, but the program suffered from slow bureaucratic responses, a lack of community participation, and changing political fortunes. The Pakistan government has increased social sector spending with a view to improving income poverty, education, and health.

Both countries lack many basic provisions that might encourage more manufacturing industry to locate there. Power supplies are subject to shortages, although Pakistan hopes to double its power provision with the aid of private investment. In the 1990s, it completed major hydroelectricity projects in the Himalayas, while in 1997 a thermal power station was completed with foreign capital and expertise at Hub near Karachi.

Mountain and Island Rim

The countries of the northern mountains—Afghanistan, Nepal, and Bhutan—have environments that isolated them from global connections and fostered internal strife through tribal rivalries. Events in these mountain countries are now influenced not only by what happens in the rest of South Asia, but also by interactions with neighboring countries in Central Asia, Iran, Russia, and China. The island countries of Sri Lanka and the Maldives have easier ocean connections with the global economic system but experience other problems that restrict their fuller development.

In 2006, Afghanistan, Nepal, and Sri Lanka each had 20 million to 31 million people. Bhutan and the Maldives had tiny populations of fewer than 1 million people between them. The mountain countries are among the world's poorest, while the Maldives and Sri Lanka have better living conditions. Nearly all these countries suffered internal conflict since the 1990s.

Afghanistan

In the late 1800s the British and the Russians delineated the borders of the state of Afghanistan that lay between their empires. The northern border was seen by the British as a buffer to Russian expansion. The country brought together groups fragmented by language, creed, geography, and historic cultural traditions. The Pashtuns formed the majority in the southern parts, but Tajik and Uzbek people dominated the north (see Figure 6.8b). King Abdur Rahman (1880–1901), backed by British money and weapons, tried to create the foundations of a centralized Afghanistan. This and later attempts to build and unify the country were often ruthless, resented by the intensely independent groups, and generally unsuccessful.

During the early decades of the 1900s, internal feuding and external influences held back attempts to modernize Afghanistan. Tensions with Pakistan emerged over the northwest frontier lands that Pakistan swiftly annexed after its own independence in 1947. Afghanistan then linked more closely to the Soviet Union until 1979, when the latter invaded and occupied it. During the period of Soviet occupation, local warlords, or mujahideen, supplied with weapons by the United States through Pakistan, fought the Soviet armies and established their own military units. After the Soviet Union withdrew in 1992, the local mujahideen and their armies fought each other for control of the country. Most government systems broke down. Education became irregular, while public health and health care were weakened.

In the late 1990s, the Taliban group conquered virtually all Afghanistan, although fiercely opposed by northern tribes who resisted the imposition of hard-line Sunni Islamic tenets. From the west, Iran supported Shia groups, including Hizbollah, but they failed to overthrow the Taliban. The rest of the country lived under the repressive Taliban order, which expelled most international aid agencies. Strong opposition to the Taliban among the Afghani people resulted in their accepting and using U.S. military intervention and military aid to depose the Taliban government. A new government was established and the first "democratic" elections held in 2004 (though marked by local coercion).

Nomadic herding of sheep and the cultivation of wheat, rice, and other cereals are important to Afghanistan's agricultural base. In a politically motivated act in 1998, the Taliban destroyed the fertile central farming region of Afghanistan, the Shardi Plain near Kabul. They cut fruit and nut trees, burned villages and wheat fields, blew up irrigation channels, and left crops to wither. Kabul's industrial district had 200 factories in 1992, the number was reduced to 40 under the Taliban, and only 6 still operated in late 2001. The 1992 fighting destroyed most of the Jangalok steelworks outside Kabul, and the machinery was taken to Pakistan. Even the Army Factory that made cannons, guns, and ammunition is largely abandoned.

Opium poppies became an important crop for Afghanistan's beleaguered farmers. In the 1990s Afghanistan supplied 70 percent of the world's heroin as the country's other farm products dwindled through war and Turkey, Iran, and Pakistan enforced strict drug control laws. Afghanistan's isolation, disrupted government, lack of alternative commercial products, and warlord control of the drug trade make it difficult to stop the production of opium and heroin and their entry to illegal drug traffic routes. Although Afghanistan was threatened with the withdrawal of aid because of its increasing production of these drugs, payments continued for fear of angering the warlords. Planting continues, with the farmers becoming indebted to the opium traders, who take their land and even their women.

Afghanistan faces a difficult and uncertain future in the light of its history of factional strife, oppressive rulers, devastated economy, and poor relationships with surrounding countries. Money from the sale of opium continues to fund local groups and undermines overall political stability. The Afghan International Office for Migration encourages the return of qualified Afghans who fled the Russian- or Taliban-dominated country but has difficulties in attracting professionals who settled comfortably in the United States, Europe, or Australia, where they have higher salaries and prospects of asylum or citizenship.

Nepal and Bhutan

From the 1850s, the Hindu kingdom of Nepal had a close relationship with Britain, and many Nepalese served in the British Indian army. Following British attacks in the 1800s, Nepal remained independent under pro-British rulers who hired out Gurkha soldiers (from different Nepalese ethnic groups) to the British. Nepal became a democracy in the 1980s. Despite this political orientation, alternative political parties were banned until 1990 while left-wing (Maoist) guerrillas caused major disruption in rural areas and attacked remote army and police barracks. In 2001, Nepal's monarchy was almost all murdered by the heir to the throne in a family argument. In 2005, the embattled king suspended all constitutional freedoms. In May 2006, following weeks of street protests and strikes, a cease-fire was declared between Nepal's monarchy and the Maoist rebels and the king agreed to cede power to a multiparty alliance. Still, the future of the monarchy and laying down of arms by the rebels continue to be contentious issues.

In Bhutan, the main group of people, the Bhote, drove out Indian peoples in the AD 800s and built impressive fortified Buddhist monasteries in the 1500s. The isolated kingdom of Bhutan was annexed by Britain in 1826 and given autonomy in 1907. Today, it remains a tiny buffer state between India and China. It has few cross-border contacts, preferring to maintain its own culture and to limit Western tourists and influences. Bhutan's official policy emphasizes "Gross National Happiness," placing cultural, spiritual and environmental well-being and the happiness of its citizens ahead of material wealth.

Sri Lanka

Sri Lanka's (formerly Ceylon's) proximity to the Indian peninsula and its openness to ocean-borne trade and conquest attracted competing groups. Indo-Aryan peoples with Buddhist beliefs from northern India developed a civilization based on irrigated rice agriculture in the north-center of the island that lasted from around 100 BC to the AD 1200s. During the later part of this period, Tamils—Dravidian peoples from southern India—established kingdoms in northern Sri Lanka around Jaffna. As Tamils expanded their influence, the majority Sinhalese peoples drifted southward, abandoning the irrigation systems and relying on rain-fed agriculture. They grew spices such as cinnamon. Arab traders settled in the country and controlled the overseas spice trade.

The Portuguese who were early traders and colonists in Ceylon left a strong legacy of Roman Catholic missions and of Portuguese as a trading language. In 1795, the British took Ceylon, which remained a crown colony until its 1948 independence. British companies set up plantations for growing tea, rubber, and coconuts, instead of the traditional spices (Figure 6.17). The emphasis on export crops, however, was accompanied by the neglect of traditional agriculture, leading to a decline in the rice crop and a need to import half the island's food needs.

After independence, Sri Lanka's government by the dominant Sinhalese people instituted reforms that were disliked by the Tamil minority, leading to a civil war. From a time when Sinhalese and Tamils lived side-by-side in Colombo, the capital, there were greater separations of people, over 60,000 deaths in the fighting and urban terrorism, and the emigration of many Tamils to India, Europe, and Canada.

Sri Lanka has tea, rubber, and coconut commercial plantations. These provide 35 percent of the country's exports, down from over 90 percent in 1950 as a result of diversification in the economy since independence. Sri Lanka's adoption of new rice varieties in the late 1960s converted the major food deficit—which arose out of the colonial focus on export crops—into a virtual self-sufficiency in rice today. Sri Lanka is an important source of gemstones and also has graphite reserves. Local agricultural processing together with the making of carpets and footwear are most common industries. Much manufacturing is in small-scale factories or cottage industries based on the dispersal of production processes in homes. Government policies since independence resulted in a diversified manufacturing base in Sri Lanka, which also produces steel, textiles, tires,

FIGURE 6.17 Sri Lanka: tea plantation. How are the tea bushes arranged on the hills?

and electrical equipment. By the 1990s, however, expansion placed pressure on existing transportation and telecommunications facilities, while power shortages reduced output and exports of garments and other textiles. In early 2004, tourism accounted for 10 percent of Sri Lanka's GDP, but the industry is still reeling from the devastating effects of the 2004 Indian Ocean tsunami.

Maldives

Fishing and tourism are the mainstays of the economy of the Maldives. Here, tuna accounts for 60 percent of exports. The islands also have commercial coconut plantations. A popular tourist destination, the Maldives saw a dramatic increase from a few thousand visitors in 1980 to nearly half a million by the early 2000s. Almost 80 percent of the economy of the 1,190 islands in the Maldives depends on tourism. The tourist industry employs over 25 percent of the working population, up from only 4 percent of the total labor force in 1985. The limits of expansion are however being reached, signaled by coastal pollution, groundwater contamination, and the need to bring in labor to cope with the increasing demand. The industry also suffered a big blow due to the Indian Ocean tsunami, but recovered speedily with international aid and government investments.

Contemporary Geographic Issues

Population Growth and Patterns

South Asia's total population of 1.5 billion in 2006 is predicted to rise to nearly 2 billion by 2025, compared to less than 1.7 billion in East Asia. The region has a high average population density of 186 persons per square kilometer (Figure 6.18) and a high proportion (39 percent) of arable land This means that additional growth will place severe pressures on the environment when the region has few current or prospective means of supporting its people at improved levels of life quality. India's population of 1.12 billion in 2006 is the world's second largest after China's, and its rate of increase is greater. Estimates for 2025 raise India's population to 1.36 billion. Growth continues because of the huge number of people of childbearing age (Figure 6.19) and increased life expectancy. In the demographic transition process, India remains in the population increase phase. The rapid increases of population in South Asia are caused by birth rates that remain higher than death rates. Birth rates are highest among the materially poorer groups, especially in rural areas, leading to significant changes in the social balance and political pressures from hitherto disadvantaged groups.

Population Dynamics and Comparisons

Bangladesh had a slightly higher population than Pakistan until 1980. By 2006, however, Pakistan's population exceeded that of Bangladesh (166 million compared to 147 million), and estimates for 2025 suggest a greater gap, with Pakistan having 250 million and Bangladesh 180 million people. The greater rate of

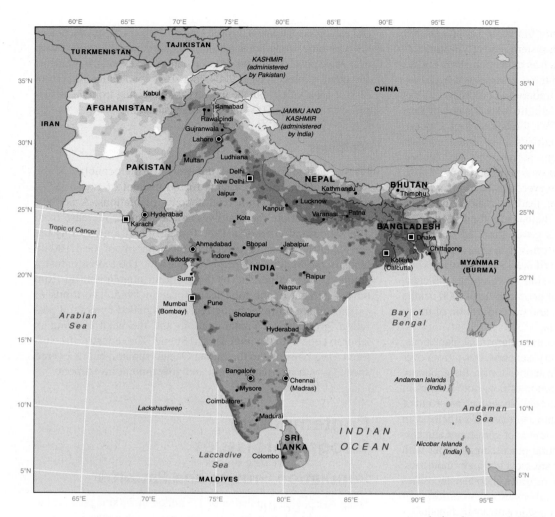

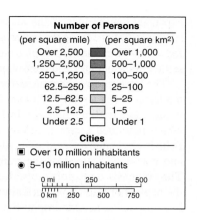

FIGURE 6.18 South Asia: distribution of population. Note the contrasts within South Asia: high population densities occur in the major river valleys and low population densities occur in high mountains and arid areas.

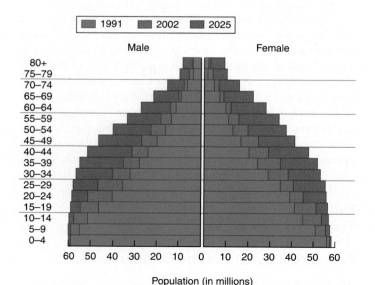

FIGURE 6.19 India: age-sex diagram. Even if fertility declines, the large numbers of young people will produce continuing population growth for many years to come when they reach childbearing age.
Source: U.S. Census Bureau; International Data Bank.

population increase in Pakistan is the result of a continuing high total fertility rate of nearly 5, compared to 3 in Bangladesh. Both have life expectancies in the upper fifties.

Given its smaller area, Bangladesh has greater problems of feeding its growing population, an issue that stimulated its government to institute effective family planning programs in the 1970s and significant reductions in fertility. By the late 1990s, half the married women used contraception, up from 8 percent in 1975. Despite a continuing subordinate role for women, new policies encourage their social and economic development.

Pakistan's need for an educated work force is undermined by the combination of influences from Islamic leaders, feudal landlords, and poor government that kept Pakistan's school population low until the late 1990s. Its enrollment rates in primary and secondary education were only 30 percent, compared to 70 percent in India and over 80 percent in Indonesia and China. In the mid-1990s, aid donors' demands led to new national policies for education, health, and population. Government funding shifted from elite university education toward universal primary education.

Of the countries on the margins of this world region, Afghanistan's population grows most rapidly at around 3 percent per year. Bhutan's, the Maldives', and Sri Lanka's populations increase at just over 1 percent. In most countries, total fertility rates are declining but remain over 6 in Afghanistan and almost 3 in the Maldives and Bhutan. All the countries remain at a stage when the population increases rapidly as life expectation increases, and there are many young people. For example, Nepal has a strong family planning program, widely disseminated, but only 29 percent of women and 2 percent of men use birth control methods; fertility rates decline slowly. This is linked to the Nepalese female adult literacy rate, which was 3.7 percent in 1971 but rose to only 15 percent by 2000.

Distribution and Density

As a relatively small world region that is home to the world's largest population, South Asia's population numbers and distribution reflect interactions between global and local influences. Apart from a few almost uninhabited areas, this region has high densities of population. The greatest concentrations are in the lowlands of the Indus, Ganges, and Brahmaputra Rivers and around the southwest coasts of Sri Lanka (see Figure 6.18).

Areas of very low population density include the desert areas of Pakistan and the western area of India, the mountainous areas of western and northern Pakistan, Afghanistan, the swampy Rann of Kutch that is mostly within India, and the Himalayan region in northernmost India, Nepal, and Bhutan. These areas hold little prospect for future population expansion. Such areas of very low population form breaks along the region's political borders and help form a clear definition of South Asia.

The rest of the region has moderate population densities, which often reflect the availability of resources that support rural employment. Over most of the region, improving water supplies for irrigation scarcely keep up with rising numbers of people, and so these areas of moderate densities offer few prospects for supporting more people.

Around 15 million ethnic Indians live abroad, with the largest groups in Nepal, South Africa, Malaysia, Sri Lanka, the Persian Gulf states, the United Kingdom, and the United States. The Indian diaspora has introduced elements of Indian cultures in these countries in the form of Indian food, Indian films, and even ethnic religions such as Hinduism and Sikhism through the erection of temples. In recent years, flows of workers to the oil-producing countries of the Persian Gulf and of professionals to the United Kingdom and United States for better-paid opportunities have created a significant "brain drain." Many Pakistanis, Bangladeshis, and Sri Lankans also live and work abroad and there was a major period of migration to the United Kingdom from the 1960s when former colonial people were allowed a British passport. More recently, workers from these countries moved to jobs in the Persian Gulf and Southeast Asia. Their remittances to families in South Asia form important income contributions.

Birth Control, Gender Inequalities, and HIV/AIDS in India

By 1961, India's population was 440 million and Indian government estimates of future population growth caused it to institute policies to slow population increase. Efforts to spread the use of birth control methods gained acceptance slowly among a largely illiterate people. The threat of forced imposition of male sterilization was a major political issue in the 1977 elections that ousted

Indira Gandhi's government until 1980. Greater effectiveness of family planning in the 1980s and 1990s built on rising levels of literacy, increased urbanization, and improvements in the status of women. Hindu women of higher castes use family planning more than those among the poorer castes and poorer, less-educated Muslims. The new urban middle class provides signs of the Indian population stabilizing around fertility rates of two to three children. More women adopt family planning, including sterilization, once they have given birth to a boy, and this is the most successful aspect of the program. Many families in rural areas, however, continue to want large families to provide extra manual labor and care of parents in later life. Moreover, many of the 120 million Muslims living in India disapprove of birth control.

The Indian age-sex diagram records fewer females than males (see Figure 6.19). The 2001 Indian census of population showed that India has a sex ratio of 107 males per 100 females, slightly lower than in 1991 but higher than that of its neighbors. Within India, sharp differences exist across the states, from 95 males per 100 females in Kerala to 116 in Haryana. Furthermore, in the under-7 age population, the sex ratio increased from 105.8 in 1991 to 107.8 in 2001, with the sharpest rises in the more prosperous states. It has been suggested that there are millions of missing females in India because of female infanticide, excused by some as saving the girl from a lifetime of suffering. In 1997, the *Times of India* estimated that mothers or village midwives kill 16 million girl babies a year. In 1994, after discovering that most aborted fetuses were female, the Indian government outlawed tests that could determine the sex of a fetus. Laws now prohibit doctors from revealing the sex of an unborn baby, although they are flouted among wealthier families. The poor cannot afford surgical abortions and carry out infanticides. In part of Rajasthan State, there are only 550 women to every 1,000 men, and newborn girls are generally buried without question.

The escalation of the practice of dowry in all sections of Indian society is causing a further devaluation of girls and women. For example, a man with a steady job in Tamil Nadu State expects a dowry of US$1,000, 100 grams of gold, household goods, and a car. Wives may be badly treated if their dowries are not up to expectations, and there are cases where the wife has been murdered by being burned alive so the husband can remarry for a better dowry. It is also a wifely duty to have sons; families with more daughters than sons are often implicated in dowry murders. Families with daughters start saving at birth but still require loans.

From the 1990s, HIV/AIDS emerged as a major health and development threat. An estimated 1 percent of India's adult population (or at least 5 million—but probably many more) is affected by the disease—the highest number for a country after South Africa. Worst affected are the southern Indian states (apart from Kerala), and Manipur in the northeast next to the Myanmar boundary. Together, these states account for 80 percent of the reported cases of HIV/AIDS in the country. However, rates in some of the northern Indian states that do not have checks may be just as high. The cities of Mumbai, Chennai, and Delhi also have high prevalence of the disease. HIV/AIDS is spreading fast in India's rural areas, where health care and knowledge of the disease are poor.

Alarmed by the predicted effects of an unchecked epidemic on its population and economic development, the government launched an HIV/AIDS prevention campaign. Its mission is to educate the Indian people on the means of transmission of the virus and offer support for prevention measures. To date most action comes from private organizations and funds from abroad, although the Indian government also is funding prevention measures. Many of the reactions typify Indian social situations. While vulnerable groups such as prostitutes and truck drivers are particularly liable to be infected, the low status of women in families assists transmission and those infected by husbands are cast out by their in-laws. Many HIV-positive men lose their jobs. Indian pharmaceutical companies Cipla and Ranbaxy produce generic anti-retroviral drugs for the Indian and developing country markets, selling them at a fraction of what they usually cost in developed countries.

Urbanization

Five of the world's twenty largest cities are found in predominantly rural South Asia. The metropolitan areas of Mumbai, Delhi, Kolkata, Dhaka, and Karachi each house over 10 million people (Table 6.2). The largest cities in all countries are immense and growing. Attracted by new jobs in factories and offices, better access to education and health services, and opportunities in the informal sector of the urban economy, migrants from rural areas and small towns pour into cities.

Urban Populations

Indian cities continue to house under 30 percent of the country's growing population. However, the total population of urban areas is projected to rise from 250 million people in the mid-1990s to over 600 million by 2025. From 1980 to 2003, the number of Indian cities with over a million people jumped from 12 to 37. By 2015, Mumbai (Bombay), Delhi, and Kolkata (Calcutta) will be three of the world's 10 largest cities (Table 6.2). Several Indian cities, including Mumbai and Delhi, have increasing global corporate services such as accounting, advertising, banking, and law, while Bengaluru and Hyderabad have global connections through their high tech industries.

Like India, Bangladesh and Pakistan remain largely rural countries. Pakistan's urban population was 34 percent in 2006, while Bangladesh's towns and cities account for 23 percent of the country's population. The major cities in both countries struggle to cope with the arrivals of rural people and the shantytowns and social tensions that this influx creates.

Karachi is the largest city in Pakistan and its major port and commercial center. Violence in the 1990s among rival ethnic Muslim groups here was made worse by the glaring contrasts between rich and poor, the large numbers of unemployed teenagers, and the availability of guns following years of warfare in Afghanistan. Other major Pakistani cities are inland, where Lahore is the major center of Muslim culture and Faisalabad is the cotton industry center. Islamabad, the new capital, is close to the

TABLE 6.2	Populations of Major Urban Centers in South Asia (in millions)	
City, Country	**2003 Population**	**2015* Projected**
INDIA		
Mumbai (Bombay, India)	17.4	22.6
Delhi, India	14.1	20.9
Kolkata (Calcutta), India	13.8	16.8
Chennai (Madras), India	6.7	8.1
Bengaluru (Bangalore), India	6.0	8.4
Hyderabad, India	5.9	7.5
Ahmadabad, India	5.0	6.6
BANGLADESH AND PAKISTAN		
Dhaka, Bangladesh	11.6	17.9
Chittagong, Bangladesh	3.8	6.2
Karachi, Pakistan	11.1	6.2
Lahore, Pakistan	6.0	8.7
Faisalabad, Pakistan	2.4	3.5
MOUNTAIN AND ISLAND RIM		
Kabul, Afghanistan	3.0	5.4

*estimated

Source: United Nations Urban Agglomerations 2003, with estimates for 2015 (2003).

Kashmir border, while Peshawar is the commercial center of the far north at the eastern end of the Khyber Pass into Afghanistan.

Bangladesh has fewer large cities. Dhaka is the capital, Chittagong the main port, and Khulna is a growing industrial center. The pressure of people on scarce farming land and lack of farm jobs for adults forced migration into cities, where jobs demand educational qualifications that most migrants do not have. From the 1980s, the shortage of jobs in Bangladesh resulted in many laborers seeking employment outside the country, such as in the Persian Gulf countries and Malaysia.

In the mountainous countries, main concentrations of people are in lower areas that are accessible to major transportation routes. Kabul (Afghanistan), Colombo (Sri Lanka), and Kathmandu, capital of Nepal, are the largest cities of the subregion and contain extensive shantytowns. The Himalayan states have most of their people near the Indian border. Sri Lanka's population is focused on the rainier southwestern coastlands around Colombo. In 2006 Nepal was only 14 percent urbanized, while the other mountain and island countries were just over 20 percent.

Changing Urban Landscapes

The urban landscapes of South Asia reflect the waves of cultural influences that washed across the region. Many ordinary buildings were constructed of materials such as wood that do not have a long life, but most towns have more permanent religious, royal, or military buildings that have lasted for centuries. Hindu temples are often very ornate (Figure 6.20). Muslim mosques and tomb gardens also provide distinctive landscape elements.

FIGURE 6.20 South Asia: contrasts. An Internet café is close to a Hindu temple in Bengalaru (Bangalore) India.

Older precolonial sections of towns left a heritage of high housing densities with poor transportation access along winding alleys. Shops and artisan workshops encroach on walkways. Few buildings are more than two stories and often there is no clear distinction of residential, commercial, and industrial functions. People of similar caste or trade live and work together in localities.

Sections added to towns during the British rule from the 1700s to the mid-1900s included a central market, administrative offices, and frequently a clock tower. New residential areas were built on European plans with wide streets and open places separating business and residential functions. Britain built and developed major port cities such as Kolkata, Mumbai, Chennai, Karachi, and Colombo initially as a means of establishing trading security. The ports were then joined to inland centers by railroads that remain the major mode of transportation available to most Indians. Some smaller cities, such as Bhopal, have no planned infrastructure, and, even in major cities, funds for maintaining infrastructure are scarce. Shantytowns are common.

Modern cities have major industrial areas, including steelworks and aluminum plants, chemical factories, textile mills and garment-assembly factories, vehicle makers, and electrical goods production. Concentrations of factories are most common around the metropolises of Mumbai, Delhi, Kolkata, and Chennai and in urban areas located close to raw material sources.

Up to the mid-1990s, the wealthy lived in the city centers and the poorer people toward the outskirts, but many of the wealthy have moved to new suburbs on the fringes of cities. Slums and shantytowns spring up to accommodate the continuous flow of poor urban migrants. Some slums have brick or block dwellings with windows, doors, and corrugated roofs; others are shacks made of any available material, often at risk from fire. The poorest ragpicker settlements adjoin and merge into refuse heaps. Where slums are recognized by municipalities, they have piped water, legal electricity connections, and shared sanitation. Many are not. There is no street lighting, cleaning, or repairs. Municipal policy is generally to control and raid the slums rather than protect them. Water is supplied by tankers and public standpipes, and there are few toilets apart from open spaces and the gutters. Yet homes are kept clean. Disease (diarrhea, tuberculosis) is endemic or occurs in rampant outbreaks of cholera, hepatitis, and typhoid. HIV/AIDS is a major scourge, as are drug dealing and drinking illicit alcohol.

The contrasts between the wealthy and the poor are clearly evident in cities such as Mumbai, India's primary commercial and industrial center. The city accounts for over one-fourth of India's manufacturing output and is home to over half of India's top 100 companies. Mumbai handles one-fourth of India's trade in its port and 70 percent of its stock exchange transactions. It is the headquarters of nearly all of India's commercial banks and has factories making a wide range of products from textiles to pharmaceuticals. India's $1.5 billion film industry is the largest in the world, making close to 1,000 movies a year, five times Hollywood's output. Mumbai is the center of India's movie industry (often called "Bollywood"), with half of all films made in India being filmed here. Real estate prices in Mumbai rival those of New York City, making it difficult for all but the wealthy to own their own homes.

Dharavi, allegedly the largest slum in Asia, is also found in Mumbai (see Figure 6.12). Poor housing, very high densities, limited access to water, sanitation, and other infrastructural facilities characterize the slum. However, over 100,000 people produce $500 million in goods each year from small businesses such as bakeries, metal workshops, recycling, tanneries, and potteries in Dharavi. Families who profited and moved out continued the business connections. The Mumbai government plans to turn the area into a showpiece development by 2010, although some suspect the motivation is to clear an eyesore.

Pollution due to improper and inadequate disposal of industrial effluents and domestic and industrial wastes greatly detract from the quality of life in South Asian cities. For example, Delhi generates nearly 4,000 tons of trash each day but clears only 2,500 tons, leaving the rest on the streets. Of 1,800 tons of sewage, only two-thirds are collected.

Air pollution in cities like Delhi and Kolkata are at their worst during the winter months when toxic gases, hydrocarbons, and particulate matter are trapped near the surface by cold air. Contributors to poor air quality are vehicular traffic, coal and kerosene used as cooking fuels, coal-fired power plants, and the poor condition of city streets. According to the WHO, more than 11,000 people die each year in Kolkata due to air pollution-related illnesses and at least half the city's children have excessive amounts of lead in their blood. The city struggles to improve its air quality. Delhi, which in the mid-1990s had the most air pollution of any Indian city, has succeeded in dramatically improving its air quality by mandating a switch to cleaner fuels like liquid petroleum gas and compressed natural gas for city buses.

Liberalization and the Rise of High Tech Cities in India

Consequent to greater openness to foreign investments and internal private enterprise following structural adjustment and liberalization in the 1990s, India emerged as one of the world's fastest growing economies. In 2005, a report by America's National Intelligence Council likened the emergence of India and China in the early twenty-first century to the rise of Germany in the nineteenth and the United States in the twentieth, with "impacts potentially as dramatic."

An important component of this economic revolution is the growth of high tech industries in India. The rise of this sector was assisted by a large youthful population, many with excellent scientific and technical training, a work force that speaks English, and legal systems that offered good copyright and contract law protection. The total output of India's modern high tech industry rose from US$2 billion in 1995 to US$12 billion in 2001. India is moving from established services such as chip design, back-office work such as call centers (Figure 6.21), medical transcription and information technology (IT) consulting to high-end ones such as research for banks and other financial agencies and research in drug development and engineering. High tech industries are found not only in large established metropolises such as Mumbai and Delhi, but also spurred growth in smaller regional centers such as Bengaluru and Hyderabad in southern India.

Well into the 1970s, Bengaluru (Bangalore) was known as the "Garden City" for its open areas, parks, tree-lined streets, and single-story residences. The city's economy was largely based on public sector industries such as munitions and aerospace and some private engineering and textile firms even in the 1980s. In the early 1980s Texas Instruments set up a factory here, but it was with the opening up of the Indian economy in the 1990s that Bengaluru began to attract numerous multinational corporations. The main new industries here are electronics, computer engineering, software and services, telecommunications, aeronautics, and machine tools. Multinational corporations in the city include 3M, AT&T, Digital, Ericsson, Hewlett-Packard, IBM, and Motorola. Bengaluru's attraction as a center for computer software and hardware development was complemented by a growing pool of skilled labor from its various engineering colleges and from migration to the city from other parts of India. Information technology and IT-enabled services accounted for over 60,000 jobs in the Bengaluru area by the late 1990s. More recently, biotechnology industries have joined the IT sector firms.

As its population and industries burgeoned, Bengaluru expanded in area. Technology parks such as Electronic City (Figure 6.22) and the Whitefield International Tech Park that house many electronics, telecommunications, and computer software and services firms emerged on the outskirts of the city where land was available. These planned industrial parks have facilities that are similar to those found in their counterparts in Western countries. Electronic City has over 100 firms including Motorola, Siemens, ITI, and Indian software companies Infosys and Wipro, set on

Figure 6.21 India: call center. A call center in Bengaluru (Bangalore). This service industry is a major part of Indian development of IT outsourcing. Jobs in this area are set to rise to over 1 million and sales to $14 billion by 2007.

FIGURE 6.22 India: modern technology. The Infosys Technology Campus, Electronics City, Bengaluru. Infosys employs over 14,000 staff and is India's top software exporter with clients like Bank of America and Citigroup.

330 landscaped acres. IT companies seeking more accessible city locations even set up offices in buildings zoned for residential use, in violation of city codes. Over 1,000 IT companies with an investment of $1.3 billion are based in Bengaluru, which is now considered India's Silicon Valley and IT capital.

New townships and residential areas (some of which are gated communities) have developed to house the city's growing IT worker population. International construction companies are vying with each other to develop new housing in the area. A 25 km long IT corridor, anchored by Electronic City and International Tech Park at either end, is designed to cater to a million people by the year 2021. The corridor will include business parks, commercial centers, six townships, two universities, hospitals and polyclinics, and two golf courses.

While Bengaluru's status as India's premier high tech city is widely acknowledged, other urban centers such as Hyderabad, Pune, and Noida are also developing their IT sectors. As is the case in Bengaluru, new technology parks and industrial campuses such as Hitech City are planned and developed in Hyderabad, a fast-growing IT hub. A progressive state government that promoted the establishment of new centers of technical, scientific, and business education, provided the infrastructural support to help fledgling firms and institutions of higher education, and initiated the development of infrastructure such as highways and a new international airport has transformed the former capital of a princely state into a modern city. Companies such as IBM Global, Google Online, Mindtree Consulting, iSoft, and Adea International all have operations in Hyderabad.

Ethnic Conflicts: Intra-State and Inter-State

All South Asian countries have sizeable ethnic and other social groups whose antagonism toward each other spills over into conflicts. Local rivalries often escalate into conflicts that affect wider areas. Tensions between Hindus and Muslims that led to the partition of British India into India and Pakistan continue. In secular India these came to the fore when in 1992, militant Hindus pulled down a Muslim mosque at Ayodhya, a site that is thought by some to be the major historic center of Hinduism and the birthplace of the god Rama. In 2002, Hindu-Muslim tensions erupted in violence in Gujarat as Hindu mobs burned Muslim areas after 58 Hindu pilgrims were killed in a train fire. These and other events led to the rise of *Hindutva* (Hindu View) as a political movement that is fertile ground for right-wing Hindu chauvinism and anti-Muslim and anti-Christian sentiments. The Rashtriya Swayamsevak Sangh (RSS, National Volunteers Association), founded in 1925 with an agenda for national unity on the basis of Hindu supremacy, and its political arm the Bharatiya Janata Party (BJP) have built a strong base in India.

In Nepal, after eight years of insurgency, Maoist guerrillas control the rural areas and threaten to take over the government in Kathmandu. Similar Maoist groups known as "Naxalites" operate within India and, although they talk to state govern-

ments as in Andhra Pradesh, refuse to disarm. Bangladesh is becoming involved with the spread of Islamic extremism linked to attacks in northeastern India. A plan to dredge the Sethusamudram Ship Canal through the shallow Palk Strait between Sri Lanka and India is causing fresh tensions following India's decision to go ahead despite Sri Lankan opposition. This canal would save over a day's sailing for ships between east and west Indian ports, but would harm local fisherman, Colombo's transshipment trade, and local environments.

South Asia may be turning into a **"shatter belt"** of conflict and disruption linked to the troubled areas of southwestern Asia (from Iran, through Iraq to Israel-Palestine) through the Pakistan-Afghanistan linkage. The interregional conflicts generated by such events hold back the type of cooperation and regional trading agreements that have become common in other world regions. The South Asian Association for Regional Cooperation (SAARC) set up in 1985 generated little economic interaction apart from some academic studies of the potential. Moreover, external and internal rivalries raise defense expenditures and affect development. Additionally, the region has many of the ingredients for nationalism and ethnic conflict. The geographic concentration of different ethnic groups, strong feelings of cultural pride, and perceptions of being excluded and discriminated against politically, culturally, and economically by a dominant group all form a potent mix that can lead to nationalistic fervor and ethnic strife.

Civil War in Sri Lanka

Sri Lanka was a politically stable country with a growing economy after independence and until the 1980s when northern minority Tamil people began to fight the dominant Sinhalese for their rights and a defined Tamil territory, Tamil Eelam (*eelam* means "homeland"). The Tamils comprise two distinct groups—the "Sri Lankan Tamils" who have lived on the island since before the Christian era, and "Indian Tamils" who are descendants of plantation laborers brought in the 1800s and early 1900s by the British colonizers. Sri Lankan Tamils are concentrated in the northern Jaffna area and the districts along the eastern coast, while the Indian Tamils are found mostly in the tea plantation areas of the Central Province. (Figure 6.23).

Problems between the majority Sinhalese and the minority Tamils arose after independence due to government policies that allegedly favored the Sinhalese. Attempts were made to repatriate the Indian Tamils, many of whom were denied citizenship in post-independence Sri Lanka. The Sinhala Only Act of 1956 made Sinhala the only official language of the island nation, restricted many government jobs to Sinhala speakers, and changed university admission policies in favor of the dominant group, causing the Tamils to lose educational opportunities as well.

An attack on the military by the Liberation Tigers of Tamil Eelam (LTTE) in 1983 and retributive anti-Tamil rioting in Colombo and other Sinhalese-dominated areas, in which up to 2,000 Tamils lost their lives, can be considered the beginning of the civil war. Peace talks between the Tamils and the government in 1985 failed. In 1987, although the Sri Lankan

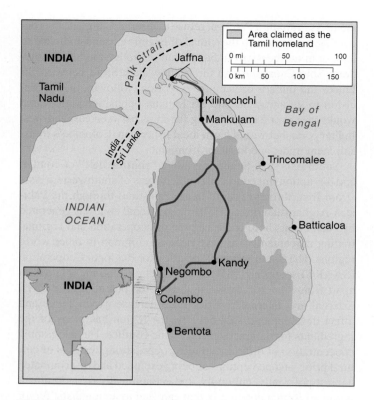

FIGURE 6.23 Sri Lanka: basis of civil war. The Tamil population lives mainly in the north and northeast of the island. *Source: Data from The Economist 1998.*

government succeeded in pushing rebels to the northern city of Jaffna, the LTTE and other Tamil guerilla groups set off several bombs in Colombo, causing hundreds of deaths. This landmark year also saw the signing of the Indo-Sri Lanka Peace Accord under which the Sri Lankan government conceded to some Tamil demands including greater power to the provinces and making Tamil an official language. The Indian Peace-Keeping Force (IPKF) was established to oversee the cease-fire and help maintain order in the Tamil-dominant north and east, but met with resistance from both Sinhalese and rebel Tamils. Hostility towards the IPKF and its brutality culminated in the 1991 assassination of former Indian Prime Minister Rajiv Gandhi, under whose leadership the IPKF was established.

Following the IPKF withdrawal in 1989, the LTTE recaptured significant areas of the north. Both sides intensified their attacks in a struggle for control of territory and power. Government forces attacked civilian buildings such as Tamil temples, churches, and schools that were used as safe havens by refugees fleeing war-torn areas, causing many Tamil casualties. The LTTE stepped up its attacks in public places and on important political figures. A Tamil suicide bomber killed Sri Lankan President Ranasinghe Premadasa in 1993. In 1996, the Central Bank in Colombo was bombed and in successive years so were the Sri Lankan World Trade Center and the Temple of the Tooth, a holy Buddhist shrine in Kandy. In 1999, the LTTE attempted to kill Chandrika Kumaratunga, Sri Lanka's president at that time.

However, the 1990s also saw greater attempts to bring peace to the island through domestic and international efforts. The LTTE, which had been banned as a terrorist organization by governments of India, the United Kingdom, and the United States, announced a unilateral cease-fire in December 2000 in a show of willingness to explore strategies to end the war while safeguarding Tamil rights. But they attacked Bandaranaike International Airport the following year, destroying planes and wreaking havoc on the country's tourist trade.

In 2001 a fragile cease-fire was brokered by the government of Norway. Since then, the island has been in an uneasy state of truce, complicated by accusations of covert operations by both the LTTE and the government. In 2005, newly elected President Mahinda Rajapakse promised to pursue peace talks with the Tamil rebels. However, in his election campaign he had called for renegotiation of the cease-fire and a tougher stance against the LTTE. By the end of 2005, the Tamil rebel group had renewed its attacks on the Sri Lankan Army and Navy, while a parliamentarian linked to the Tigers was assassinated during a Christmas mass. These and other assaults do not bode well for peace.

Casualties on both sides continue to occur and an estimated 65,000 lives, both Tamil and Sinhalese, have been lost in over two decades of fighting. Pressure from donor countries of the European Union as well as Norway and Japan may persuade the warring factions to return to serious negotiations. The international community is against the division of the island into separate states. However, many of the 700,000 Tamils who left Sri Lanka in the aftermath of the 1983 riots are fervent Eelamists, and politically and monetarily support the formation of a Tamil state. Without stronger commitment from both sides to end the war, Sri Lanka may prolong its history of violent ethnic clashes.

The Indo-Pakistan Dispute over Kashmir

The events of 9/11 and the subsequent attacks on Afghanistan brought to light many links of agreement and conflict across sets of countries. One of these was the long-lasting dispute between India and Pakistan over the Kashmir Province at the northern end of their common border (Figure 6.24). By early 2002, India and Pakistan were sending military forces to the border and war looked possible. As in its relationships with the former Taliban government in Afghanistan, however, the "War on Terrorism" caused the Pakistan government to hold back and even repudiate groups of Islamic radicals it supported and trained for many years. The situation along the Indian-Pakistani border remained tense—as it had since the partition adopted at independence in 1947.

Some history of the Kashmir situation explains present events. In the mid-1800s, the British sold the rulership of the Muslim state of Kashmir to a Hindu maharajah for around US$50 million at today's prices. At independence, Kashmir opted to be part of India. Although the maharajah tried to avoid the need to join India or Pakistan, an invasion by Pakistani tribesmen forced his hand, and he chose India in return for military help. Indian Prime Minister Jawaharlal Nehru, in his eagerness to get popular ratifi-

From 1989, separatist violence that India alleges is backed by Pakistan was matched by Indian military repression.

By the late 1980s, Pakistan looked friendlier to Kashmiris than India despite having a very different culture based on Islamic rules and military dictatorships. Indians believe their democratic constitution can provide liberation for Kashmiris. In fact, most Kashmiris see themselves as prisoners of either country and are weary of continuous fighting. They prefer independence.

It is difficult to see who speaks for Kashmiris. The state government based in Srinagar is very unpopular. Some Kashmiris look to the United States and its allies as potential saviors that can push India into a settlement. But India and Pakistan are the real arbiters. India shifted from inaction to attempts to talk with Pakistan or separatists. The current Pakistani crackdown on terrorists causes Indian cynicism about Pakistan's promises when banned groups such as Lashkar-e-Taiba continue to act after merely changing their names. India, however, keeps a security force of 400,000 men in Kashmir and hopes the crackdown will moderate the militancy. There is speculation that Pakistan will return support to more moderate Kashmiri groups such as Hizbul Mujahideen.

Any election or referendum faces major problems. If separatists who deeply distrust India participate, they must declare themselves Indian citizens to qualify to vote. Without the separatists, however, Kashmiri Muslims will mock the result as worthless. The Indian government rejects proposals for foreign election observers or outside mediation as questioning its authority. This attitude continues a high-handed and superior approach based on the original partition decision, Indian democratic institutions, and resistance to terrorist attacks but avoids discussion of how India alienated the Kashmiri people.

Partition of Kashmir between India and Pakistan with a degree of autonomy for both parts of Kashmir might then occur. Pakistan conceded much after 9/11, withdrawing support for Taliban in Afghanistan and shutting down a potential war against India after terrorists attacked the Indian parliament in Delhi. It remains to be seen whether Pakistan can accept a more democratic and humane version of the status quo that Pakistan fought against for over 50 years.

Pakistan's President Pervez Musharraf is a military dictator who gained international disapproval when he dismissed the previous democratic (but corrupt) government in 1997, ordered a nuclear bomb test in 1998, and oversaw training camps for Islamic extremists. On the other hand, he ended ties with the Afghanistan Taliban and banned Islamist groups likely to annoy India, showing that he is at least preparing the ground to act on his stated wishes to modernize Pakistan and bring in a democratic government again. Most Pakistanis see the winning of Kashmir and supporting the "freedom fighters" there as more important than backing the former Taliban government and al-Qaeda in Afghanistan.

Both India and Pakistan took advantage of U.S. policies arising from the war on terrorism from late 2001, but the Kashmir question threatened to undermine the uneasy peace. The buildup of Indian troops along its border with Pakistan took

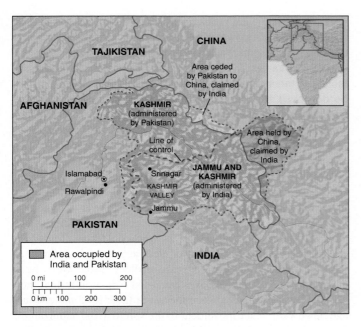

FIGURE 6.24 Kashmir. This map shows a complex geography. A line of control separates Indian and Pakistani areas of present occupation. India also has border disputes with China.

cation for the accession, brought Pakistan's 1947 invasion to the notice of the United Nations and urged Pakistan forces to withdraw until a referendum allowed Kashmiris to choose between India and Pakistan. After the 1948 cease-fire, Pakistan held onto the one-third it conquered, much of it close to the city of Rawalpindi and later the new capital, Islamabad. The Pakistanis never withdrew, and Nehru never put the matter to a vote. At first the maharajah, who only handed over control of defense, foreign affairs, and communications to India, and left Kashmir with its own prime minister until 1965, encouraged local politics.

India progressively integrated Kashmir, moving many of its Hindu population to Delhi and setting them up with craft industry employment making rugs, woolen clothes, and decorated plates and selling them in tourist outlets. In 1964, India extended to Kashmir the right of the federal government to dismiss state governments. Each decision, however, seemed to thwart the popular will of Kashmiris. India and Pakistan fought over Kashmir in 1965–1966, when India accused Pakistan of backing insurgents.

War broke out again in 1971, when Bangladesh (East Pakistan) became independent, backed by India. Kashmir was another source of fighting in that war, after which the 1972 Simla Agreement divided Kashmir into the Pakistan state of Kashmir and the Indian state of Jammu and Kashmir. This agreement established today's line of control.

Any goodwill generated by the first genuinely free Jammu-Kashmir elections in 1977 that brought back Kashmiri leader Sheikh Abdullah collapsed with the ousting of his son from power in 1984. Antigovernment Muslim groups contested the 1987 state elections, but Delhi annulled some of their successes.

Pakistani units from their patrols along the Afghanistan border, allowing Taliban and al-Qaeda elements to escape more freely. India used the situation to defuse Islamic militant terrorism in Kashmir and Delhi, getting U.S. President George W. Bush to freeze assets of Pakistan-based groups. The war on terrorism broadened to include cooling antagonisms between two nuclear powers who became allies.

Environmental Problems in a Developing Region

The effective use of the natural resources of South Asia, such as its waters, soils, and plant and animal life, is critical to the development of the region. In their quest for economic progress, governments of South Asian countries have dammed rivers for hydroelectric power, diverted channels and tapped underground aquifers for irrigation and industrial and domestic use, and harvested forests and wildlife. Religions and traditions in South Asia emphasize the interdependency of humans and the environment. They consider humans a part of the natural system rather than in charge of it. However, rapid population growth and industrialization have taken their toll on the environment, which has become increasingly polluted and degraded. Humans and nature do not coexist in harmony in this world region.

Problems of Water Use

A critical issue for agricultural and industrial development and growing settlements is the availability of water. Seventy percent of India's agricultural land is rain fed. The amount, timing, and distribution of rainfall control the success or failure of crop production. Water availability during the dry season has been increased through canal irrigation, tube wells, and storage tanks. However, water management remains a critical problem. The overuse of irrigation water in dry areas can lead to salinization as in prime agricultural lands in Pakistan and Uttar Pradesh in India. Disputes between countries or states within a country over the use of water from rivers that flow through different jurisdictions add to complications of water management.

The building of dams to regulate water flow to irrigation canals and generate hydroelectric power is a strategy favored by planners in India, who viewed the dams also as symbols of progress. The country has more than 1,500 dams, making it one of the world's largest dam-building nations. Many of India's large-scale dam projects were modeled on the Tennessee Valley Authority in the United States, with its multiple goals of increased agricultural productivity through irrigated farming, flood control, generating hydroelectric power, and constructing sites for recreation. Huge investments and foreign technical aid mark these megaprojects. However, problems such as rapidly silting reservoirs, leaking canals, and loss of the natural environment and traditional ways of life accompany the expected positive outcomes of such dams.

The environmental problems associated with water-based irrigation and hydropower projects occur in both the source and use areas. Upstream water storage requires building dams and flooding vast areas. Even where dams are constructed in sparsely inhabited mountains, they eventually fill with sediment or may be destroyed by earthquakes. Downstream, the extraction of water for irrigation alters river flow patterns. For example, Ganges River floods are reduced, but dry season flow is much lower. In areas using irrigation, poor management of water sometimes causes soils to become waterlogged or saline. Wells already provide nearly half the irrigation water used and remain the main source of irrigation growth. Tube wells drilled by modern rigs increased in numbers from 200,000 in 1960 to over 4 million by the 2000s. However, the sinking of deeper wells lowered water tables, cutting out smaller farmers who could not afford deep wells. Water is lifted by diesel or electric pumps, in contrast to the human or animal power used for older wells. Wells have environmental advantages over canal-fed irrigation in that they use local water supplies, keep the water table low, and encourage downward water movement in the soil to avoid salinization. They are increasingly combined with canal irrigation so that local groundwater sources can be replenished.

The Narmada Valley Project

The Narmada Valley Project is one of the largest of its kind undertaken in South Asia. The idea of damming the Narmada, which flows from east to west through arid regions in central and western India (see Figure 6.1), was considered in the late 1800s and later dismissed. Post-independence plans for the giant project on this sacred river were formulated in the 1960s.

From its inception, the project was subject to controversy as governments of Gujarat, Madhya Pradesh, and Maharashtra, the three states the river traversed on its route to the Arabian Sea, clashed over the division of the costs and benefits of the project. Madhya Pradesh claimed greater benefits as much of the land submerged due to dam construction was in its territory, while Gujarat raised the issue of water needs in its drought-prone areas to make its case for a larger share of the water. The entire scheme called for 30 major dams, 135 medium dams, and more than 3,000 small irrigation systems.

The Bagri Dam, the first to be completed, flooded 162 villages, displacing their populations while providing no plans for their resettlement. The cause of the largely tribal displaced people was taken on by activists led by Medha Patkar, forming an action committee in 1986 called the Narmada Bachao Andolan (NBA; Save the Narmada Movement). The next two dams to be built, the massive Sardar Sarovar dam and the Maheshwar dam, became the focus of the NBA. The Sardar Sarovar Project is a joint venture of the states of Madhya Pradesh, Gujarat, Maharashtra, and Rajasthan. The storage dam is expected to submerge over 37,000 hectares of land and result in the displacement of more than 1.5 million people. On the positive side, the dam projects are expected to generate 1,450 megawatts of power, irrigate some 1.8 million hectares, particularly in the arid Barmer and Jalore districts of Rajasthan and in drought-prone areas of western Gujarat, and provide water to between 20 and 40 million people.

Social activists argue that the benefits of the projects are spatially and demographically skewed. A disproportionate number of those displaced due to flooding are indigenous tribal people (Adivasis) and dalits in Madhya Pradesh and Maharashtra, while those who benefit are largely settled agriculturists and urban dwellers in the lower Narmada valley in the state of Gujarat. Using nonviolent Gandhian practices such as hunger strikes, passive resistance (such as refusing to evacuate as flood waters rose), and petitioning and mobilizing international environmental support, the NBA called attention to the human and social costs of building the huge dams (Figure 6.25). Protesters demonstrated that such projects were environmentally unsustainable. International support for the movement caused the World Bank to withdraw its funding for the project in 1992. In 2000, however, work resumed on some of the major sections of this project as the National Hydro Power Corporation became involved. Ninety thousand people being displaced received improved compensation.

FIGURE 6.25 **India: Narmada Valley Project.** Protesters oppose the building of the Sardar Sarovar dam in the Narmada Valley as it will submerge many villages and displace thousands of people.

GEOGRAPHY AT WORK

Battling Infectious Diseases

Mohammed Ali (Figure 6.26) uses his skills in geographic information systems (GIS) to study the spatiotemporal patterns of infectious diseases such as cholera, dengue fever, and acute lower respiratory infection in developing countries. People's environments and their health behaviors vary in space. The population groups to which individuals belong and the neighborhoods in which they live influence their lifestyle, health, and health-seeking behaviors. Epidemiologists and medical geographers study environmental risk factors at different geographical/ecological scales to understand health problems.

A native of Bangladesh, Dr. Mohammed Ali works at the International Vaccine Institute (IVI) in Seoul, South Korea. He and his colleagues at the IVI have conducted vaccine evaluation studies in research sites in China, Vietnam, Pakistan, and India. They created household GIS databases to investigate the ecological factors that influence vaccine coverage and to map geographic variations in vaccine use. The IVI team use GIS, satellite remote sensing, and spatial modeling techniques in research that is grounded in geographic theories of human-environment interaction. The team modeled the emergence and fluctuation of cholera in central Vietnam and in Matlab, Bangladesh. By integrating spatial data sets on sea surface temperature, rainfall,

FIGURE 6.26 **Dr. Mohammed Ali** working with colleagues in Guanxi Province, China.

vegetation, land use, climatic variables such as monthly temperature and monthly rainfall, and socio-demographic data such as population distribution and socioeconomic status, they examine the effects of individual and ecological factors on the emergence of disease. They believe that associations between these variables and cholera incidence in Vietnam and Bangladesh can be used to predict future epidemics in other parts of the world.

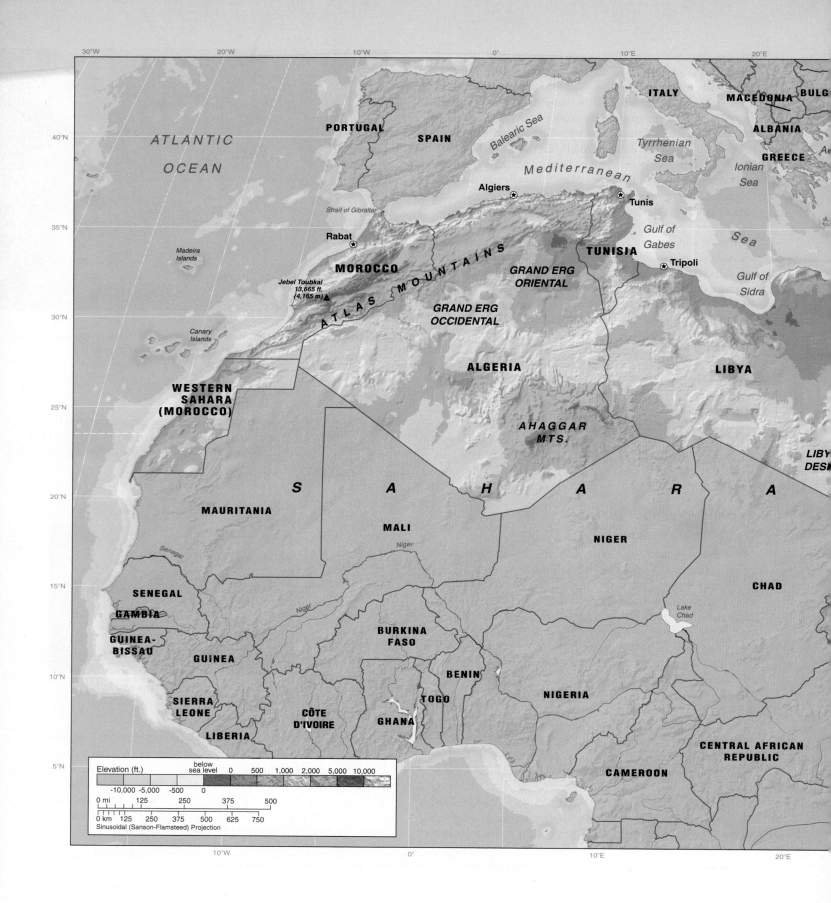

ATLANTIC
OCEAN

PORTUGAL

SPAIN

Balearic Sea

ITALY

MACEDONIA

BULG

Tyrrhenian Sea

ALBANIA

GREECE

Ionian Sea

Mediterranean

Algiers ⊛

⊛ Tunis

Strait of Gibraltar

Madeira
Islands

Rabat ⊛

MOROCCO

Jebel Toubkal
13,665 ft.
(4,165 m) ▲

A T L A S   M O U N T A I N S

Sea

TUNISIA

*Gulf of
Gabes*

⊛ Tripoli

GRAND ERG
ORIENTAL

*Gulf of
Sidra*

GRAND ERG
OCCIDENTAL

Canary
Islands

WESTERN
SAHARA
(MOROCCO)

ALGERIA

LIBYA

AHAGGAR
MTS.

LIBY
DESE

S   A   H   A   R   A   A

MAURITANIA

Senegal

MALI

Niger

NIGER

CHAD

SENEGAL

Niger

Lake
Chad

GAMBIA

GUINEA-
BISSAU

GUINEA

BURKINA
FASO

BENIN

NIGERIA

SIERRA
LEONE

CÔTE
D'IVOIRE

TOGO

GHANA

LIBERIA

CENTRAL AFRICAN
REPUBLIC

CAMEROON

Elevation (ft.)

below
sea level   0   500   1,000   2,000   5,000   10,000

-10,000 -5,000   -500   0

0 mi   125   250   375   500

0 km 125   250   375   500   625   750
Sinusoidal (Sanson-Flamsteed) Projection

192

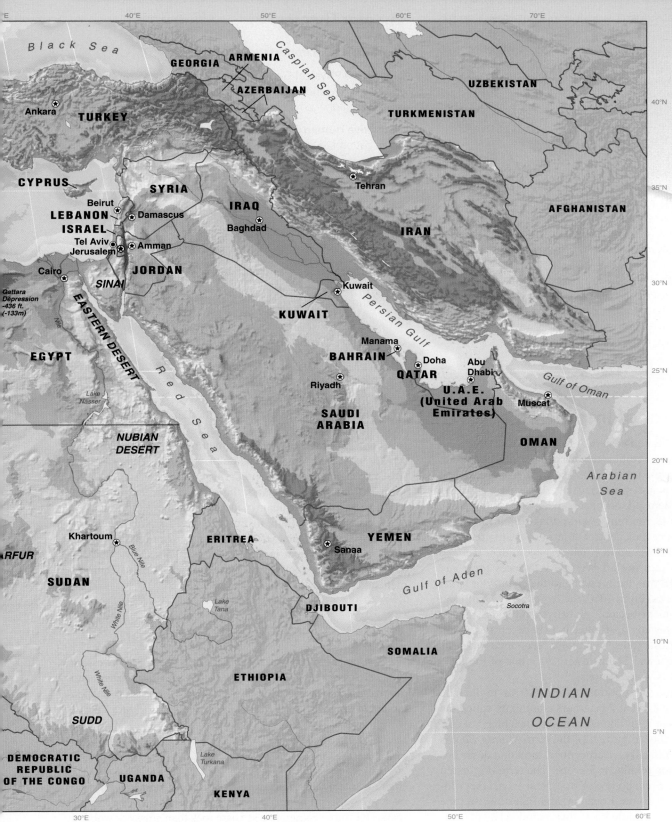

FIGURE 7.1 **Northern Africa and Southwestern Asia: physical features, country boundaries, and capital cities.** Northern Africa is bounded to the north by the Mediterranean Sea and to the south by the almost empty Sahara Desert area, except in the east, where the Nile River joins Sudan to Egypt. Southwestern Asia is almost surrounded by seas, except in the narrow Sinai Peninsula at the northern end of the Red Sea and the northern mountain boundaries between Turkey, Iran, the Caucasus countries, Turkmenistan, Afghanistan, and Pakistan.

Ahmed is a medical student, studying in what is regarded as the "cream" of Omani professional courses. He is one of seven children, and going to university in Muscat was his first time away from home. He began the medical training directly out of high school. It took some adjustment and several weekend trips home—a three-hour drive taken in relatives' cars or taxis. In Muscat itself, public minibuses called *biaza* are cheap and stop anywhere along the main routes.

In his first year, Ahmed lived in a hostel for men. He enjoyed sports and a variety of clubs in mixed groups on campus. In Oman, men and women are taught together in universities. At Sultan Qaboos University, the men enter lecture rooms from the front and the women from the back. The men have walkways at ground level, the women on the level above. All students wear national dress, except when engaged in sports. The women at university may not wear face coverings for security and identity reasons.

The facilities for study are as good as anywhere in the world, with modern laboratories and huge libraries. Everything is free. There are no tuition fees, room and board costs, or charges for books. Ahmed now lives off the university campus and receives 120 rials (US$300) a month toward his expenses.

Chapter Themes

Global Center of Importance

Distinctive Physical Geography: clashing plates; dry climates and desert vegetation; natural resources: water; environmental problems; global environmental politics.

Distinctive Human Geography: cultural diversity; population; urban patterns; evolving political geographies; global economics.

Geographic Diversity
- North Africa.
- Nile River valley.
- Arab Southwest Asia.
- Israel and the Palestinian territories.
- Turkey and Iran.

Contemporary Geographic Issues
- Israelis versus Palestinians.
- Human rights.
- Iraq.

Geography at Work: Operation Iraqi Freedom

Global Center of Importance

Northern Africa and Southwestern Asia is one of the most influential world regions. Prominence began early in history with some of the world's first urban civilizations (Mesopotamia and Egypt). The three major monotheistic religions of Judaism, Christianity, and Islam all originated here before spreading worldwide, and the region remains important to all three religions today. Europeans took control of most of the region in the 1920s and created new countries that reshaped identities and relationships. Hostilities toward Western influences grew and intensified after 1948 with the founding of the country of Israel as a homeland for Jews. From 1950, the region's oil wealth and then resurgent Islamism became globally significant, though material wealth is unevenly distributed both between and within the region's countries. Today's strategic economic interests make Northern Africa and Southwestern Asia one of the most politically contentious world regions.

This world region incorporates North Africa to include much of the Sahara and what Westerners commonly called the "Middle East" (Figure 7.1). Though it has its own blend of cultures and environments (Figure 7.2), the term *Middle East* is an ethnocentric, specifically a Eurocentric, term that describes lands by their location relative to Europe. The term originated from an early 1900s French view of the "Near East," incorporating modern Turkey, Syria, Lebanon, and Palestine, and the British concept of the "Middle East" that included Egypt, Arabia, and the Persian Gulf area. The "Far East" (Pakistan and beyond) indicated greater distance from Europe. Rather than

apply the Eurocentric term of "Middle East," the more global and culturally sensitive term "Southwestern Asia" is used in this book. In this chapter, the defined boundaries of Northern Africa and Southwestern Asia are primarily coastlines, with a line drawn along country boundaries through the Sahara in the south. Mountain ranges mark the northern and eastern limits of Turkey and Iran. Sudan is included in the region because of its strong connections to Egypt along the Nile River valley. The region has a largely arid natural environment, making fresh water a political issue. Nevertheless, despite extensive deserts, more productive mountain and river plain environments are significant as well.

Distinctive Physical Geography

The natural environments of Northern Africa and Southwestern Asia help to explain the region's geographic distinctiveness and internal diversity. Hot, dry plains exist alongside snow-capped mountain ranges and well-watered river valleys (Figure 7.3).

Clashing Plates

High mountain ranges and extensive plateaus contrast with wide river valleys (see Figure 7.1) and stand out where vegetation is sparse. It is more common to see bare rock than grassy or wooded slopes. Such landform differences are related to geologic activity along tectonic plate boundaries of the African, Arabian, and Eur-

(a)

(b)

(c)

(d)

FIGURE 7.2 Northern Africa and Southwestern Asia: diverse landscapes and peoples. (a) McDonald's in central Cairo, Egypt, is one example of global connections. (b) Hassan Mosque in Casablanca, Morocco, was built for the sixtieth birthday of former Moroccan king Hassan II in 1993. It is the second largest religious monument in the world after Mecca. It has space for 25,000 worshippers inside and another 80,000 outside. The 210-meter minaret is the tallest in the world. (c) Weekly animal market at Sinaw, Oman. Although some camels are sold, most of the animals traded are goats. Oman is an oil-producing country, but many Omanis still depend on agriculture and pastoralism for their livelihoods. (d) Village of Bilad Sayt in the Hajar Mountains of Oman. Although all of Oman's cities and towns are connected by good roads, this and many other villages can still only be reached by dirt roads. Note that the village is built on a bluff, as good farmland is scarce.

asian plates (see Figure 1.5). Such rigid sections of Earth's crust underlie the extensive plateaus of northern Africa and Arabia. The ancient rocks are often covered by flat layers of sedimentary rocks, such as the limestones and sandstones forming the plateaus on either side of the Nile River valley.

Collisions of the African and Arabian plates with the Eurasian plate formed the highest mountains: the Atlas in North Africa and the Taurus, Zagros, and Elburz ranges in Turkey and Iran. The Atlas Mountains rise to nearly 4,000 m (12,500

ft.) and the Elburz to over 5,500 m (17,000 ft.). Both are parts of the young folded mountain ranges that extend from North Africa and Southern and Alpine Europe, through Southwestern Asia, to the Himalayas in South Asia. Rocks, which were forming in the seas that filled the basins between the plates, were caught up as the collision closed the seas and raised them as parts of the mountains. Volcanic eruptions along the geologically active belt injected mineralized fluids that solidified as veins in the rocks. Earthquakes are common and threaten

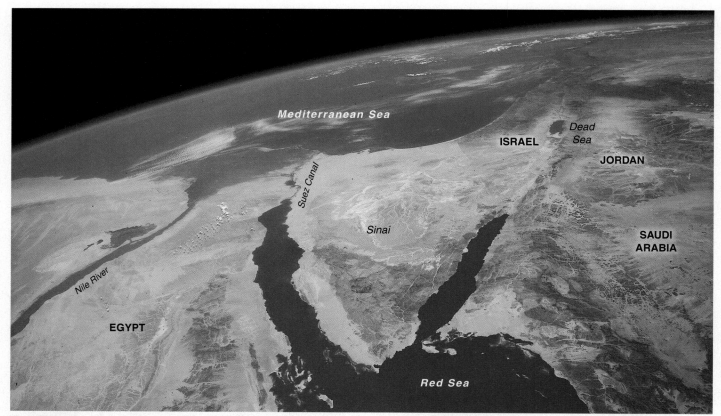

FIGURE 7.3 **Arid lands: parts of Egypt, Israel, Jordan, and Saudi Arabia.** A space shuttle view taken over the northern Red Sea across desert lands to the Mediterranean Sea. On the left, the irrigated valley and the delta mouth of the Nile River and the Suez Canal; in the center, the Sinai Peninsula; on the right, the Gulf of Aqaba leading northward to the Dead Sea.

populations. Compression folded and lifted up these rocks, and large quantities of eroded rock were deposited in downwarped sections of the crust, where oil and natural gas could be trapped. The Persian Gulf area became the environment for the world's largest concentration of these fuels.

The Red Sea is in a rift (the Great Rift Valley of Africa) where two tectonic plates are pulling apart, causing the Red Sea to widen. At its northern end, the rift splits in two around the Sinai Peninsula (see Figure 7.3). The eastern branch forms the valley into which the Jordan River flows from the uplands around the Sea of Galilee toward the Dead Sea, some 400 m (1,312 ft.) below sea level. The western branch falls short of entering the Mediterranean Sea but was extended to that sea by the construction of the Suez Canal, which opened in 1869.

Dry Climates and Desert Vegetation

Arid climatic environments, in which evaporation rates are greater than precipitation, dominate virtually the whole region (see foldout world climate map inside back cover). The region boasts the world's highest recorded shade temperature (58°C,

or 136°F, at Al'Aziziyah, Libya), but the lack of cloud cover makes nights cool or cold, and freezes are possible in winter.

Most places have some rain, but it falls irregularly and with increasing uncertainty as aridity increases. The rainy winters of the coasts of North Africa and the eastern Mediterranean lands are caused by the seasonal southward shift of midlatitude frontal weather systems bringing Atlantic moisture. The mountains of Turkey and Iran force air to rise, often adding lift to the frontal zones, and precipitating rain and snow. Winter snowfall provides a source of meltwater that feeds the Tigris-Euphrates River system in spring. In southern Sudan, summer rains come from the northward movement of the equatorial rainy belt.

The arid climatic environments support only drought-resistant desert plants that partially cover the ground. Large areas of desert have little vegetation and are gravel-strewn, rocky, or sand-covered: sand seas cover one-fourth of the Sahara, which has a mostly rocky or gravel-covered surface. The vegetation cover thickens and becomes denser on uplands, where there are higher precipitation totals and lower rates of evaporation. In the uplands of North Africa, Lebanon, western Syria, Turkey, and Iran, grassland and woodland vegetation increase with altitude as temperatures and evaporation levels fall and make the low-to-moderate precipitation more effective.

Soils are poor and undeveloped through most of the region: the best occur in the rainy coastal areas and along the valley floors of rivers where annual floods deposit fertile alluvium. Soil erosion resulting from plowing for grain and cultivation on steep slopes is a major problem in most of the region.

Climate change left its mark on the region and affected the history of human settlement. The present aridity began some 5,000 years ago, forcing many people into the watered valleys of the Nile and Tigris and Euphrates rivers, thus concentrating populations. The desert margins continue to fluctuate. They retreat in series of wetter years and advance in drier years—or as human actions remove vegetation and lower the groundwater levels.

Natural Resources: Water

The arid climate makes rivers unusual in Northern Africa and Southwestern Asia. Rivers that exist often receive their water from surrounding mountains or from rainy areas outside the region. For example, the Tigris-Euphrates River system is fed by snowmelt on the high mountain ranges of Turkey and Iran. The Nile River system has two major headwater branches that begin in the rainy equatorial area around Lake Victoria (White Nile) and the seasonal rainy area of the Ethiopian Highlands (Blue Nile). The two major Nile River branches join near Khartoum (Figure 7.4). The White Nile River, flowing from Lake Victoria on the equator, has tributaries that are fed by rains through the year and supply a fairly constant flow of water. The Blue Nile River flow is more important because it is less subject to evaporation and produces the annual September Nile River flood in Sudan and Egypt.

The renewable supplies that come from rain and snowfall feeding rivers are meager over most of this region. Other sources include underground reservoirs that accumulated over millennia (virtually nonrenewable) and **desalination plants** that make seawater usable (expensive). Water demands are mainly from irrigation farming, the oil industry, and growing urban centers.

Political issues arise when precipitation falls in one country and feeds rivers flowing through another. Turkey, Syria, and Iraq compete for the waters of the Tigris and Euphrates. Israel and Jordan must share the Jordan River. Egypt and Sudan have conflicts with each other over the Nile River and must now take into account increasing use in Ethiopia and the other headwater countries, including Uganda, Kenya, and Tanzania (see Chapter 8). By 1988, the expectation that the Aswan High Dam (completed in 1970) would ensure Egypt's water needs for 50 years was proved wrong, and discharges below Aswan had to be reduced to maintain flow through the year.

In Saudi Arabia, most water comes from underground sources. The growth of the oil industry and urbanization in the arid areas placed greater demands on water resources, leading to increased groundwater extraction. Coastal settlements have built costly desalination plants to provide fresh water. A large proportion of the land is without fresh water, however, and has little potential for economic development.

Environmental Problems

Ecosystems in arid environments are particularly fragile, as they are easily destroyed by human activities and do not recover quickly. As this region's population rapidly grows, new industries and cities are built in sensitive arid environments, and higher living standards raise the demand for water. Irrigation farming in arid areas requires good management to prevent the high rates of evaporation from drawing so much salt to the surface soil that crop productivity is reduced or ended. This process is known as **salinization.** Under careful management, maintaining good drainage allows water to flush the salts downward. Adding too much water waterlogs the soil and concentrates salts at the surface. In ancient irrigation schemes, 60 percent of the land in the Tigris-Euphrates River lowlands became unusable because of poor management. Modern usage led to loss of farmland for similar reasons.

The oil industry is a major polluter throughout the world, and the concentrations of production and distribution centers in the Persian Gulf area pollute the atmosphere and waters around oil wells, surface seepages, and where unwanted gases are flared off. Leakages pollute the waters near ocean terminals. Particles in the atmosphere cause fogs that worsen respiratory ailments. The Persian Gulf had lost most of its plant and animal life as a result of pollution since oil production first began there in the early 1900s. At the end of the Gulf War in 1991 the retreating Iraqis set oil wells on fire, adding carbon gases and particles to the atmosphere, but the particles fell on desert sands. They also released a huge oil slick, damaging plant and animal life, but it had less impact than it would have if the Persian Gulf had not already been heavily polluted.

Global Environmental Politics

Iran and Saudi Arabia ranked 18th and 19th in the world respectively in their total CO_2 emissions from fuel combustion in 1997, the year of the Kyoto Protocol. The top emitters per capita in the world in 1997 were Qatar, Bahrain, United Arab Emirates, and Kuwait, followed by the United States in 7th place and Saudi Arabia in 10th place. Though the oil-producing countries of Northern Africa and Southwestern Asia are contributing significantly to global CO_2 emissions, none of the countries are on the Kyoto Protocol's list of Annex I countries, industrialized countries needed to implement the Protocol. Nevertheless, countries of the region have shown their support for the Kyoto Protocol. Morocco was the first country of the region to ratify the Protocol (47th country overall to ratify), followed by Jordan (103rd), Israel (121st), and Yemen (125th). Russia ratified the Kyoto Protocol in November 2004, thus satisfying the minimal requirements for it to go into effect. Qatar, Egypt, Oman, United Arab Emirates, and Saudi Arabia ratified the Protocol soon after.

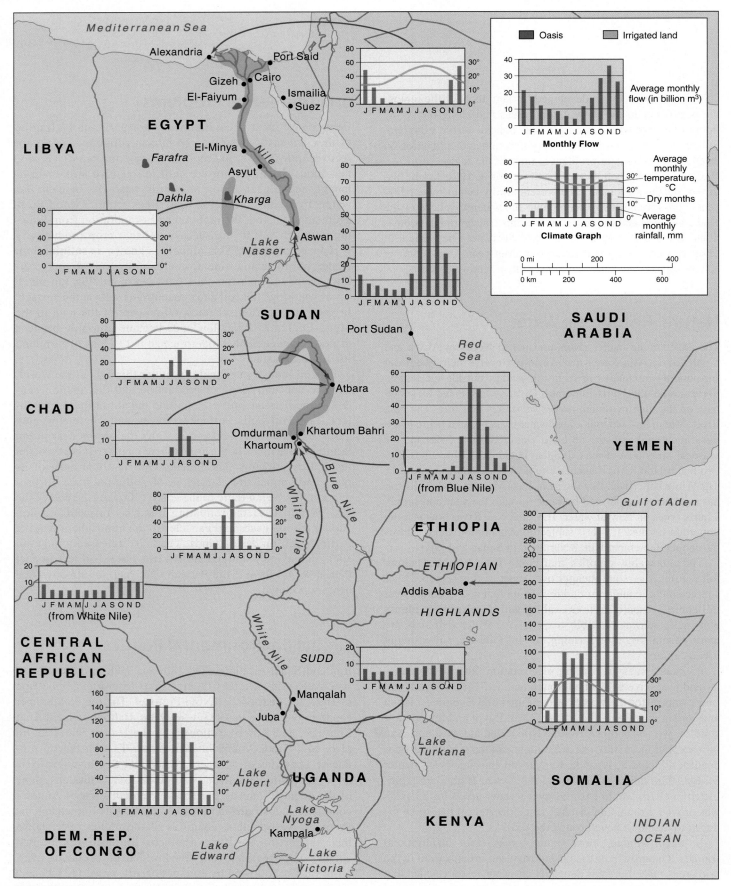

FIGURE 7.4 **The Nile River valley: rainfall and river flow compared.** The relatively small areas of irrigated land are shown in green. The climate graphs (yellow background) and the river flow graphs (blue background) show how the main flow comes from rainfall in the tropical south, and that the arid area through northern Sudan and Egypt receives little rain. The White Nile flow is moderate, but consistent; the Blue Nile annual flood dominates flows downstream of the confluence.

Distinctive Human Geography

This world region is culturally diverse, and a historic center for artistic, scientific, and technological innovations. Along with having differing languages, the region is the hearth for Judaism, Christianity, and Islam, related to one another through worshipping the God of Abraham. Muslim empires flourished in the Middle Ages before this world region became a focus of geopolitical strategies by major world powers in the 1800s and 1900s. Its location at the junction of other world regions meant that this region acted as a source of diffusion outward and as an inward-directed focus of external influences. The discovery of oil has provided wealth and influence to some, but has also become a geopolitical issue. Growing populations live mainly in urban areas in arid areas and also stress water resources.

Cultural Diversity

Languages

Northern Africa and Southwestern Asia is linguistically very diverse with numerous languages among four major language families spoken (Figure 7.5). The most common are Arabic, Berber, Farsi (Persian), Turkish, Kurdish, and Hebrew. Arabic is spoken by just under 50 percent of the people in the region. It is also the preferred language in the Qu'ran and Muslim prayers. "Arabs," who were once defined as living in the Arabian Peninsula, now include all those using Arabic as their first language. Berber is linguistically connected to Arabic but older and found primarily in Morocco and Algeria, especially in the Atlas Mountains and the Sahara Desert where it best resisted the spread of Arabic (Figure 7.6). Farsi (Persian) is the official language of Iran and is linked to the Shiite Islamic beliefs of Iran's leaders and majority population. Kurdish speakers straddle Turkey, Syria, Iraq, and Iran. Hebrew is the official language of Israel. The language of the Hebrew Bible and religious services is an archaic form, superseded in everyday usage by modern Israeli Hebrew. The movement devoted to creating a Jewish state from the 1800s, known as Zionism, revived the use of Hebrew, modernizing it for secular usage to provide a common religious and secular element among Jews from many countries.

Religions

Religion and related traditions contribute greatly to peoples' identities in Northern Africa and Southwestern Asia. The early animist religions had many gods linked to natural phenomena. Some human emperors were treated as gods. After approximately 1000

FIGURE 7.5 Language families and languages of Northern Africa and Southwestern Asia.

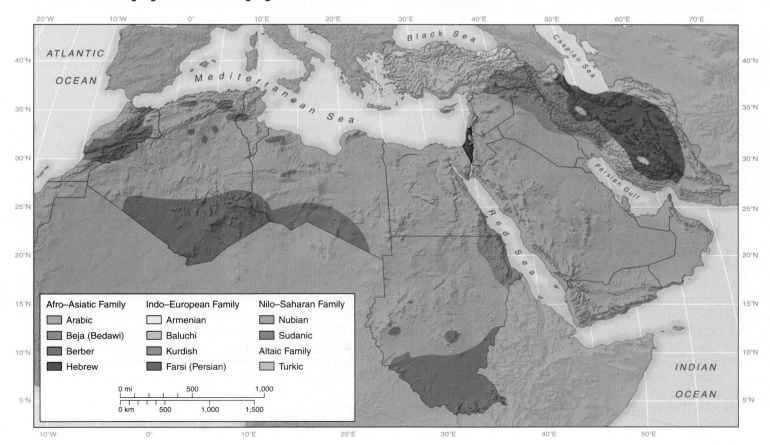

FIGURE 7.6 Berbers in Morocco. These two Berber men guide tourists into the Sahara Desert close to Erfoud, Morocco.

BC, dominance by religions with many gods gave way to religions based on a single god: **monotheism.** Judaism was the first monotheistic religion and was followed by Christianity and then Islam. These three monotheistic religions share common religious figures and religious texts. All three diffused from their hearths in Southwestern Asia to Europe, Africa, and Asia (Figure 7.7).

Judaism is a religion whose adherents worship Yahweh, seen as the only God, creator, and lawgiver. It began in the area now known as Israel and the Palestinian territories some 2000 years BC, where Abraham and his descendants settled after moving from Mesopotamia. Jewish beliefs focus on the historic role of family based on the line from Abraham, persecution beginning with slavery in Egypt, redemption as Moses led the people out of Egypt, and their occupation of the promised land. Yahweh intervened to support and punish through times of trouble and deportation, promising a messiah to save Jews from domination by others. In AD 70, the Roman army destroyed the holy city of Jerusalem and dispersed Jews through Southwestern Asia, northern Africa, and Europe to form a major **diaspora** (dispersed community).

Christianity stemmed from new interpretations of the beliefs and teachings of Judaism in the early years AD, focusing on the teachings of Jesus of Nazareth. Christians began as Jews who believed that Jesus of Nazareth was the messiah, which translates to "Christ" in Greek. Christian churches spread across Southwestern Asia and into Africa and Europe. Controversies over how much Jesus was man and how much God resulted in divisions among the churches of this region. A later division occurred between the eastern (Orthodox) and western (Catholic) groups of churches in Europe. From AD 395 to 1453, the eastern church was centered in Constantinople (modern Istanbul, Turkey). The largest Christian sect in this region today is the Coptic Church, with a pope who resides in Alexandria, Egypt.

Muhammad founded Islam in Arabia, notably in Mecca and Medina, during the early AD 600s. **Islam** means "submission to the will of Allah (God)" and the followers of the religion, **Muslims,** are "those who submit to Allah." Muhammad carried on many Jewish and Christian beliefs, such as monotheism. Seen as the prophet of Allah, Muhammad is believed to succeed earlier prophets such as Moses, David, and Jesus. The **Qu'ran** (holy book) is believed by Muslims to be the word of Allah revealed to Muhammed.

After Muhammad's death in 632, Arabs spread Islam rapidly westward to northern Africa and Spain as well as eastward into central Asia, uniting the Arab peoples and creating a series of empires that went on to convert Persians, Turks, and people in India through conquest and trade. It led to a Muslim "Golden Age" of artistic and scientific achievements alongside further military expansions from the 800s to the 1100s. New forms of art and architecture developed, and Muslim mosques remain dominant features of town landscapes, often doubling as centers of religious and secular activities (Figure 7.8).

Early divisions within Islam continue to be significant today. **Sunni Muslims,** or Sunnites, are the majority Muslim group and base their way of life on the Qu'ran, supplemented by local traditions. Political power was at first given to a succession of leaders, or caliphs, descended from historic Muslim leaders. Many Sunni Muslims are moderates in maintaining traditional Islamic practices. Exceptions include the Wahhabi sect that dominates Saudi Arabia, and the Taliban faction that ruled Afghanistan from 1996 to 2001.

A minority of Muslims also follow the Qu'ran but dispute the caliph leadership succession accepted by Sunnis. They look to descendants of Ali, the fourth caliph of Islam and a cousin and son-in-law of Muhammad. His son, Hussein, was defeated and killed by the Sunni caliph of Damascus in 680. Supporters see Ali as the only *imam*—authoritative interpreter of the Qu'ran—and are known as *Shi'at 'Ali* ("partisans of Ali"). For most of their history, these **Shia Muslims** (Shiites) were known for commemorating and lamenting Hussein's death by flagellation and withdrawal from the world. Today, Shias comprise large proportions of the population in a few countries, including 90 percent of the Iranian population and a majority in Iraq.

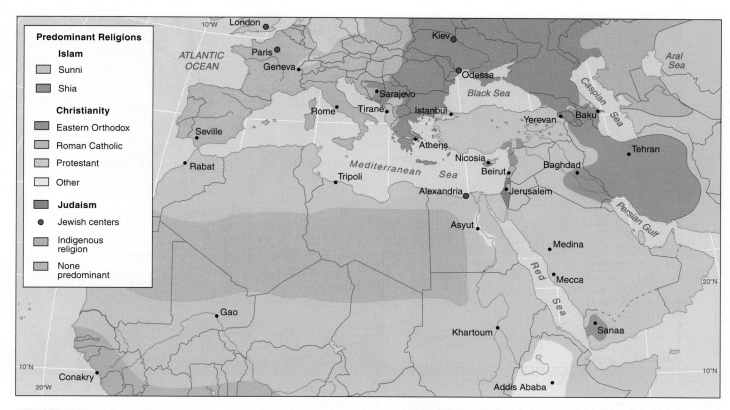

FIGURE 7.7 Major religions in Northern Africa and Southwestern Asia. Beginning in Mecca and Medina, Muslim Arabs conquered the area shown in green after Muhammad emerged as a new religious leader around AD 600. Islam also spread along trade routes. Christianity became the religion of the Roman Empire in the AD 300s, spreading through Europe. Jews were dispersed at the fall of Jerusalem in AD 70, and the modern country of Israel was not formed until 1948.

FIGURE 7.8 Islam and architecture. Mosques are a common feature of urban landscapes in Muslim countries. Older (1300s) and more recent (1800s) mosques tower over Cairo, Egypt.

Until recent times, they shunned politics and looked to clerics, who were supported by alms, for leadership. The Iranian Revolution in 1979 marked a new era of Shiite political activism as Shiites took control of Iran's government. The religious government of Iran promotes Islamic isolationism from Western ways and supports extremist groups such as Hamas (Palestine) and Hezbollah (Iran, Syria, Lebanon).

Population

The population of Northern Africa and Southwestern Asia tends to be concentrated near the Mediterranean and Red seas, the Persian Gulf, and in the fertile valleys of the Nile and the Tigris-Euphrates rivers (Figure 7.9). Moderate densities are also found in the rest of Turkey and western Iran. Much of the rest of the region is almost empty of people.

With high birth rates and low death rates, the region's population will continue to grow rapidly over the next 20 years, with the total population increasing by almost 50 percent (Table 7.1). As the age-sex diagram in Figure 7.10a (Egypt) indicates, the large numbers of young people will produce continuing high rates of population growth as they reach childbearing age. Countries in Arab Southwest Asia in particular, namely Yemen,

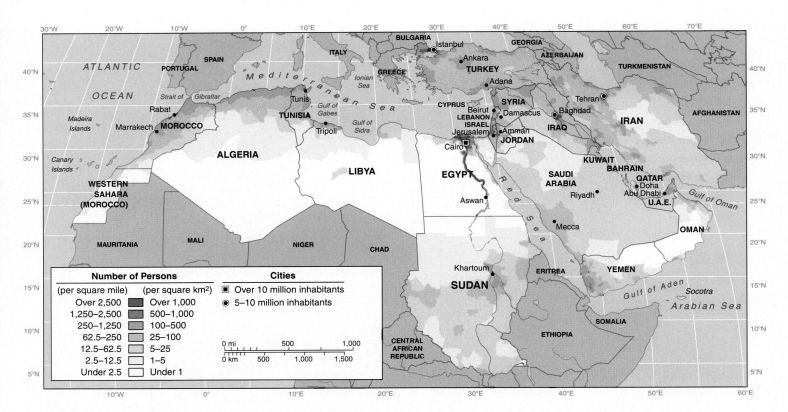

FIGURE 7.9 **Northern Africa and Southwestern Asia: population distribution.** Note where the highest and lowest densities of population occur.

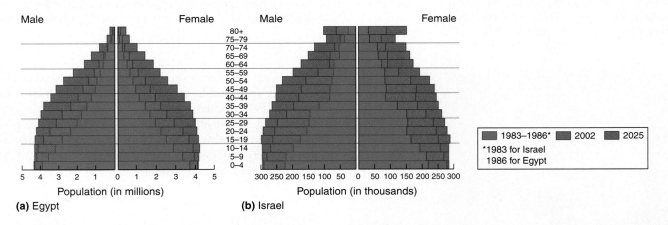

FIGURE 7.10 **Age-sex diagrams.** (a) Egypt. (b) Israel. Account for the different patterns of change. *Source: U.S. Census Bureau; International Data Bank.*

Saudi Arabia, and the Palestinian territories, have high total fertility rates with over four children per woman. In most of the smaller and richer Gulf countries—Bahrain, Kuwait, Qatar, and UAE—the 2005 fertility rate was under 4. When Iranian religious leaders gained power in 1979, they encouraged more births, but later reversed the policy.

In recent years, many of the region's countries have implemented policies to slow birth rates. For example, Tunisia set a minimum age for marriage and instituted a successful family planning program soon after it obtained independence. Morocco and Algeria implemented family planning and maternal and child services. Egypt's support for family

planning resulted in a drop of that country's total fertility rate from over 5 to 3.1 and to under 3 in cities between the 1970s and 2006.

Variations in immigration make Israel's population growth irregular. The annual growth rate fell from 1965 to 1980 but rose again in the mid-1990s as a result of immigration. In the early 1990s, over 1 million Russian Jews moved to Israel. The 2006 resident Israeli population of 7.1 million included 20 percent non-Jews, most of whom were Arabs. The fluctuating number of immigrants and the higher proportion of older people means that Israel's population structure differs from that of the region's Arab countries (Figure 7.10b (Israel)).

TABLE 7.1   NORTHERN AFRICA AND SOUTHWESTERN ASIA: Data by country, area, population, urbanization, income (Gross National Income Purchasing Power Parity), ethnic groups

Country	Land Area (km²)	Population (millions)		% Urban	GNI PPP 2005	2005	Ethnic Groups (%)
		mid 2006 Total	2025 est. Total	2006	Total (US$ billions)	Per Capita (US$)	
NORTH AFRICA							
Algeria, Democratic and Popular Rep. of	2,381,740	33.5	43.1	48.6	226.8	6,770	Arab 83%, Berber 16%
Socialist People's LIBYAN Arab Jamahirya	1,759,540	5.9	8.3	86.0	—	—	Arab-Berber 97%
Morocco, Kingdom of	446,550	31.7	38.8	55.1	138.3	4,360	Arab-Berber 99%
Tunisia, Republic of	163,610	10.1	11.6	64.9	79.9	7,900	Arab-Berber 98%
Totals/Averages	**4,751,440**	**81**	**102**	**64**	**445**	**6,343**	
NILE RIVER VALLEY							
Egypt, Arab Republic of	1,001,450	75.4	101.1	42.5	334.9	4,440	Egyptian-Bedouin-Berber 99%
Sudan, Republic of	2,505,810	41.2	61.3	36.1	82.5	2,000	Black African 49%, Arab 39%, Nubian 8%
Totals/Averages	**3,507,260**	**117**	**162**	**39**	**417**	**3,220**	
ARAB SOUTHWEST ASIA							
Bahrain, State of	680	0.7	1.0	100.0	15.8	21,290	Bahraini 63%, Asian 13%, other Arab 10%, Iran 8%
Iraq, Republic of	438,320	29.6	44.7	67.9	—	—	Arab 75–80%, Kurd 15–20%, Assyrian, Turkmen
Jordan, Hashemite Kingdom of	89,210	5.6	7.9	82.3	29.8	5,280	Arab 98%, Circassian, Armenian
Kuwait, State of	17,820	2.7	3.9	96.0	63.9	24,010	Kuwaiti 45%, other Arab 35%, Indian, Pakistani 9%, Iranian 4%
Lebanese Republic	10,400	3.9	4.6	86.6	22.2	5,740	Arab 93%, Armenian 5%
Oman, Sultanate of	212,460	2.6	3.1	71.5	37.8	14,680	Omani Arab 75%, Indian, Pakistani 21%
Qatar, State of	11,000	0.8	1.2	100.0	—	—	Arab 40%, Pakistani 18%, Indian 18%, Iranian 10%
Saudi Arabia, Kingdom of	2,149,690	24.1	35.6	86.2	355.5	14,740	Arab 82%, Yemeni 13%
Syrian Arab Republic	185,180	19.5	28.1	50.2	72.9	3,740	Arab 90%, Kurd, Armenian, Turkmen and others 10%
United Arab Emirates	83,600	4.9	7.1	74.3	118.9	24,090	UAE Arab 19%, other Arab 23%, South Asians 50%, other expatriates 8%
Yemen, Republic of	527,970	21.6	38.8	26.3	19.9	920	Mainly Arab, with African-Arab and South Asian
Totals/Averages	**3,726,330**	**116**	**176**	**76**	**737**	**12,721**	
ISRAEL AND THE PALESTINIAN TERRITORIES							
Israel, West Bank, Gaza	**21,060**	**7.1**	**9.3**	**92**	**128**	**19,200**	Jewish 82% (born in Israel 62%; white 26%, African 7%, Asian 5%) non-Jew (mainly Arab) 18%
IRAN AND TURKEY							
Iran	1,648,000	70.3	89.0	66.7	566.1	8,050	Persian 60%, Azerbaijani, Turkic 25%, Kurd 7%
Turkey	779,450	73.7	86.0	59.3	620.3	8,420	Turkish 80%, Kurd 17%
Totals/Averages	**2,427,450**	**144**	**175**	**63**	**1,186**	**8,235**	
Region Totals/Averages	**14,433,540**	**465**	**624**	**67**	**2,914**	**9,944**	

Source: World Population Data Sheet 2006, Population Reference Bureau. Microsoft Encarta 2005.

Immigration also plays a significant role in many Arab Southwest Asian countries. The booming oil business requires more labor than many of the oil-producing countries possess. Imported labor comes from other parts of the region (e.g., Jordan, Lebanon, and Yemen) and from South and East Asia (e.g., India, Pakistan, Bangladesh). In 2005, foreign workers comprised large percentages of the labor force in Saudi Arabia (35 percent), Bahrain (44 percent), United Arab Emirates (74 percent), and Kuwait (80 percent).

Most of the region's countries have life expectancies around 70 years. Israel has the highest with almost 80 years. Life expectancy in Yemen remains lower, at 60 years. In Iraq, it fell below 60 after years of war and deprivation.

Urban Patterns

Arid environments, coupled with the growth in oil industry and government employment, make most countries in the region very highly urbanized (Table 7.2). For example, between 1950 and 2006, Kuwait and Qatar went from 50 percent to over 95 percent urban, while Saudi Arabia went from 10 to 86 percent urban. The rate of urban expansion means that many cities are dominated by new buildings (Figure 7.11). However, the expanding demand for housing by poorer people was often more than governments could meet. Shantytowns are features of cities across the subregion, known as *bidonvilles* in Casablanca, Morocco, and as *gourbivilles* in Tunis, Tunisia.

The Fertile Crescent, which runs along the Tigris-Euphrates Rivers, along the eastern shore of the Mediterranean, and up the Nile River valley, is the location of some of the world's oldest and most historically significant cities. These older cities that were central to agricultural and trading economies survive as enclaves within today's expanded cities. Some are much changed by the clearing of crowded buildings that made way for new highways. High densities of homes, commercial premises, and public buildings inside city walls marked the traditional small towns of the

region and their central **medinas** (Figure 7.12). Medinas, named after the sacred Muslim city in Saudi Arabia, are historic sectors of cities, valued for their distinctive structure and social fabric. Their labyrinthine alleys, *souks* (commercial areas) (Figure 7.13), and artisan shops relate them to the past and attract tourists. Within these sectors existed a rigid pattern of land use. Prestigious craft

TABLE 7.2	Population of Major Urban Centers in Northern Africa and Southwestern Asia (in millions)	
City, Country	2003 Population	2015* Projection
NORTH AFRICA		
Casablanca, Morocco	3.6	4.6
Algiers, Algeria	3.1	4.2
Tripoli, Libya	2.0	2.5
Tunis, Tunisia	2.0	2.4
Rabat, Morocco	1.8	2.3
NILE RIVER VALLEY		
Cairo, Egypt	10.8	13.1
Khartoum, Egypt	4.3	5.6
Alexandria, Egypt	3.7	4.5
ARAB SOUTHWEST ASIA		
Baghdad, Iraq	5.6	7.4
Riyadh, Saudi Arabia	5.1	7.2
Jeddah (Jiddah), Saudi Arabia	3.6	4.9
Aleppo (Halab), Syria	2.4	3.1
Damascus, Syria	2.2	2.8
Beirut, Lebanon	1.8	2.2
Sana'a, Yemen	1.5	2.7
Mecca, Saudi Arabia	1.4	2.0
Mosul, Iraq	1.2	1.6
Amman, Jordan	1.2	1.5
Kuwait City, Kuwait	1.2	1.4
Basra, Iraq	1.1	1.4
ISRAEL AND THE PALESTINIAN TERRITORIES		
Tel Aviv-Jaffa	2.9	3.5
IRAN AND TURKEY		
Istanbul, Turkey	9.4	11.3
Tehran, Iran	7.2	8.5
Ankara, Turkey	3.4	4.2
Izmir, Turkey	2.4	3.0
Meshed (Mashhad), Iran	2.1	2.5
Isfahan (Esfahan), Iran	1.5	1.9
Tabriz, Iran	1.3	1.7
Bursa, Turkey	1.3	1.8
Adana, Turkey	1.2	1.5
Karaj, Iran	1.2	1.6
Shiraz, Iran	1.2	1.5

*estimated

Source: United Nations Urban Agglomerations 2003, with estimates for 2015 (2003).

FIGURE 7.11 Southwestern Asia: Dubai. New buildings along Sheikh Zayed Road.

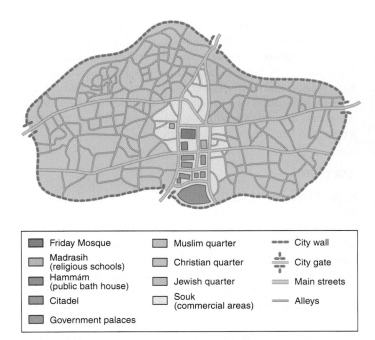

■ Friday Mosque	■ Muslim quarter	---- City wall
■ Madrasih (religious schools)	■ Christian quarter	⊹ City gate
■ Hammám (public bath house)	■ Jewish quarter	══ Main streets
■ Citadel	■ Souk (commercial areas)	═══ Alleys
■ Government palaces		

FIGURE 7.12 **The typical layout of a medina.** *Source: Data from Kheirabadi.*

FIGURE 7.13 **In the medina.** The souk of Marrakech, Morocco, Northern Africa.

workers, such as religious artisans, had premises close to a central mosque, castle, and square; those in lowlier occupations, such as leather workers, located on the edges. The residential population tended to concentrate in "quarters," usually based on religion so that separate Muslim, Jewish, and Christian quarters were common. The World Heritage list of historic cities contains many medinas across the whole region, from Fez and Marrakech in the west to the tall, brown mudhouses of Yemen in Southwest Asia. Today, city walls remain only in cities where tourist interests are economically important, such as Fez and Marrakech (Morocco).

FIGURE 7.14 **Nile River Valley: Cairo, Egypt.** Cairo is a modernizing city, the political and economic center of Egypt.

Just outside the old centers are newer high-rise apartments, hotels, and offices. Beyond are extensive suburban areas of housing, factories, and shopping facilities. Many older town areas were abandoned to poorer people when merchants and businesspeople moved to more spacious homes and commercial premises in the new suburbs. The factories and supermarkets changed patterns of working and social movements. Segregation by income and social group is common within the cities, especially in oil-rich Arab Southwest Asia where Arabs often live apart from immigrant workers.

Historically, Fez, Cairo, and Constantinople (Istanbul) were three important global cities and ranked as the third, fourth, and fifth largest cities in the world as recently as the AD 1200s. Though having fallen in global significance, Istanbul (Turkey) and Cairo (Egypt) are the two largest cities in the region today and have many global city-region characteristics (Figure 7.14). Abu Dhabi, Dubai (both in UAE), Riyadh (Saudi Arabia), and Tel Aviv-Jaffa (Israel) are considerably smaller in size but have many global connections. On the other hand, the isolation of Iran from much of the global economy leaves Tehran with few global city features though it is the third largest city in the region.

Evolving Political Geographies

Empires to Countries

The Tigris-Euphrates River valley of Mesopotamia (modern Iraq) and the Nile River valley formed two of the world's early cultural hearths. From these places, many early human achievements diffused to the surrounding continents. Even in ancient times of slow and limited transportation, this region acted as a hub with constant movements of people to and from Northern Africa, Europe, and China. Over the centuries, a series of empires originating from both inside and outside the region dominated this part of the world. Some of the most noteworthy are the Assyrian, Babylonian, Egyptian, Persian, Roman, Byzantine, Mongol, Umayyad, and Abbasid empires. The last great empire of the region was that of the Ottoman Turks, which began in 1299 and lasted until 1923.

The Ottomans flourished in the 1500s and 1600s but turned inward-looking after the 1700s and did little to encourage modernization or involvement in the expanding global economic system. By then, European empires were encroaching on the region and eventually took control in many places, including France (Tunisia, Morocco, Algeria), Italy (Libya), and Britain (Egypt, Sudan, and much of the southern Arabian Peninsula and Gulf coasts after the 1869 opening of the Suez Canal). Russia took land in the north. Some of the more isolated tribes in Arabia became largely independent under their own kings or sultans. During World War I, the British and French fueled Arab nationalism to expel the Turks from the areas that became Palestine, Syria, Lebanon, Jordan, Iraq, and the Arabian Peninsula.

The Arab Southwest Asian countries gained independence from their European colonial overlords in different ways. After the defeat of the Ottoman Empire in World War I, the League of Nations made Syria and Lebanon French protectorates and placed much of the rest of the area under British protection. The French encouraged republican governments. The British guaranteed the survival of the kingdoms they established, including Jordan (still a monarchy today), Iraq (a monarchy until 1958), and the small emirates along the Persian Gulf under traditional local rulers, or emirs. In the 1970s, seven emirs joined to form the United Arab Emirates (UAE). A homeland for the Jews was also established and eventually became the modern state of Israel in 1948.

Pan-Arabism

In the mid- and late 1900s, post-independence movements, generally opposed to Western economic colonialism, sought to unite Arab peoples on the basis of (mostly secular) nationalism into a single country with increased world influence. None succeeded in uniting the Arab world though their efforts are noteworthy.

The **Arab League** was created in 1945 to encourage the united opposition of Arab countries to the establishment of Israel. Its seven founding members were the only independent Arab countries at the time, but its membership increased to 21 as more gained independence. Members of the Arab League eventually included the **Palestine Liberation Organization (PLO)**, a political organization providing an umbrella for many smaller groups that demand a country for Palestinians, the people living in the lands used to create the state of Israel. After Egypt's 1979 accord with Israel, the Arab League expelled Egypt and transferred Arab League headquarters from Cairo to Tunis. From 1958 to 1961, Egypt and Syria joined as the United Arab Republic under the leadership of Colonel Jamal Nasser, with the intent of persuading other countries to commit their futures to a single **Pan-Arab country.** It did not attract others, and soon broke up.

Disunity, military defeats, and tensions over the Israel-Palestine issue and among countries with different resource bases weakened the Arab League and its ability to foster unity among Arab countries in the 1980s and 1990s. A major blow to the Arab League came during the Gulf War of 1990–1991, when one Arab country (Iraq) invaded another (Kuwait) and was defeated by a coalition of other Arab countries backed by the United States and other Western countries. Many Arabs felt betrayed by both Saddam Hussein's invasion of another Arab country and their own need to rely on outside help. It was clear that the interests of individual countries remained more significant than an overriding Pan-Arabism, feelings that were repeated in later Arab economic summits.

Islamism

In the 1970s and 1980s, as individual countries preferred their independence to a Pan-Arab identity, religious affiliation became more significant. In 1970, foreign ministers of Muslim countries set up the **Organization of the Islamic Conference (OIC),** which now has 45 members including such countries as Pakistan, Indonesia, and Nigeria. However, it is more successful in advancing individual member countries' interests than in defining and pursuing a common agenda to rival Western-dominated globalization. One of OIC's most important affiliates is the Islamic Development Bank, which is dedicated to economic development among OIC members.

In the 1970s, Islamic political groups also came to the fore, basing their ideology on the Qu'ran, interpretations of *jihad* as holy war, and references to past Islamic triumphs. These ideas contradicted many long-held Islamic beliefs and practices that were peaceful in nature. However, persecuted under nationalistic Arabism and secular governments, Islamic political groups called for the reestablishment of a single Islamic country and rejected the traditional view that involvement in politics should not be a concern of good Islamic practice. They hated the fragmentation of Islamic lands into separate countries, in which the religious establishment was subservient to those educated in Westernized ways.

Political Islam was fueled by the 1967 and 1973 Arab-Israeli wars, but its greatest success was the 1979 Islamic revolution in Iran under Ayatollah Khomeini. Religion further mixed with politics when Islamic law (*sharia*) became official in Iran in 1983. Although the success in Iran stimulated Muslims in other countries to consider similar action, most Muslim countries, including Libya, Syria, and many small Gulf countries, suppressed such ideas and imprisoned religious activists. At the same time, Iran and Saudi Arabia struggled for dominance through the 1980s. As Iran tried to export revolution, the Saudi princes wished to maintain the system by which they had grown rich. When Saddam Hussein of Iraq declared war on Iran in 1981, he had support from the Saudis and other Gulf countries as well as Western countries. However, the war stalled, and after huge losses of life, an armistice was reached between Iran and Iraq in 1989.

The 1979 invasion of Afghanistan by the former Soviet Union generated another jihad financed by the Gulf and Western countries. This unified Islamists around the world as international brigades from Egypt, Algeria, the Arabian Peninsula countries,

Pakistan, and Southeast Asia worked together in guerrilla warfare as part of the Islamist armed struggle. In 1989, the Soviet Army withdrew from Afghanistan, causing Islamists to declare a great victory and gain immense self-confidence.

The Gulf War of 1990–1991 and subsequent failures to impose military solutions in Algeria and Egypt were major blows for Islamists. A radical Islamic fringe turned against the Arab kingdoms with rulers who repressed them and their international networks. Their hostility toward the United States increased.

Islamist failures caused most countries that had supported guerrillas to withdraw from extremist actions. In Iran, for example, moderate candidates were elected president during the 1990s and in 2001. Turkey and Pakistan turned from Islamist leaders to moderate military or secular leaders. In Jordan, radicals lost their parliamentary seats when Jordan began negotiating with Israel after the Gulf War.

Extreme Islamists frustrated by political impasse continued and expanded terrorist actions. Attention reverted to Israel and Arab resentment of the lack of movement by the U.S.-supported Israeli government to accommodate the wishes of Palestinians. Attention also focused on the U.S. occupation of Iraq following the fall of Saddam Hussein in 2003. But terrorism, despite the events of 9/11 and suicide bombings in Israel, has yet to yield any results. Many Muslim countries held memorial services for the victims of 9/11, including Iran in its capital, Tehran. The election of a conservative hard-liner as Iran's president in 2005 can be seen as a recent victory for Islamists, but the victory resulted more from voter dissatisfaction with corruption allowed by reformers, and reformers who did not redistribute the wealth to poorer segments of society as they had promised. Indeed, the new conservative president redistributed wealth to the poor when he was mayor of Tehran.

Global Economics

The discovery of oil and the founding of new countries after World War I brought the region into world prominence again. The 1973 war with Israel triggered massive increases in the price of oil, and the oil-producing countries in this region have built huge financial surpluses. Subsequently, many of the region's countries have become wealthier than other formerly poor countries in the world (Figure 7.15). Ownership of consumer goods (Figure 7.16) reflects a range of material wealth across the region.

Oil Resources

The huge oil and gas production from Saudi Arabia and the Gulf states amounted to 22 percent of world oil production from the 1960s to the early 2000s. Despite such extraction and the addition of new producers in Latin America, Asia, and Africa, the proportion of total known world reserves located around the Persian Gulf increased (Figures 7.17 and 7.18). Continued discov-

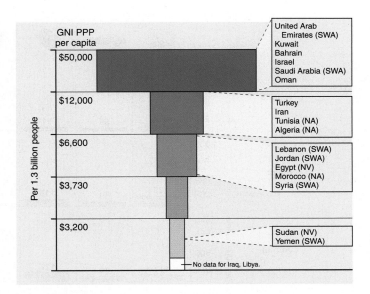

FIGURE 7.15 **Northern Africa and Southwestern Asia: country incomes compared.** The countries are listed in the order of their GNI PPP per capita. *Source: Data (for 2005) from* World Development Indicators, *World Bank; and Population Reference Bureau.*

eries increased reserves from 61 (1965) to 65 (2000) percent of world total reserves, despite increased usage (from 215 billion to 560 billion barrels per year). These figures do not include Iran, Libya, Egypt, or Algeria, which are also major oil and natural gas producers in this region. Algeria, for example, has the fifth-largest natural gas reserves in the world and is the second-largest natural gas exporter. Libya was politically shunned by most countries in the 1980s and 1990s because its government supported international terrorists. However, after 2004 the Libyan government became more cooperative, and foreign investment in its oil industry subsequently increased. New discoveries of oil fields in the Caspian Sea area (see Chapter 3) may rival the Gulf in output, and Iran and Turkey have attracted oil pipeline traffic across their territories from this new source. Oil sources and movements remain a major geographic feature of the region.

Not all the countries of Northern Africa and Southwestern Asia, however, are major oil producers. Morocco, Turkey, Israel, and Jordan produce no oil. Tunisia, Sudan, Syria, and Yemen produce and export modest amounts. Countries that import oil face the burden of purchasing oil, whatever the price, and many went into debt during times of high prices.

Organization of Petroleum Exporting Countries

From the early 1900s, international oil companies kept oil prices low for consumers in the world's wealthiest countries by paying little to the producing countries. In 1960, the producers around the Persian Gulf, together with Venezuela, formed the **Organization of Petroleum Exporting Countries** (OPEC). Today, Arab countries still account for 75 percent of OPEC membership. OPEC's main purpose is to work as a **cartel**—an

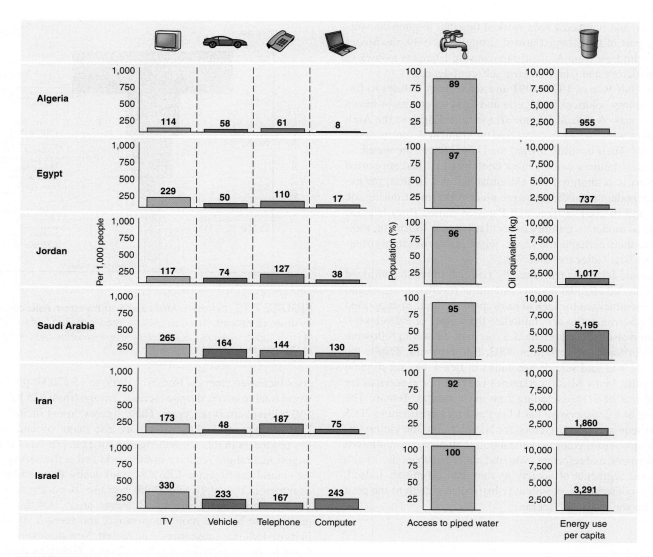

FIGURE 7.16 **Northern Africa and Southwestern Asia: ownership of consumer goods, access to piped water, and energy usage.** Compare these levels with other countries in Africa and Asia. *Source: Data (for 2002) from* World Development Indicators, *World Bank, 2004.*

organization that coordinates the interests of producing countries by regulating oil prices.

OPEC's oil embargo on the West during the Arab-Israeli War in 1973 was successful and allowed OPEC to control world oil distribution for a few years. As prices rose fourfold, revenues of the oil-producing countries boomed in the mid- and late 1970s. However, higher oil prices caused economic recession in the wealthier Western countries. Western oil companies opened new oil fields outside the OPEC area, including those in the North Sea (Europe) that had previously been too expensive to develop. Beginning in the 1990s, Russia exported increasing quantities of oil and natural gas as its main source of currency. A world oil glut brought very low market prices and financial problems to the Arab producers. By the late 1990s, the OPEC oil producers and major industrial countries saw that their interests lay in a moderately high, but stable, oil price. However, in the early

2000s, war, other crises, and demand from growing economies such as China and India pushed oil prices to record highs.

Water Politics

Much attention is given to this region's oil but the issue of water is crucial in this arid region of the world. Like oil, however, water is not evenly distributed. Eighty percent of the region's fresh water is found in Nile and Tigris-Euphrates river basins. The Jordan River, though much smaller, is crucial to countries that depend on it. However, each of these rivers flow through more than one country (see Figures 7.1 and 7.4) and require cooperation among governments, creating and sometimes exacerbating political conflict where agreements cannot be reached. Conflict may increase as fresh water scarcity is becoming more of an issue as populations grow rapidly. Between 1975 and

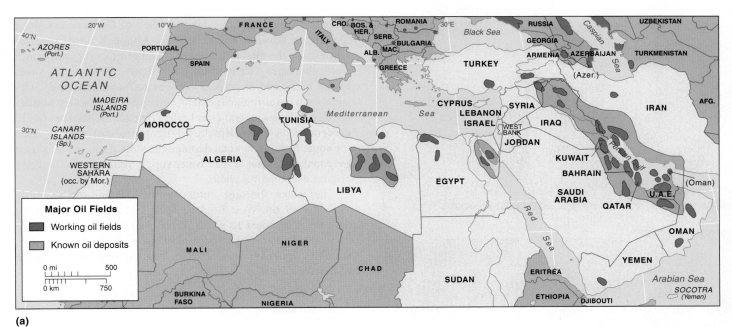

(a)

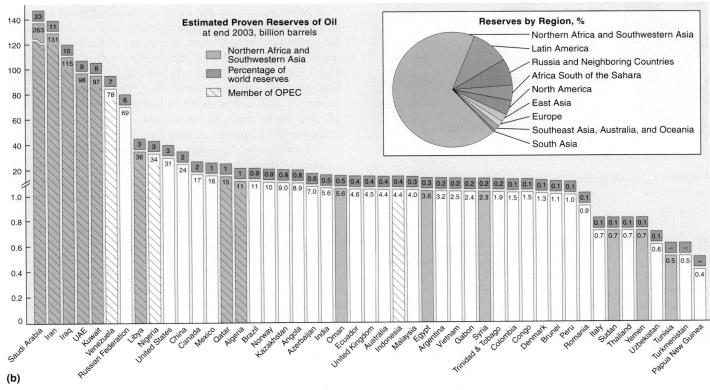

(b)

FIGURE 7.17 Oil resources in Northern Africa and Southwestern Asia. (a) Map of major oil fields. (b) World oil resources at the end of 2003, showing the continuing significance of the region. The share, 67.8 percent of world reserves in 2003, is a little greater than the previous decade, and reserves in the region are being used more slowly. Note which countries in this region have major oil reserves, and which have little or no oil. *Sources: (a) Data from* The Economist; *(b) Data from British Petroleum.*

2001, the amount of fresh water available to each individual dropped by more than half.

In 1959, the Egyptian and Sudanese governments signed the **Nile Waters Agreement,** which led to the building of the Aswan High Dam to store sufficient water and generate electricity. On completion of the dam in 1970, Lake Nasser behind the dam stored three times Egypt's annual water usage. Sudan receives only 13 percent of the annual flow. Along with

FIGURE 7.18 Oil in Iraq. The Shaiba oil refinery, 20 kilometers south of Basra.

disputes they may continue to have with each other, Egypt and Sudan also now contest the use of the water with those in the upper Nile River watershed, such as Ethiopia, Uganda, and Tanzania.

The Turkish government constructed a series of dams in Southeastern Turkey on the upper reaches of the Tigris and Euphrates rivers as a part of a greater plan to irrigate lands to increase agricultural production (Figure 7.19). Because these irrigation projects divert water from Iraq and Syria downstream, tensions have increased between Turkey and these two countries.

To provide fresh water, the wealthier countries of the Arabian Peninsula built desalinization plants to turn seawater into drinking water but such efforts are extremely expensive. In fact, 60 percent of the world's desalinization capacity is in the Arabian Peninsula with 30 percent of the world's total in Saudi Arabia alone.

Israel, Gaza, and the West Bank occupy a small area of land between the Mediterranean coast and the Jordan River valley. The terrain includes a coastal plain and mostly hilly land with lower areas around Lake Tiberias (Sea of Galilee) and along the Jordan River to the Dead Sea. Although Israel receives winter rain, the total rainfall is low, and summer drought brings water shortages. Great efforts made it possible to supply water to dry areas and to manage the environment efficiently. Despite such careful management, internal groundwater sources are now fully used, and Israel relies on external sources for 25 percent of its water. Brackish (slightly saline) water is used for some farming but can damage the soils if too much is used over many seasons. Urban water supply relies increasingly on coastal desalination plants.

The water shortage enters politics. Israel retains jurisdiction over the West Bank and the Golan Heights both for defense and access to water. Peace negotiations with Jordan included agreement on the use of Jordan River basin waters, which have become central in the negotiations over the future of the West Bank.

Agriculture

The arid lands of Northern Africa and Southwestern Asia provide little opportunity for farming, requiring most countries to import foodstuffs to adequately feed their citizens. Most of the arable land extends from the shores of the Mediterranean (Figure 7.20) and Red seas, and the Persian Gulf. The Nile and Tigris-Euphrates rivers provide for irrigation. Interior deserts have given rise to nomadic herding. The most agriculturally productive areas are on the fringes of the region such as in Turkey and the southern Sudan. Indeed, the building of dams in southeastern Turkey has enabled Turkey to increase its production of cotton, soybeans, grains, fruits, and vegetables.

Many of the problems in the region's agricultural production stem from the type of economy established by colonial countries. The colonial system in North Africa involved land appropriation for European settlers and local elites, who used irrigation water for intensive commercial farming tied to markets in Europe. Consequently, agriculture is geared in many areas to export crops such as citrus fruits, olive oil, and cotton. Iran is still known for its export of nuts, especially pistachios. In recent decades, governments have encouraged greater cultivation of domestically needed crops such as cereal grains, vegetables, dairy, and poultry products. Productivity has increased with the introduction of mechanization, the application of fertilizers, and double cropping, which allows some areas to harvest two crops within a year. As noted previously (see "Environmental Problems" section, p. 197), improper irrigation techniques in places lead to continual problems of salinization.

Geographic Diversity

Northern Africa and Southwestern Asia is divided into five subregions: North Africa, Nile River valley, Arab Southwest Asia, Israel and the Palestinian Territories, and Iran and Turkey (Figure 7.21).

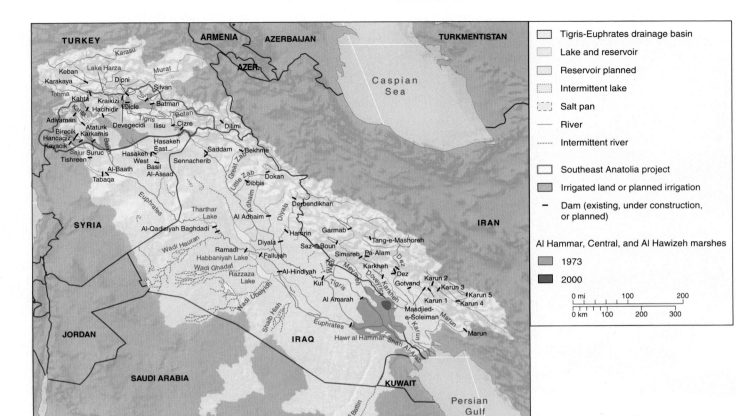

(a)

(b)

FIGURE 7.19 Tigris-Euphrates drainage basin. (a) Turkey's southeast Anatolia region has more than twenty dams to divert water from the Tigris and Euphrates rivers for irrigation projects, depriving Syria and Iraq of water. Dam building in Iran also has reduced the flow of water into wetlands of the lower Tigris River. In the early 1990s, Saddam Hussein's regime launched a program to drain marshes, partially to drill for oil and partly to eradicate the Marsh Arabs who opposed him. In southern Iraq, the Central Marshes and Al Hammar Marshes have completely dried up, and the Al Hawizeh Marsh is a fraction of its former size. (b) The Ataturk Dam on the Upper Euphrates River as it neared completion in 1992. *Source: (a) Data from UNEP.*

North Africa

The four countries of North Africa are Algeria, Libya, Morocco (with Western Sahara), and Tunisia (Figure 7.22), each with differing population sizes (see Table 7.1). Over 80 percent of Algeria's

and Libya's territories are desert, but Morocco and Tunisia do not extend so far into the arid Saharan environment. The northern parts of Morocco, Algeria, and Tunisia are dominated by the Atlas Mountains, and that area is known as the Maghreb. It includes high ranges (Mount Toubkal in Morocco is 4,165 m, or 13,665 ft.,

FIGURE 7.20 North Africa: rural landscape. Bedouin women taking goods to market near Foudouk al Aouerab, Tunisia. The cultivated valley behind them and the bare hillsides above are typical of Northern Africa and much of the wider region.

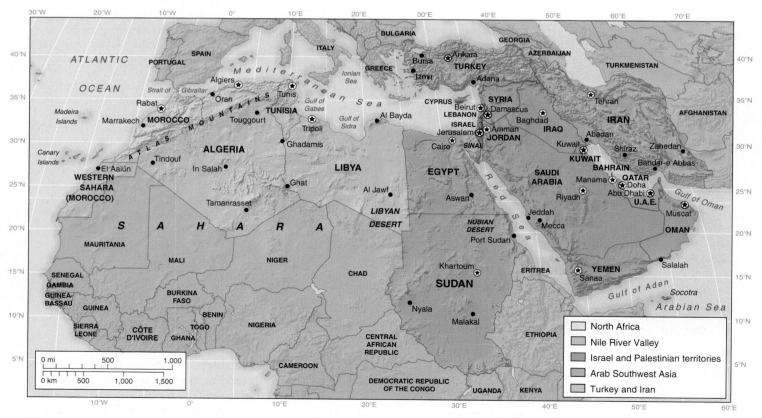

FIGURE 7.21 Northern Africa and Southwestern Asia: subregions and aspects of geography.

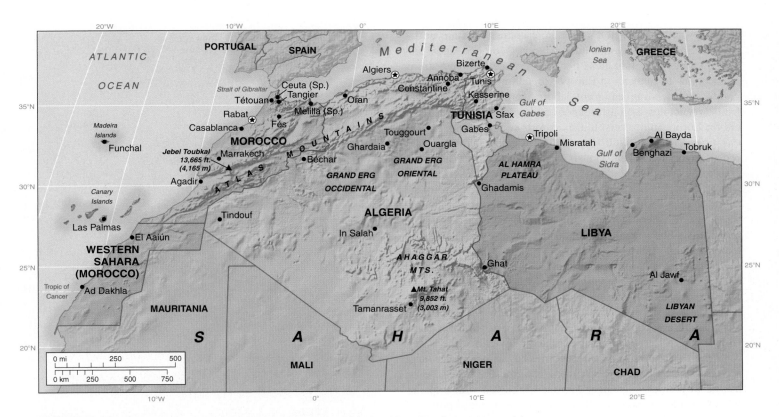

FIGURE 7.22 **North Africa: countries, physical features, and main cities.** "Ergs" are major sand dune areas.

in altitude), broken by internal plateaus and river valleys. The harsh, largely arid, and often mountainous natural environments of the North African countries restrict agriculture and most human settlement to a small percentage of the territory along the northern coasts and in the immediate mountainous hinterland (Figure 7.23). Problems of water supply affect all these countries.

Arabs are dominant but the Berber people help to make this subregion unique. European colonization, which began in the early 1800s, explains why much of the educated middle classes in the Maghreb countries speak French and those in Libya speak Italian. After World War II, nationalist groups fought for and obtained independence—in 1956 in Morocco and Tunisia, and in 1962 in Algeria. Morocco is politically stable under its moderate king, Muhammed VI. Tunisia had 30 years of one-party rule when Islamic extremists were repressed and women's rights were established. Women in Tunisia have a better position than those in other Arab countries, resulting in lower adult illiteracy (26 percent in 2004). Algeria has been ruled by demo-cratically elected governments with socialist policies based on central planning. The first-round election success of the funda-mentalist Islamic Salvation Front (FIS) in the 1992 elections led to an army takeover of the government and civil war with the dispossessed Islamic militants. Terrorist activities and army repression through the 1990s devastated Algeria's economy and people, with over 100,000 deaths and the army's destruction of the FIS as a political party.

Libya was a mainly desert area of little economic or political outside interest until Italy occupied it in 1911. After World War II, Italy was replaced by a British-French protectorate until Lib-yan independence was achieved in 1952. Colonel Muammar al Qadhafi seized power in 1969 and runs the country as a mixed military, socialist, and Islamic republic based on oil wealth.

FIGURE 7.23 **North Africa: mountainous environment.** The town of Moulay Idriss (north of Fez), named after a Moroccan saint who is descendant of the Prophet Muhammad, is located in a dry mountainous environment where orchards are planted among the desert vegetation.

Many of the subregion's economic connections are northward to Spain, France, and Italy. Algeria, Morocco, and Tunisia, in particular, retain close ties with France and have strong connections to markets in Europe for selling products, buying goods, and sending emigrant labor. Morocco exports citrus fruits, vegetables such as tomatoes and potatoes, and cut flowers for European markets. It also exports cork from the bark of oak trees and fish: squid (for export to Japan) and tuna. Morocco, Algeria, and Tunisia also export phosphate for fertilizer. Half of Morocco's population is still dependent on agriculture. Morocco's manufacturing sector is substantially craft industries. Algeria also has light industries, including the manufacture of electrical components. Libya makes steel and aluminum, and Tunisia has a small steel industry. Algeria and Libya also export oil and gas.

North African countries have growing government bureaucracies and service occupations in education and health care. In Morocco and Tunisia, tourism is a major source of income, based on their sunshine, coastal locations, historic and cultural sites, and stable political environments. Most tourists come from northern Europe, with its cool and rainy climate.

In the 1990s, North African countries began to privatize large sections of their economies. Tunisia is farthest ahead and even has its own stock exchange. Morocco followed with unparalleled sales of state holdings that reversed the previous policy of "Moroccanization."

Nile River Valley

The Nile River connects Egypt and Sudan (Figure 7.24) and provides them with water that has sustained a human presence in the dry eastern Sahara since the early days of farming and civilization (see Figure 7.4). However, the Nile River waters are finite, and pressures from growing populations make it difficult for Egypt and Sudan to continue to modernize and diversify their economies. Neither produces sufficient export income to pay for the imported needs of its people.

Egypt and Sudan are similar in some ways, but their geographic positions, political environments, and products make them different too. Egypt is the largest Arab country in population (75 million in 2006). Though Muslim and Arab, Egyptians constitute a discrete subgroup within the larger Arab population spread throughout the region. Egypt also plays a strong role in international relations. It retains control over the major global choke point of the Suez Canal and the Sinai Peninsula, which is the land route from Africa into Southwest Asia.

In 1952, Egypt became fully independent after centuries of Ottoman and then British domination. It became a socialist state focused on its own internal needs. Under the leadership of President Gamel Abdul Nasser, Egypt developed more rapidly than the rest of Africa in the 1960s and 1970s. Nasser died in 1970, and by the late 1970s, Egypt's foreign policy shifted by recognizing Israel. Then generous aid came from

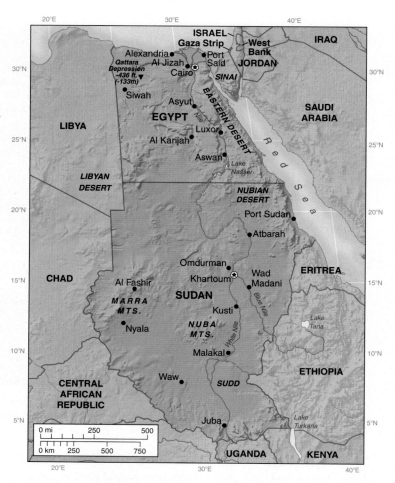

FIGURE 7.24 Nile River valley: countries, physical features, and main cities.

the United States, but Egypt was shunned by the rest of the Arab world.

Sudan has twice the area of Egypt, but only half the population. Sudan is very diverse with over 600 ethnic groups and 400 languages spoken (see Figure 7.5). Muslim Arabs dominate in the north and black Africans practicing traditional animistic religions or Christianity dominate in the south. Overall, southern Sudan has conditions similar to those of its neighbors Chad and Ethiopia, and resembles other African countries more than those of northern Africa.

In 1956, Sudan gained its independence from the United Kingdom despite Egyptian claims. The tensions resulting from the period of Egyptian occupation of Sudan and fears of future Egyptian expansion remain in the minds of many Sudanese and affect negotiations over the use of Nile River water. Internally, the Arabic Muslim peoples have controlled the government in a series of military regimes that have tried to impose Arabic and Islamic culture and traditions upon all the peoples of the country. Consequently, a civil war erupted from 1972–1982. A ruling military junta composed of a mixture of the military and Islamists took control in 1989. Sporadic violence continued but began to die down in 1999 when Sudan's government moved toward multiparty politics internally and better international relations. Peace talks in 2002 and 2003 led to the signing of peace accords and a cease-fire, but episodes of violence have not completely ceased, with the latest violence erupting in the western Sudan (see "Human Rights" section, p. 222).

Arab Southwest Asia

Arab Southwest Asia is the heart of the Arab and Islamic worlds. It comprises the Arabian Peninsula and the Fertile Crescent that includes the Tigris-Euphrates River basin and the Lebanon coast (Figure 7.25). Despite small populations, the countries of this subregion play a major part in world affairs because of their oil wealth and involvement in the Arab-Israeli peace process.

The countries of Arab Southwest Asia are differentiated by forms of government, emphases within Islam, and natural resources of oil and water. The differences in natural resources and economic management produce a wide range of economic status—from countries that remain materially poor to those that rival the world's wealthiest (see Table 7.1).

The governments of countries in this subregion mostly move slowly, if at all, toward democracy. Many of the oil-rich countries remain dominated by ruling families or dictators; Saudi Arabia is the only country named after a family, the Sauds, who still rule with the title of king and support a huge range of related princes. Other Gulf countries have sheikhs, emirs, and sultans with similar roles, and Jordan has a king. Syria has an authoritarian military-dominated regime. Until the fall of Saddam Hussein, Iraq had an authoritarian, military-dominated regime but is now moving toward democracy. In Lebanon, a greater degree of democracy now exists after a destructive civil war in the 1980s and 1990s between the Muslim and Christian groups spurred on by Islamic extremists, although Syria still exercises a major influence.

Arab Southwest Asia has major economic differences between countries with high oil revenues and those that produce little or no oil. The richer countries tend to border the Persian Gulf and sit on huge oil reserves (see Figure 7.17). The Saudi Arabian economy produces about three times the total income of any other country in the subregion. These countries have small total populations and rely on immigrant labor. Wealth is not distributed widely or evenly among their populations. Menial, low-wage jobs and much of the commerce, especially in retailing, are left to Indians and other Asian immigrants. Lebanon, Jordan, and especially Yemen are much less developed economically. Iraq and Syria are the exceptions to this rich-poor duality, since until the mid-1980s they both had oil revenues, significant water resources and agriculture, and large labor forces. The Iran-Iraq War, Gulf War and subsequent UN sanctions, and the U.S. invasion in 2003, however, almost destroyed Iraq's economy, though the United States is attempting to rebuild it.

In oil-rich countries, income benefits the ruling elites and is used partly to generate industrialization, intensify agricultural output, provide more and better roads, airports, health services, and education, and increase living standards. Up to one-third of revenues in some countries go to purchasing military hardware. The oil income makes it possible to shift economies toward sustainable development based on diversification. A

FIGURE 7.25 Arab Southwest Asia and Israel: countries, physical features, and main cities. The Fertile Crescent stretches from the Mediterranean coastlands through Syria and Iraq to the Persian Gulf. Israel declared Jerusalem its capital in 1950, but most countries of the world do not recognize this declaration and have their embassies in Tel Aviv.

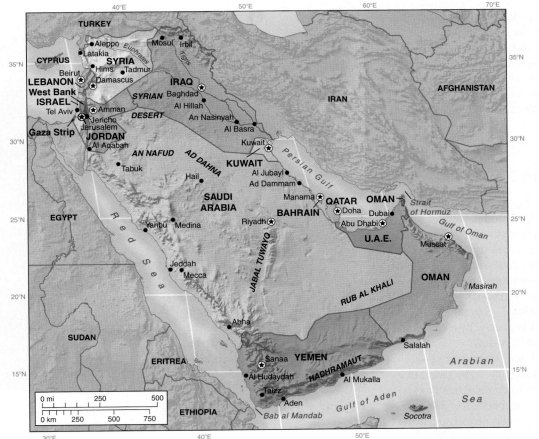

diversified economy is one in which manufactured goods are more important than primary products, and a variety of manufactured products is joined by a growing service sector. For example, Bahrain and UAE now have aluminum smelters and Oman a copper smelter. Qatar and Saudi Arabia produce steel. Several countries also manufacture construction materials, including cement, to provide the materials for upgrading roads and building new housing and factories. Countries developed food and consumer goods industries. Services also became important in many of these countries where banking and government employment are increasing rapidly.

Since the 1980s, the Gulf countries have invested some of their oil revenues in higher education facilities, and the many graduates are now employed in banking, education, health care, government, and new industries, including the media and information technology. In 2002, Dubai's Internet City had 200 firms that employ 4,000 workers and is linked to similar centers from Europe to Bangalore, India. The adjacent Media City, Festival City, conference center, expanding major airport and booming Emirates airline, luxury housing, and high-quality tourist facilities all signal Dubai's investment in the future (Figure 7.26).

In 1981 the Gulf oil countries, except for excluded Iraq, formed the **Gulf Cooperation Council** in the context of the Iran-Iraqi War. It focused on common economic and political interests. Saudi Arabia dominates the organization, which was effective in bringing together other countries to resist the 1990 Iraqi invasion of Kuwait.

Arab Southwest Asia is also rich in history and culture, and is the location of some of the world's earliest known towns. The historical sites and hot, sunny climate attract many tourists. The rock-hewn city of Petra in Jordan is one such example. Every year, millions travel—mostly as religious pilgrims—to Saudi Arabia to visit the Islamic holy sites in Mecca and Medina (Figure 7.27). Mecca is the birthplace of Muhammad, Islam's founder, and Medina became Muhammad's power base after he was expelled from Mecca.

Israel and the Palestinian Territories

Israel stands out in Southwestern Asia. It is a unique example of a country created by the United Nations for a particular ethnic or religious group, despite opposition from those living in and around it (see Figure 7.25). The Palestinian territories of the West Bank and Gaza are Israeli-occupied territories following conquests in the 1967 war, although the United Nations ruled that they should be returned to Syria, Egypt, and Jordan. Territorial disputes, terrorism, and refugee groups have major local geographic impacts and global implications. This subregion is marked by a dual society of different opportunities for Jew and non-Jew.

Israel's population is composed of Jews, Israeli Arabs (which include Druze and Bedouin), and Palestinian Arabs (see Table 7.1). The Arabs and some of the Jews trace their family existence in the country to before Israel's independence. The Sephardic Jews from southern Europe form the majority, but the Ashkenazi Jews of Central and Eastern European origin play

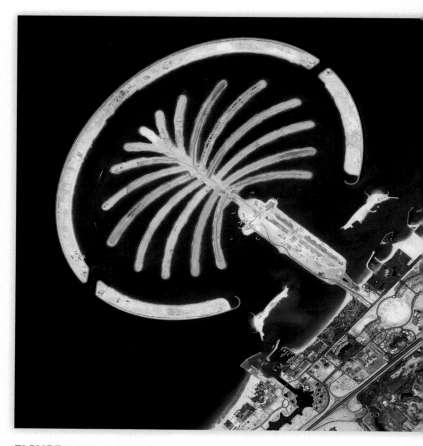

FIGURE 7.26 **Arab Southwest Asia: Dubai.** Palm-shaped artificial islands in the Persian Gulf built from 130 million cubic yards of rock and sand create 75 miles of waterfront and new houses, hotels, and a marine park.

FIGURE 7.27 **Arab Southwest Asia: Mecca, Saudi Arabia.** Al-Haram mosque, the main pilgrimage site for Muslims, who seek to visit it at least once in their lifetimes.

important roles in politics. Other Jews are of Asiatic origin or arrived from Russia in the 1990s.

Israeli Arabs generally accept the Israeli state and rarely conflict with Israeli Jews. On the other hand, Palestinian Arabs conflict greatly with Israelis because they reject the state of Israel and would prefer to have their own Palestinian state (see "Israelis versus Palestinians" section p. 219). Palestinian Arabs are most often simply called Palestinians, referring to the name of the land before the Israeli state was carved out of it in 1948. Interestingly, it was common to also call Jews Palestinians before Israel was created. Palestinian Arabs compose 90 percent of the population in the West Bank and Gaza and resent not having full governmental powers.

Israel's economy places its per capita income in the same league as countries of southern Europe. Ownership of consumer goods is high (see Figure 7.16), and Israel's economy is diversified. The agricultural sector, using intensive reclamation and irrigation methods, produces fruits, vegetables (Figure 7.28), and flowers for export to Europe but now constitutes only 5 percent of total GDP.

Israel possesses a well-educated workforce and access to foreign aid and investments. Providing a sophisticated defense capability prepared a generation of engineers for work in Israel's high technology industries. Manufactures make up 44 percent of exports and include diamonds, machinery, military equipment, and chemicals. In the 1990s Israel became a major center and leader of high technology development in manufacturing areas such as telecommunications, electronic printing, diagnostic imaging systems for medicine, and data communications. Such products account for 50 percent of industrial output, compared to 15 percent in 1990. Most of the manufacturing industries are based around the coastal cities such as Tel Aviv and Haifa. Industrial estates along the borders with Gaza and the West Bank employ cheap Arab labor.

Tourism is a major industry that attracted 1.9 million visitors in 2005 but is often interrupted by conflict and war. Many visitors combined visits to, for example, Jerusalem, Petra (Jordan),

FIGURE 7.28 Israel: agriculture. Sprinkler irrigation on a farm in the Negev Desert of southern Israel. Assess the importance of the presence or absence of water for landscapes in this region.

and the pyramids (Egypt), but the 2002 invasions of Palestinian areas caused both Jordan and Egypt to close their borders with Israel and suspend diplomatic relations, effectively ending such combined itineraries.

While Israel has a growing economy that places it ahead of its neighbors in development and lifestyles, the Palestinian areas of Gaza and the West Bank continue to have poorer conditions for human development. Palestinians accuse the Israelis of paying unequal attention to the needs of Palestinians in these territories compared to Israelis—a form of apartheid. Israelis accuse Palestinians of harboring terrorists. Though billions of dollars in international aid was given to Palestinians in the late 1990s, the intense and very destructive conflict between Palestinian terrorists and the Israeli army in the 2000s has destroyed much of the infrastructure for Palestinians in both Gaza and the West Bank, in turn ruining the economy and leading to a social crisis. In Gaza, for example, 50 percent of the work force is unemployed and 40 percent of its population lives in refugee camps. Many families depend on emergency food supplies.

Turkey and Iran

Turkey and Iran occupy the northern and eastern margins of this region (Figure 7.29). They are largely mountainous countries lying along fault lines that make them subject to frequent and devastating earthquakes. Their mountains receive precipitation, much of which falls as winter snow, feeding rivers and providing fresh water for urban areas and farming.

The two countries are influential in Southwestern Asia and the wider world, sharing economic leadership of the region with Saudi Arabia, Egypt, and Israel. They have crucial strategic positions between the southern boundary of Russia and Neighboring Countries (see Chapter 3) and the Persian Gulf oil fields (Figure 7.30). Together their populations comprise nearly one-third of the total population of the entire region (see Table 7.1). Both countries have Kurdish minorities who desire independence, a problem they share with Iraq and Syria. Governments feel threatened by the Kurds, especially the Turkish government (see "Human Rights" section, p. 222).

Some differences distinguish Iran and Turkey. For example, Iran stands out in the region as a country where Shiite Islam dominates. Iranians (Persians) are a distinct people from Turks, and both groups are unrelated to Arabic peoples who dominate this world region. Separate histories of empire building and differing forms of government and economies add to both countries' distinctiveness.

In the early 1900s, Iran was ruled by shahs who kept the country largely under military control but allowed for the adoption of Western education and the development of a wide range of economic activities. Though Iran's economy grew, the last shah was repressive and resentment toward him increased. In 1979, nationalist religious leaders, led by the Ayatollah Khomeini, seized political power. The shah fled to the United States but the U.S. government, which had long supported the shah, refused to turn him over to stand trial. Consequently, anti-American feelings grew in Iran, culminating with the taking of hostages at the

FIGURE 7.29 Turkey and Iran: countries, physical features, and main cities.

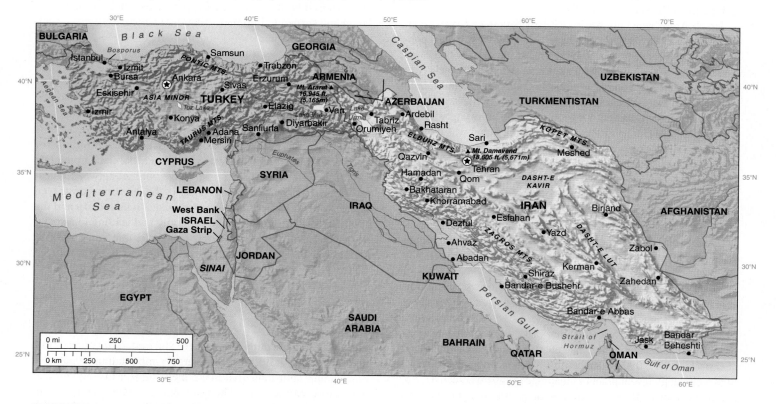

FIGURE 7.30 Iran and the Persian Gulf. A space shuttle view of the Strait of Hormuz with the Zagros Mountains of Iran to the left (north) and the United Arab Emirates and Oman on the south side of the strait. All marine traffic into and out of the Gulf must pass through the strait, which has great strategic significance.

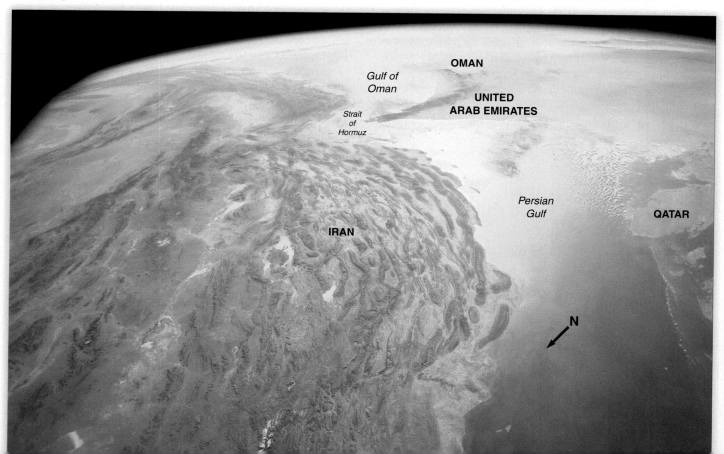

American embassy in Tehran. At the same time, Iran shifted to Islamic religious leadership and isolationism. The new and current constitution of the Islamic republic in Iran depends on a Shiite interpretation of Islamic government. Clergy are expected to establish a just social system and implement Islamic laws.

Turkey's political system is very different from that of Iran. After the disastrous defeat of the Ottoman Empire in World War I, Turkey became a nationalist and secular republic, putting the country before religion in questions of government. From the 1920s until 1936, the new leader, Mustafa Kemal Ataturk, ruled with a single-party government. Turkey gradually modernized and became increasingly involved in the global economy. Neutral during World War II, Turkey became a member of NATO (see Chapter 2) during the Cold War.

In recent times, both Iran and Turkey face tensions between liberalizing Western influences and pressures from Islamist political groups. Many young Iranians dislike the strict Islamic rules imposed by a fundamentalist few on a moderate majority. In contrast, Turkey faces continuing challenges from Islamic political groups to its long-term secular state principles. For example, many citizens do not like the ban on the religious practice of wearing headscarves.

Iran and Turkey also have different types of economic development. With 10 percent of world oil reserves (see Figure 7.17), Iran has the potential of earning great export income like Saudi Arabia. However, the takeover of the country by religious leaders, U.S. sanctions, and the war with Iraq in the 1980s has inhibited economic development. The Iranian government's decision to invest in nuclear power plants in the 2000s was condemned by the United States, which fears that nuclear technology employed to generate electricity also can be easily used to develop nuclear weapons.

Turkey has little oil but invested heavily in the development of its water resources for more agricultural output and hydroelectricity. Industrial expansion increased from 1950 and especially in the 1970s and 1980s. Turkey's real income rose steadily to exceed that of Iran until the late 1990s when Turkey's government implemented new policies that put Turkey's economy into recession. Nevertheless, Turkey has a more developed services sector than Iran. Government employment is very important in the economy. International tourism grew and annually brings in over a billion dollars. In 2005, Turkey attracted 20.3 million visitors (up from 1 million in 1980). The development of tourist resorts along its sunny coasts and the availability of historic, often religious, sites made Turkey a major venue for Europeans. Turkey is the only country in Northern Africa and Southwestern Asia (outside of Israel) with such a diversified economy.

Contemporary Geographic Issues

Israelis versus Palestinians

Perhaps the best known world conflict today is between the Jews living in Israel and the Arab Palestinians living in the occupied Palestinian Territories of the West Bank and Gaza. Although the

United States plays a significant role in trying to reconcile the two sides, the Arab countries see U.S. positions and policies as unfairly favoring Israel. The conflict's origins go back in history, before Israel was established as a new country in 1948. The continuing resistance of Palestinians to the heavily armed Israelis suggests an irreconcilable conflict, at least one that will not likely be resolved until an independent homeland is created for the Palestinian Arabs.

The land in question is a small part of what is known as the "Fertile Crescent" that cradled and connected the early civilizations of Mesopotamia and Lower (northern) Egypt. The hilly coastal lands provided a home for the ancestors of the Israeli nation (Hebrews). Roman armies subdued their lands, which were then governed as the Roman province of "Palestine." Eventually, the Romans tired of Hebrew rebellions, sacked Jerusalem, and dispersed most of the people in AD 70 to create the Jewish diaspora. Long after the fall of the Roman Empire, Arabs inhabiting the area quickly converted to Islam in the AD 600s, and Muslim armies spread the Islamic faith. Jerusalem became the third most holy site for Muslim pilgrimages after Mecca and Medina.

The Turkish Ottoman Empire governed "Palestine" from medieval times to the early 1900s. In the 1800s, the idea of a separate country for Jews arose out of Zionism, a movement that began in Europe and Russia which called for the creation of a separate Jewish nation-state. Anti-Semitism in Europe and Russia, including the imprisonment and murder of Jews, caused waves of Jewish settlers, numbering 60,000 from 1880 to 1914, to migrate to Palestine. The settlers taught the modern Hebrew language in schools and established socialist institutions such as labor organizations and farming in **kibbutzim** (singular: kibbutz). In the kibbutzim, land is communally owned and decisions are made collectively. Settlers planned for mass migration into a new, independent country as a safe haven from a persecuting world.

During World War I, the Ottoman Empire sided with Germany and against the United Kingdom and France. To combat the Ottoman Empire, the British and French fanned nationalistic feelings among the Arabs of the Ottoman Empire. The British also put forth the Balfour Declaration (1917), which called for "the establishment in Palestine of a National Home for the Jewish people." After World War I in 1923, the Ottoman Empire was dismantled. Much of Southwestern Asia was granted to the United Kingdom and France as protectorates, including Palestine as the Jewish National Homeland.

British authorities tried to address Arab discontent over the creation of a Jewish homeland in their midst by restricting the number of Jews migrating to the territory. However, numbers had been building since the late 1800s. Then Nazi persecution in Europe in the 1930s caused another 350,000 Jews to move to Palestine by 1940, despite prohibition of such movement under the British mandate (Figure 7.31a). After World War II and the genocide of the Nazi Holocaust, a million more European Jews joined them. Violence among Jews, Arabs, and British forces increased. Pressure mounted for the creation of an independent Jewish state, while Arabs resisted turning over lands to such

(a) Palestine was a UK protectorate after World War I. Jewish immigrants made up 11 percent of the population by 1922, 29 percent by 1936, and 32 percent by 1946. Palestinian Arabs objected to the influx and rioted in 1920, 1921, 1929, and 1935–1939 in the face of land purchases and exclusive labor policy by Jews.

(b) The 1947 UN Partition Plan envisaged the orange-shaded area as Israel, but was rejected by Arabs. It was a UN initiative after the British withdrew, finalized in May 1948, when Israel declared itself to be a country.

(c) Palestinian Arabs supported by Egypt, Jordan, and Syria, invaded the newly declared country of Israel, but were repulsed as Israel extended its land area to that outlined in red. Two-thirds of the Palestinians became refugees. Further wars took place in 1956, 1967, and 1973.

Israel in 1967

Controlled by Israel after Six Day War

(d) After the 1967 war, Israel extended its frontiers to include the West Bank (from Jordan), the Sinai Peninsula (from Egypt), and the Golan Heights (from Syria). In 1973 a Syrian and Egyptian attack failed, leading to a desire for a diplomatic settlement: the 1978 Egypt–Israeli Camp David Peace Accord in which Israel returned Sinai.

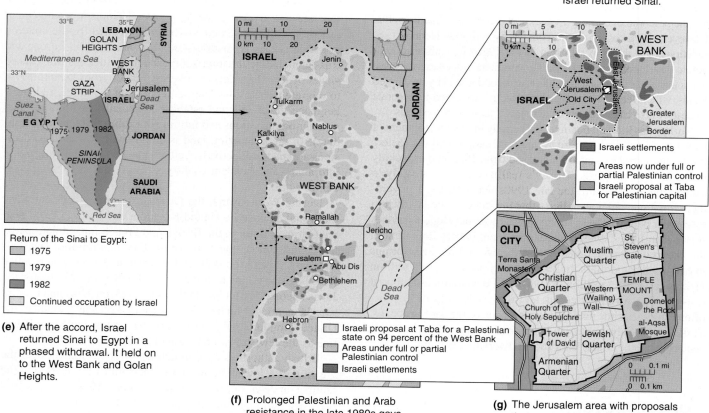

Return of the Sinai to Egypt:
- 1975
- 1979
- 1982
- Continued occupation by Israel

(e) After the accord, Israel returned Sinai to Egypt in a phased withdrawal. It held on to the West Bank and Golan Heights.

Israeli proposal at Taba for a Palestinian state on 94 percent of the West Bank

Areas under full or partial Palestinian control

Israeli settlements

(f) Prolonged Palestinian and Arab resistance in the late 1980s gave way to diplomacy after the Gulf War (1990–91). From 1993, the United States fostered talks that led to partial Palestinian control of West Bank and Gaza. Israel expected to return over 90 percent of these lands to form a Palestinian country.

Israeli settlements

Areas now under full or partial Palestinian control

Israeli proposal at Taba for Palestinian capital

(g) The Jerusalem area with proposals that were moving forward in 2000, before the later hard-line position of Ariel Sharon's Israeli government ended diplomatic means.

FIGURE 7.31 History of modern Israel: A timeline of changing maps (a) to (g).
Source: Data from The Economist.

a state. The United Kingdom could no longer control its protectorate and handed jurisdiction over to the United Nations, which in 1947 proposed a Partition Plan that would have created separate lands for Jews and Arabs (Figure 7.31b). Though the Arabs rejected the plan, the British withdrew their forces in 1948. Jews took matters into their own hands and declared the establishment of a new independent state which they called Israel. Egypt, Syria, Jordan, Palestine, and Iraq declared war, but the new state of Israel repulsed attacks and also expanded the country's original territory by the end of the war in 1949 (Figure 7.31c). During the conflict, some 600,000 Palestinian Arab refugees were forced out or fled for their lives and moved mainly into Jordan and Lebanon. By 2000, some 3.6 million Palestinian Arabs were living in Jordan (1.5 million), West Bank and Gaza (1.4 million), Lebanon and Syria (600,000 together).

Arab countries refused to accept the existence of independent Israel and fought it unsuccessfully again in 1956, 1967, and 1973. During the 1967 war, Israel extended its territory southward into the Gaza Strip and Egypt's Sinai Peninsula, eastward across the West Bank area of Jordan, and northward to the Golan Heights of Syria (Figure 7.31d, e). Despite UN resolutions, Israel refused to give up the occupied lands, apart from Sinai, for security reasons. Israel wanted surrounding countries to accept the existence of Israel and to renounce military activity against it. Syria would not acknowledge Israel's existence, so Israel held on to the Golan Heights. Unsuccessful in open warfare, Arab groups, both secular and religious, turned to terrorist methods from the 1970s.

In the 1970s and 1990s, U.S. presidents made attempts to reconcile differences. After the 1967 war, Israel returned the Sinai Peninsula to Egypt but retained control of the Israeli-occupied Palestinian territories of the West Bank and Gaza. The Jordanian and Egyptian governments gave up their rights to the West Bank and Gaza lands (which were previously parts of their territories) as a basis for creating a new Palestinian country by negotiation.

In 1994, following the Oslo Accord between Israel and the Palestine Liberation Organization (PLO), a Palestinian Authority was established and given limited jurisdiction and autonomy in Gaza and the West Bank. Israeli settlements in these areas (Figure 7.31f, g) were maintained, however, and their populations expanded. After changes in the Israeli government occurred in 1996, progress toward the development of a Palestinian country halted as Israel resisted further devolvement of power and transfers of land to Palestinians. Terrorist activity by a Palestinian group known as Hamas increased. The Israeli military entered Palestinian territories to punish people for atrocities, such as suicide bombings, committed against Israelis.

A major source of tension and a serious problem for Israeli-Palestinian negotiations relates to how the Israelis use Jerusalem, Gaza, and the West Bank. Both Israelis and Palestinians claim Jerusalem as their rightful capital for their respective countries (Figure 7.32). Israel asserted its rights over Jerusalem by proclaiming the city to be its capital in 1950 and building the Gnesset (parliament) in the city's suburbs, though the United States and most other countries do not recognize Jerusalem as

(a)

(b)

FIGURE 7.32 Jerusalem: aspects of conflict. (a) Dome of the Rock and the Western Wall. The Western Wall (sometimes referred to as the "Wailing Wall") is the holiest site for Jews because it is all that remains of Herod's Temple complex destroyed in AD 70 by the Romans. The gold dome of the Dome of the Rock, the holiest Islamic site in Jerusalem, is just 150 meters away on the Temple Mount (Haren al-Sharif) complex. (b) Israeli soldiers relaxing in the Old City.

Israeli's capital and maintain embassies in Tel Aviv. After Israel took East Jerusalem in the 1967 war, it confiscated land in the occupied area and made Palestinians second-class citizens by denying them property rights.

Another problem stems from the Israeli government's decision to build Jewish settlements in Palestinian areas. The resulting violence has led to increased geographical separation as the Israeli government attempts to provide security to Jewish settlers. Because Israelis are in the position of power, they dictate the course of separation. To protect their settlements in the West Bank, they have been constructing a wall that runs through the territory (Figure 7.33a). The wall prevents Palestinians from easily entering Israeli settlements, but the land for the wall comes from Palestinians who are then displaced. Moreover, while the wall is constructed to give Israelis easy access to the rest of Israel, many Palestinians find themselves cut off from their communities and workplaces. Palestinians are mainly restricted to low-wage jobs and denied access to higher-earning professions. Palestinians who wish to become doctors, for example, have to train abroad. The wall has resulted in much international criticism and has even been rejected by Israel's highest court, causing the Israeli government to modify its placement in many locations though the government is forging ahead with its construction in the hopes that it will provide security for Jewish settlers. In 2005, the Israeli government decided to remove Jewish settlers from the Gaza strip (Figure 7.33b). Many Israelis supported the move in the hopes that it would foster peace. Many thought it too costly to provide security for 9,000 settlers occupying more than 25 percent of the Gaza strip, overall inhabited by more than a million Palestinians mostly hostile to the settlers.

In 2006, an armed conflict erupted between Israel and Hezbollah, a Shia Islamic militia and political organization in Lebanon supported by Iran and Syria. It began on July 12 when Hezbollah launched missiles across Israel's border, captured two Israeli soldiers, and killed three others. The Israeli military responded with a military campaign aimed at rescuing the soldiers and destroying Hezbollah's military capability.

FIGURE 7.33 **Israeli barrier in the West Bank and Gaza.** (a) Israeli barrier in the West Bank. (b) Jewish settlements in Gaza abandoned in 2005. *Sources: (a) Data from* The New York Times *2004; (b) Data from* The Economist *2005.*

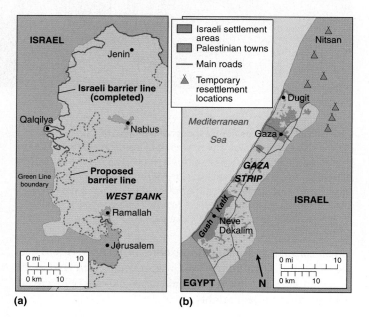

(a) (b)

Hezbollah responded by launching 100–200 missiles per day into Israel. The conflict lasted until a UN brokered cease-fire took effect on August 14. It was destructive and bloody. For example, 119 Israeli soldiers and approximately 400 Hezbollah fighters were killed. More than 1,191 Lebanese civilians and 43 Israeli civilians also died, and many more soldiers/fighters and civilians were injured on both sides. Nearly 1 million Lebanese and 300,000 Israelis were forced to flee their homes. Lebanon's infrastructure was largely destroyed, totaling US$7–15 billion, and Israel suffered $1.6–3.0 billion in damage. Both sides claimed victory but neither achieved many of their goals, and losses clearly overshadowed gains. The only clear outcome was that both sides hardened their negative views of each other.

While Israeli Jews and Palestinian Arabs contest each others' rights to their claimed homelands, it should be remembered that neither group is homogeneous or speaks with one voice. Both groups are very diverse and contain members with wide ranges of opinions. Each side has those who are conciliatory and willing to give up land for peace and each side has hard-liners, perhaps one in five inhabitants, unwilling to make any compromises and grant any concessions. The views in Table 7.3 are those that might be expressed by an Israeli Jew and Palestinian Arab.

Human Rights

Basic human rights for all citizens are a concern in some countries of Northern Africa and Southwestern Asia. For example, while no countries have abolished the death penalty, countries like Turkey only use it in exceptional circumstances and countries like Algeria and Tunisia have not imposed the death penalty in more than 10 years though it is still legal. However, all the other countries retain the death penalty. According to Amnesty International, Iran ranks number two in the world for the number of its citizens that it executes, behind China and just ahead of the United States and Vietnam.

A wider variety of human rights abuses are found in the region. The role of women in society is a major concern and so are specific armed conflicts within individual countries. The ones that have created the greatest number of deaths and refugees include the Israeli-Palestinian conflict (see previous section), the civil war in Sudan, and repression involving the Kurds, who are found in Turkey, Syria, Iraq, and Iran.

Women in Society

Gender inequalities are a continuing issue in Muslim countries. Male attitudes and the law work against equal opportunities for women. As with other aspects of Muslim culture, local interpretations and outcomes vary. Considerable debate exists in Muslim countries over what the Qu'ran says on these matters. Saudi Arabia boycotted the 1995 Fourth World Conference on Women in Beijing on the grounds it was anti-Islamic. In Iran, it is a criminal offense for a woman not to wear a headscarf. And yet, Muslim-dominated countries such as Turkey, Pakistan, Bangladesh, and Indonesia have elected women leaders in recent years. While

TABLE 7.3   DEBATE: ISRAELIS VERSUS PALESTINIANS

A Hard-Line Jewish View	A Hard-Line Arab (Palestinian) View
We were here first and have a longer history of occupying this land: it is ours, as claimed by Abraham and Joshua.	Our ancestors were here from the time of Abraham, whom we also recognize as a father of our people.
Our religion started first in this area. Our holy sites include Hebron (where Abraham is buried) and Jerusalem, both in, or partly in, the West Bank.	Jerusalem and Hebron are sacred to Muslims.
The United Nations agreed to the partition of Palestine and recognizes Israel with its Jewish majority.	The decision resulted from a combination of weak Arab support and strong U.S. and British pressure. We were ignored, although we made up most of the population here in 1948. Israel has taken large areas of land from us that were not part of the UN plan.
We have returned some territories we took in the 1967 war but hold on to the rest as a matter of national security and survival.	After the 1967 war, the United Nations ordered Israel to hand back the occupied territories, but it has not done so more than 40 years later.
Arabs deny our rights to exist and want to wipe us off the map; we are exercising our right to defend ourselves. When joined, the surrounding Arab countries outnumber us, so we have to ensure strong security.	We were forced out of our land, we lived in crowded camps without amenities, and we now live in poverty. The Jews close checkpoints with no notice and interrupt our lives. We cannot argue with them because they are supported by the United States, and we now hate that country as well.
We regard them all as possible terrorists who blow up our restaurants and nightclubs, kill our athletes, assassinate our leaders, and drive suicide bombs into our neighborhoods.	They refuse to recognize our presence and nationality, suppressing our language, religion, and culture. Most of us want to live peaceful lives, but they treat us all as spies and criminals, abusing our human rights.
Jerusalem is our real capital city and is central to the Jewish faith. Some Jews want to remove the mosque on the holy mount.	Jerusalem is sacred to us, and any moves to destroy the mosque would be a declaration of all-out war that would unite Muslims.
No strong Palestinian nationality was expressed here before 1948. This area of land was merely a British protectorate carved out of the former Ottoman Empire, and most people knew of the intention to create a land for Jews. The present Palestinians are Arabs who should have been taken in by existing Arab countries. They have invented Palestinian nationalism as part of a plot to eliminate Israel.	We want our own lands and independence from Israel.
They are poor workers and earn only low wages. We go out of our way to employ them, but it would be better to employ only Jews.	They are well fed and materially wealthy. If we want to study for the qualifications that would earn us better jobs, we cannot do so in our country and have to go elsewhere. A doctor friend of mine, who works in a Jerusalem hospital, had to go to Greece to qualify.
The Palestinian Arabs have not repaid all the help we have given to them, raising their well-being above that of other Arabs in this region.	We can do nothing that is legal to improve our lot, so it is not surprising that some of us take to the gun and bomb.

Saudi Arabia excludes women from all public activity, Tunisia promoted women's rights after its 1957 independence.

Literacy and education are a measure of human rights and women's empowerment. In general, a great discrepancy exists in the region between female and male literacy. For example, in 2005, only 51 percent of women in Yemen were literate compared to 84 percent of men. However, literacy rates are not uniform throughout the region. In United Arab Emirates, the literacy rates are 95 percent and 88 percent respectively. Overall, literacy rates have improved dramatically for women in the region over the last 30 years (Figure 7.34).

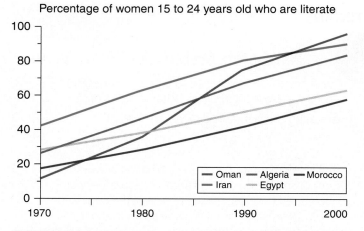

FIGURE 7.34 **Literacy rates among young women in selected countries, 1970–2000.** *Source: Data from UNESCO.*

Sudan

Not long after Sudan became an independent country in 1960, internal conflict ensued with the country experiencing numerous civil wars that have extended well into the 2000s. Much of the conflict has been between the Arab Muslim north and the black Christian and animistic south. The warfare created famine and misery in this drought prone country. As many as 2 million people died. Many others were displaced and left homeless. Arab Muslims have controlled the government in recent times. However, years of war weakened the country's economy and made it difficult for the government to assert effective control over the country. The Sudanese government relies on undisciplined troops and paramilitary groups to fight those who oppose it. Consequently, human rights abuses during campaigns have been widespread. Northern soldiers taking southern captives as slaves are just one example. Many Christian organizations in the United States have raised funds to buy the freedom of southern Sudanese Christians. It is a controversial practice because it may actually encourage more slave-taking.

In the early 2000s, armed conflict erupted between the government and non-Arabic tribes of western Sudan. To put down the rebellion, the government armed local Arab tribes who were steadily moving into the western province known as Darfur (meaning "land of the Fur people"), a source of anger for the Fur and a partial cause of the rebellion. These armed individuals earned the name Janjaweed, an Arabic word meaning "devil on horseback with a gun." The name reflects the fact that these nonprofessional troops, though acting on behalf of the government, terrorize the local population, burning villages and raping women in what the international community labels as genocide. Thousands of refugees have fled into neighboring Chad. The international community condemned the Sudanese government, and some international peacekeepers have been sent to the region. However, the Sudanese government denies its support of the Janjaweed despite deputizing many Janjaweed as police with the assignment to restore order in the western region. This policy encourages human rights violations.

Kurds

Many Kurds have long desired their own nation-state called Kurdistan and have struggled and been persecuted for trying to establish their own state (see Figure 7.5). In 1984, the Kurds of southeastern Turkey launched a military campaign to free the Kurdish areas of Turkey. The Turkish military responded with force, leading to many human rights abuses. After the capture of a Kurdish leader who was then sentenced to life in prison, the Kurdish fighters changed their military struggle to a political one. In the early 2000s, the Turkish government, desiring membership in the European Union, succumbed to pressure from the European Union to grant more rights to its citizens. Kurds now hold political office and are allowed to speak Kurdish in schools.

While the situation of the Kurds in Turkey is much better today, life for Kurds in Iraq is far less certain. In the early years of Saddam Hussein's reign, Kurds enjoyed a number of rights. However, after a number of Iraqi Kurds joined with the Iranians in the Iran-Iraq war in the 1980s, Saddam Hussein persecuted Kurds, even dropping lethal mustard gas on them. After the Gulf War in 1991, the United States and its allies set up a no-fly zone in northern Iraq to prevent the Iraqi airforce from bombing Kurdish areas. The act allowed the Kurds of northern Iraq to effectively set up their own governmental structures. Following the ouster of Saddam Hussein in 2003, the United States encouraged Kurds to participate in a new federally organized Iraq.

Kurds in Iran have suffered less persecution in recent years and have not engaged in armed conflict. The last armed struggle led by Iranian Kurds occurred right after the Iranian Revolution in 1979. Though Iran's Kurds have engaged in fewer armed struggles than Kurds elsewhere, they succeeded in creating the only modern Kurdish nation-state, the State Republic of Kurdistan with the capital in Mahabad. This nation-state was short-lived as it was founded and destroyed in 1946.

Kurdish rights have been increasingly violated in Syria since 1963 when the Baath Party took power. The Baath Party is similar to the party by the same name in neighboring Iraq that governed under the leadership of Saddam Hussein. The Syrian Arabization program deprives Kurds of Syrian citizenship, ownership of their land, and since 1992, the right of children to use their Kurdish names when registering for school. False imprisonment, torture, and mass arrests of Kurds are common in Syria.

Iraq

Iraq is a source of major tension in Southwestern Asia and the world today. The U.S. attempt to build a democratic country is challenged by complex ethno-religious and political geographies of the region (Figure 7.35 and see Figures 7.5 and 7.7). To understand these challenges, it is useful to examine the circumstances that created Iraq and its various internal geographies.

Iraq occupies much of the land that was part of ancient Mesopotamia (meaning "land between the rivers," namely the Tigris and Euphrates). Until the end of World War I, it was within the Ottoman Empire and did not exist as a distinct political unit. In 1916, British military forces began occupying the area. At the

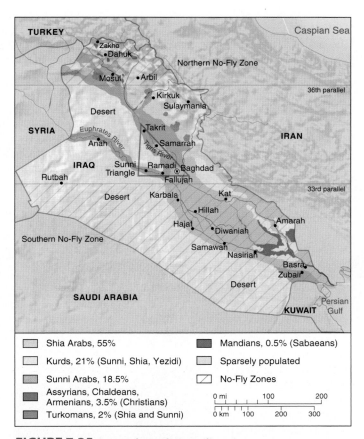

Legend:
- Shia Arabs, 55%
- Kurds, 21% (Sunni, Shia, Yezidi)
- Sunni Arabs, 18.5%
- Assyrians, Chaldeans, Armenians, 3.5% (Christians)
- Turkomans, 2% (Shia and Sunni)
- Mandians, 0.5% (Sabaeans)
- Sparsely populated
- No-Fly Zones

0 mi   100   200
0 km   100   200   300

FIGURE 7.35 **Iraq: ethnoreligious diversity.** *Source: Data from Wikipedia.*

end of the war, modern-day Iraq came into existence when the three Ottoman provinces of Baghdad, Mosul, and Basra were joined together and became a British mandate.

As a European creation, Iraq has little internal homogeneity. The north is occupied by Sunni Kurds, the north-central portion by Sunni Arabs, and the southeast by Shia Arabs. All three groups have as much in common with peoples in neighboring countries than with one another. Ethnic and religious diversity does not necessarily undermine the functioning and well-being of countries. However, well-functioning countries usually were created by their own people and have established a governing tradition that the country's citizens respect.

When Iraq first became a British mandate, the British tried to create a governing tradition by installing a king from the Hashemite clan, which traces its roots back to the prophet Muhammad, and controlled the holy cities of Mecca and Medina for a time as well; the clan also rules Jordan (formally known as the Hashemite Kingdom of Jordan). Iraq seemed stable when it was granted independence in 1932, but the Hashemite king was overthrown in 1958. Iraq then was governed by a series of military coups until 1968, when the Arab Socialist Baath Party took power. The Baath Party combined Arab nationalism with socialism and pan-Arabism. With branches in other countries, Baathists also came to power in Syria in 1968. As a member of the Baath Party, Saddam Hussein served first as vice president of Iraq and then became president in 1979.

Seeing himself as a social revolutionary and modernizer, Saddam Hussein abolished most of the *sharia* (traditional Islamic law) courts in Iraq and established a Western-style legal system, the first in Southwestern Asia. Hussein also granted women greater freedom and the opportunity to rise in government and business. Secular in outlook, Hussein disliked fundamentalist Islam, which in turn made him distrust the Shias of southeastern Iraq. A pan-Arabist, he was suspicious of the Kurds in northern Iraq. He believed that Iraq's ethnic and religious diversity was a threat to Iraq and subsequently believed that he must maintain unity through force.

Saddam Hussein received his support from the Sunnis in north-central Iraq, primarily in an area known as the Sunni Triangle, marked by Baghdad in the southeast, Ramadi in the southwest, and Tikrit (Hussein's hometown) in the north. As Hussein's reign progressed, he built an oppressive police state to promote unity. Kurds and Shias suffered greatly. At one point, Hussein had lethal mustard gas dropped on Kurds as part of a broader campaign often described as genocide. After he was deposed, it was discovered that the ratio of men to women among the Shias in the south was particularly low, indicating that Hussein's regime had targeted and killed many Shia men.

In 1979, an Islamic revolution brought Shia fundamentalists to power in neighboring Iran under the leadership of Ayatollah Khomeini. Saddam Hussein feared that Iran's Shia Muslims would incite Iraq's Shia population. In 1980, Hussein launched an invasion of Iran, allegedly to settle a border dispute. He claimed that the adjacent oil-rich province of Khuzestan rightfully belonged to Iraq because it had been within past Arab empires. The invasion was initially successful, but then stalled and turned into a long war of attrition that claimed more than a million casualties before ending in 1988.

Relations between Iraq and Kuwait were poor too. The Iraqi government never recognized Kuwait's sovereignty, believing for historical reasons that Kuwait properly belonged to Iraq. Hussein regularly referred to Kuwait as Iraq's nineteenth province. He also accused Kuwait of slanting drilling underneath the border between Iraq and Kuwait to illegally obtaining Iraq's oil and used this as a justification to invade Kuwait in August 1990. Iraqi forces quickly conquered Kuwait. However, UN economic sanctions followed, and a coalition of military forces largely comprised of American troops drove Iraqi forces out of Kuwait and southern Iraq in early 1991 in the Gulf War (also known as Operation Desert Storm).

Iraq lost the Gulf War but its military remained formidable, and Hussein was seen by the international community as a continued threat. The United Nations continued its economic sanctions with the intent of forcing Hussein to disarm. The United States and the United Kingdom established no-fly zones (see Figure 7.35) over northern and southern Iraq that forbade the Iraqi airforce from launching campaigns against the Kurds and Shias. Hussein was uncooperative and campaigned to have the sanctions removed. The Iraqi military periodically tested American and British military resolve, resulting in brief shooting and bombing incidents.

U.S. President Bush accused Hussein of pursuing a nuclear weapons program and not cooperating with UN weapons inspectors. In 2003, an American-led military force invaded Iraq and quickly toppled Saddam Hussein's regime. At the end of the year, Hussein was captured and put on trial before a special

Iraqi tribunal, charged with crimes against humanity. He was found guilty and executed on December 30, 2006. In the meantime, the United States and its allies initially set up a Coalition Provisional Authority to govern Iraq. It was replaced by an Iraqi interim government in 2004. Elections in May 2005 created a transitional government until elections in December 2005 established a permanent government to serve from 2006 to 2010.

Though many Iraqis were freed from Hussein's rule, the history of favoritism and oppression within Iraq has left deep divisions within Iraqi society. Sunni Kurd, Sunni Arab, and Shia Arab differences only represent the main cleavages. Tribal, social, and other layers of identity represent additional divisions.

The lack of a strong government in Iraq has become an attraction for extremists and terrorists from other countries, making Iraq a frontline battleground in the global war of ideologies. The ensuing violence is widening and hardening sectarian differences as insecurity increases and the basic provision of food, water, and electricity decreases. As the U.S. military clamps down on terrorist activity, it is perceived by many as an occupational power. The longer it takes U.S. military force to subdue violence, it ironically creates a sense of Iraqi nationalism aimed at driving foreigners out of Iraq. The challenge for the United States is to provide security and services to Iraqis until the new Iraqi government becomes effective enough to offer such provisions.

GEOGRAPHY AT WORK

Operation Iraqi Freedom

Ask the soldiers in the 450th Movement Control Battalion (MCB) about their experience in Operation Iraqi Freedom (OIF) and they'll tell you with a smile that "It's all about geography." Their commander, Lieutenant Colonel (LTC) Mark Corson (Figure 7.36), told and showed them time and again that geography mattered in everything they did. When not active as a U.S. Army Reserve officer, Mark is a geography professor at Northwest Missouri State University in Maryville, Missouri.

Mark notes that military transportation has three parts: terminal operators, mode operators, and movement controllers. Terminal operators work the airports, seaports, and trailer transfer points to stage cargo for movement. Mode operators drive trucks, fly helicopters, and operate railroads and watercraft. Movement controllers are the brains of the operation, telling the mode operators what to haul, where, and to whom (they also control the roadways to prevent traffic jams). The 450th MCB served as the movement controllers in southern Iraq during the opening phases of the war. They later moved to control transportation operations in Kuwait, where they oversaw movements for the major airports, seaport, and staging bases.

Understanding the geography of Iraq helped soldiers of the 450th MCB a lot. Their knowledge of physical geography prepared them for the weather, climate, and terrain of the desert. Their well-designed uniforms and equipment allowed them to brave searing heat, sandstorms, and desert sand so that they could accomplish their mission. Understanding the terrain, the location of key cities, and the nature of the connecting roads allowed them to site their logistics bases and pick the best transportation routes.

Knowledge of human geography was also very valuable to Mark and his soldiers. Understanding Iraq's three major cultural/religious groups, ensured that Mark and his soldiers respected these peoples' traditions. Their knowledge of the political geography and history of the region was very helpful in negotiating sensitive diplomatic agreements to allow their trucks to cross the Kuwait-Iraq border.

Geography provides some very powerful technical tools such as cartography, remote sensing, geographic information systems (GIS), and the global positioning system (GPS). Mark and his soldiers used these geographic tools extensively. Mark's Highway Regulation Teams conducted route surveys using GPS to get exact coordinates of facilities, bridges, intersections, and key terrain. His operations section then created a simplified strip-map that they distributed widely. Anyone who needed to go anywhere in their area referred to that map so he or she could travel safely and without getting lost. Mark and his soldiers used a sophisticated navigation and communications sys-

FIGURE 7.36 **Geography at War.** LTC Mark Corson (right) and 1LT B. J. Vincent of the 450th Movement Control Battalion at Saddam International Airport (now Baghdad International Airport) shortly after its capture by the U.S. Third Infantry Division in April 2003.

tem called the Military Tracking System (MTS) in their vehicles and in their headquarters. MTS has a digital map display linked to a GPS that shows where the user and other MTS users are on the map. Mark's soldiers used the satellite text-messaging capability to communicate over very long distances. In many cases the soldiers saved lives by text-messaging medical evacuation requests that could not be passed by radio. Mark's unit also maintained in-transit visibility of convoys and cargo using the Joint Distribution Logistics Model, or JDLM, which is a type of geographic information system (GIS).

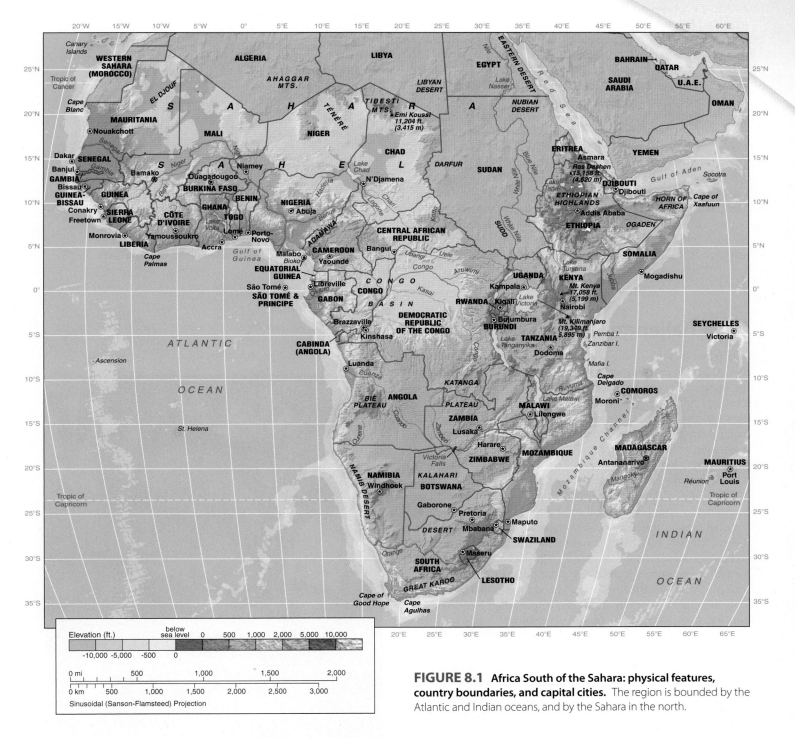

FIGURE 8.1 **Africa South of the Sahara: physical features, country boundaries, and capital cities.** The region is bounded by the Atlantic and Indian oceans, and by the Sahara in the north.

"Growing up in Ghana, I, Yaa Baodi, had a village-based childhood. I hauled water from the village well and was raised by my grandparents in a family compound. An uncle encouraged me to achieve at school and I got to the university at Kumasi, one of only four women to enroll for civil engineering. As the only woman to complete that course, I had to hold my own with male classmates from privileged families and prestigious high schools. When I graduated, I became the first woman hired by an Accra engineering company, working on highway projects. As a Ford Foundation scholar, I studied for a masters degree in the United Kingdom to prepare me for work on infrastructure projects in neglected parts of Ghana. I am happy with my life and achievements."

"I, Ntwari, lived in Rwanda until the 1994 troubles. My parents had a small farm and I was one of the few in my village to get through elementary and high school and study at university. I became a teacher and married. However, in 1994 the Hutu people revolted against Tutsis and rampaged through the country, killing thousands. My wife and I fled as I taught in a Tutsi-headed school and her family were Tutsis. We reached eastern Congo, where I could live in a Hutu camp, but she had to hide. I walked and caught trains to Nairobi in Kenya and then returned to bring out my wife. The United Nations airlifted us out of Congo since we could not return to Rwanda. We now live abroad, hoping to return to Africa and help others to have fulfilled lives."

Chapter Themes

The Challenge

Defining Africa South of the Sahara

Distinctive Physical Geography: plateaus and river valleys; major rivers; tropical climates; cooler southern margin; forests, savannas, deserts, and their soils.

Distinctive Human Geography: African ethnic diversity; shifting control; growing and mobile populations; political and economic pressures.

Geographic Diversity
- Central Africa.
- Western Africa.
- Eastern Africa.
- Southern Africa.

Contemporary Geographic Issues
- HIV/AIDS pandemic.
- Exploding cities.
- Global intrusions and local responses.
- Culture shock.
- Can Africa claim the twenty-first century?

Geography at Work: Mapmakers and GIS Analysts

The Challenge

Africa South of the Sahara (Figure 8.1) was the cradle of humanity and people from the region spread around the world. Within the region, Africans developed their own distinctive social systems and cultural expressions (Figure 8.2). The natural environments of the region are endowed with huge mineral resources, major rivers and water resources, extensive areas where farming is possible, and the tourism resources of wildlife and landscape.

However, today it is the world's poorest region, and many of its 42 countries are affected by internal armed conflicts. In a 2006 *Foreign Policy* Failed States Index, based on political, economic, military, and social indicators of instability, half of the countries in this region were regarded as "critical" or "in danger" (Figure 8.3). Many countries have political leaders who siphon off scarce funds into their own pockets. The region's natural environments include expanding deserts, impoverished soils, and diseases affecting humans and their animals.

In this chapter we examine the challenges posed by a world region where so many current human development trends are negative. After years of colonialism, serving as political footballs during the Cold War, and then being largely forgotten by the world's wealthier countries, new attention is being directed at the major task of helping African countries to improve their well-being. We start by assessing the physical and human forces acting on the region over time to produce today's geographic diversity.

FIGURE 8.2 Ancient Africa. The ruins of Great Zimbabwe record the existence of a major trading center from around AD1100 preceding European intrusions. A symbol of African historic achievement.

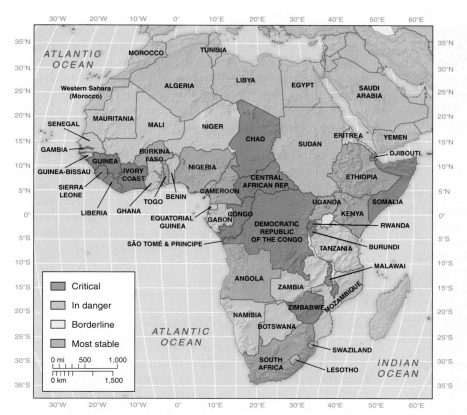

FIGURE 8.3 Failed Countries. Ranked by indicators of instability including high population growth, refugees, uneven development, economic decline, poor security, human rights issues, factionalized elites, and external interference. Half the African countries are in the "critical" or "in danger" groups. *Source: Data from Foreign Policy, 2006.*

Defining Africa South of the Sahara

This world region is mostly defined by its coasts, from the border of Sudan and Eritrea in the east to the southern Cape of Good Hope and along the west coast to the northwestern corner of Mauritania. Some of the islands in the Indian Ocean to the east (Madagascar, Reunion, the Seychelles) and in the Atlantic Ocean to the west (Cape Verde, St. Helena) are commonly included.

The main question mark is the region's northern boundary. Although it crosses the largely uninhabited Sahara, some countries include territory in this desert. We include Mauritania, Mali, Niger, Chad, Eritrea, and Ethiopia in this region since their human activities are directed southward. We omit Sudan, despite its links with Ethiopia and its southern area that has many features like the countries on either side, because the Nile River links it northward to Egypt.

Distinctive Physical Geography

The natural environments of Africa South of the Sahara provided stages for changing human activities over time, offered resources on which to base a set of choices, and imposed some restrictions at each stage of human development. These environments remain significant factors in today's geography.

Plateaus and River Valleys

The African stage is dominated by plateaus (see Figure 8.1), which are underlain by some of Earth's oldest rocks. The plateau edges rise steeply from the oceans in many places and the interior is marked by steps from one level to another—steps that result in waterfalls or rapids on the major rivers. In eastern Africa the plateau surface was raised to domes, where the top cracked, causing crustal rocks to drop and form **rift valleys.** These are often floored by deep lakes. Between two of the rift valleys, Lake Victoria and the source streams of the Nile River are located in a down-warped section of plateau.

Volcanic peaks such as Mount Kilimanjaro, the highest point in Africa, rise above the plateaus and form this region's main mountains. Other volcanic mountains include the Ethiopian Highlands and Mount Kenya along the eastern rift valley and parts of the Cameroon Highlands on the border of Cameroon with Nigeria.

The ancient rocks of African plateaus contain a wealth of mineral resources, which were one basis for European interest and local economic development from the late 1800s. However, the exploitation of some large deposits awaits peaceful conditions or the building of transportation links. Coastal and offshore oil occur in some of the newest rocks and delta deposits. Beginning in Nigeria, but now extending southward to Angola, coastal oil and natural gas resources are making the continent a major world source (see Figure 7.17). In South Africa layers of sedimentary rocks contain coal deposits, of which the country is the world's sixth largest producer.

Major Rivers

The largest river valleys include those of the Niger, Nile, Congo, and Zambezi. Each provides a major source of water and a transportation system, although all have courses that interrupt navigation. For example, the largest river, the Congo, has a series of rapids from source to mouth. It enters the ocean through narrows with more rapids, preventing ocean-going ships from sailing upstream. Short lengths of road and rail are needed to transfer goods around the rapids. The next largest river, the Niger, has a large marshy delta at its mouth and its northern loop flows through dry lands where the river remains shallow. In the south, Zambezi River navigation is also interrupted, including by the Victoria Falls.

Unlike Europe, Africa does not have deep ocean inlets and there are few natural port sites. Access from the coast to the interior relied on rivers until modern transportation modes were introduced. Even today, there are only a few railroads or all-weather roads, mostly built to link interior mines and cities to ports. Air transportation overcame many of these difficulties, but depends on the availability of airports and landing strips.

Tropical Climates

Almost all of Africa South of the Sahara lies within the tropics (see foldout map inside back cover). Climates range from the hot and rainy all-year type at the equator to hot arid areas in the Sahara and Namibia. In the former, temperatures average 30°C (80°F) and annual rainfalls 3,500 mm (140 in.). In the latter, summer temperatures exceed 30°C (80°F) and winter temperatures reach 25°C (70°F), but rainfalls are under 100 mm (4 in.) Between these extremes, the temperatures remain high all year, but rains are concentrated in the summer season, totaling around 500 mm (20 in.). The extensive plateau uplands provide cooler, but often drier, conditions than the tropical latitudes suggest.

Tropical climates allow the production of a wider range of crops than in the world's temperate regions: rubber, cocoa, coffee, cotton, tropical fruits and vegetables. They also attract tourists looking for sun and big animals. However, such tropical environments harbor many diseases that restrict the productive ability of people, crops, and cattle. Malaria, sleeping sickness, and river blindness still affect millions, while the tsetse fly prevents cattle raising over large areas. Poor people and poor countries do not attract the wider application of medical and veterinary treatments.

Cooler Southern Margin

On the southern edge of the continent, in the Republic of South Africa, subtropical rainy climates with cooler winters provide some of the world's most pleasant living conditions, with the hottest periods ameliorated by ocean breezes. Summer temperatures average 25°C (70°F) and winter temperatures 10°C (50°F). Rains come in winter (Cape Town area) or summer (Durban area), with totals of 500–700 mm (20–30 in.).

Forests, Savannas, Deserts, and Their Soils

The natural vegetation regions of Africa South of the Sahara (Figure 8.4) correspond closely to the climatic regions, but are becoming increasingly modified by human actions. The dense **tropical rain forest** of the equatorial rainy climate regions (Figure 8.5a) is marked by a huge variety of broadleaf, evergreen tree and plant species. The great variety of plants pro-

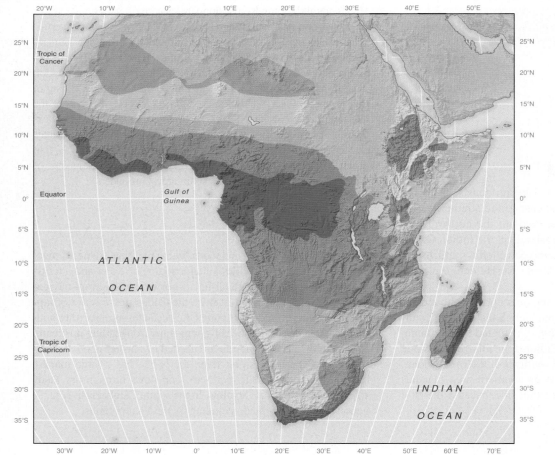

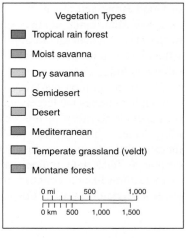

FIGURE 8.4 Africa South of the Sahara: vegetation regions. Major natural vegetation types. Compare this map with the foldout world climate map inside the back cover.

Vegetation Types

- Tropical rain forest
- Moist savanna
- Dry savanna
- Semidesert
- Desert
- Mediterranean
- Temperate grassland (veldt)
- Montane forest

vides a stock of biodiversity that many see as vital to the future of human existence on Earth. However, the variety of insects and microbes in these biomes present challenges of diseases that affect humans, animals, and plants. Much of the rain forest has been cut for timber or to make way for commercial farming. Slash-and-burn forest clearance for short-term crop production is followed on abandonment by a less-rich secondary forest. Forest soil quality deteriorates rapidly, or erosion is rapid, when the trees are cut.

The **savanna** grasslands (Figure 8.5b) of the seasonally rainy areas have more or less tree cover depending mainly on the length of the dry season. It is thought that the extension of this biome into wetter areas was caused by dry season burning by humans to reduce the dense forest in favor of grasses that encourage greater numbers of large herbivore animals (such as elephants, giraffes, zebras, antelopes, and springboks, many of which provide meat) and their carnivore predators (such as lions, leopards, and wild dogs). Today, viewing these animals is a focus for tourist industry. Poaching of some animals in some countries reduces numbers there, but wildlife populations in other countries expand and exceed available food resources.

Soils beneath the grassland are often workable and many contain good proportions of plant nutrients that support agriculture. However, once exposed following cultivation or overgrazing, soils containing high quantities of clay and iron minerals become cemented into red **laterite.** Laterite forms an almost impenetrable layer; its main use is as rough blocks in house building. Furthermore, some diseases such as tsetse fly infestations affecting cattle, as well as malaria, sleeping sickness, and river blindness in humans are still prevalent. Cures are available, but expensive.

The grasslands and seasonal rains grade into arid **deserts** (Figure 8.5c), the margins of which move back and forth as a result of climate change and human destruction or planting of vegetation. The **Sahel** zone along the southern margin of the Sahara suffers periodic droughts that are made worse by overgrazing and firewood cutting. The **desertification** of the Sahara and the dry margins in Southern Africa causes famines and destitution in countries that have few resources. Drought affected wider areas in the early 2000s. Not only is Lake Chad in the Sahel shrinking, but also the major lakes in Eastern Africa such as Lake Victoria and Lake Tanganyika (see Figure 8.1) have falling water levels. The fall of 2 m (6.5 ft.) since the independence of the Eastern African countries is due to increased abstractions, deforestation, and global warming and has adverse effects on fishing and hydroelectricity production.

Distinctive Human Geography

Homo sapiens first emerged in this region, from which modern humans pushed outward to populate the rest of the world. Indigenous societies developed characteristic cultures over thousands of years before Arab and European interventions and incursions resulted in a diverse set of geographies today.

(a)

(b)

(c)

FIGURE 8.5 **Africa South of the Sahara: contrasting ecosystems.**
(a) Congo rain forest, based on plentiful rain. (b) East African savanna close to Mount Kilimanjaro in an area of seasonal rains. (c) Rocky desert with vegetation supported by subsurface water.

African Ethnic Diversity

For thousands of years after the early dispersal of *Homo sapiens* to other world regions, population changes within Africa were mainly a matter of internal movements. The great variety of African peoples mixed and migrated. Cattle herders from the upper Nile River valley, such as the Masai, lived in tension with farming Bantu groups such as the Kikuyu, who migrated from Central and Western Africa. Bantu groups also reduced the populations of pygmy groups in the tropical rain forest and displaced the San and Khoikhoi hunters in Southern Africa. In the south, Great Zimbabwe (see Figure 8.2) became the center of a widespread trading empire in the AD 1100s. By the early 1800s, Bantu peoples, who had moved into southernmost Africa—the Swazis, Zulus, Zhosas, and Sothos—established local kingdoms, fighting each other for territory.

Members of groups (called "tribes" by European colonists) linked by kinship, language, and territory formed the basic social and political units (Figure 8.6a,b). Such groups are often distinguished by appearance, as in the tall cattle herders or the short pygmies. However, individuals and families changed allegiance as their economic and social circumstances altered.

Common constituents of African culture emerged from the long history of social development. Human, natural, and spiritual forces were contained in religious **animism** that sees human success or failure dependent on gods and spirits in rivers, tree

FIGURE 8.6 Africa South of the Sahara: cultural features. (a) Ethnic group areas (red lines) and colonially imposed country boundaries (gray lines). (b) Ethnic groups, languages, and precolonial kingdoms. The transition zone contains a great diversity of languages. *Sources: (a) Data from English and Miller 1989; (b) Data from Grove 1993.*

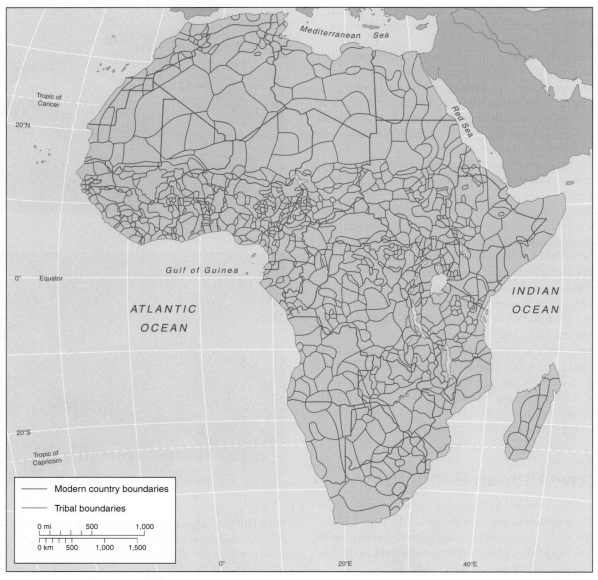

(a)

groves, and rock outcrops. Land is traditionally a communal possession, inheritance, and responsibility in a still largely rural economy based on cultivating, herding, or hunting. Each person is part of a continuing life chain, revering ancestral spirits and the family unit. Large families are a blessing and childlessness a tragedy. Such beliefs cause the wider family to support members through difficult times. The chief of a group is respected as a strong, trusted leader acting for the common welfare. These cultural foundations have been challenged by new belief systems, colonization, urbanization, and technological changes.

Shifting Control
Muslim Influences and African Empires

From the AD 600s Islam spread from Arabia into the northern and eastern parts of Africa. Introducing camels, Muslim Arabs traded via routes from the Mediterranean across the Sahara: in the west from Fez and Marrakesh (modern Morocco) to the middle Niger River valley, centrally from Tripoli to the Lake Chad area, and in the east along the Nile River valley. As it moved into Africa, Islam accommodated many local practices, such as polygamy and the use of the African drum.

Trade in salt, gold, ivory, and slaves underpinned the sophisticated Western African empires of Ghana (AD 700–1240), Mali (1050–1500), and Songhai (1350–1600). Timbuktu, on the northernmost bend of the Niger River, began as a market for crops and cattle, but by the AD 700s traded salt and gold. Universities were established at Timbuktu and Djenne, employing scholars from Greece, Egypt, and Arabia and boasting libraries stocked with imported books. When the Mali emperor, Mansa Musa, undertook his Muslim pilgrimage to Mecca in 1324, he demonstrated his wealth with 500 porters each carrying a golden staff. Later Islamic expansion led to Fulani

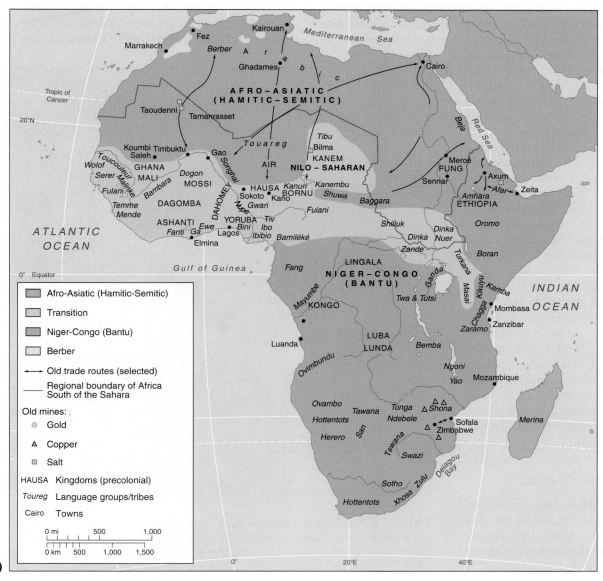

(b)

warriors and zealots settling grazing lands along the southern Sahara margins.

In Eastern Africa, Arab traders established African ports such as Zanzibar and Pemba to exchange goods with other ports around the Indian Ocean. From the 700s to the 1800s, ivory, gold, and around 5 million slaves were exported to Arabia, Persia, and even China. Traders used the **creole** language of Swahili, which incorporated elements of African, Indian, and Arabic languages. The Ethiopian monarchy resisted the Muslim invasions, although repeated wars devastated much of the land. In 1523 Muslims established coastal settlements around Ethiopia in modern Eritrea, Djibouti, and Somalia.

European Traders

From the mid-1400s, European influences in the western and southern parts of Africa rivaled the Arab intrusions in the north and east. At first European interests were limited to coastal trading, exchanging alcohol, guns, and sugar for slaves, gold, ivory, and palm products. Parts of the Western African coasts were labeled "Ivory Coast," "Gold Coast," and "Slave Coast." The demand for slave labor on the North and Latin American plantations growing tobacco, sugar, and later cotton resulted in a triangular pattern of trade (Figure 8.7) that forced over 12 million Africans across the Atlantic and brought great wealth to European ship owners, merchants, and port cities. In Africa, the loss of so many people and the acquisition of guns and luxury goods by the slave-trading aristocracies inside the continent resulted in long-lasting conflicts and underdevelopment.

Antislavery movements beginning in the late 1700s led the United Kingdom to abolish Atlantic slave shipments in 1808, although some countries did not officially abolish slavery until the 1880s. Humanitarian efforts in Europe and the United States during the 1800s led to the return of some freed slaves to the new country of Liberia, and the ports of Freetown (modern Sierra Leone) and Libreville (modern Gabon). However, the returnees seldom integrated with local populations, fueling later ethnic conflicts.

The Colonial Period

By the late 1800s, European explorations of the African interior, the opening of the Suez Canal (1869), and the demands of European manufacturers for raw materials, led to changes. The United Kingdom, France, Belgium, Germany, Spain, and Portugal decided that they needed to control such developments and shared out African lands at the Berlin Congress in 1884–1885 (Figure 8.8). There was no consultation with Africans, but the boundaries agreed upon and colonial policies played major roles in determining the shapes and characters of modern African countries. Plantations in the tropical rain forest environment grew tree crops such as cocoa and palm oil close to the coast and ports; mines in the interior were connected to the ports by railroads. Most colonists—the administrators, plantation and mine owners, and missionaries—stayed for short periods in Africa before returning home.

In some areas, however, Europeans set up home, investing their lives and long-term prospects in the new lands. South Africa, with its warm temperate climate, received the largest numbers of European settlers, beginning with the Dutch, who built the port of Cape Town in the mid-1600s to supply ships trading with the Dutch East Indies. Farmers in the immediate hinterland provided the port and ships with meat and crops. They took land from the indigenous Khoikhoi, whom they called Hottentots, often enslaved, and forced into the desert margins. When the United Kingdom bought Cape Colony in 1814, British colonists clashed with the Boers (Dutch farmers), forcing the latter to

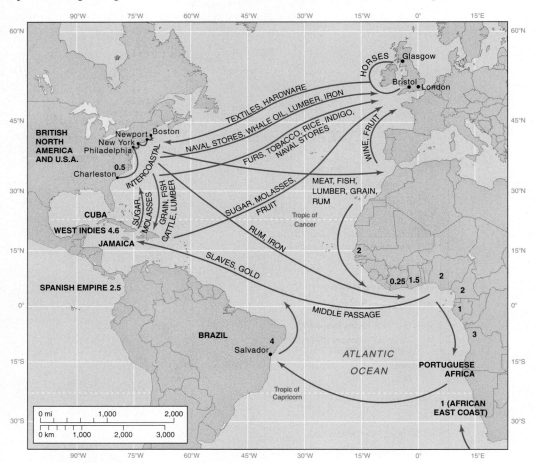

FIGURE 8.7 **Atlantic Ocean: slave trade.** The 1700s Atlantic economy traded colonial commodities with the home countries and brought slaves from Africa to the Americas. Numbers in millions are estimates of slave movements from Africa. The British ended most of this trade in the early 1800s. The country boundaries are those of today. *Source: Data from Thomas 1997.*

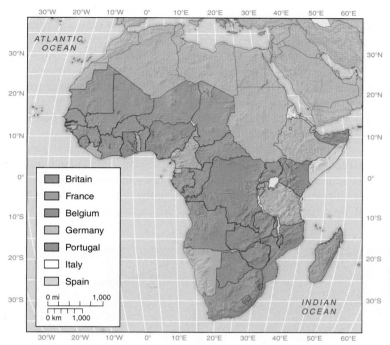

FIGURE 8.8 Africa south of the Sahara: colonial territories around 1900. The colonies of European countries established at the 1884 Berlin Congress and before World War I.

undertake the "Great Trek" to areas where they could govern themselves and use their Dutch-derived Afrikaans language. These Boer movements, together with Zulu aggression, pushed the Ndebele group northward into Shona lands.

Conflicting interests reached a high pitch when diamonds and gold were discovered in the Boer Transvaal and claimed by British venturers such as Cecil Rhodes. Following British incitements, the Boers declared war in 1899. The British army soon took the major towns, but failed to conquer the rural areas. In 1910, the four colonies—Cape, Natal, Orange Free State, and Transvaal joined to form the Union of South Africa, a self-governing dominion within the British Empire. However, the rights of black Africans were again not considered and it was over 80 years before they gained control of their country.

British commercial farmers settled in other colonies from Rhodesia to Kenya. After Germany colonized South West Africa (modern Namibia) in 1884, local chiefs to the east opted for British overlordship in the colonies of Bechuanaland (modern Botswana), North and South Rhodesia (Zambia and Zimbabwe), and Nyasaland (Malawi). Cecil Rhodes' mining ventures and the construction of railroad connections northward from South Africa also supported such developments.

In Eastern Africa, the British settlers in the pleasant upland climates of British East Africa (Kenya and Uganda) faced little resistance. After World War I the assignment of German East Africa (Tanganyika) to British protectorate status attracted less settlement. In Western Africa, the combination of less attractive climate and local resistance resulted in fewer permanent British settlers. The British met considerable armed resistance in the early

1900s when they tried to combine government of the coastal crop growers and plantations in Gold Coast (Ghana) and Nigeria with the drier inland areas of largely Muslim herding groups. In general, British rule was based on informal control and tax collection through local African chiefs and village headmen.

The French made considerable investments in their West and Equatorial Africa colonies, although the lands were mainly either rain forest or dry savanna and supported few people. Attempting to make their African lands and people part of France, they allocated a few seats in the French parliament. However, France's policy of differentiating between a few full French citizens and the majority of Africans led to conflicts. The government of these two colonies, so large in geographical extent, focused on local units, which later became the centers of independent countries.

The other colonial countries—Belgium, Germany, Portugal, Spain, and Italy—regarded their African lands as sources of minerals and other wealth. Germany lost its colonies during World War I, but it was only after World War II that Portugal tried to introduce settler farmers to Angola and Mozambique.

Political Independence

After gaining its independence within the British Empire in 1910, South Africa's white population, dominated by the Boer element, maintained a separation of black African, Asian, and European populations. In 1948 the **apartheid** policy was passed into law, including a "petit apartheid" of separated ethnic groups and a "grand apartheid" of relocating black Africans to homeland areas. The black Africans were regarded merely as laborers for the whites and each was assigned to a homeland or temporary urban location.

Independence came gradually to the rest of the region, beginning with Gold Coast/Ghana in 1957 and ending with the Portuguese colonies in 1975 and Namibia in 1990. Local agitation since the 1920s and virtual abandonment by the European countries preoccupied in World War II stirred nationalisms within the colonial territories. In some countries, such as Ghana, the colonial power transferred control reluctantly but peacefully. In others, such as Kenya, Mozambique, and Angola, there were years of guerrilla warfare. In Southern Rhodesia (Zimbabwe) the European settler population declared independence in 1965, but this was not accepted by the United Kingdom or other countries; external trade sanctions and internal guerrilla conflicts ended with independence under an elected African government in 1979.

Few new African countries were prepared for independence. Conflicts between chiefs and educated and commercial elites, who had frequently led the push to independence, mirrored the contrasts between rural areas dominated by the chiefs and the growing urban centers, where chiefs had little power and lawyers were prominent in politics. The authoritarian colonial experience foreshadowed the centralization of political power and dictatorships in the new countries. Oppositions were removed "in the national interest." Military interventions and takeovers replaced poor governments and self-serving leaders, but often continued corrupt practices. Such internal conflicts over political power sapped local and international confidence in economic development.

Growing and Mobile Populations

Population Distribution and Dynamics

The population distribution of Africa South of the Sahara (Figure 8.9) reflects cultural history, environmental challenges, and external influences. The highest densities are often found in lands where traditional empire organization was followed by colonial product extraction. Coastal areas and some inland places, such as central Ethiopia, northern Nigeria, interior Kenya and Uganda, and the Johannesburg region of South Africa, have the highest densities. The desert margins of the northern edges and the southwest of the region support few people.

Two-thirds of the region's population remain rural and exist by traditional subsistence farming. Urban populations increase (see "Contemporary Geographic Issues, Exploding Cities" p. 254) as people move to cities for better employment prospects in a money-based economy, although many find that there are few job opportunities in the formal sector. Better educational and health facilities also attract migrants to cities. Many Africans see urban centers as safer in times of civil war.

At independence, African countries had low population densities. Assisted by modern medical treatments, population numbers rose as a result of falling death rates and continuing high birth rates. The 380 million people in the region in 1980 rose to over 700 million by 2006 and could reach 1 billion by 2025.

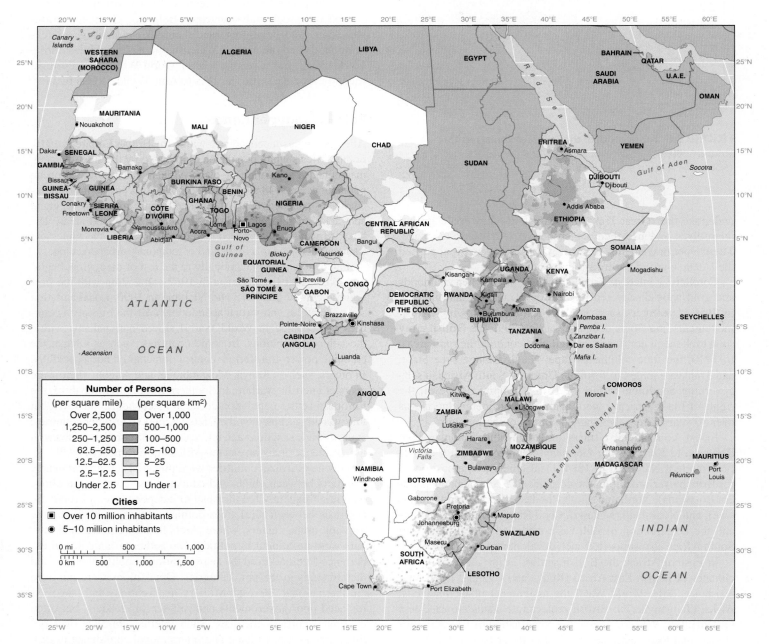

FIGURE 8.9   Africa South of the Sahara: population distribution. Relate the information on this map to Figure 8.1 and the foldout world climate map inside the back cover.

However, the annual rate of population increase of 3 percent was paralleled by agricultural production growth of only 2 percent. African countries did not take part in the Green Revolution that enabled other materially poor countries to escape undernutrition and periodic famines. The age-sex diagram for Nigeria (Figure 8.10) shows the continuing dominance of young age groups. Reducing population growth is a major key to Africa's future, but efforts to encourage smaller families seldom work. Large families are a sign of male virility and many women prefer them as a basis for security in old age.

African population growth continues despite the widespread occurrence of long-term diseases such as malaria, sleeping sickness, and river blindness, the more recent onset of HIV/AIDS (see "Contemporary Geographic Issues," p. 252), and the prevalence of civil wars. Although life expectancies increased in many countries, they were reduced in the countries with the worst incidence of HIV/AIDS.

Migration

Among world regions, Africa South of the Sahara has the largest proportion of its people living outside their country of birth. Some migrants move for economic reasons to places where they may be more likely to obtain paid employment. They include Western Africans from other countries moving into and out of Nigeria and Côte d'Ivoire, and millions moving to work in South Africa from the surrounding countries.

However, increasing numbers of migrants flee from violence against political oppositions or ethnic minorities. Movements across the borders of Rwanda and Burundi with the Democratic Republic of the Congo, Uganda, and Tanzania in the 1990s were associated with mobile militia groups, many deaths, and property destruction. Movements from Somalia

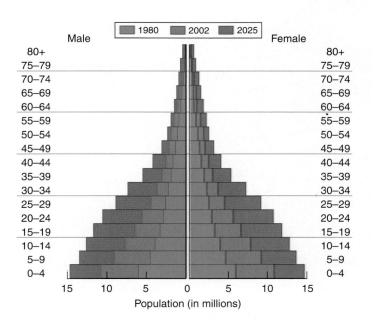

FIGURE 8.10 **Nigeria: age-sex pyramid.** This is typical of most African countries, where the populations are dominated by young age groups—a challenge for the present and resource for the future. *Source: U.S. Census Bureau; International Data Bank.*

into Ethiopia and from Ethiopia into Sudan and Uganda are also composed of people displaced by conflict. Civil war in Liberia spread into Sierra Leone, Guinea, and Côte d'Ivoire since 1989 and caused over 1 million people to flee their home towns and villages, to become internally displaced persons or refugees in other countries.

As apartheid ended, many whites moved from South Africa to the United States, the United Kingdom (1.4 million), or Australia. Around half have higher education degrees and many were leaders in business. However, there are probably as many black Africans who left South Africa as part of a brain drain of abilities. Many African countries also lose educated and skilled personnel to better-paying jobs in the materially wealthy countries. For example, the loss of doctors and nurses affects African health care. Ten percent of Canadian hospital doctors are South Africans, and in 2005 there were more Ethiopian doctors practicing in Chicago than in Ethiopia. Few doctors who were trained in Zambia since independence have stayed. The positive side of this outward migration includes remittances that bolster local economies and business links, and a few returning experienced professionals.

Political and Economic Pressures
Independence and the Cold War

During the Cold War, the United States and Soviet Union superpowers vied for control of other countries. They either propped up friendly governments or encouraged rebel groups against less friendly ones. At independence many African countries had economies that rivaled materially poor countries in Asia and Latin America, but levels of internal conflict increased within the countries, delaying economic development. For example, in Ethiopia the Soviet Union supported an unpopular Communist government, which oppressed some groups. In Angola, Soviet and Cuban troops supported government forces, while South Africa and the United States supported guerrilla oppositions. In Mozambique the guerrilla groups had Soviet support. In the Democratic Republic of the Congo, the United States supported the military dictator, attempting to ensure the continued production of important minerals from the interior. Supported leaders often used much of the financial assistance to buy armaments for protecting themselves. Many invested funds in personal accounts abroad. When the Cold War ended, internal conflicts fueled by the widespread availability of weaponry, and even by the adoption of combative multiparty politics, often turned into ethnic strife. At the same time, the ending of the superpower rivalries also resulted in African issues of poverty and conflict being ignored until they further deteriorated.

Global Outsiders

By the early 2000s almost every African country faced major economic and political problems. Heavy debts and internal conflicts made it difficult for them to attract new investment or aid, except for short periods when humanitarian crises directed world attention to specific countries. Of the foreign investment going to developing countries, 50 percent went to Asian countries, 20 percent to Latin American countries, and under 3 percent to African countries.

African countries continue to face many disadvantages in global trading. They produce and export primary products: metallic minerals (copper, bauxite, diamonds, gold, platinum), oil, commercial farm products (cocoa, coffee, tea, rubber, palm oil, vegetables, flowers, cotton), timber, and fish. These products are subject to price swings, most commonly downward as world competition increases. Much of the improving African country income levels in the 2000s derives from the export of such raw materials subject to higher prices at a time of increasing world demand. However, the most significant financial advantage comes when such items are processed and marketed, mainly in the materially wealthy countries. Apart from the local manufacture of bulky goods from cement to bottled drinks and some assembly of automobiles and trucks, only South Africa has a diversity of more sophisticated manufacturing.

In the early 2000s, political leaders of the world's materially wealthy countries in Europe, North America, and some Asian countries belatedly offered more support, including debt forgiveness and longer-term investments. But a large number of African countries with the greatest need could not meet criteria of good government and peaceful business environments. Most of the offered assistance was relatively short-term and of small significance compared to the level of need. The lack of agreement among the wealthier countries on such aspects as reducing their own tariffs on imported food products and commodities such as cotton and sugar also held back development prospects for African countries.

Geographic Diversity

The 42 countries of this world region are grouped in four subregions (Figure 8.11). Each subregion contains a large country of over 50 million people; these four countries made up half of the region's population in 2006 and produced almost 60 percent of the region's total GNI PPP. By contrast, over half the

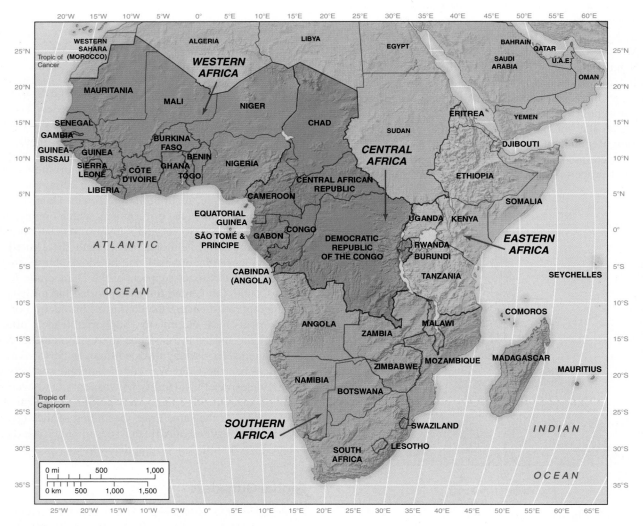

FIGURE 8.11   Africa South of the Sahara: subregions and country boundaries.

FIGURE 8.12 Central Africa. (a) The subregion, its countries, and main cities. (b) Democratic Republic of the Congo (DRC). (c) Rwanda.
Sources: (b) & (c) Data from Microsoft Encarta 2005.

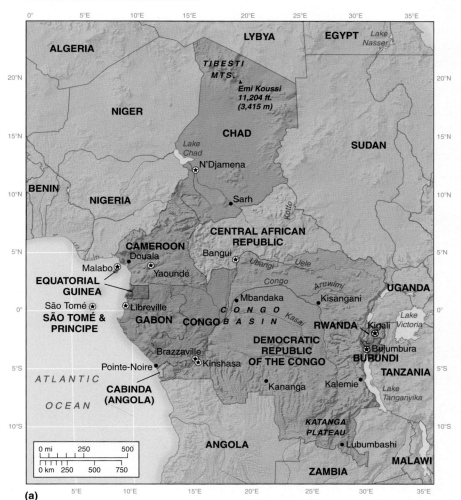

(a)

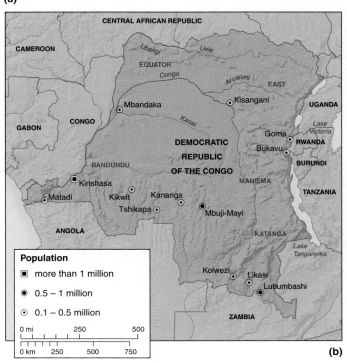

(b)

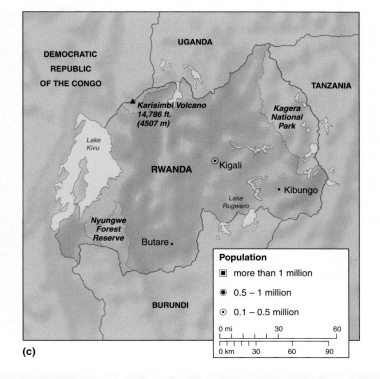

(c)

countries in the region had fewer than 10 million people each. Small countries and internal country regions illustrate further the geographical diversity within the region.

Subregion: Central Africa

In 2006 Central Africa (Figure 8.12a) had the fewest people (Table 8.1) and lowest population density of the African subregions, although rapid population increases may take it above Southern Africa by 2025. The subregion's development was slowed by its natural environment. The tropical rainy climate, tropical rain forest vegetation, and plateaus crossed by the Congo River and its tributaries with many rapids form a difficult environment. Humans, from precolonial times, through colonization, to political independence, always struggled with these conditions. Before European intrusions, these forested lands supported few people, and they were subject to external domination from all sides.

Colonial development of mineral and crop exports was less extensive than in other subregions and, apart from the coastal countries of Gabon, Equatorial Guinea, Cameroon, and southern Democratic Republic of the Congo, economies developed slowly. Contacts with the rest of the world are still slowed by poor transportation, mainly because the major Congo River artery is interrupted by rapids. Internally, conflict divides and disrupts all countries. There is little intraregional cooperation.

TABLE 8.1

AFRICA SOUTH OF THE SAHARA: Data by country, area, population, urbanization, income (Gross National Income Purchasing Power Parity), ethnic groups

Country	Land Area (km²)	Population (millions)		%Urban*	GNI PPP 2005	2005	Ethnic Groups (%)
		mid 2006 Total	2025 est. Total	2006	Total (US$ billions)	Per Capita* (US$)	
CENTRAL AFRICA							
Burundi, Republic of	28,000	7.8	14.0	9.0	5.0	640	Hutu (Bantu) 79%, Tutsi (Hamitic) 20%
Cameroon, Republic of	475,440	17.3	24.3	52.8	37.3	2,150	Over 200 groups: Fang, Barnileke, Fulani, Pahouin
Central African Republic	622,980	4.3	5.5	41.2	4.9	1,140	Baya 34%, Banda 27%, Mandjia 21%, Sara 10%
Chad, Republic of	1,284,000	10.0	17.2	23.8	14.7	1,470	Many groups: Muslim in north, non-Muslim in south
Congo, Republic of the	342,000	3.7	5.9	52.2	3.0	810	Kongo 48%, Sangha 20%, Teke 17%, Mboshe 12%
Congo, Dem. Rep. of the	2,344,860	62.7	108.0	30.3	45.1	720	Over 200 African groups: Bakongo, Mongo, and Luba 45%
Equatorial Guinea, Rep. of	28,050	0.5	0.8	38.8	3.8	7,580	Fang 80%, Bubi 15%
Gabonese Republic	267,670	1.4	1.8	81.4	8.3	5,890	Fang 60%, Mpongwe 15%, M'bete 14%, Punu 12%
Rwanda, Republic of	26,340	9.1	13.8	16.7	11.9	1,320	Hutu 90%, Tutsi 9%, Twa 1%
Totals/*Averages	**5,419,340**	**117**	**191**	**38**	**134**	**2,413**	
WESTERN AFRICA							
Benin, Republic of	112,620	8.7	14.3	40.4	9.7	1,110	42 groups: Fon, Aja, Yoruba main
Burkina Faso, Dem. Rep. of	274,000	13.6	23.2	16.3	16.6	1,220	Mossi 25%, Gourounsi, Senufo, Lobi, Bobo, Mande
Côte d'Ivoire, Republic of	322,460	19.7	27.1	46.7	29.3	1,490	60 groups: (Côte d'Ivoire Akan 42%, Goer 18% (GUR) Mandes 27% Krous 11%)
Gambia, Republic of the	11,300	1.5	2.4	50.0	2.8	1,920	Mandinke 42%, Fulani 18%, Woluf 16%, Jola 10%
Ghana, Republic of	238,540	22.6	32.7	43.8	53.5	2,370	(Ghana) Akan 44%, Moshi-Dagomba 16%, Ewe 13%, Ga 8%
Guinea, Republic of	245,860	9.8	15.2	29.6	22.0	2,240	Fulani 35%, Malinke 30%, Susu 20%
Guinea-Bissau, Rep. of	36,120	1.4	2.4	47.7	0.9	700	Balante 27%, Fulani 23%, many others
Liberia, Republic of	97,750	3.4	5.8	45.4	—	—	Bassa, Gio, Kpelle, Kru 95%, U. S. Liberians 5%
Mali, Republic of	1,240,190	13.9	24.0	30.2	13.9	1,000	Mande 50%, Peul 17%, Voltaic 12%, Tuareg/Moor 10%
Mauritania, Islamic Rep. of	1,025,520	3.2	5.0	40.0	6.8	2,150	Mixed Moor/Black 40%, Moor 30%, Fulani, Wolof
Niger, Republic of	1,267,000	14.4	26.4	20.6	11.5	800	Hausa 56%, Djerma 22%, Fulani 8.5%, Tuareg 8%
Nigeria, Federal Republic of	923,770	134.5	199.5	44.1	139.9	1,040	Hausa, Fulani, Yoruba, Ibo total 71%
Senegal, Republic of	196,720	11.9	17.3	45.1	21.1	1,770	Wolof 44%, Fulani and Tutulor 24%, Serer 15%
Sierra Leone, Republic of	71,740	5.7	8.7	36.1	4.4	780	Mende, Temne, Limba, Creoles, others
Togo, Republic of	56,790	6.3	9.6	33.4	9.8	1,550	37 groups: Ewe, kabye main
Totals/*Averages	**6,120,380**	**270**	**413**	**38**	**342**	**1,439**	

(continued)

Country	Land Area (km²)	Population (millions) mid 2006 Total	2025 est. Total	%Urban* 2006	GNI PPP 2005 Total (US$ billions)	2005 Per Capita* (US$)	Ethnic Groups (%)
EASTERN AFRICA							
Djibouti, Republic of	23,200	0.8	1.1	82.0	1.8	2,240	Somali 60%, Ethiopian 35%
Eritrea, Republic of	121,140	4.6	7.4	18.7	4.6	1,010	Tigrinya 50%, Tigre-Kunama 40%, Afar 4%
Ethiopia, Fed. Dem. Rep. of	1,100,760	74.8	107.8	15.3	74.8	1,000	Oromo 40%, Amhara, Tigrean 32%, Sidamo 9%, others
Kenya, Republic of	580,370	34.7	49.4	35.9	40.6	1,170	Kikuyu 21%, Luhya 14%, Luo 12%, Kalenjin 11%, Kamba 11%
Somali Republic	637,600	8.9	14.9	34.0	—	—	Somali 85%
Tanzania, United Rep. of	945,090	37.9	53.6	32.3	27.6	730	Over 120 groups: Sukuma, Haya, Nyakyusa, Nyamwezi, Chagga
Uganda, Republic of	235,880	27.7	55.5	12.2	41.5	1,500	Ganda 18%, Nyankole 10%, Kiga, Soga, Iteso, Langi
Totals/*Averages	**3,644,040**	**189**	**290**	**33**	**191**	**1,275**	
SOUTHERN AFRICA							
Angola, Republic of	1,246,700	15.8	25.9	33.4	35.0	2,210	Ovimbundu 37%, Mbundu 25%, Bakongo 15%, others
Botswana, Republic of	581,730	1.8	1.7	54.2	18.0	10,250	Tswana 75%, others
Lesotho, Kingdom of	30,350	1.8	1.7	13.4	6.1	3,410	Basotho 79%, Nguni 20%
Madagascar, Republic of	587,040	17.8	28.2	26.0	15.6	880	Merina 27%, Betsimisaraka 15%, Betsileo 12%, others
Malawi, Republic of	118,480	12.8	23.8	14.4	8.3	650	Chewa, Nyanja main
Mozambique, Republic of	801,590	19.9	27.6	32.1	23.3	1,170	Makua-Lomwe, Sena, Shona, Tsonga
Namibia, Republic of	824,290	2.1	2.5	33.0	16.2	7,910	Orambo 50%, other black 36%, white 6%, mixed 7%
South Africa, Republic of	1,221,040	47.3	48.0	52.9	573.5	12,120	Black 76%, white 13%, mixed 9%, Indian 2%
Swaziland, Kingdom of	17,360	1.1	1.0	23.1	5.9	5,190	Black 97%, white 3%
Zambia, Republic of	752,610	11.9	16.4	35.1	11.3	950	Black (over 70 ethnic groups) 98.7%, white 1.1%
Zimbabwe, Republic of	390,760	13.1	14.4	33.6	25.4	1,940	Shona 71%, Ndebele 16%, others, white 2%
Totals/*Averages	**6,571,950**	**145**	**191**	**32**	**739**	**4,244**	
Region Totals/ Averages	**21,755,710**	**722**	**1,085**	**35.3**	**1,406**	**2,343**	

Source: World Population Data Sheet 2006, Population Reference Bureau. World Development Indicators, World Bank (2006). Microsoft Encarta 2005.

Democratic Republic of the Congo

The combination of large size (largest country in the region) and large population (over half of the subregion total), centrality in the Congo River basin, many mineral resources, and huge hydroelectricity potential might be expected to make the Democratic Republic of the Congo (DRC) (Figure 8.12b), with over 60 million people in 2006, a focal country in Central Africa, or even in the continent. However, instead of forming a central driving force for development in Central Africa, DRC was largely taken over by feuding private armies, making government of the whole country impossible. These events are central to understanding the geography of the country today. It is esti-

mated that thousands were killed in the fighting. From 1998 to 2004, up to 4 million people died of war-induced starvation or disease, and millions more were displaced from their homes. Even huge aid income failed to lift GNI PPP per capita income above US$720 in 2005.

Soon after independence in 1960, the Democratic Republic of the Congo army under Mobutu Sese Seko seized control from the Communist Prime Minister Patrice Lumumba. Supported by the United States, Mobutu remained in power, building up his family positions and private bank accounts abroad until a 1997 rebellion deposed him. Mobutu's robbery of the national wealth resulted in a description of his government as "kleptocratic" ("klepto" infers stealing).

Laurent-Désiré Kabila led the 1997 rebellion from his base in eastern DRC, using exiled Tutsi and local related groups with the backing of Rwanda. The demoralized DRC armies offered little opposition, Mobutu went into exile, and Kabila became president.

However, Kabila could not control DRC from the capital, Kinshasa (Figure 8.13). Other countries intervened, widening the impact of the internal conflict instead of calming it. Old Marxist ties led Zimbabwe and Namibia to send troops and equipment to support Kabila. Uganda and Rwanda backed eastern (Tutsi) militias against the Kabila regime, which now used Hutu men as the basis of the Congolese army. Angolan troops entered from the west, outwardly to support Kabila but mainly to cut off the supply routes to Angolan rebels. Once there, all the countries extracted DRC's mineral wealth, from diamonds to gold and copper, but by 2003 stresses at home caused them to withdraw.

After Kabila was assassinated in 2001, his son Joseph Kabila replaced him. Ethnic rivalries again erupted into local wars, such as the Lunda fighting the Luba in the south (Katanga). By 2004, ceasefires became more common between flare-ups in the fighting. At least 16,000 United Nations peacekeepers were having an effect. In 2003 a coalition of rebels formed a DRC "government." However, many local warlords and militia groups resisted invitations to join a government of national unity. A catalog of confusion, corruption, and atrocities continued.

Elections set for June 2005 were postponed and funds designated for development were redirected to military hardware. An election took place in mid-2006, surprising many that it could be held under such difficulties: the last census was in 1984, there is no experience of democracy, no functioning civil service, no registered voters, few electoral workers, and virtually no roads.

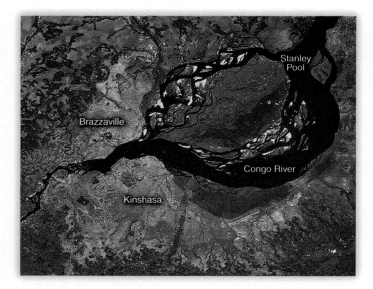

FIGURE 8.13 **Central Africa: Congo River at Kinshasa, DRC, and Brazzaville, Republic of Congo.** This satellite image illustrates some of the problems of navigating the river.

Rwanda

In 1994, the genocide of nearly 1 million people brought Rwanda and Burundi, two countries in the heart of Africa, to the world's notice. More than 10 years later, Rwanda is quiet under the virtual dictatorship of Paul Kagame, who ended the massacres by force.

Rwanda is one of Africa's smallest countries (Figure 8.12c), although its 9.1 million people in 2006 gave it one of the highest population densities. It is known locally as the "Land of a Thousand Hills" with the highest points (often volcanic) in the west, where the steep slopes lead down to Lake Kivu in the rift valley. Eastward is a rolling hilly upland and there are swampy plains on the border with Tanzania. Although just south of the equator, the hilly terrain and altitudes create a pleasant climate—warm temperate with temperatures around 70°F (25°C) throughout the year. Rain falling from frequent thunderstorms is heaviest in the west.

Landlocked, with poor surface transportation, some 90 percent of the population is dependent on self-sufficient **subsistence agriculture.** Most families produce a little coffee or tea for cash and export (in contracts, for example, with Starbucks), but there are few natural resources for a manufacturing base. Incomes are very low; few people live in towns.

Before the colonial period, the original population of Twa Pygmies (now only 1 percent of the total) was displaced by Bantu peoples—the majority Hutu cultivators, and the taller, thinner Tutsi cattle and sheep herders. They skirmished with each other. However, tribal priority was never an issue, intermarriage was common, and many Hutus socially upgraded to become Tutsis. Unfortunately, the Belgian colonists created separation and resentment by deposing Hutu chiefs in favor of Tutsis and issuing ethnic identity cards. At independence in 1962, the Hutu majority won the elections and imposed a 9 percent quota on Tutsis (their proportion of the population) for salaried jobs. General Juvenal Habyarimana seized power in 1973 as one of "the majority people" and further discriminated against the Tutsis—many of whom left the country. In 1990, Kagame gathered these exiles and invaded from Uganda. Despite a 1993 peace accord, some members of the previous Hutu regime recruited and indoctrinated thousands of Hutu militiamen and armed them with machetes to meet the Tutsi "threat." When Habyarimana's plane was shot down in 1994, his most bigoted associates took control and set off the genocide of Tutsis through local gatherings. The property of those killed was given to enthusiastic murderers.

Kagame's Rwanda Patriotic Front (RPF), formed of exiled Tutsis, won the short war that ensued, killing many of the murderers and chasing them into the Zaire (now DRC) rain forests. The RPF continues to govern Rwanda, using massive foreign aid to rebuild the school and health care systems. Hundreds of thousands of Tutsis returned, bringing cash and skills to replace the lost middle class. Many had been born abroad and often returned with little knowledge of the local language. Rwandans returning home or to visit surviving friends and families find improved housing, shopping facilities, and transportation links without feelings of fear.

As part of the RPF's main objective is to maintain peace and involve all Rwandans, it tries to re-educate the murderers who survived. However, security is tight, with no press freedom or freedom of association, party loyalty is imposed, and Rwandan prisons, pronounced as "life-threatening" by the U.S. State Department, provide a major deterrent to opposition. In 2007, Kagame, in a gesture of defiance against the French forces, whom he accuses of being a bad influence in 1994, attended the British Commonwealth of Nations conference in Nairobi—also a reflection of increasing links with Eastern Africa.

Subregion: Western Africa

Western Africa's geography (Figure 8.14a) contrasts with that of Central Africa and makes it more open to the rest of the world. Instead of an interior focus on a major river system and a short section of coast, it has a long coastline and several rivers connecting the interior to the coast. European exploration and trade began in the mid-1400s, but colonial political control began in the late 1800s. France, the United Kingdom, Germany, and Spain were the colonial powers, who left a legacy of country boundaries and common languages still used in commerce and government.

The colonial development of the subregion was based on local mineral, forest, and climate resources. The humid tropical rain forest in the south was cut for hardwood lumber, while plantations were established for cocoa, palm oil, rubber, and tropical fruits by multinational corporations, and the smaller plots of local farmers. These produced exports and income. Inland, cattle were raised in the savannas (Figure 8.15a). Mining for gold, diamonds, and bauxite was mainly by international corporations and depended on world markets. Railroads joined inland centers and mines to ports. After independence, single-party government and military dictatorships became common, often tearing apart the countries on ethnic lines. Ports and inland historic cities experienced rapid population growth (Figure 8.15b).

Arising from their history as French colonies, many countries of Central and Western Africa continue financial links with France (Figure 8.16). After independence, the **Communauté Financière Africaine** (CFA) franc was shared by 14 African countries. Its value was tied to the French franc and guaranteed by the Bank of France. High exchange rates allowed French companies to retain dominant positions in local contracting, but made it difficult for French-speaking African countries to export goods, used up their foreign exchange reserves on imports, and made them increasingly dependent on France. In January 1994, the French government devalued the CFA franc by 50 percent. This raised the cost of imports, but France wrote off the debts of the poorer countries, and the World Bank made increased grants available. By the early 2000s, most CFA countries saw some benefits in lower inflation and better foreign currency reserves.

The **Economic Community of West African States** (ECOWAS), established in 1975, is Africa South of the Saha-

ra's most active regional organization. It has been prominent since the early 1990s through peacekeeping operations in Liberia, Sierra Leone, and Côte d'Ivoire and opposition to a coup in Guinea-Bissau. The consultative ECOWAS parliament with 115 members was established in 2002. A Court of Justice arbitrates individual, corporate, and country-level complaints. ECOWAS has also promoted integration among countries, and many workers cross borders daily. Tariff barriers are being removed to allow free trade, an investment bank supports private sector enterprises, and a gas pipeline connects Nigerian oilfields to Benin, Togo, and Ghana. There is also scope for rationalizing air routes to make internal connections easy. A common currency, the "eco," is being developed for member countries. It will meet strict criteria, but will compete with the CFA franc in former French colonies. The problems of open borders include easier arms traffic across them and con artists taking advantage of a travelers' check initiative.

Nigeria

With half the subregion's population, Nigeria (Figure 8.14b) exemplifies many of the problems facing the other countries of Western Africa. Nigeria's geography includes coastal tropical rain forest and inland savannas, drained by the Niger River and its main tributary, the Benue River. The Muslim Hausa and Fulani people in the north are more numerous than the combined Christian and traditionalist Yoruba groups in the southwest and Igbo in the southeast. After the initial division of the independent federal country into three states, the conflicts resulting from the threefold division and the presence of other ethnic groups in the center of the country led to the designation of 36 states within Nigeria.

Like DRC, Nigeria failed to become the dominant force in its subregion. It experienced so much bad government from dictator generals in the 1980s and 1990s that it moved from an oil-rich, middle-income country to become one of the world's poorest. A lot of the problems arose from the continuing north-south ethnic and religious conflict. The northern Hausa and Fulani took control of the military and government, with the southern Yoruba being outvoted and the Igbos being widely disregarded after losing the 1960s civil war in which they had tried to establish their independence.

While the 1966–1979 military rule brought life to the economy and united peoples with very different interests inside the country, the military rule from 1984 to 1999 was a disaster. Despite huge controversy within the country, the northern Muslims made Nigeria a member of the Organization of the Islamic Conference, annulled the 1993 presidential election, and repressed the peoples of the oil-producing Niger River delta region. Cultural divisions deepened and the economy faltered through neglect and corruption. In 1998, it was estimated that three-fourths of official GDP was generated by the informal economy. Some US$12.4 billion of government funds was paid out without proper accounting, while the military dictator in the mid-1990s stole $3 billion.

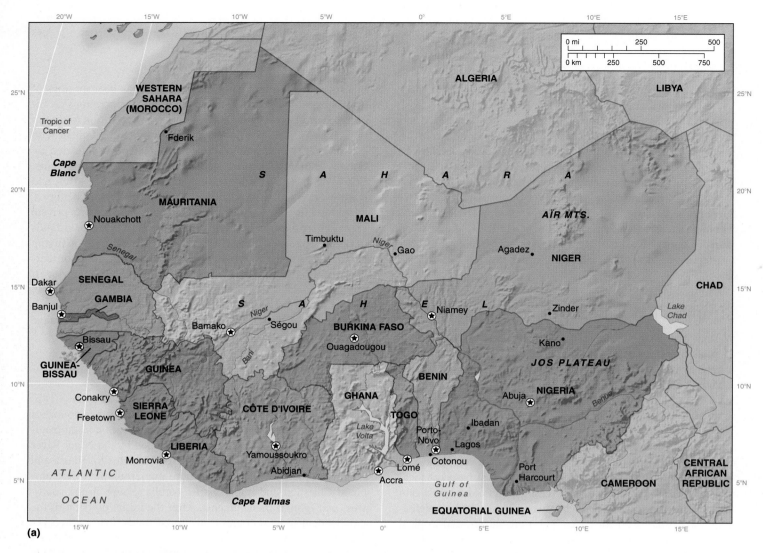

(a)

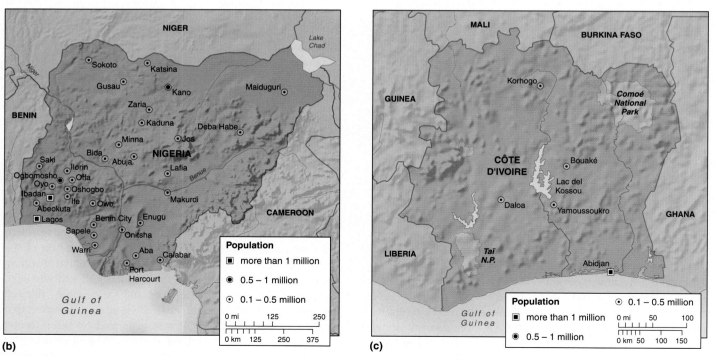

(b)

(c)

FIGURE 8.14 Western Africa. (a) The subregion, its countries, and main cities. (b) Nigeria. (c) Côte d'Ivoire. *Sources: (b) & (c) Data from Microsoft Encarta 2005.*

(a)

(b)

FIGURE 8.15 **Western Africa.** (a) Cattle in the dry Sahel region of western Niger. (b) Oshodi Market, Lagos, Nigeria.

(c)

FIGURE 8.16 **Western and Central Africa: the French influence.** Former French colonies had a currency depending on the French franc and used French loans. The French government devalued the CFA franc in 1994 and adopted the euro in 2002, resulting in some confusion in the African countries, which were also part of the ECOWAS (Economic Community of West African States) zone.

After the 1999 elections, Nigeria made a rapid transition from military to democratic government. The newly elected democratic government of Nigeria led by Olusegun Obasanjo faced the difficulties of a hastily devised new constitution and the legacies of a fragile federal system, an undermined judicial system, and a police force that was reduced in numbers in case it competed with the previous military. In 1999 and 2000, several states in the Muslim north challenged the new government by declaring that Islamic religious (sharia) law should take precedence over the common law order inherited from the colonial era. This raised questions about the relationships between public institutions and religious traditions throughout Nigeria.

A second challenge to the government comes from the delta of the Niger River, where most of Nigeria's oil is produced, and which is inhabited by several small, poor ethnic groups. The oil revenues, however, go to the federal government, which takes the largest share and divides the rest among the 36 states on the basis of population and area. The oil-producing areas demand that more funds be returned to them, and local people cause pipeline disruptions. The federal government scarcely listens and replies with repressive measures such as the murder of the activist Ken Saro-Wiwa, who protested oil company and government attitudes. Demands from the delta peoples increased when world oil prices rose sharply in 2004.

A third challenge arose when the 17 southern states complained that the federal government made most of the important decisions without listening to the states. The states want more local power over the police force, education, revenues from resource exploitation, and infrastructure provision. But the 19 northern states resist this proposal.

By 2005, Obasanjo had achieved more democracy and had appointed young technocrats to key posts. To reduce the temptations of corruption, civil servants' benefits were increased and advertised on the Internet; there was more public bidding for government contracts.

Commercial Farming, Southern Côte d'Ivoire

For much of the late 1900s the people of southern Côte d'Ivoire (Figure 8.14c) enjoyed political stability and rising well-being based on the exports of mainly tree crops, including cocoa, coffee, tropical fruits, rubber, and palm oil, from the tropical rain forest coastal area. In the 1970s the country's economic development based on this region earned the title "Ivoirian miracle."

Settled late by many of the modern African ethnic groups, the dense tropical forest of the Ivory Coast suffered little from the slave trade because of its poor harbors. From the 1840s France persuaded local chiefs to grant French traders a monopoly. The southern tropical rain forest environment was developed by French settlers using a hated forced-labor system on their plantations to produce cocoa, coffee, and palm oil. The French built a naval base and undertook a long war to conquer the interior, lasting into the 1890s.

In the 1940s, Félix Houphouët-Boigny, son of a chief, formed a cocoa farmer trade union that enabled local farmers to compete with the French plantations by recruiting migrant workers. He was elected to the French parliament in Paris, where he became the first African minister. At independence in 1960, Côte d'Ivoire was the most prosperous part of French West Africa. Houphouët-Boigny's government stimulated cocoa and coffee production helped by good world prices. By 1979, the country was the world's largest cocoa producer and third in coffee, Africa's main exporter of pineapples and palm oil, and its rain forests produced hardwood timber. The country enjoyed annual economic growth of 10 percent for 20 years. Economic expansion was made possible by French technical expertise and encouraged a rise in French settlers from 10,000 to 50,000, mainly as teachers and advisers. The infrastructure of roads, railroad, and the port of Abidjan were the best in Western Africa. The need for more workers on the Côte d'Ivoire plantations attracted immigrants from poverty-stricken Burkina Faso and Mali and strife-torn Sierra Leone.

By 1990, the country was troubled. Markets for the tree crops worsened at a time of world recession in the 1980s, droughts affected the northern lands, the overcutting of timber led to declining income, and overspending on show projects led to huge debts. The projects included expansion of the port of Abidjan and the building of the new capital inland at Houphouët-Boigny's home village of Yamoussoukro. Most foreign delegations remained in Abidjan.

When Houphouët-Boigny died in 1993, his successor, Henri Bédié, emphasized the concept of "Ivorité" that excluded the many immigrants—and also his political rivals in the north. Ethnic and regional tensions rose. Although there was a period of improved economy with better commodity prices and better trading conditions after the CFA franc revaluation in 1994, a 1999 coup removed Bédié. For a year public order improved, but during the October 2000 elections civil and military unrest increased. The Supreme Court disqualification of a candidate because of his nationality sparked riots. In 2002, troops in the poorer Muslim northern areas mutinied and took control, supported by people who felt marginalized inside the wealthy south. Northern migrants into the southern areas began fighting over the prime cocoa-growing areas. That drew in militias from lawless Liberia and Sierra Leone. French troops arrived to restore order, but ended up fighting both sides after some of their number were killed by a government air strike and they responded by destroy-

ing the planes involved. The French were accused of fanning the flames of conflict to maintain their own presence in the country. Despite a January 2003 accord and an attempt at a government of national unity, Côte d'Ivoire's political stability and economic viability were badly sapped.

Subregion: Eastern Africa

Eastern Africa (Figure 8.17a) is dominated by plateaus and the Ethiopian highlands in the north, both broken by rift valleys. The subregion has mostly summer rainfall, heaviest in the Ethiopian highlands, but increasingly arid on the coasts toward the north. The subregion became a crossroads where African groups from the Nile River valley met those from Western and Central Africa. Facing the Red Sea and Indian Ocean, Arab influences were strong. Europeans (mainly British, but also Italians and Germans) brought new pressures after the 1869 opening of the Suez Canal. Superimposed on the African ethnic areas with their emphases on cropping or animal husbandry, British colonial commercial farming in Kenya and Uganda produced tea, coffee, and cotton. People from South Asia, introduced to construct and manage transportation, factories, and shops, added to the subregion's ethnic diversity.

The **East African Community** (EAC), linking Kenya, Tanzania, and Uganda, was launched in 1967 but soon collapsed. Revived in 2000, it began operations in 2005, with its headquarters at Arusha in northern Tanzania. The EAC faces concerns from those now used to their own country sovereignty: the Kenyans seek to dominate, the Ugandans see it as a source of conflict, and the Tanzanians are worried about keeping their distinctive identity. Progress has been slow toward a common parliament and political union contemplated for 2010.

Ethiopia

Ethiopia (Figure 8.17b) has the distinction of being the only large African country that was not colonized by a European power. Although occupied by the Italian military from 1936 to 1941, it was soon returned to the Emperor Haille Selassie's rule by British forces at the outset of World War II. Ethiopia remained an isolated feudal country, and Selassie was the last in a line of Ethiopian emperors dating back to the Coptic Church conversion in the AD 300s. He reigned from 1930 to 1974, when he was deposed and murdered in a military coup that brought to power a Communist government with Soviet Union support. After further coups, devastating droughts, and movements of refugees in and out of the country, rebel groups replaced that regime in 1991. A new constitution in 1994 led to multiparty elections.

After World War II, the United Nations federated Eritrea with Ethiopia. Disappointed at being reduced to a province within Ethiopia, Eritreans fought Ethiopian and Soviet Union forces. After the Communist government in Ethiopia was deposed and the Soviet forces departed, Eritrea gained independence in 1993. Ethiopia is now landlocked and has difficult relations with surrounding countries. Tensions continue between

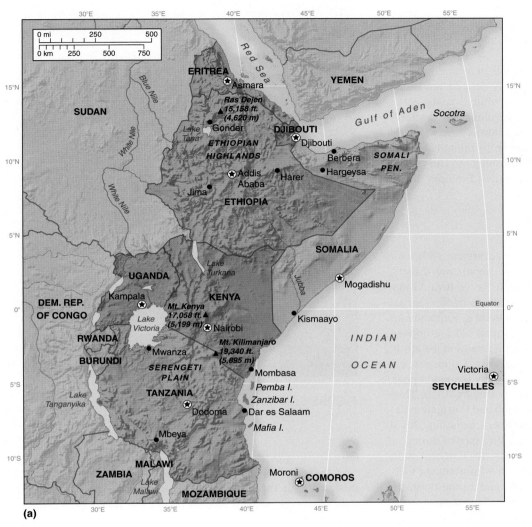

FIGURE 8.17 Eastern Africa.
(a) The subregion, its countries, and main cities. (b) Ethiopia. (c) Interior Tanzania. *Sources: (b) & (c) Data from Microsoft Encarta 2005.*

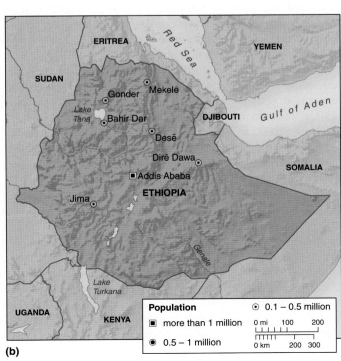

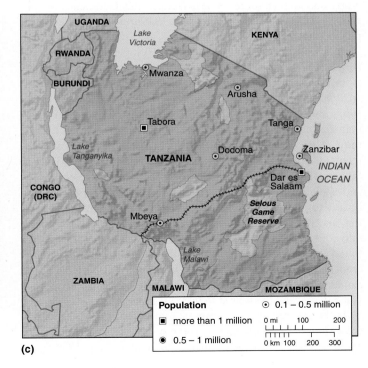

Ethiopia and Eritrea over the placing of the border and access to an ocean port. Costly wars from 1998 into the early 2000s sapped both countries' resources.

Ethiopia is greatly concerned with events in neighboring Somalia. The international boundary is poorly defined and Ethiopia seeks to influence events in Somalia, while Somalis act as opposition groups within Ethiopia. A loose alliance of warlords based in Baidou, supported by Ethiopia, controls most of southern Somalia, but it was under pressure from militant Islamic groups who controlled the capital, Mogadishu, until early 2007. There are frequent Somali incursions into east Ethiopia, where many ethnic Somalis live.

Interior Tanzania

Inland Tanzania has a tropical plateau environment with variable moderate summer rains, mostly low population density, and low income (Figure 8.17c). Although the Olduvai Gorge in northern Tanzania is the site of many finds of the earliest humans, most of this area's history is unknown. Remnants of Khoisan (click-tongue) languages suggest that older groups of people were displaced by Bantu and Nilotic groups in the north. Some of the interior groups had well organized societies, especially in the northwestern lakes area; others remained poor, combating a semiarid natural environment (Figure 8.18). They were largely untroubled by the coastal Arab and Portuguese developments. In the 1800s, European explorers, including David Livingstone (whose last mission was at Ujiji), passed through.

Germany indicated its colonial interest in Tanganyika in 1884 and made treaties with tribal chiefs, who accepted German protection, and with the British colonies to the north. In 1891 the German government took control of the territory with headquarters at Dar es Salaam and intentions to develop cash crops and transportation lines. Neither the Germans nor the subsequent British rulers in the post-1919 trust territory accomplished much development.

Julius Nyerere, a school teacher, led Tanganyika at independence in 1961, soon (1964) to be united with the island of Zanzibar as The United Republic of Tanzania. Nyerere introduced African socialism (*Ujamaa*) to emphasize justice and equality. While most economic activity still focused on the coastal area, major attempts were made to develop the interior, including moving the country's capital to Dodoma in 1996. However, many government offices remain in the former capital of Dar es Salaam. The Nyerere policies failed economically, politically, and socially. In particular, the poor performance of national boards to market the commercial crops of coffee, tea, and cotton, and the attempts to move traditional village-based groups into collective farms, were disasters. Combined with the failure to develop manufacturing, such policies led to increased poverty, famine, and indebtedness. The egalitarian socialist Tanzanian government resisted building large luxury hotels in the game parks, and the tourist industry remained small.

In the 1990s, a new government moved away from the Nyerere policies. In an attempt to forge a sense of national unity, Nyerere's successor detribalized the country's politics and

FIGURE 8.18   Eastern Africa: the dry plateau. A straight road in Tanzania, but potholed and dusty in the dry season, changing to a river in the wet season.

made Swahili the official language. A more open economic policy now encouraged foreign investment and improvements in farming technology. However, the bad feelings between the mainland and Zanzibar remain; people on the mainland think Zanzibar has too much representation for its 1 million people, outmoded politics, and poor human rights, while Zanzibarians complain about the erosion of their previous sovereignty.

Having resisted the development of tourism for wealthy foreigners in Nyerere's equality-driven society, there was a major expansion of park resorts in the Serengeti area in the mid-1990s. Tanzania received over half a million visitors in 2005, up from 80,000 in 1980. Tourism made up over 60 percent of export earnings, replacing the crop sector as a dominant revenue source. The future depends on further improvements in government to restrict overspending and corruption.

Tanzania is known to possess large deposits of minerals, including gold and natural gas. Road construction is increasing to bring the interior into the orbit of global trade. The Tazara Railway connecting Zambia across Tanzania to Dar es Salaam was completed in 1975 with Chinese funding, in order to provide an outlet to the sea for countries that did not wish to trade through South Africa. The railway continues to function since the end of apartheid, and there is a tourist train from Cape Town to Dar es Salaam. However, competitive rates attracted Zambian copper and other trade southward again. Furthermore, the railway was built in a direct line that missed most Tanzanian development opportunities except for the Selous Game Reserve, passed through just before reaching Dar es Salaam.

Subregion: Southern Africa

Southern Africa includes the lands south of the Democratic Republic of the Congo and Tanzania (Figure 8.19a). The tropical summer rain climates grade into the southwestern arid and southern warm temperate conditions. The dominant plateaus with their pleasant climate provided many good farming areas for colonial Europeans, who displaced African groups. This subregion experienced the development of some of the world's largest deposits of valuable minerals (diamonds, gold, platinum, and important steel constituents). In the later 1900s, the apartheid policies of the white-ruled Republic of South Africa made enemies of the newly independent, black African-ruled countries in this subregion. However, since the mid-1990s the new black African-dominated majority government in South Africa improved the links of this most advanced African country with its neighbors.

The **Southern African Development Community** (SADC), with its economic policies taking over from its anti-apartheid stance in the early 1990s, includes all the countries in this subregion, plus DRC and Tanzania. Although trade is freer, South Africa, Botswana, and Namibia resist allowing free movements of people. Other countries fear a two-tier trading system as South Africa looks to the United States and the EU for trading opportunities without including its SADC partners.

Republic of South Africa

South Africa (Figure 8.19b) has 7 percent of the population of Africa South of the Sahara, but its GNI PPP is 43 percent of the total. Its diverse economy of mining, manufacturing, and increasing services built up during the apartheid years, often in response to external sanctions, continues to prosper following the peaceful transition from the apartheid regime to majority black government after the 1994 elections. The Government of National Unity, and the ability of the Truth and Reconciliation Commission to deal with sensitive interethnic concerns, made this possible. The flight of some white professionals

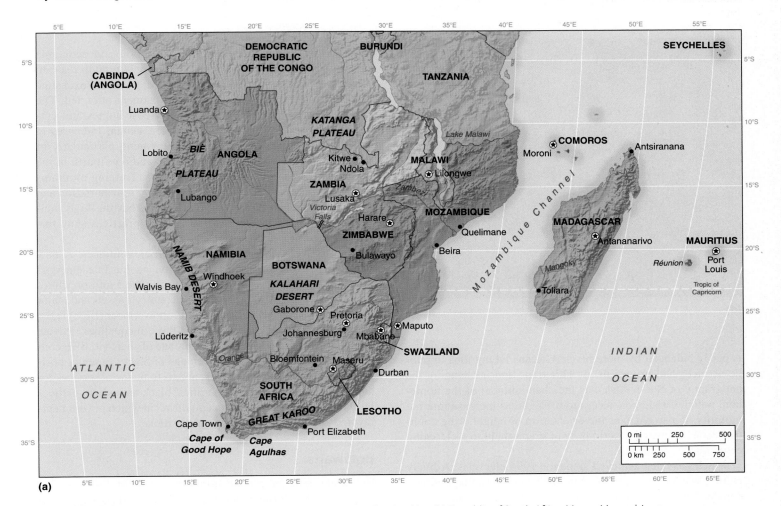

(a)

FIGURE 8.19 Southern Africa. (a) The subregion, its countries, and main cities. (b) Republic of South Africa. Moz. = Mozambique; Swz. = Swaziland; Les. = Lesotho. (c) Botswana. (d) An aerial view of the Johannesburg area. *Sources: (b) & (c) Data from Microsoft Encarta 2005.*

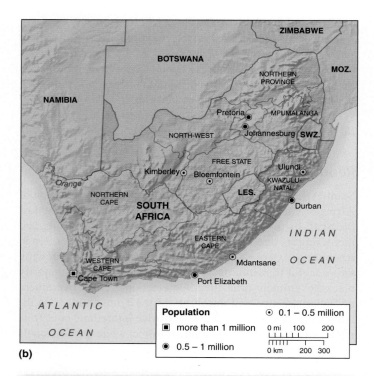

(b)

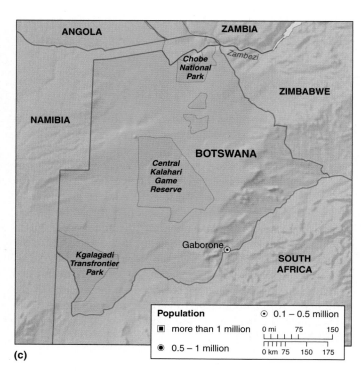

(c)

(d)

heid deprived nonwhite people of education and economic opportunities. In the 2000s, annual economic growth reached almost 4 percent, inflation was controlled, trade and exchange controls relaxed, and new external investment attracted. The government built nearly 2 million low-cost homes and created over 1 million new jobs. However, annual increases in the youthful labor force keep unemployment, particularly of unskilled people, high (30–40 percent), and one-fourth of the population lives on government handouts. Training programs and major public works projects to improve transportation and power supplies are helping the economy, but the high incidence of HIV/AIDS and restrictive labor practices slow further increases in well-being.

Debate over who controls South Africa continues. A new black middle class of some 15 million people that emerged after 1994 is fueling a consumer boom, but the 5 million whites who remain control 90 percent of the country's assets and major corporations. The Black Economic Empowerment (BEE) legislation of 2003 attempted to redress these inequalities. Private commercial companies give a percentage of shares to BEE consortia, contract to employ a percentage of blacks in senior positions, develop skills, and provide access to financial services for small companies. BEE has its critics, black and white, particularly for suspicions of "crony capitalism" in which government-favored people are placed in good positions.

Botswana

Botswana, situated on the northern border of South Africa (Figure 8.19c), forms a contrast to other small African countries such as Rwanda. It has 20 times the land area, but only

is partly balanced by more black Africans taking jobs in public services and private firms. This transition is changing the geography of South Africa as the abandonment of the homeland policies and open access to urban areas increase internal mobility. It also changes the orientation of neighboring countries from imposing sanctions against apartheid to welcoming economic association.

Much has been achieved since the 1994 elections. A lively culture was liberated, with open media and vigorous political debate. Through the 1980s South Africa's distorted and underperforming economy grew at only 1 percent per year. Apart-

1.5 million people, because around 70 percent of the land is dominated by the Kalahari Desert. Formed as the British colonial protectorate of Bechuanaland, following a request by the Tswana people in the face of a perceived threat from Boer farmers in the Transvaal, the country took its present name at independence in 1966.

Botswana is a stable democratic country, electing its National Assembly. An advisory House of Chiefs represents the main ethnic groups, and the chiefs preside over traditional courts (although citizens have the option to be judged under the British-based legal system).

Traditionally the people are cattle herders, but Botswana has experienced major economic growth based on diamond mining. The government owns half of the Debswana mining company and uses the income to build infrastructure, diversify employment opportunities, and build foreign exchange reserves. However, in the early 2000s spending on economic development was cut in response to the rise in health care expenditure. Botswana has one the world's highest HIV infection rates (22 percent of the population aged 15 and over in 2005) and is spending money on free anti-retroviral treatments and a program to reduce the mother-to-child transmission. Botswana also maintains a relatively high level of military spending despite the low probability of internal or external conflict involvements.

Industrialization in the Johannesburg-Rand Hub

With 8 million people, Greater Johannesburg is Africa's third largest city (after Lagos, Nigeria, and Kinshasa, DRC) and this region's main economic hub (Figure 8.19d). Unlike many other major world cities, it is not situated on the coast or a major river. Nor is it the political capital of South Africa, although it is the capital of Gauteng, South Africa's wealthiest province.

Johannesburg occupies an area that yielded some of the oldest human remains. The long-term occupation by nomadic Bushmen gave way to Bantu farmer incursions from the north around AD 1060 and Boer farmers in the 1800s. The area was totally rural until the discovery of gold in the 1880s. The first gold rush was in Barberton, but further exploration located the riches of the Witwatersrand Mountains, which became the center of development. Following the Boer War (1899–1902) and the declaration of the Union of South Africa (1910), a harsh racial system barred blacks and Asians from skilled jobs and instituted the migrant labor system. From the 1940s, people of non-European descent were removed to specified areas, such as the South Western Townships (Soweto), which often became sprawling shantytowns.

The abandonment of apartheid in 1990 and the 1994 elections led to freedom from discriminatory laws and the integration of black townships into the municipal government system. Although the suburbs became more multiracial, many businesses moved out of Johannesburg's Central Business District (Figure 8.20) to the northern suburbs following perceptions of increased crime rates and serious traffic congestion. The north and northwestern suburbs are the wealthiest with high-end retail

FIGURE 8.20 Southern Africa; central Johannesburg. The CBD with suburbs and old mine workings beyond. Relate this to Figure 8.19d.

shops and residential areas, including Houghton and the mostly black Sophiatown. The southern suburbs are low-income residential, including old townships such as Soweto.

The mining basis of Johannesburg's economy gave way to manufacturing and service industries. Today there is no mining within the city limits. Sanctions during the apartheid period led both to losses of multinational corporations and their replacement by local firms in many fields. Post-apartheid economic renewal brought back the multinationals. Steel and cement manufacture are paralleled by many medium and smaller concerns, making engineering products, chemicals, armaments, and food and drink. There are government branch offices and consular offices. Many new shopping centers are opening in the suburbs. Johannesburg is the center of South African media from newspapers (leading national Afrikaans, English language, and others aimed at black readers) to broadcasting. This city is also a center for higher education with the newly consolidated University of Johannesburg and the longer-standing University of Witwatersrand building on its reputation of resisting apartheid.

As an inland center, Johannesburg was dependent on surface transport until the age of air transport. As a rail and road center it has the City Deep container "dry port" that takes 60 percent of cargoes arriving at Durban for further distribution. Although the large numbers of low-income people depend on bus and informal minibus taxi transportation, the many cars and trucks cause huge traffic jams, even on the 12-lane sections of the ring road. Johannesburg has South Africa's premier international airport, which acts as a hub for the subregion, and is the center of tourist arrivals despite the area's industrial and crime-ridden reputation. Visits to Soweto, the Apartheid Museum, the world-famous zoo, and museums are often part of packages that use the airport as an exchange point for visits to Kruger National Park, Cape Town, and Namibia.

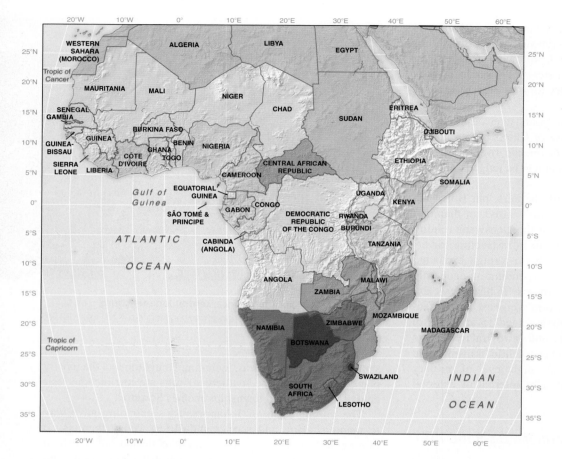

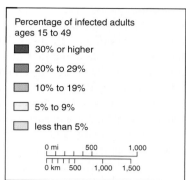

Percentage of infected adults ages 15 to 49

- ■ 30% or higher
- ■ 20% to 29%
- □ 10% to 19%
- □ 5% to 9%
- □ less than 5%

FIGURE 8.21 Africa South of the Sahara: HIV/AIDS. The percentage of adults aged 15 to 49 years infected with HIV/AIDS in 2003. Some countries are unlikely to report the full extent. What could be the causes of the higher prevalence in Southern Africa?
Source: Data from United Nations and Population Reference Bureau.

Contemporary Geographic Issues

HIV/AIDS Pandemic

Africa South of the Sahara has 10 percent of world population, but in 2005 had 60 percent of people living with HIV (Figure 8.21). In 2005 over 3 million people became newly infected, while 2.5 million adults and children died of AIDS. **HIV/AIDS** is a major threat to world health and especially to millions of people in poorer countries, where 90 percent of infections occur. The World Health Organization lists AIDS as the third main cause of global deaths. Its occurrence is of **pandemic** status—a disease that has a long-term presence around the world. For geographers, one of the major issues arising from a study of the HIV/AIDS pandemic is the differences in approach between materially wealthy and poor countries (Table 8.2).

People contract HIV through unprotected sexual contact with HIV carriers or through contact with HIV-contaminated blood or body fluids. Current medical research does not indicate transmission by other contacts with HIV carriers. HIV infection can be passed from mother to baby. Patients become prone to many other sexually transmitted diseases and to other serious illnesses such as tuberculosis (TB), pneumonia, toxoplasmosis, fungal infections, and cancers. Available medical treatments are complex and expensive, and require close monitoring (Figure 8.22). These do not cure HIV but

FIGURE 8.22 Africa South of the Sahara: treated HIV/AIDS. An HIV-positive man visits a health clinic in Cape Town, South Africa. After being close to death, he received antiretroviral drugs in a trial project and is well enough to run his own business.

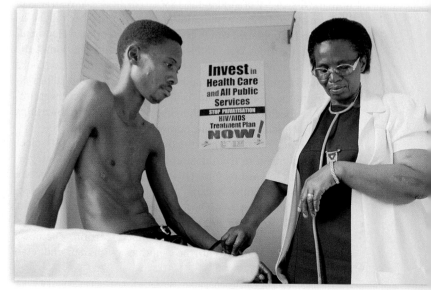

TABLE 8.2 DEBATE: HIV/AIDS IN MATERIALLY WEALTHY AND POOR COUNTRIES

Materially Wealthy Country	Materially Poor Country
Retroviral drugs are having a significant effect in delaying the onset of AIDS, giving a sense that the disease is on the wane.	The drugs are designed in materially wealthy countries for the strains of HIV that are common there, and those produced by multinational pharmaceutical companies are too expensive for wider use in poorer countries. However, Brazilian and Indian producers sell them more cheaply and the MNCs reduced the price when faced with the terrible outcome of no action. From the mid-1990s to 2004, drug prices had been reduced by 97 percent.
The greater investment in health care facilities makes vital monitoring centers widely available.	Monitoring centers are few and far between, making the distribution of drugs difficult. Access is being improved in middle-income countries such as Botswana and South Africa, but not in the poorest countries.
HIV rates are low, but are rising again after a period of careful control. In particular, the behavior of sexually active males is tending toward less protection, while some of the drugs are less effective.	HIV rates of infection and deaths from AIDS are high, although recent surveys show that previous estimates on less evidence were often too high.
Men have been affected more than women because of the connection with homosexual activities.	Women are affected more than men. Reasons include: male circumcision helps prevent infection; women are vulnerable to rape; older male sexual partners pick up infections; mobile miners and military personnel are most likely to be infected from prostitutes and pass on infection to marriage partners and the resultant children.
Education programs are effective, together with provision of syringes for drug users, although few countries impose HIV/AIDS tests.	Effectiveness of education varies. Some countries did not take the threat seriously until the later 1990s, by which time HIV/AIDS was a major cause of death and social disruption. Even when testing is available, people avoid voluntary involvement because of stigma and social penalties. Botswana, with the highest rate of infection, reduced such high-profile testing to a routine action during doctor visits.
HIV/AIDS is not regarded as a major socioeconomic threat, although more significance is given to the worldwide situation. Cash invested in HIV/AIDS programs increased by 20 times from 1996 to 2003 (although UNAIDS wants twice that in 2005 and even more by 2007).	The high incidence could lead to economic collapse in South Africa, which has more cases than any other country (5.3 million HIV-positive citizens out of a total population of 45 million). Arising from this threat, more countries are devising national plans to combat the disease. However, much of the funding available has strings attached (e.g., the United States does not support family planning programs that "promote" abortions or some NGOs such as the Global Fund to Fight AIDS, TB, and malaria).
Most countries are now open in reporting cases of HIV/AIDS.	Many countries resist full reporting, giving a false view of the total picture. There is a particular difficulty in many Arab countries, where activities contributing to the spread of HIV/AIDS are illegal or not admitted and so there is little detailed monitoring. However, it is becoming clear that there is a high incidence among sex workers and drug users, and that the many migrant workers pose a considerable threat.

can prolong life. First recognized in wealthier countries, HIV/AIDS is now a major plague in Africa South of the Sahara, where most countries have increasing HIV prevalence and few are showing a decline. Southern Africa is the "epicenter" of the global HIV/AIDS pandemic, with the main countries having over 20 percent of the adult population infected. Western and Central Africa have lower HIV prevalence, under 10 percent of adults. Eastern Africa has most evidence of declining prevalence, particularly in Uganda and Kenya, but HIV rates in other countries in that subregion remain at high levels.

The causes of the high levels of HIV/AIDS in Africa include poverty, the breakdown of traditional family support systems, the apartheid policy in South Africa that brought miners into male-only camps serviced by prostitutes, continuing promiscuity at a time when traditional polygamy gives way to the taking of sexual partners outside monogamous marriages, and mistaken government policies. HIV/AIDS spreads quickly in cultures that value male sexual prowess.

Although reduced (for a time) in Europe and North America in the 1990s by expensive triple-drug therapy monitored at special clinics, the disease diffused rapidly through Africa. The adoption of such antiretroviral drugs in African countries is patchy. By mid-2005, a third of those needing the drugs received them in Botswana and Uganda, with up to 20 percent receiving them in Cameroon, Côte d'Ivoire, Kenya, Malawi, and Zambia. Low levels of provision occurred in South Africa, Ethiopia, Nigeria, and Zimbabwe. Progress requires the well managed use of these drugs, but the high levels of HIV/AIDS are expected to continue in Eastern and Southern Africa.

As well as the demographic impacts, HIV/AIDS has geographic social impacts that will take many years to change. Half the miners in South Africa are HIV carriers, and millions of orphans, often carriers themselves, constitute a growing need for help in the region. By 2010, orphans with HIV/AIDS will rise from 1 to 2 million in Nigeria and from under 4 million to nearly 6 million in Eastern Africa. Throughout Africa South of the Sahara, the lives of young workers and their families are being shattered. Military personnel, migrant miners and their wives, and prostitutes have the highest proportions of infection. The lack of medical understanding and panic at not having the funds or expertise to do anything about the disease generate false taboos and myths, such as the one that men can cure themselves by having sex with a virgin girl, a basis for many child rapes. Although HIV/AIDS mainly occurs in urban areas, it also threatens rural communities in Southern Africa, where one-fourth of farm workers are infected.

HIV/AIDS requires global action that is as vital as controlling terrorism, the armaments and drugs trades, and slavery. However, current United States laws prohibit programs to support condom distribution or abortions, cutting off a major funding source.

Exploding Cities

Two-thirds of the populations of most countries in Africa South of the Sahara remain rural, linked to the dominance of subsistence farming in the economy. However, Africa is now the fastest urbanizing continent in the world: the growth of urban centers (Table 8.3) reflects the rising influence of global connections. Around 15 percent of Africans lived in towns in 1950, 28 percent in 1980, and 50 percent are expected to do so in 2020. The 300 million African urbanites in 2004 could rise to 500 million by 2015. In 1960, Johannesburg was the only city in the region with more than 1 million people. There were 4 such centers in 1970, 12 in 1990, and 24 in 2003.

Urban places account for high proportions of economic activities. They are perceived to contain more prospects for waged employment and provide better educational and health facilities than rural villages. Urban populations also grow as people migrate into towns for perceived safety in times of civil war. However, the informal urban economy becomes the only means of livelihood for many people who are cut off from their home village and subsistence food production. In the absence of the wealth common in Western countries, many Africans create ways of living that enable them to survive and even enjoy life. Items that might be regarded as waste become children's toys; bicycles or walking are common modes of travel, while cheap rides on crowded vans and buses enable wider circulation; and loads such as water and many goods and possessions are often carried on people's (especially women's) heads.

When placed alongside the fact that Africa is the only world region where extreme poverty is expected to increase (from 100 million people in 2000 to an estimated 345 million in 2015, mainly in towns), it is clear that there is an urban crisis. Although the rate of urban population increase is similar to that in Europe and North America in the late 1800s, with similar problems, few African cities benefit from the economic growth experienced in those regions.

African cities have distinctive geographies with current changes imposed on past patterns. Accra in Ghana illustrates some common features (Figure 8.23). Many of the oldest African urban landscapes occur in Western Africa. They include the trading centers of Timbuktu, Sokoto, and Kano at the southern end of trade routes across the Sahara. These Islamic cities typically have central markets, mosques, citadels, and public baths, and are still dominated by craft workshops rather than modern industry. Ibadan, the center of the Yoruba culture in southwestern Nigeria, preserves another type of pre-European urban landscape. The central palace and nearby market are at the focus of streets radiating toward other towns. Fortifications were added later to resist Muslim attacks. The modern populations of these older towns remain much more ethnically unified than those in many of the newer urban areas.

The colonial era was a time of building ports and interior control centers that left often grandiose buildings of European

TABLE 8.3	Populations of Major Urban Centers in Africa South of the Sahara (in millions)	
City, Country	2003 Population	2015* Projection
CENTRAL AFRICA		
Kinshasa, DRC	5.3	8.7
Lubumbashi, DRC	1.0	1.7
Yaoundé, Cameroon	1.6	2.2
Douala, Cameroon	1.9	2.5
Brazzaville, Congo	1.1	1.6
WESTERN AFRICA		
Lagos, Nigeria	10.1	17.0
Abidjan, Côte d'Ivoire	3.3	4.4
Dakar, Senegal	2.2	3.1
Accra, Ghana	1.8	2.6
Bamako, Mali	1.3	2.2
EASTERN AFRICA		
Addis Ababa, Ethiopia	2.7	4.1
Nairobi, Kenya	2.6	4.0
Kampala, Uganda	1.2	2.0
SOUTHERN AFRICA		
Cape Town, South Africa	3.0	3.2
Johannesburg, South Africa	3.1	3.7
Durban, South Africa	2.6	2.7
Pretoria, South Africa	1.2	1.4
Maputo, Mozambique	1.2	1.9
Luanda, Angola	2.6	4.3
Harare, Zimbabwe	1.5	1.8
Lusaka, Zambia	1.4	1.8
Antananarivo, Madagascar	1.6	2.6

* estimated

Source: United Nations Urban Agglomerations 2003, with estimates for 2015 (2003).

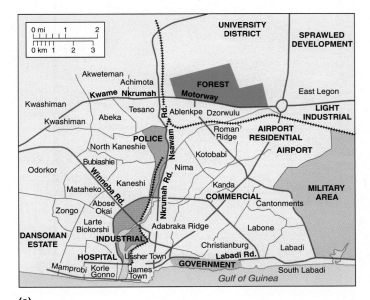

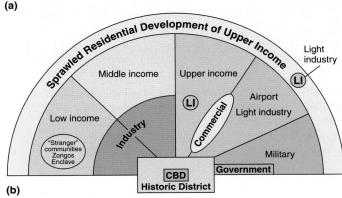

FIGURE 8.23 African urban landscapes: Accra, Ghana. (a) Land use in the port city of Accra, which expanded from a village through colonial trade and now has a variety of urban sectors. (b) A generalized pattern of urban land uses in Western African ports. *Source: (a) Data from S. Aryeetey-Attuh 1997.*

design and separated economic zones. After independence local elites took over colonial properties. Many new homes, offices, and factories were built, but planning was often at a low level and such city expansion was haphazard.

Shantytowns are slum areas that occur in all African cities, housing over 70 percent of the urban population. Shantytowns are unplanned, constructed of any materials that come to hand—from wooden crates to cement blocks and corrugated iron—and basic in their services. Few are linked to piped water or sewerage systems, leading to high infant and child mortality. HIV/AIDS incidence in South Africa is twice as high in slums as in the rest of urban areas, and three times as high as in rural areas. Some shantytowns condemn families to a hopeless future of poverty. Their inhabitants are often involved in the informal economy. The high incidence

of such poor housing, unemployment, street children, and the wide availability of small arms are linked to high levels of urban crime.

In many cases, however, people move into slums on arriving from a rural area but eventually find better accommodations. Governments may supply utilities and build schools, hospitals, and roads to integrate shantytowns with the rest of a large urban area, usually after a considerable period while these become established. Shantytowns were encouraged under apartheid in South Africa (Figure 8.24), and the long process of improving housing conditions there is now under way.

A combination of the United Nations Habitat Agenda's Sustainable Cities Project, the Millenium Development Goals, and the New Partnership for Africa's Development (NEPAD) is beginning to address this region's urban problems. There is

FIGURE 8.24 South Africa: Soweto Township, Johannesburg in 2002. Contrasts within Soweto include the squatter shacks in the foreground and the more permanent housing in the rear. Once a symbol of apartheid, the township is now a tourist venue.

an emphasis on improving water, sanitations, and housing in NEPAD Cities. The first international ministerial meeting on land, housing, and urban development was held in Durban, South Africa, in early 2005. Improved governance, security of tenure, and better urban-rural linked regional planning are also on the agenda. Geographers are involved in producing local poverty maps for the UN-HABITAT Global Urban Observatory that will provide a basis for focused policies.

Global Intrusions and Local Responses

Global Connections and China's Role

Much of Africa South of the Sahara remains a plantation or quarry, for long providing raw materials to the materially wealthy Western world—often a hangover from trade established in colonial times. Some African countries and cities are more integrated in the global economy than others because their ports and airports link commercial farms and mines to markets in wealthier countries. Cities such as Nairobi (Kenya), Johannesburg and Cape Town (South Africa), and Lagos (Nigeria) exhibit some signs of becoming global city-regions as they provide headquarters for the subregional centers of commerce, multinational corporations, and NGOs.

However, many African countries are part of the global economic system as dependent debtors and recipients of aid. In their post-independence focus on self-sufficiency to develop the well-being of their people, they often spent their small resources on what they perceived to be important in raising their country's identity, through lavish new capital cities and military purchases. In some countries, poorly managed projects, together with the corrupt siphoning of funds into personal bank accounts abroad, slowed economic opportunities and led to international lenders and bankers imposing conditions that made it difficult to obtain further loans.

In the 1980s, the World Bank and the International Monetary Fund—the two major lending institutions for developing countries—established new guidelines for grants and loans. This was a response to the low rates of success achieved with previous loans and the large proportions of loans that were absorbed by high exchange rates, the cost of internal government bureaucracies, and the corrupt mismanagement of funds. To date, these structural adjustment policies have had little success in the countries of Africa South of the Sahara, although at times Ghana, Tanzania, Burkino Faso, Nigeria, and Zimbabwe came close to adopting the guidelines. The policies relied too much on exporting commercial crops instead of growing food for local use. The reduction in government employees reduced health care and education provision. At present, countries in Africa South of the Sahara seem to lose out whether they decide to follow structural adjustment policies or not. If they do not adopt them, they lose access to funds from the World Bank, International Monetary Fund, and aid agencies. If they adopt the stringent policies, they often alienate their people.

For many African countries, the immediate future looks brightest through association with China. Africa may become a prime trade battlefield between China and the West. In 2004 the Chinese president, Hu Jintao, made Africa one of his first foreign visits. The Chinese keep human rights and politics out of business arrangements and so gain vital resources and new markets in this region. China trades with countries, such as Ethiopia, Somalia, Equatorial Guinea, and Zimbabwe, which are shunned by European Union countries over human rights issues. It purchases oil (30 percent of China's oil imports), iron ore, and other commodities, and invests billions of dollars in producing the commodities, along with tourism, agricultural, infrastructure, and health (such as anti-HIV/AIDS) projects. China is a major supplier of military hardware to the region. In 2005 African trade with China increased by 30 percent to $40 billion. China is offering a prospect of free trade for African countries.

Local and Global Connections: Cell Phones

The development of telecommunications is indispensable to Africa's growth and domestic stability. In 2005, the United Nations launched its "Digital Divide Fund" to help reduce the technology gap between richer and poorer worlds. However, the establishment of rural telecenters (communal telephone, Internet, and other computing facilities) is an example of top-down development requiring capital and often government funding. The bottom-up spread of cell phones is faster because their use is a matter of private demand. The world's poorest people often share or rent cell phones by the call and so overcome the high cost and shortage of landlines and electricity supplies. Cell phones have the greatest impact on development, raising long-term growth rates. Cell phones cut transaction costs, widen trading networks, and reduce travel costs. They also access the Internet and provide a clock, camera, and music download source—bringing an information revolution. An extra 10 phones per 100 people in a developing country increases annual economic growth by up to 1 percent.

In Africa, while only 3 percent of the population had access to a landline telephone in 2001, there were already 50 million cell phone subscribers, with numbers increasing by one-third each year. By 2005 South Africa, the main hub, had over 60 cell phones per 100 people, Botswana 40, and Kenya 15 (the top three in the region), and South African providers were expanding their facilities across the region. Cell phone growth rates doubled in 2005 in Nigeria, Angola, Ghana, Liberia, Tanzania, Zambia, Guinea-Bissau, and Chad. Urban areas are well provided and rural users are being targeted. Ethiopia has the lowest usage with just one cell phone network under the prevailing state telecommunication monopoly.

Multinationals and the World Trade Organization

At the time of independence in the 1950s and 1960s, Western economists claimed that the new countries would grow economically by following the world's wealthier countries in moving from the primary sector into manufacturing and services. Few African countries achieved this progression.

Multinational mining companies and makers of coffee, tea, and chocolate products continue to buy African raw materials. For example, multinational aluminum manufacturers helped arrange funding for the Volta River project in Ghana that generated hydroelectricity to refine bauxite, the ore of aluminum, as cheaply as possible. However, reliance on producing raw materials and low levels of processing keeps local incomes low.

Many countries reliant on primary product exports are kept poor by low world prices and restrictive practices in the wealthier countries over farm products. Thus Nigerian and Ghanaian coffee producers receive around US50¢ per pound. Each cappuccino served in U.S. coffee chains takes about one ounce (4.5c) of coffee but customers pay over $2. Most of the markup goes to the coffee traders, blenders, grinders, and retailers in the United States (and other wealthier countries). Purchasers of African raw materials often maintained low world prices by opening up new areas of production in other parts of the world as growing markets absorbed what established areas produced.

Soaring raw material prices in the 1970s (minerals) and in the early 1980s (beverages) were short-lived, but often enabled the producer countries to take out loans for economic development projects. When the prices fell, these countries faced debts that they could not repay. In 2004, the World Trade Organization championed the cause of African products to gain wider access to world markets.

Tourism

Tourism is the world's largest industry, with international tourist arrivals in 2005 exceeding 800 million people. African visitors were up from 33 to 37 million, also growing by 10 percent. Tourism provides many African countries, particularly those in parts of Eastern and Southern Africa with a major potential for earning foreign currency. In 2005, 7 percent of all employment in the region was in tourism. The International Council of Tourism Partners aims to triple the numbers of visitors to Africa by 2015.

The commitment of some governments to conservation in designated game and national parks resulted in Africa South of the Sahara having a higher proportion of such land uses than any other continent. Unfortunately, the governments have little money to spend on maintaining the parks and their wildlife, and cannot finance realistic management policies. Other tourist attractions include the historic slave-trading centers and Robben Island in South Africa, where Nelson Mandela was a long-term inmate.

An example of the growing significance of tourism in Africa is provided by the Great Limpopo Transfrontier Park, connecting the Kruger National Park (South Africa) with the extended Limpopo National Park (Mozambique) and the Gonarezhou National Park (Zimbabwe). Formalized in 2002, the three-country initiative was taken forward in 2006 by the presidents of the three countries opening a border post. The combined park area of over 41,000 km^2 (16,000 mi.^{2}) has both ecological and economic goals: animals will be able to range widely and the already prosperous tourist industry will be extended. In particular, the park will be a draw to the many thousands of visitors expected in South Africa for the 2010 soccer World Cup.

Tourism does have negative impacts, since tourist revenues fluctuate with global security crises and changes in demand. It puts pressure on land, water, and power resources, diverting demand from often poorly provided residential and industrial consumers. However, Africa's unique cultural and natural environments, along with the rewards from well-planned tourist facilities, will continue to make this industry significant.

Local Emphasis in a Globalizing World

Globalization places an emphasis on export goods and foreign trade, but most people in the countries of Africa South of the Sahara rely on the local economy. Consumer goods remain unusual or communal (Figure 8.25). Many rural Africans live their lives with little reference to the global economy, even though it brings the use of motor vehicles or clothes made of synthetic fibers into the remotest villages. Many small towns experience a growing mobility of people, increased levels of commercial exchange, and rising demands for the consumer goods advertised on global TV channels. Some villages mushroom into small service centers with rapidly built shops and market stalls where food and consumer goods are sold and buses bring people from surrounding rural areas.

In some countries, a trend toward interaction with the global economy has been reversed. For example in Zimbabwe, which was one of the main African growth countries in the early 1990s,

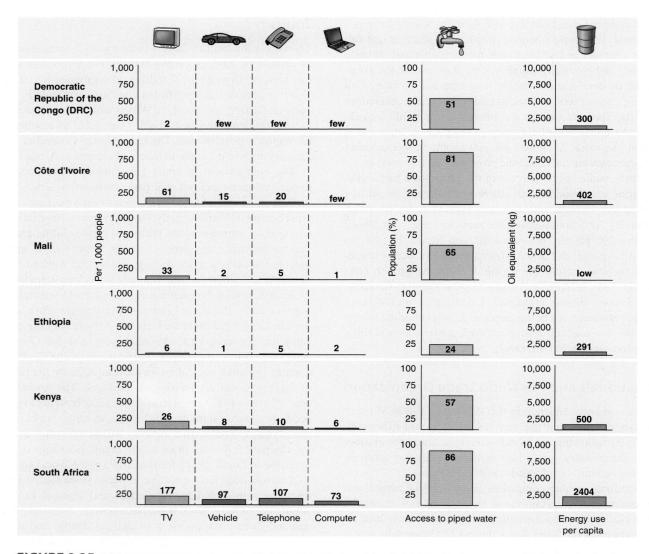

FIGURE 8.25 Africa South of the Sahara: indicators of well-being. The ownership of consumer goods is low in almost all countries and access to piped water and energy varies greatly among the countries. Compare this diagram with those for other regions. Details are provided in Figure 1.19. *Source: Data (for 2002) from* World Development Indicators, *World Bank, 2004.*

President Mugabe encouraged the takeover of large, white-owned commercial farms by war veterans. This changed the farming emphasis from tobacco and vegetable crops for export to the growing of corn for local consumption. This, in turn, cut off much of Zimbabwe's foreign exchange and reduced many black former farm workers to unemployed poverty.

Culture Shock

Human Rights and Women's Roles

Africa South of the Sahara has a poor record on human rights. Few people can earn what Westerners classify as a decent standard of living. Many Africans with formal employment are paid low wages and exploited by their firms. Personal security is often threatened by arbitrary arrest and violence. Injustice is widespread, illustrated by the apartheid culture that affected South Africa for many years and is proving difficult to change. Discrimination by gender, race, ethnicity, and religion is rife, as examples earlier in the chapter show.

Women in Africa are expected to bring up children, draw water, collect wood (Figure 8.26), raise crops, and cook meals. Few, such as Baodi (see p. 228), avoid such a life and hardly any attain high political office. Only 13 percent of African members of parliament are women. Before the colonial period, ethnic groups such as the Kongo people in Central Africa had matriarchal inheritance, but the colonial powers ended many female institutions and reduced women's rights.

Unequal access to education has been a major deprivation of rights for many women, especially in the northern, Muslim parts

FIGURE 8.26 Women in Nigeria. Govari women carrying firewood across a highway in Gwagwalada village in the middle belt. The increases in kerosene prices in 2004 led to increased firewood usage.

of this region, although efforts are being made at present to put that right. Female genital mutilation is common. It is estimated that 100 million women are affected, mainly in the northern countries such as Nigeria (25 percent of women) and Mali (90 percent). Complications from the cutting include bleeding that encourages HIV/AIDS, painful intercourse, and childbirth difficulties.

In contrast to the lot of most African women, a few women wield immense power. For example, wives of dictators often make their presence felt. The wives of Nigerian dictators in the 1980s and early 1990s built their own personal fortunes. In Rwanda, the wife of the dictator from 1994 is suspected of links to the groups who carried out the genocide. In Gabon and Zambia, estranged leaders' wives returned to embarrass their husbands as pop stars or in court cases. The wife of the leader of Liberia in 2003 claimed that she, not her husband, was in charge.

Religion: A Controversial Role

Religion has always been important in Africa and continues to be so, often in new ways. Many people profess membership in a religious organization, traditional, Muslim, Christian, or other.

For Africans, traditional religious belief is about an invisible world linked to the visible; spiritual beings or forces communicate with them and influence their daily lives; people relate to a spirit world, linking them to other people and the land they cultivate. Traditional beliefs loom large in rural societies about the common ownership of land, the rights of chiefs to grant and revoke uses of the land, and taboos about women owning land or farm implements. Such beliefs often conflict with Western ideas of individual human rights and personal ownership.

In the 2000s the advances of Islam in the north and of evangelical Christianity elsewhere in the region, and the reinvigoration of traditional beliefs as part of ethnic identity, both aided and

hindered personal well-being and development. Peace is basic to human development. In many parts of Africa, religious groups work together to establish or maintain it. In South Africa, the Truth and Reconciliation Commission was instituted and led by Archbishop Tutu and closely linked to the country's faith communities. But in other parts of Africa strong religious involvements led to violence as extreme Islamists persecuted Christians or traditional beliefs become the basis of militia initiation rites.

While revenue collection is a major problem for African governments, which rely on foreign aid and are often deeply in debt, religious communities survive on monies donated by members. Health and education are the most conspicuous provisions by religious groups, often begun by Christian missions and more recently developed by Muslims, that contribute toward human development. In many areas outside major towns that are largely abandoned by country governments, there is a trend toward religious communities assuming some of the functions of government. These functions include education, health care, and vigilante protection.

Better Education

After independence, all countries increased educational enrollments and achievements. By the early 2000s, countries such as Botswana, Cameroon, Kenya, South Africa, Zambia, and Zimbabwe had virtually total enrollment, male and female, in elementary schools, up from 50 percent in 1965. Burundi, Chad, and Mauritania made major strides by increasing primary education from under 20 percent in 1965 to between 70 and 80 percent. However, Burkina Faso, Guinea, Mali, and Niger in the northern Muslim belt and Ethiopia still have only half of their children in elementary school, and female education lags behind male.

Increasing numbers of people in the region also have opportunities to become fully literate in secondary school and to earn higher academic qualifications. It is often disappointing to many who gain higher qualifications that there are few jobs available in their home countries. African doctors, lawyers, and airline pilots are increasing in numbers, but many enter the brain drain and find their employment abroad in the world's wealthiest countries. Emigration from Africa to the United States more than doubled since the 1990s, disproportionately in the professional, managerial, and technical occupations, and continues to grow. Many Africans living in the wealthier countries may benefit their home countries by sending money to their families. Some of the skilled and experienced personnel return. Meanwhile, African countries pay expatriates from wealthier countries high salaries to carry out professional jobs.

Can Africa Claim the Twenty-First Century?

On the assumption that there is a wish to end poverty around the world, Africa South of the Sahara presents a huge challenge (Figure 8.27). Many countries are short of the basic human

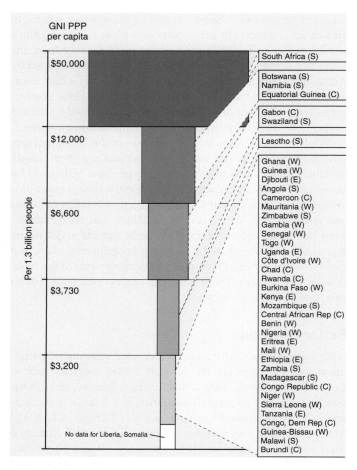

FIGURE 8.27 **Africa South of the Sahara, 2005: average country incomes per capita compared.** GNI PPP figures. C = Central Africa, W = Western Africa, E = Eastern Africa, S = Southern Africa. Compare this diagram with Figure 1.20 and those for other regions (e.g. North America, Figure 10.11a). *Source: Data (for 2005) from World Development Indicators, World Bank; and Population Reference Bureau.*

resources and infrastructure needed to increase economic development, slow population growth, and encourage political democracy. This is a local problem with global implications.

The debate about the future of this region is posed in Table 8.4.

External Pressures

Throughout 2005, world leaders—politicians and global organizations—proclaimed that they would focus on poverty reduction with special reference to Africa. The Millennium Development Goals (see Table 1.1), increased aid, debt forgiving, and WTO trade liberalization that could make it possible for poor African countries to emulate countries in Asia.

External (France, the United Kingdom, the World Trade Organization, and the World Bank) critiques of Africa's problems highlighted four areas of action.

1. ***Improved governance and conflict resolution*** is the most basic need. To date, reductions in poverty have resulted as much from better domestic government in a few countries,

as from external aid. Civil conflicts impose huge costs at home and in neighboring countries through deaths, maimings, property destruction, and refugee migrations.

2. ***Investment in people.*** The vicious circle of high fertility and mortality, low enrolments in education (especially of girls in some countries), high numbers of dependent children and youths, slow action against HIV/AIDS, and low family savings lie behind much of Africa's static or declining development.

3. ***Economic diversification*** makes countries more competitive in world markets. To date, Africa South of the Sahara has been a loser in the global economy. The countries of this region produce only 1 percent of global GNI PPP. Most people have little access to the consumer goods that are the signs of material well-being, or lack the financial ability to develop entrepreneurial skills or engage with global connections. The region's industries need new products and better terms of trade, new incentives, and wider access to markets in wealthier countries. The perceived risks of investing and doing business in Africa make job creation slow.

Internal reforms needed include reducing corruption, improving infrastructure and financial services, and providing better access to the information economy. Countries do not have sufficient all-weather roads or other forms of internal transportation. Ports are poorly equipped and expensive. People lack clean water supplies and adequate sanitation. There are shortages of electricity and telecommunications.

4. ***Reduced aid dependence, debt, and stronger intraregional partnerships.*** Africa remains the world's most aid-dependent and indebted region. Programs of debt relief have become more significant since the 1990s. By the end of 2005, 25 African countries had debt burdens eased by $35 billion under the Heavily Indebted Poor Countries initiative begun in 1996. Aid donors still insisted on approved development policies to avoid corruption. The World Trade Organization tries to help African and other developing countries improve their access to markets for agricultural products and reduce farm subsidies in wealthier countries, but the United States and the European Union make small concessions and increase their own farm subsidies to maintain their farming communities.

Internal Efforts:
African Union and Regional Links

In July 2001, the heads of African governments meeting in Lusaka, Zambia, changed the name of the Organization of African Unity (OAU) to the **African Union (AU)** with headquarters in Addis Ababa, Ethiopia. The purpose was to enable African countries to compete better in a tough global environment by creating strong African institutions—including an executive assembly, a fixed parliament, a central bank with a single currency, and a court. By 2005, little progress had been made and few countries had contributed their share of costs; the central bank was seen as "decades away." The main contribution of the AU is through its Peace and Security Council involvements in conflicts within countries. There are also moves to bring the independent **New Partnership for Africa's Development (NEPAD)**, at present headquartered in South Africa, within the AU.

At the initial conference, President Thabo Mbeki of South Africa proposed the Millennium Action Plan with a twofold thrust. First, it restated the policies previously urged by Western countries and institutions: better government, more democracy, respect for human rights, market reforms, and recognition of the advantages of globalization. Second, the plan highlighted the need to reduce poverty by improving education and public health. It asked for continuing, more accountable aid together with the removal of trade barriers and agricultural subsidies in richer countries. However, subsequent actions suggest that the world's wealthier countries demand the first part but contribute little to the second. When the wealthiest (G8) countries met in 2002 and discussed African needs, they offered $1 billion of the $64 billion requested at a time when the United States increased its own farm subsidies by $190 billion.

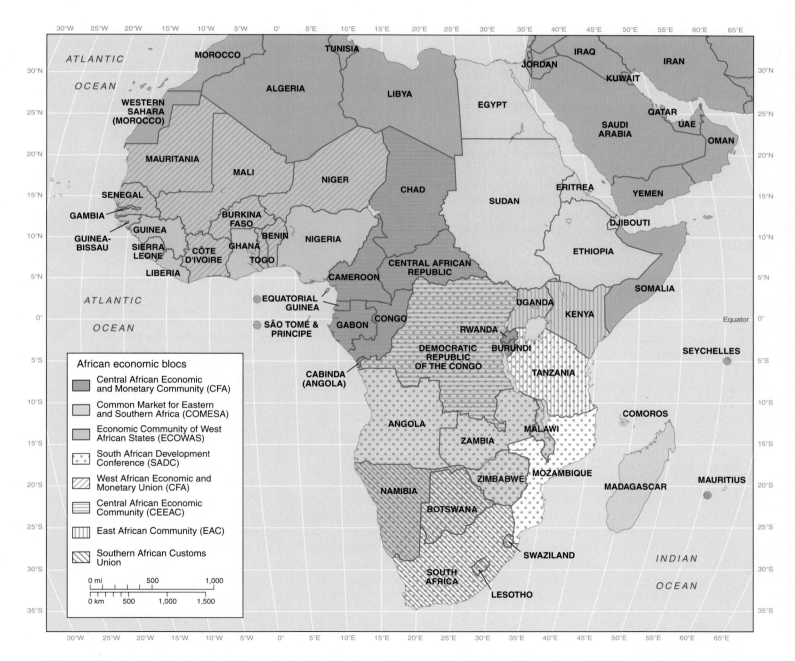

FIGURE 8.28 Africa South of the Sahara: economic blocs. Few influence world markets. Most have achieved little, but may be at the start of greater acceptance. If the map is somewhat confusing, it is because many of the regional blocs overlap. *Source: Data from The Economist Newspaper Group, Inc.*

TABLE 8.4   **DEBATE: THE FUTURE OF AFRICA SOUTH OF THE SAHARA**

Africans Can Do It	No Hope for Africa's Future
There are signs that Africans are moving toward a better future. The peaceful transition in South Africa, Uganda's success with AIDS, fewer wars, and increasing democracy are examples of good trends.	Such signs are few and temporary. Most things get worse. Civil strife springs up in new places and most democracy is a façade. These are basic reasons today for crippling African development.
Younger "born frees" (since independence) are better educated and more inclined to expect good leadership from the current leaders rather than blaming the past.	The present problems stem from the colonizers and their racism. Whatever Africans do to put them right is futile. (This is a common complaint of older people born into colonial rule.)
African countries can produce goods in addition to the crops and minerals that others want to buy if trading terms with the wealthier countries are improved. Some countries are growing rapidly, some as a result of oil windfalls, and a few others such as Mozambique, Rwanda, and Uganda have seen a decade of economic growth after decades of strife and poverty.	Few leaders place much importance on sustained economic growth and their actions tend to reduce the ability of governments to develop education, health, and employment prospects. Few African leaders allow widespread involvement in government that reduces their powers of patronage and ability to make arbitrary decisions.
Smart businessfolk can do well, as in the oil companies on the west coast, the mining corporations, and those making cheap luxuries such as bottled drinks and soap powder prosper. Cell phones increased from almost none 10 years ago to nearly 100 million today. The business climate could improve with more privatization in areas such as utilities (telephones, electricity) and with shorter periods of business registration.	Too many countries place difficulties in the way of foreign corporations wishing to do business in African countries, including the continuation of the bribe culture and lawlessness. The past has left too many examples of squandered opportunities.
The rise in the numbers of urban Africans is leading to wider political participation and, hopefully, to demands for better government. Better government is particularly important at this time, but has to be widely wanted and supported. Power is based in country governments. But improved political involvement has seldom been linked to greater prosperity.	There is still too much rule and control by a few "big men" who turn the law and finances to their own ends. President Mugabe of Zimbabwe is a well-known example. His once fairly prosperous country is now among the poorest in the world. Too many governments are predatory and few are competent. There are still few leaders who leave after electoral defeat, compared with those overthrown by war or coup.
Aid agencies are now putting more research into pre-funding activities. More philanthropic donors are needed, overseeing their giving in relation to criteria such as "saving the maximum number of lives at minimum cost" (Bill Gates).	In the past, too many aid agencies left behind more harm than good. Dependency on outside resources, and spending aid such as World Bank grants in profligate ways, do not lead to local entrepreneurial actions.
Land reform is occurring, although there is a need to ensure that people who can farm get title to farming land and government favorites do not take over productive land.	Most countries suffer from a lack of security in property rights. People with communal or tenant rights cannot use that land to underpin bank financing.
African countries are at last realizing that they have to deal with the HIV/AIDS problem and there are encouraging signs of government intentions.	HIV/AIDS is out of control, with rapid spread of devastating social problems including a generation of orphaned and infected children. Even temporary palliatives are too expensive and there is little infrastructure to monitor their use.
South Africa acts as an example of democracy, modernization, involvement of black talent in major corporations, free press, strong labor unions, independent judiciary, and large middle class.	South Africa does little to help other African countries improve their systems, partly because it is also struggling. Its recent dominance by a single political party (African National Congress) could result in greater corruption.
Africans are beginning to pay more attention to marrying local and global trends.	The change from traditional approaches, ways of doing things, and expectations to the trappings of modernization has been too great.

African countries also attempt to work with each other through regional trading groups (Figure 8.28) along subregional lines. Unlike the EU (see Chapter 2) or NAFTA (see Chapters 9 and 10), however, the African groups are loosely organized and often overlap. They often set out with enthusiasm but then become dormant or achieve little for want of political support from members, credibility among the wealthier countries, or difficulties in administrating their activities.

By 2006, soaring prices for oil and minerals gave Africa South of the Sahara's encouraging annual economic growth a boost, and not only in the mineral-rich countries. Most country economies are better with lower inflation, tighter government accounting, and less violent politics. The improvements attracted foreign direct investment, trebling since the low point of the 1990s. Yet the real income per person has scarcely risen—by only one-fourth from 1960 to 2005, whereas it multiplied by over 30 times in Asian countries. Half the population continues to live on less than a dollar a day, compared to reductions in this criterion to 30 percent in South Asia and

17 percent in East Asia. Aid and investment in Africa South of the Sahara have been uneven in the past and the region, apart from South Africa, continues to rely on commodity exports, paying less attention to diversifying economies and developing human resources.

Amid the complexities of the challenges facing Africans today, it is clear that many of the materially poor are not poor in the broadest sense. Particularly in rural areas, there are self-contained communities with clear leadership, community festivals, music, songs and stories, herbalist "doctors," practical education, and the satisfaction of basic needs. Conflicts cause more rural poverty than droughts. The real poverty occurs in cities.

Africa faces major difficulties in the transition from a self-sufficient and communal, largely rural economy to a modern industrial and urban one, where the money economy dominates. Countries such as China, India, and Malaysia might be better qualified to advise the region than the affluent industrialized West.

GEOGRAPHY AT WORK

Mapmakers and GIS Analysts

Elio Spinello and Steve Lackow are Californians who are the authors and managers of AtlasGIS, a software package that combines mapping with a geographic information systems approach. In the 1990s, the huge GIS software company, ESRI, took over AtlasGIS, but has recently placed marketing and development back with Elio Espinello and Steve Lackow's company, RPM Consulting.

As an example of their work, Elio completed a project that was an epidemiological assessment of river blindness (onchocerciasis) in Mozambique. He was commissioned by Aircare International, which provides small aircraft to help local groups in Mozambique and wanted to know the most important areas for delivering medical care. The disease does not kill, but is chronic and widespread and affects the lives of many people. The African blackfly inhabits areas with fast-flowing streams and the female carries the river blindness from an infected person to an uninfected one. Fibrous nodules form in the infected persons, producing microfilariae that attack skin pigmentation, causing skin atrophy and blindness. Some surgery is

needed to remove the nodules, but new medicines make it possible to treat many more people without bad side effects.

Elio began by mapping the factors that encourage a concentration of African blackflies: the density of population, the concentration of rivers, and the occurrence of steep slopes that give faster river flow. He put together a composite index of these factors, giving greater weight to the population and waterway density than slope steepness (Figure 8.29). The results highlighted districts of high risk (yellow), making it possible for Aircare International to concentrate its delivery of medical support.

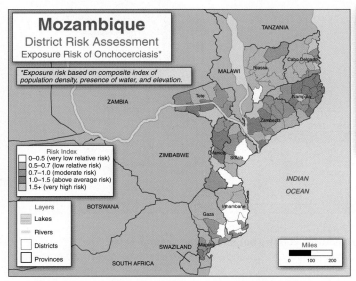

FIGURE 8.29 Mapmakers and GIS. (a) Mapping the risk of river blindness in Mozambique. (b) Elio Spinello. *Source: (a) Data from Elio Spinello, RPM Consulting.*

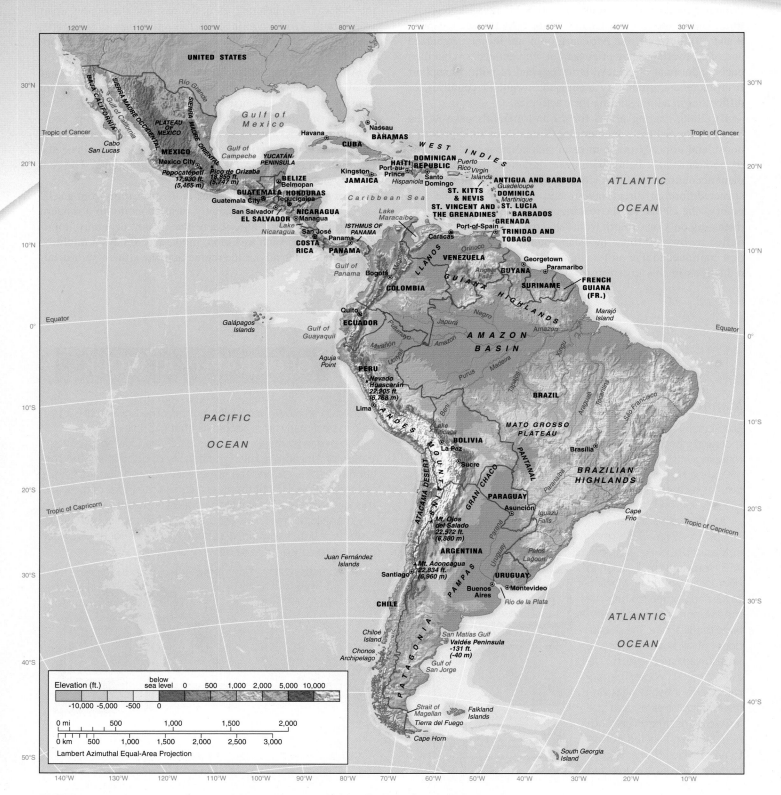

FIGURE 9.1 Latin America: physical features, country boundaries, and capital cities. The region is dominated by the almost-island of South America, linked by the Panamanian isthmus to Central America and Mexico. The islands of the Caribbean add a further element of diversity. The major boundaries are the Atlantic and Pacific Ocean. The short land boundary in the north is with the United States, much of it along the Rio Grande.

Eighteen-year-old Jorge works as a bellman at a business hotel in the Amador neighborhood in Panama City, Panama. Jorge hopes to attend college in a year or two, but first wants to save money from his job to pay off the used Toyota sport utility vehicle he bought last year. The hotel where he works is situated along the eastern bank of the Pacific entrance to the Panama Canal. The Amador neighborhood (and nearby "Causeway") of Panama City has large areas of waterfront parks, restaurants, and shops and is very popular with locals, business travelers, and tourists. Jorge earns $US4 per hour plus tips. He lives with his parents and siblings in the San Francisco neighborhood on the eastern side of Panama City. Jorge drives his SUV to work each day. If he leaves by 6:30 a.m., he is able to avoid the worst of Panama City's traffic, and be at work early after a half-hour's drive. If he oversleeps and does not leave his father's house until after 7:00 a.m., then it may take him an hour and a half to arrive at work. Jorge's official work at the hotel ends at 5:00 p.m. each day. Several days per week he agrees to stay late to taxi hotel guests around the city. Sometimes he uses the hotel's van and other times he chauffeurs them in his SUV. Some days he earns significantly more money from this informal driving service than he does from his bellman position (he may earn from $US30 to $US100 more per night). Jorge thinks the money is great, but is challenged by the hours. Often he will taxi people, after a long days work, until 2:00 or 3:00 a.m. Then, after an hour or two of sleep at home, he rises early to start his workday again at the hotel. Jorge enjoys two days off each week. He spends his free time playing guitar in a band in clubs around Panama City. Like most Panamanian youth his age, Jorge likes to go out to Panama City's numerous and diverse clubs, walk around with friends in one of the city's several modern shopping malls, go to the movies, or just hang out with relatives or close friends. Jorge loves the cosmopolitan nature of Panama City and could not imagine living anywhere else.

Chapter Themes

Latin America: Dramatic Contrasts

Distinctive Physical Geography: landforms of Latin America; tropical to polar climates; major river basins; natural vegetation; natural resources; environmental problems.

Distinctive Human Geography: ethno-cultural diversity; Latin American population patterns; regional urban geography; political geography, economic geography.

Geographic Diversity
- Mexico.
- Central America.
- The Caribbean Basin and Environs.
- Northern Andes.
- Brazil.
- Southern South America.

Contemporary Geographic Issues
- Tropical forests and deforestation.
- Urban pressures in Mexico and Brazil.
- The Northern Andes and the international drug trade.

Geography at Work: Floodplain Erosion in Mexico

Latin America: Dramatic Contrasts

Latin America is a vast region situated from temperate latitudes north of the equator in northern Mexico, to the subarctic conditions of southern South America (Figure 9.1). Latin America is examined here as a world region based on the lasting human geographic impact of Iberian colonization and the influence of the global economy on Latin America's human inhabitants and physical geography. The countries of Latin America range in size from population giants such as Brazil (186.77 million people in 2006) and Mexico (108.33 million), to the Caribbean Basin, where many countries have fewer than 100,000 people (Table 9.1). Latin American countries range from high to low incomes, from dependence on a single economic product to a diverse and integrated economic base, and from highest levels of involvement in, to isolation from, the global economic system.

The region is one of dramatic physical contrasts. The Andes Mountains, the second-highest mountain range in the world, contrast with the huge, low-lying basin of the world's largest river system, the Amazon. The Earth's largest tropical rain forest, the Amazon rain forest, is situated across the mountains from one of the world's driest deserts, the Atacama. The region's great range in latitude produces a variety of climate regimes, which are further altered dramatically by changes in elevation and proximity to mountain ranges. Some countries, such as Mexico, have both tropical beaches and snow-capped mountain peaks.

The contrasts present in Latin America also exist within the region's human geography and are especially evident inside Latin America's large cities and between the growing cities and their hinterland rural areas. Mexico, Brazil, Argentina, Peru, and Chile contain large urban-industrial areas around their major cities. The São Paulo and Mexico City metropolitan areas each contain more than 20 million people and are two of the world's largest urban centers. Extreme contrasts between the region's materially wealthy and materially poor are most dramatic within Latin America's cities.

Distinctive Physical Geography

Although the Latin America region extends through almost 90 degrees of latitude—the greatest north-south distance of any major world region—the majority of Latin America lies within the tropical latitudes. When coupled with high mountain

TABLE 9.1   LATIN AMERICA: Data by country, area, population, urbanization, income (Gross National Income Purchasing Power Parity), ethnic groups

Country	Land Area (km²)	Population (millions) mid 2006 Total	Population (millions) 2025 est. Total	%Urban 2006	GNI PPP 2005 Total (US$ billions)	2005 Per Capita (US$)	Ethnic Groups (%)
MEXICO, CENTRAL AMERICA							
United Mexican States	1,956,200	108.33	129.38	75.1	1,086.5	10,030	Mestizo 60%, Native American 30%, white 9%
Belize	22,960	0.30	0.40	50.2	2.0	6,740	Mestizo 44%, Creole 30%, Maya 11%
Costa Rica, Republic of	51,000	4.27	5.57	59.0	41.4	9,680	Mestizo and white 95%
El Salvador, Republic of	21,040	7.00	9.05	59.0	35.8	5,120	Mestizo 90%, Native American 9%
Guatemala, Republic of	108,890	13.02	19.96	39.4	57.4	4,410	Mestizo 56%, Native American 44%
Honduras, Republic of	112,090	7.36	10.70	46.9	21.3	2,900	Mestizo 90%, Native American 7%
Nicaragua, Republic of	130,000	5.60	7.67	58.6	20.4	3,650	Mestizo 69%, white 17%, black 9%, Native American 5%
Panama, Republic of	75,520	3.28	4.24	62.2	24.0	7,310	Mestizo 70%, Native American, mixed black 14%, white 10%
Totals/Averages	**2,477,700**	**149.16**	**186.98**	**56**	**1,289**	**6,230**	
THE CARIBBEAN BASIN							
Antigua and Barbuda	440	0.07	0.08	36.8	0.8	11,700	Black 96%, white 3%
Bahamas, Commonwealth of	13,880	0.30	0.33	88.5	—	—	Black 85%, white 12%
Barbados	430	0.27	0.28	50.0	—	—	Black 90%, mixed 4%, white, other 6%
Cuba, Republic of	110,860	11.27	11.82	75.6	—	—	Mixed white/black 51%, white 37%, black 11%
Dominica, Commonwealth of	750	0.07	0.08	71.0	0.4	5,560	Black, Carib native
Dominican Republic	48,730	9.02	11.59	63.6	64.5	7,150	Mixed 73%, white 16%, black 11%
French Guiana	91,000	0.20	0.29	75.1	—	—	Mixed white, Native American, black, Arawak natives
Grenada	340	0.10	0.11	38.5	0.7	7,260	Black, South Asian, white
Guyana, Cooperative Republic of	214,970	0.75	0.70	36.3	3.2	4,230	Asian Indian 51%, black and mixed 43%, Native American 4%
Haiti, Republic of	27,750	8.52	12.95	36.0	15.7	1,840	Black 95%, mixed and white 5%
Jamaica	10,990	2.67	3.02	52.1	11.0	4,110	Black 76%, mixed black/white 15%, Asian Indian 3%
St. Kitts and Nevis, Federation of	360	0.05	0.06	32.8	0.6	12,500	Black
St. Lucia	620	0.17	0.21	27.6	1.0	5,980	Black 90%, mixed 6%, South Asian 3%
St. Vincent and The Grenadines	390	0.11	0.12	45.0	0.7	6,460	Black 82%, mixed 14%, whites, South Asian, Carib native
Suriname, Republic of	163,270	0.50	0.53	74.1	—	—	South Asian 37%, Creole white/black 31%, Indonesian 15%, Maroon 10%
Trinidad and Tobago, Republic of	5,130	1.31	1.34	74.1	17.2	13,170	Black 41%, South Asian 40%, mixed 17%
Totals/Averages	**689,910**	**35.36**	**43.50**	**55**	**116**	**7,269**	
NORTHERN ANDES							
Bolivia, Republic of	1,098,580	9.12	12.05	62.7	25.0	2,740	Quechua 30%, mestizo 25–30%, Aymara 25%, white 5–15%
Colombia, Republic of	1,138,910	46.77	58.29	74.9	347.0	7,420	Mestizo 58%, white 20%, mulatto (white/black) 14%
Ecuador, Republic of	283,560	13.26	17.47	61.0	54.0	4,070	Mestizo 55%, Native American 25%, white 10%, black 10%
Peru, Republic of	1,285,000	28.38	34.11	72.6	165.5	5,830	Native American 45%, mestizo 37%, white 15%

(Continued)

Country	Land Area (km²)	Population (millions)		%Urban	GNI PPP 2005	2005	Ethnic Groups (%)
		mid 2006 Total	2025 est. Total	2006	Total (US$ billions)	Per Capita (US$)	
Venezuela, Republic of	912,050	27.03	35.17	87.9	174.1	6,440	Mestizo 67%, white 21%, black 10%, Native American 2%
Totals/Averages	**4,718,100**	**124.56**	**157.10**	**72**	**766**	**5,300**	
BRAIZIL							
Brazil, Federative Republic of	8,511,970	186.77	228.87	81.3	1,537.1	8,230	White 55%, Black 11%, mixed white/black 32%
SOUTHERN SOUTH AMERICA							
Argentine Republic	2,766,890	38.97	46.42	89.3	542.5	13,920	White 85%, mestizo, Native American, other 15%
Chile, Republic of	756,950	16.43	19.13	86.8	188.5	11,470	European and mestizo 95%, Native American 3%
Paraguay, Republic of	406,750	6.30	8.57	56.7	31.3	4,970	Mestizo 95%
Uruguay, Eastern Republic of	177,410	3.31	3.52	92.5	32.5	9,810	European descent 88%, mestizo 8%, black 4%
Totals/Averages	**4,108,000**	**65.02**	**77.64**	**81**	**795**	**10,043**	
Region Totals/ Averages	**20,505,680**	**560.88**	**694.09**	**69**	**4,502**	**7,414**	

Source: World Population Data Sheet 2006, Population Reference Bureau. Microsoft Encarta 2005.

altitudes, the latitudinal expanse of the region results in a wide variety of climates, natural vegetation types, and soils.

Landforms of Latin America

The major relief features of Latin America were formed by a combination of tectonic plate interaction and of precipitation runoff in major rivers and mountain glaciers. Along the west coast, the South American plate overrides the Nazca plate (see Figure 1.5). The convergent plate margin marks the line of the Andes Mountains and causes earthquakes and volcanic eruptions. The tectonic pattern is more complex in Middle America (Mexico, Central America, and the Caribbean Basin), where the North American, South American, Caribbean, and Cocos plates meet.

Insular and Mainland Middle America

High-altitude plateau lands between the eastern and western Sierra Madres dominate northern Mexico. The plateau rises over 2,000 m (6,000 ft.) and contains shallow basins. The western slopes facing the Pacific Ocean are steep, but those on the east are less steep with wide coastal plains (see "Geography at Work," p. 301). Prominent mountains continue southward from Mexico through the Isthmus of Panama.

South of the Tehuántepec isthmus in Mexico, the Caribbean plate collides with the Cocos plate, forming a single spine of mountains along and parallel to the Pacific coast. There are very narrow coastal plains and some areas without any flat land between mountain and ocean. A large limestone platform emerged from the seafloor to form the Yucatán Peninsula.

Middle America is subject to earthquakes and volcanic eruptions where tectonic plates collide. A major earthquake centered off the west coast of Mexico devastated Mexico City in 1985 (Figure 9.2), and earthquakes twice leveled the Nicaraguan capital city of Managua in the 1900s. There are 25 active volcanoes between northern Mexico and Colombia that periodically spew lava and ash on surrounding areas.

Insular Middle America is also affected by the clashes of tectonic plates. The North American plate drove into the Caribbean plate, producing intense volcanic activity along the plate margin and forming the Lesser Antilles arc. The eruption of lava and ash continued through the 1900s at Martinique (1902), Mount Soufrière on St. Vincent (1979), and the Montserrat eruption

FIGURE 9.2 Mexico City: earthquake damage. Rescue workers search for victims in a building completely demolished during an earthquake that devastated parts of Mexico City in 1985.

FIGURE 9.3 Caribbean Basin: coral islands. The Exuma Cays, Bahama Islands, are formed of coral reef limestone and rim the Great Bahama Bank for over 150 km (100 mi.).

of 1995, which forced the evacuation of the island and heavily damaged the island's capital city. The winter of 2007 saw new volcanic activity on Montserrat, once again threatening the island's population. In the Greater Antilles, the plate movements caused a mass of continental rock to founder, leaving only the highest points above sea level. The Bahamas and an outer group of islands including Anguilla, Barbuda, and Barbados are flat limestone islands constructed of coral reefs on top of subsiding former volcanic peaks (Figure 9.3).

Andes Mountains

The Andes Mountains have an impact on most of the physical environment of South America. The collision of the South American and Nazca plates produced a volcanic and earthquake-prone western mountain range and a folded and faulted eastern range. The Andes rise to over 6,500 m (20,000 ft.) in Argentina, Chile, Peru, and Ecuador. In Bolivia, Peru, and Ecuador, the central Andes have two main ranges, the Cordillera Occidental (west) and Cordillera Oriental (east) (Figure 9.4).

Between the two ranges, a high plateau, the **altiplano,** is widest in Bolivia and narrows northward into Peru. In Peru, rivers cut deep gorges as they flow northward and eastward to join the Amazon River tributaries.

In Colombia, the Andean ranges splay out northward into three cordilleras—the Occidental, Central, and Oriental. In northern Venezuela, the Cordillera Oriental branches into a further series of lower ranges. Along the Caribbean coast, the mouths of the Colombian rivers, Lago de Maracaibo, and the Orinoco River delta provide limited areas of lower coastal land between the ranges. Islands such as Trinidad and the Dutch Antilles are extensions of mainland geologic structures.

The southern Andes Mountains dominate the landscapes of Chile and the western parts of Argentina, with their highest points constituting the border between the two countries for most of its length. The snow line on the Andes gets lower toward the southern tip of the continent, where glaciers descend to sea level (Figure 9.5).

On the Chilean side, the Andes come close to the Pacific Ocean in the north. Southward, a coastal range is separated from the main Andes ranges by a series of basins and then a wide continuous valley south of the Chilean capital, Santiago. The coastal range and the valley get lower in height alongside the main Andes range and are drowned by the ocean south of Puerto Montt. On the Argentine side of the Andes, deep, river-carved valleys break the front ranges in the north. Their eastern margins have large alluvial fans formed by the deposition of rock material eroded from the mountains. These fans mark both sides of the Andes throughout Chile and Argentina.

Broad Plateaus

Broad plateaus and wide river valleys dominate Brazil's physical environment. Locally, relief is sharp near physical transition zones, as when one travels from the coast to the first plateau level or from one plateau level to the next. The main relief features of Brazil consist of the ancient rocks of the Brazilian Highlands, which are topped in the southeast by layers of lava flows, and the similar ancient rocks of the Guiana Highlands

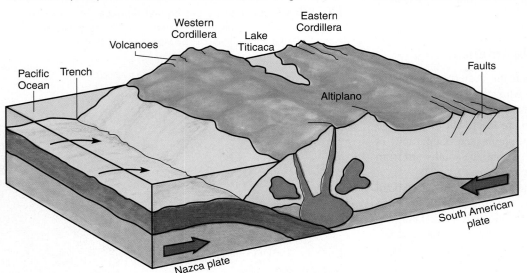

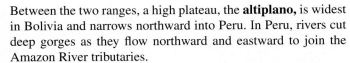

FIGURE 9.4 South America: Andes Mountains. The major features of the central Andes Mountains related to their formation along a destructive plate margin. The Nazca plate plunges beneath the South American plate, causing volcanic activity in the Western Cordillera along the coast and uplift of the Eastern Cordillera and the high plateau (altiplano) between the two main ranges.

FIGURE 9.5 **Southern South America: Andes Mountains.** The Moreno glacier, Glacier National Park, Santa Cruz province, Argentina. The glacier is 5 km (3 mi.) wide, and its front melts on entering Lago Argentino. Such glacial features increase in significance in the southernmost parts of the Andes Mountains.

on the Venezuelan border to the north. The Guianas have low coastal plains that were formed by the deposition of sediment brought to the Atlantic Ocean by the Amazon River and then moved westward along the coast by offshore currents. Inland, these countries rise to the Guiana Highlands plateau. In southern Argentina, the Patagonia Plateau is cut deeply by rivers draining eastward from the Andes.

Tropical to Polar Climates

Middle America

Nearly all of Middle America lies within the tropics (see foldout world climate map inside back cover). East-to-northeast trade winds dominate much of the Caribbean Basin for about two-thirds of the year, contributing to the area's consistent warmth and humidity. Temperatures average around 30°C (86°F) through most of the year. The northern Caribbean islands, Mexico, and even parts of northern Central America are occasionally affected by modified cold air masses, locally referred to as "*nortes*," that move southward from continental Canada and the United States during the Northern Hemisphere's winters. In Middle America the rainfall generally increases southward, from the arid region that straddles the Mexico–U.S. border toward the coasts of Nicaragua, Costa Rica, and Panama, which receive over 2,600 mm (100 in.) of rain each year.

Most rains fall in the summer and fall, when heating of the lower atmosphere causes humid air to rise, cool, and condense into clouds. On Caribbean islands with mountains, windward slopes facing the east or northeast trade winds force the moist air upward into cooler atmospheric conditions. This increases precipitation totals. Where the air descends down the leeward slopes, it warms, becomes less humid, and stabilizes thus producing sig-

nificantly less rain in a localized, or **microclimate** area, referred to as the **rain shadow.** For example in Jamaica, the windward northeast coast receives over 3,300 mm (130 in.) a year, while the leeward southern coast requires irrigation for farming in areas where the annual rainfall is less than 750 mm (31 in.).

Middle America is a region of annual hurricane activity. The North Atlantic hurricane season runs from June through November. Early in the season, hurricanes form in the Western Caribbean and Gulf of Mexico. During the seasonal peak in August and September, hurricanes form in the eastern Atlantic near the Cape Verde Islands west of Africa. The storms move westward into the Caribbean Basin and curve toward either the northwest or north (affecting the U.S. mainland or shipping channels) or continue westward, striking the northern countries of Central America or the Mexican Yucatán Peninsula. The southernmost countries of Middle America lie outside the hurricane zone. Hurricanes cause both a dramatic loss of life and extensive damage to crops, livestock, personal property, and the vital regional tourism industry. Recovery is often slow in small countries with few resources.

Brazil

Northern Brazil experiences a tropical rainy climate in which the temperatures hover in the low 30s°C (80s°F), humidity is high, and rain falls in all seasons. On the Brazilian and Guiana Plateaus, the rains are more variable and more seasonally concentrated in the high-sun period when evaporation and rising air currents are most intense. The variability is particularly marked in the northeastern corner of Brazil, where severe and prolonged drought occurs periodically. In southernmost Brazil, the climate becomes temperate in type, with cool winters and a shorter growing season.

Andean South America and the El Niño Southern Oscillation

The climate in the southern Andes countries ranges from arid regions in northern Chile and parts of Argentina to one of the world's stormiest and wettest regions in southern Chile. The Andes ranges affect the climates of the lands on either side in differing ways.

In northern Chile, the Atacama Desert is situated along the coast between the Andes Mountains and the cold Pacific Ocean. Winds blow almost parallel to the coast or offshore, pushing the cold Peruvian current northward along the coast. Heavy, dense cold air over the cold ocean water cannot rise and cool in the atmosphere and thus it does not produce precipitation and results in the region's coastal deserts. In southern Chile, by contrast, the midlatitude westerly winds bring precipitation throughout the year. Between the desert and the stormy southern region of Chile is a transition zone of warm, dry summers when the arid climate moves south and wet winters when the storm tracks move north. Agriculture and a booming wine industry thrive in this Mediterranean type (subtropical winter rain) climate zone in Chile.

On the eastern side of the Andes is another contrast between northern and southern lands. In the north, the warm Brazilian current in the Atlantic Ocean allows high rates of evaporation offshore, and trade winds blow humid air into the continent. The Patagonia region of southern Argentina is arid. Westerly airflow descends after crossing the Andes, warms up, and becomes drier. The strong winds blowing across Patagonia, after crossing the Andes, often produce very dry rain shadow conditions at the surface.

The **El Niño Southern Oscillation (ENSO)** is regarded as a significant feature in explaining connections among worldwide climatic environments (Figure 9.6). The major 1997–1998 El Niño event brought drought to parts of Middle America and northern South America, and exceptionally heavy rains to the deserts of Peru and Chile. The same El Niño event

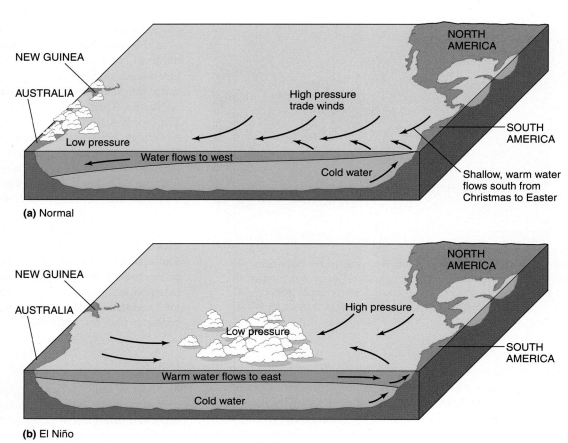

(a) Normal

(b) El Niño

FIGURE 9.6 **The El Niño effect.** During an El Niño event, strong tropical easterly winds (a) diminish, permitting warm water in the western Pacific to flow eastward (b) toward the Americas.

was blamed for unusual weather around the world. Droughts in Indonesia and Australia coupled with drought-based fires in Florida and unusually hot weather in southern Europe were indirectly linked to the event. No two El Niño events are the same, and thus the impact on global climatic conditions varies.

In general, the basic features of El Niño are understood. Every two to five years, there is a decrease in the easterly tropical trade winds that usually circulate cooler waters from the Peruvian current westward across the tropical Pacific (where it warms as it progresses westward). Warmer water from the western Pacific thus flows eastward, eventually reaching the coasts of western North, Central, and South America. Warmer and more humid air masses off the west coasts of the Americas alter the regional climate. Warm water also dramatically changes the marine ecosystem, depleting the usually cold water of its nutrients, altering the marine food chain, and producing dramatic reductions in fish catches for the countries of Andean South America.

Major River Basins

Three major river basins between the high mountains and lower plateaus dominate South America. Tributaries of the Orinoco River primarily drain the largest areas of lower land in Colombia and Venezuela. The Amazon River tributaries flowing from the Andes are muddy "white water" rivers in contrast to the "black" rivers, which contain little sediment as they flow from the plateaus. The contrast is visible where the "black" Rio Negro joins the muddy Solimões just below Manaus in the center of the Amazon basin (Figure 9.7). The Amazon River is navigable well into Peru. In Manaus, Brazil, 2,500 km (1,500 mi.) from the ocean, the Amazon River is 15 km (10 mi.) wide, and over 50 m (160 ft.) deep.

In southern Brazil, rivers drain south to the Paraná-Paraguay River system. The southward flowing rivers present great hydroelectricity potential in the waterfalls at lava plateau breaks.

Natural Vegetation

Tropical rain forest is the natural vegetation regime produced where warm humid conditions persist through most of the year, as in the Amazon River basin, along the northern Pacific and Central American coasts, and on some Caribbean islands (see "Tropical Forests and Deforestation," p. 296). Such vegetation spreads several thousand meters up the east-facing slopes of

FIGURE 9.7 **Brazil: Amazon rain forest.** A LANDSAT satellite view of the area around Manaus, Brazil, in July, 1987, approximately 150 km (100 mi.) across. Unbroken tropical rain forest is red. The wide black Rio Negro contains little silt. The blue Solimoes branch of the upper Amazon River brings large quantities of silt from its upper reaches in the Andes Mountains. Manaus is the light-colored area just west of the Negro-Solimoes confluence, with radiating roads leading to and from it north and south of the river. Small white areas with shadows to the west and northeast are clouds.

the Andes (where it becomes cloud forest). The Amazon basin contains the world's largest expanse of tropical rain forest. Soils beneath the forest vary, but the areas of good soils are small, apart from the flooded areas close to sediment-carrying rivers.

Tropical grasslands or shrub vegetation communities dominate where tropical rainfall is seasonal or significantly lower on average during the year than in rain forest areas. On the Brazilian Highlands, dense deciduous woodland gives way to more open woodland with increasing proportions of shrubs and grasses. Soils are generally poor beneath the natural vegetation and need fertilizers to support agriculture, but some of the lava flows capping the plateau in southeastern Brazil produce easily worked soils.

Cold ocean currents, arid air, and the rain shadow from easterly winds produce deserts along the central west coast of South America. Vegetation ranges from tropical plant species in the north to temperate types in the south.

In the southern part of South America, temperate conditions coupled with varying levels of annual precipitation create a range from bare desert, through semiarid bunch grasses and drought-resisting plants, to tall grasses and forest in humid areas. The plentiful precipitation of southern Chile supports natural vegetation of beech and pine forests. The pampas region of central Argentina and Uruguay is named after the tall, lush grasses that grew there before the region was plowed.

The majority of the high mountain ranges of Latin America are located within tropical latitudes. The latitudinal position of the Andes Mountains coupled with their very high altitude results in a series of vertical zones with distinctive climate and vegetation regimes. The greatest number of distinctive vegeta-

tive zones is at the equator. The **altitudinal zonation** in the region is directly linked to the variety of crops that may be cultivated (Figure 9.8). The lowest 1,000 m (3,000 ft.) have warm to hot conditions and tropical forest in the *tierra caliente.* The next 1,000 m (3,000 to 6,500 ft.) have mild to warm temperate conditions and deciduous forest in the *tierra templada.* Between 2,000 m and 3,000 m (6,500 to 10,000 ft.), the *tierra fría* has cold to mild temperature conditions and pine forests. Above this zone, there is grassland, and at the greatest heights, in the *tierra helada,* snow cover is present all year, even on the equator.

Natural Resources

The natural resources of Latin America include minerals, soils, forests, water, and marine life. The Andes Mountains and the ancient plateau rocks contain considerable resources of metal ores that were among the early attractions for European settlers. Sedimentary rock basins between the mountains and in offshore areas became petroleum and natural gas reservoirs, especially in eastern Mexico, northern Venezuela, Colombia, the offshore areas of northeastern Brazil, and along the eastern slopes of the Andes in Ecuador, Bolivia, Peru, and Argentina. The alluvial soils in parts of the Amazon River basin, the weathered lava surfaces in southern Brazil, and the pampas of Uruguay and central Argentina form large areas with predominantly good soils for farming.

The water resources of Latin America are huge, including the world's largest river by volume, the Amazon, which carries more than twice the amount of water of the next largest river, the Congo in Africa. Many Latin American countries currently generate a large proportion of their power as hydroelectricity, and they continue to explore the expansion opportunities of this power source. Marine life provides some of the world's richest fisheries along the western coast of Ecuador, Peru, and northern Chile. Other fishing grounds in the Caribbean and off southern Argentina are being developed. Fish production for North American markets is a growing source of income to many countries in Central America and the Caribbean. Such resources are vulnerable, however, to human overuse and natural phenomena such as El Niño.

Environmental Problems

Natural hazards such as earthquakes, volcanic eruptions, and hurricanes bring destruction and death to the Andean and Middle American mountain ranges and to the Caribbean islands. Many environmental problems, however, are related to European colonial patterns and subsequent decades of political instability and corruption. Two of the more serious environmental issues in Latin America are soils and air quality. Perhaps the most serious and globally controversial environmental issue centers on deforestation in the Amazon tropical rain forest (see "Tropical Forests and Deforestation," p. 296).

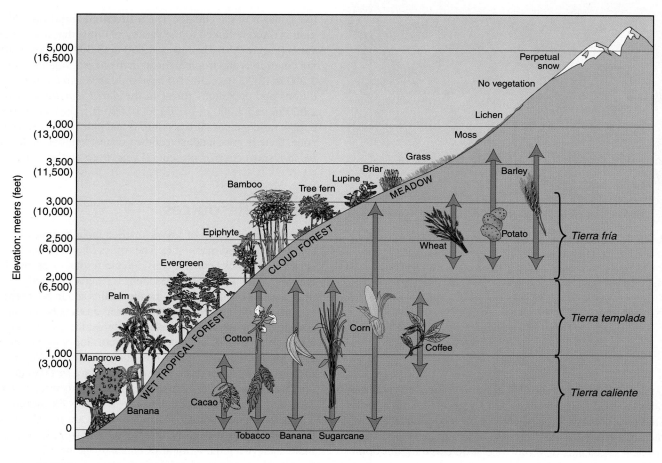

FIGURE 9.8 Northern Andes: altitudinal zoning of vegetation and crops. As they cross the equator, the Andes have maximum height and the greatest number of climatic environments from mountain foot to peak, from coastal mangroves and tropical rain forest to perpetual ice and snow cover. How do the crop zones relate to the vegetation zones?

Soil Erosion

Soil erosion is a major problem in many countries of Latin America, especially where growing populations place additional pressure on subsistence lands. The primary subsistence crop is corn, which is commonly grown in rows up and down a slope, making it easy for intense rainstorms to wash soil away. As economic and population pressures force more subsistence farmers onto the hillsides, the elimination of the natural vegetation removes one of nature's mechanisms for holding soil in place on steep slopes.

Small Caribbean islands colonized for intensive commercial agriculture became notoriously liable to soil erosion and other forms of environmental degradation. The combination of agricultural practices and hotel development along hilly areas above the coasts has accelerated soil erosion. Clearing natural vegetation on the more hilly islands of the Caribbean causes soil to wash rapidly down the hills and into the surrounding sea, resulting in the accumulation of sediment in the coral reef areas surrounding many of the islands. The **sedimentation** of a reef system may ultimately kill the coral. Fertilizers used to compensate for lost soil nutrients from rapid runoff also enter the sensitive coral reef system, further degrading the health of the reefs.

Air and Water Pollution

Air and water pollution results from mineral extraction and refining, and from the concentration of human activities in urban areas. Mexico City suffers more than any other metropolitan area in Latin America from air pollution. Mexico City is a very densely populated urban center situated in a bowl-shaped depression known as the Central Valley of Mexico. The city sits on the valley floor at approximately 2,200 m (7,000 ft.) above sea level, surrounded by much higher mountains which virtually form walls around the urban expanse. The air in and just above the city is often trapped in the valley and becomes extremely stagnant. Vehicle exhaust, coupled with industrial and domestic pollutants, collects in the valley's stagnant air mass and settles near the valley floor where millions of people breathe. Naturally occurring **temperature inversions,** or periods during the winter months when cold, dense air remains "trapped" at the surface under warmer air for several days or even weeks, further complicate the air pollution phenomenon of Mexico City. Although migration and natural population increase are now slower than they were a few decades ago in the Central Valley of Mexico, population numbers continue to challenge pollution reduction efforts.

Distinctive Human Geography

Ethno-Cultural Diversity

Three groups dominate Mexico's population: Native Americans (around 30 percent), those of European heritage (9 percent), and people of mixed Native American and European ancestry known as mestizos (60 percent) (see Table 9.1). Native Americans are particularly numerous in the southern province of Chiapas, where language dialects of Maya origin are commonly spoken in preference to Spanish, the official language. Almost 90 percent of Mexicans are nominally Roman Catholic, although some personal links to the Catholic Church are eroding as established lifestyles are disrupted by moves to urban centers.

The populations of Central American countries are comprised of racial and ethnic groups that became social classes. Native American, or indigenous peoples, remain in large numbers in western Guatemala (Figure 9.9) and comprise significant minorities in the other countries of the subregion. Many indigenous groups in Guatemala and other Central American locations do not speak Spanish as their first language. Differences in language, social customs, and livelihood marginalize such communities from those with political power and economic wealth. Several indigenous groups in Honduras and Panama have successfully achieved a measure of autonomous status within the

FIGURE 9.9 The peoples of Central America. Indigenous people comprise a significant portion of the population of Guatemala. Patterns and colors present in clothing signify the home village or region of many Guatemalans.

federal systems of those countries. Costa Rica has the smallest indigenous population of the subregion.

Along the east coast of Nicaragua and the northern and eastern coast of Honduras are many communities of English-speaking Afro-Caribbean peoples, mixtures of Miskito Indians and blacks from Jamaica and other Caribbean islands, and even pockets of English-speaking whites. The Spanish largely ignored coastal activity in this region during the colonial era, which opened the coasts to decades of influence and control by the British. The British forced the migration of blacks from the Caribbean to work primarily in agricultural production. The historical combination of African and Caribbean cultural influence coupled with the imprint of British rule created a very distinctive cultural landscape region that remains strong today. Belize also has English-speaking Afro-Caribbean peoples.

Hardly any of the indigenous population of the Caribbean Basin survived the early days of European settlement, succumbing especially to the diseases brought across the Atlantic. Although a few Caribs live on Dominica and St. Vincent and some Arawaks on Aruba and in French Guiana, the vast majority of the regional population traces its origin to the colonial occupation. Residents of the Caribbean speak a variety of languages, from French and English to regional hybrids of Patois and Papiamento. Religious practices are equally diverse.

The Andean countries contain racial and ethnic contrasts that have important economic, social, and political effects. In Colombia and Venezuela, Native American groups live primarily in highland and interior areas that were not taken over by Hispanic or mestizo groups. In Ecuador, Peru, and Bolivia, indigenous peoples also tend to be concentrated in the highlands. Smaller numbers of different Native American groups occur in the Amazon River lowlands, totally isolated from European intrusions until the 1950s. Current settlement for European descendants and mestizo groups reflects colonial occupations of upland basins and coastal locations that provided overseas links to Spain. Workers from Japan, China, and Africa immigrated to work in the coastal irrigated farmlands of Peru and established neighborhoods near their work sites. The economic and social differences between European-origin and Native American peoples continue to raise tensions within the countries of the Northern Andes.

Although the large Brazilian population is extremely varied ethnically, the citizens of the country maintain a strong sense of national pride. The original Native American population is much reduced: in the Amazon River basin, it was probably around 3.5 million in AD 1500 but is now closer to 200,000. People of Portuguese and other European heritage make up a major proportion of the current population, with highest concentrations in the southern states and elite areas of the cities. The descendants of African slaves form a significant minority proportion of the population along the northeastern coast and have spread into the southern cities. Brazilians of African decent have had a major impact on Brazilian art, food, music, and dance. Immigrants over the last 30 years include large numbers of Japanese who have easily insinuated themselves into

Brazilian business and political communities. Japanese Brazilians form the largest Japanese community living outside of Japan. Mixtures of people of African, European, and Native American descent comprise a sizable and growing portion of the Brazilian population (Figure 9.10).

The proportions of immigrant population give different emphases to the racial and ethnic mixes of the Southern South American countries. In Chile, around 40 percent of the population claims to be of European heritage, with the remainder considered to be mestizo. In Paraguay, virtually all the people are mestizos apart from a few people of European or Asian origin. Argentina and Uruguay have the largest proportions of European heritage and the smallest proportions of mixed populations.

Latin American Population Patterns
The Spatial Distribution of People

The population of Mexico is concentrated in the country's central region, from Guadalajara in the west through Mexico City to Veracruz in the east (Figure 9.11). This central plateau and the valleys cutting into it formed both an indigenous and a colonial hearth, and more recently became the center of industrial development and government functions focused on Mexico City. In Central America, the main concentrations of people are in and around the largest cities—often in the highlands and closer to the west coast, where temperatures and soils are better for cultivation. Rural to urban migration continues in Panama where the capital, Panama City is undergoing rapid growth. Haiti and Puerto Rico have very high population densities while other islands are more sparsely populated. In countries of the northern Andes Mountains, population distribution reflects the Spanish pattern of colonial settlement and the utilization of harbors and fertile valleys. The distribution of population in Brazil is a combination of both historical and contemporary regional develop-

FIGURE 9.10 Brazil: the people. Brazilians watch a World Cup soccer match between the United States and their country on TV. The crowd includes a variety of peoples with origins in Africa, Europe, Asia, and the Americas, together with those of mixed ancestral heritage.

ment goals. The highest densities are in the southeast, around and inland of São Paulo and Rio de Janeiro. Moderate densities occur in a band parallel to the coast from the southeast around Pôrto Alegre to west of Fortaleza in the north. Farther inland, the very low densities of the Amazon rain forest area create a major geographic contrast within the country. The main population centers in Southern South America are around the Río de la Plata estuary (Argentina, Uruguay) and in central Chile. Smaller centers occur in the irrigated farming oases of northern Argentina and around Asunción, capital of Paraguay.

Natural Increase and Migration Patterns

Mexico's population grew rapidly to around 70 million in 1980 and to just over 108 million in mid-2006 (see Table 9.1). Although its fertility rates and birth rates are declining, Mexico's population is projected to rise to nearly 130 million by 2025. The age-sex diagram (Figure 9.12) shows a decline of births that was achieved through a well-developed family planning program, halving total fertility rates from about 6 in 1976 to under 2.4 in 2006. The very low death rate and increasing life expectancy, however, maintain a gap between births and deaths that keeps population totals increasing.

The population of Central America is projected to grow by 60 percent in the next 30 years. By 2025, the mid-2006 total of 40.8 million may become 57.6 million people. Such growth will place increasing stress on the already pressured natural environment and the political, economic, and social systems of these countries. Very high fertility rates in the 1960s and 1970s—over 6 children per woman, apart from Costa Rica (fewer than 5)—continue to impact the population growth within the subregion. By 2006, fertility rates halved for the majority of the countries of Central America, but they remain high relative to those in the United States, Canada, and most European countries.

The total population of the Caribbean Basin more than doubled from 17 million to 35.3 million people from 1965 to 2006. Population pressure is already severe in many islands, with densities of over 600 people per km^2 (1,500 per mi.^{2}) in Barbados and over 200 in many of the other islands. Only in the Guianas does population density fall below 5 per km^2.

In the 1960s and 1970s, the Northern Andean countries had annual rates of population increase that were among the highest in the world at 3 to 4 percent. In the 1980s and 1990s, total fertility rates fell from over 5 to around 3 by 2006, except in Bolivia. These countries lag behind others in South America in bringing birth and death rates closer together and continue to have annual rates of natural population increase under 2 percent. The total population of 124.5 million people in mid-2006 may grow to nearly 158 million by 2025.

Brazil's population is moving out of a period of rapid growth, which produced the relatively young population of today and a population total that rose rapidly to the mid-2006 figure of 186.8 million people (see Table 9.1). Although Brazil is the world's largest Roman Catholic country and that church in Brazil has a conservative hierarchy, its local priests are often

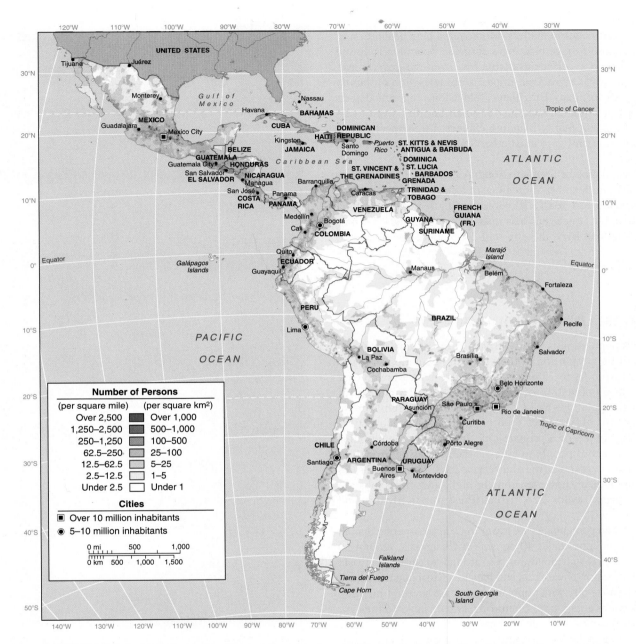

FIGURE 9.11 **Latin America: distribution of population.** Explain the distribution of heavily and lightly populated parts of this region.

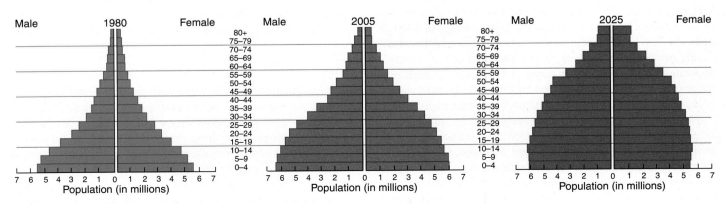

FIGURE 9.12 **Mexico: age-sex diagram.** *Source: U.S. Census Bureau; International Data Bank.*

liberal on birth control matters. Moreover, a high proportion of women choose to give birth by a cesarean operation followed by sterilization, a service available free from the government. Studies suggest a significant impact from television on family size choices for many Brazilians. Television programs depicting happy and stable situations for small families seem to influence Brazilian family planning choices. Life expectancy in Brazil remains similar to most other South American countries, at 68 years.

The population of Southern South America is growing more slowly than in the other subregions of Latin America. In 1930, this subregion had nearly 20 percent of Latin America's total population, but it now has less than 12 percent. The 65.0 million people who lived there in mid-2006 could rise to around 78 million by 2025 (see Table 9.1). Rates of population growth and fertility are low.

Regional Urban Geography

Mexico's population is highly urban. From 1970 to 2006, estimates of the proportion of the Mexican population in towns increased from 59 to 75 percent (see Table 9.1). Mexico City grew from a population of 500,000 in 1900 (2.5 percent of Mexico's population) to 18.7 million in 2003 (slightly more than one-fifth of Mexico's total population). Mexico City is a classic example of a primate city. In Mexico, the stimulus for movements of people to cities is the increasing growth of urban-based manufacturing and service jobs. The official rates of unemployment, however, remain high, and many jobs are in the informal sector (see "Urban Pressures in Mexico and Brazil," p. 299).

Total urban percentages are lower in Central America than in Mexico, remaining under 40 percent in Guatemala, around 50 percent in Belize and Honduras, and around 60 percent in Costa Rica, El Salvador, Nicaragua, and Panama. Most countries have a single primate city with populations that are several times that of the second-largest city. Central America's largest cities are the political capitals of each country.

The Northern Andean countries vary in their levels of urbanization. Venezuela was 87 percent urban in 2006, while Colombia and Peru were over 74 and 72 percent respectively (see Table 9.1). Ecuador and Bolivia, however, had lower urban percentages of 61 and 63 percent. All these figures show increases of 10 to 20 percent since 1970. Peru's primate urban expanse is the Lima-Callao metropolitan complex, where one-third of the country's population resides. This urban complex extends along the Rimac River valley from Lima to the port of Callao. In Bolivia, metropolitan La Paz has 20 percent of the country's total population.

Leading urban centers in Venezuela include Caracas, Maracaibo, and Valencia (Table 9.2). Recent economic growth in the Pacific coast city of Guayaquil, Ecuador, positions this metropolitan region ahead of the established highland capital of Quito. Colombia has a number of major cities that grew up in relative isolation from one another, producing urban regional

TABLE 9.2	Population of Major Urban Centers in Latin America (in millions)	
City, Country	**2003 Population**	**2015* Projection**
MEXICO, CENTRAL AMERICA		
Mexico City, Mexico	18.7	20.6
Guadalajara, Mexico	3.8	4.3
Monterrey, Mexico	3.4	3.9
THE CARIBBEAN BASIN		
San Juan, Puerto Rico	2.3	2.4
Havana, Cuba	2.2	2.2
Port-au-Prince, Haiti	2.0	2.8
Santo Domingo, Domincan Republic	1.9	2.2
NORTHERN ANDES		
Lima-Callao, Peru	7.9	9.4
Santa Fe de Bogotá, Colombia	7.3	8.9
Caracas, Venezuela	3.2	3.6
Medellín, Colombia	3.1	3.8
Cali, Colombia	2.5	3.1
Guayaquil, Ecuador	2.3	3.0
Valencia, Venezuela	2.2	3.0
Maracaibo, Venezuela	2.1	2.6
Baranquilla, Colombia	1.8	2.3
BRAZIL		
São Paulo, Brazil	17.9	20.0
Rio de Janeiro, Brazil	11.2	12.4
Bello Horizonte, Brazil	5.0	6.3
Pôrto Allegre, Brazil	3.7	4.2
Recife, Brazil	3.4	4.0
Salvador, Brazil	3.2	3.9
Brasilia, Brazil	3.1	4.3
Fortaleza, Brazil	3.1	4.3
Curitiba, Brazil	2.7	3.5
Campinas, Brazil	2.5	3.2
SOUTHERN SOUTH AMERICA		
Buenos Aires, Argentina	13	14.6
Santiago, Chile	5.5	6.3
Asunción, Paraguay	1.6	2.3

*estimated

Source: United Nations Urban Agglomerations 2003, with estimates for 2015 (2003).

cultural identities that challenge national cohesion. Bogotá, the capital and largest city, has extensive manufacturing industries as well as the national government bureaucracies. Medellín is another major manufacturing center for Colombia, and Cali is linked to one of Colombia's busiest port areas. Both Cali and Barranquilla are rapidly growing urban centers.

Brazil has two of the three largest metropolitan areas in Latin America and a highly urban population overall. The proportion of the population living in towns or urban settings increased from 56 percent in 1970 to 81 percent in 2006 (see Table 9.1). São Paulo is one of the largest metropolitan areas in the world with 18 million people in 2003 (see "Urban Pressures in Mexico and Brazil," p. 299). The state of São Paulo produces nearly half of Brazil's GDP and two-thirds of its manufacturing output. In the late 1990s, São Paulo's population growth slowed (from 5 percent to 0.5 percent per year). New industries tended to be sited elsewhere, often farther out in São Paulo state to avoid the high costs of traffic congestion, pollution, and strong union control (high wages and restrictive working conditions). The provision of housing, schools, and health facilities could not keep up with the metropolitan area's growth. Brazil's second-largest metropolitan center is the Rio de Janeiro complex. Rio de Janeiro first grew as a port city for early interior gold mine development, becoming Brazil's largest port and capital city in 1763. Today, Rio de Janeiro is a contemporary culture hearth for Brazilian nationals and a major draw for international visitors to the country.

Southern South America is the most urban subregion in Latin America. Except for Paraguay (57 percent urban in 2006), all the countries of Southern South America had over 85 percent of their populations living in urban environments. As in many countries of Latin America, each country has a primate metropolitan center. Buenos Aires grew from a population of 170,000 in 1870 to around 12.5 million people, one-third of the Argentine total, in 2004. Montevideo, Uruguay; Santiago, Chile, and Asunción, Paraguay all contain a significant percentage of their respective country's total population. These cities are the centers of government, manufacturing, and service industries. Few other cities in any of the four countries approach the size or important role of these primate cities.

Global Cities

Although many cities in Latin America interact and trade on a global level, the region's two giants, Mexico City and São Paulo, are the dominant forces responsible for inserting the Latin American region into globally connected world systems. Both Mexico City and São Paulo contain diverse populations, with metropolitan area residents originating from all regions of their respective countries. The cities are representative of the different dialects, social customs, religious practices, and regional pride existing throughout their respective countries. The expanse of trade and interaction with multinational corporations from all world regions, as well as relationships with government officials from various countries of the world, furthers the diverse composition of people and customs present in each urban center. Mexico City is the political, industrial, financial, and cultural capital of the country. São Paulo is the financial, industrial, and cultural capital of Brazil. Although Brasília is the political capital, São Paulo wields strong political power within the Brazilian system. Both Mexico City and São Paulo are media centers for their countries, and both serve as their countries' media connections to the world. Although

each global city-region houses representatives of all subsets of each country, the residents of each take pride in being part of the most influential metropolitan center of their country.

Political Geography
Colonial Geopolitics

Christopher Columbus embarked from Spain in 1492 on the first of four trans-Atlantic voyages. Within 50 years of the initial voyage of Columbus, much of the region was conquered and occupied. In 1494, Spain and Portugal signed the Treaty of Tordesillas, which established a demarcation line (approximately 46°W longitude) between their global spheres of interest in the Americas, giving the eastern quarter of South America to Portugal (Figure 9.13).

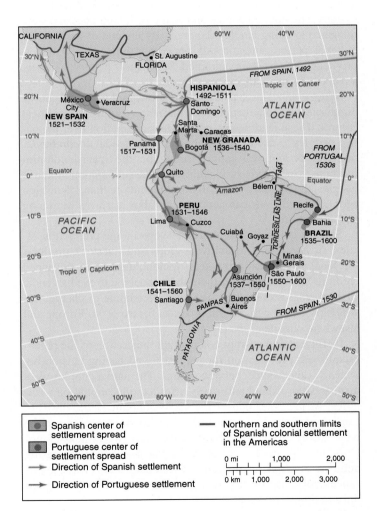

FIGURE 9.13 Latin America: colonial conquest and settlement. During the 1500s, Spanish and Portuguese adventurers established bases and rapidly conquered huge areas. The areas of initial conquest became centers of administration and development. Spain and Portugal signed the Treaty of Tordesillas in 1494, which established a demarcation line separating their colonization of Latin America.

The Spaniards conquered most of the region in the 1500s and imposed a high degree of control on their colony of New Spain by establishing an oppressive system of agricultural production and tightly connected urban settlements and ports. Several forces hindered the Native American inhabitants in forming strong resistance to Spanish intentions in the Americas. First, after the conquest, the lack of immunity to European diseases such as smallpox drastically reduced indigenous populations. Then, the Roman Catholic Church converted surviving communities. Native peoples were forced to give up their subsistence agricultural practices to provide crops and livestock for the food needs of the foreign city dwellers. Mining settlements in the Andes Mountains used various forms of slavery to coerce the local people.

The colonial period in Latin America left a legacy of underdevelopment because it primarily focused on mining for export and the imposition of large landed estates with a strict feudal system. The native peoples and many of mixed ancestry were relegated to being landless laborers. By the end of the Spanish colonial period, antagonisms between privileged and underprivileged groups were ingrained. Subsequent history is largely a record of such antagonisms at work.

Spain granted large land areas to nobles, soldiers, and church dignitaries, in conjunction with the responsibility for political control. These landowners were given jurisdiction over the native peoples to use them as laborers. This feudal *encomienda,* or tribute system, was applied throughout the Spanish colonies for more than 200 years. In addition to the subjugation of native peoples, the Spanish eventually imported more than 1.5 million people from the African continent to labor as slaves in mining and agriculture.

Spanish colonial agriculture developed large production estates, or *haciendas,* to cultivate crops and livestock products for local or domestic markets. Landholders often advanced credit to laborers, causing them to be indebted to the hacienda. Coastal areas around the Caribbean were home to plantations, which produced agricultural products for export to Europe and elsewhere outside of the region.

Inefficient and corrupt colonial administration of the region established a foundation for pervasive and lasting tension between various socioeconomic groups. The native peoples were dispossessed of land and rarely afforded educational opportunities. At the other end of the socioeconomic scale, the *Peninsulares*—Spaniards born in Spain who were working in the New World—took the highest offices and largest land grants. *Criollos*—Spaniards born in the colonies—and the increasing numbers of *Mestizos* (mixtures of European and Native American ancestry) had fewer privileges and became resentful as their links to Spain weakened.

The Portuguese installed a governor general in Brazil in 1549. The city of Salvador became the first capital and northeastern hub of Portuguese Brazil, while São Paulo was established much farther south to solidify Portuguese territorial claims and settle southern lands. The Portuguese established a settlement along Rio de Janeiro Bay after expelling French settlers from that portion of the region. The bayside settlement of Rio de Janeiro became the Portuguese colonial capital in the late 1700s. In the 1600s, the discovery of gold inland of Rio de Janeiro led to increased Portuguese immigration. This in turn set off expeditions to explore and claim the interior. In 1750, Spain agreed to permit Brazilian interior expansion westward of the Tordesillas line—which was really an expression of Spain's inability to prevent what was already occurring.

The Portuguese process of occupying Brazil resembled that of Spain in its colonies, although it had some distinctive features. Colonial Brazil was economically and socially controlled by an elite class who purchased or captured millions of African slaves and transported them across the Atlantic to work on large sugar plantations along the northeastern coast. The Portuguese were responsible for the forced migration of more people from the African continent than any other single colonial power. Estimates suggest that more than 4 million Africans were shipped to Brazil to fill the labor needs of economic development. Large numbers of present-day Brazilians trace their ancestral heritage to African slaves.

French, Dutch, and British attempts to colonize Latin America came later and were largely resisted by the Spanish and Portuguese. There were limited incursions on lands controlled by the dominant colonizers, primarily in and on the periphery of the Caribbean Basin.

Independence

The initial wealth that attracted Europeans came from exports of high-value minerals such as silver, gold, and gemstones from Latin American colonies to Europe. When the flows of such wealth slowed in the 1700s, combined with the devotion of Iberian resources to fight the Napoléonic conquest of Spain in the 1790s, the colonial powers of both Spain and Portugal weakened considerably. The dilution of Iberian strength enabled many Latin American countries to gain independence in the early 1800s. The new countries largely emerged from administrative divisions within the Spanish viceroyalties.

Countries newly independent from Spain were poorly prepared to capitalize on their sovereignty. The next 150 years after independence were marked by political instability punctuated by short periods of economic growth. Spain left behind a geographic system based on mineral exploitation, transportation to a limited number of ports, and large estates engaged in raising livestock. The postindependence period created rivalries and boundary disputes between countries. Rigid social class stratification fueled deep and lasting internal resentments. In most former Spanish colonies, government control alternated between the elite conservative landowning groups, often allied to the military, and the more liberal criollos desiring more democratic governance.

Independence came in the 1820s to Brazil, which had a stronger foundation in place for modernization than other postcolonial territories. In 1807, the Portuguese royal family evacuated from Europe to Brazil when Napoléon threatened to take their country. Rio de Janeiro became the temporary capital of Portuguese government. Although the royal family returned to Portugal after Napoléon's defeat, the prince regent returned to Brazil in 1816 at a time of flourishing revolutions in the Spanish colonies. In 1822, he proclaimed Brazil's independence from Portugal and himself king as Pedro I. Under King Pedro II (1840–1889), the Brazilian economy grew rapidly with the construction of railroads and ports and the expansion of mining in the east-central parts, commercial farming of coffee in the south, and rubber collecting in the Amazon River basin. In 1889, a military coup made Brazil a republic, but falling prices of coffee and rubber led to widespread unrest and a period of dictatorship.

Economic Colonialism

Although Spanish and Portuguese rule was broken, Europe remained the primary market for Latin America's exports. Britain in particular established a relationship of economic dependence, or **neocolonialism,** with several Latin American countries that lasted until the early 1900s. Areas targeted for economic development and control by the British were set up for the export of raw materials and products to supply Britain's industries. The British built railroad networks and port installations aimed at developing the production of minerals, cotton, beef, grain, and coffee. Elite families in Latin American countries who worked within the system often sent their children to schools in Europe and the United States and banked their wealth in those countries.

Latin America, the United States, and Regional Political Change

The United States first asserted its perceived right to influence affairs in Latin America by formulating the **Monroe Doctrine** in 1823, in which it asserted its role as the geopolitical leader of the Western Hemisphere. From the late 1800s, the United States intervened in the affairs of Cuba, the Dominican Republic, Haiti, and Panama (among other regional countries) on several occasions. Increasing economic ties throughout the 1900s culminated in the 1990s with the acceptance by Mexico, the United States, and Canada of the **North American Free Trade Agreement (NAFTA).** In the early 2000s, a unique counter-trend eroded some of the dominant unidirectional flow of influence between the United States and Latin America. Growing Latin American immigrant communities in the United States increasingly asserted their political voice while the payments, or **remittances** they sent to family in Latin American countries continued to significantly influence the economies of regional countries. In the 1990s and early 2000s, Latin American popular culture began to rival U.S. homegrown popular culture. Latin American influences in the U.S. restaurant industry thrived, and Latin American popular music topped the U.S. charts.

Numerous elections in the early 2000s in Latin American countries brought center-left to far-left political parties to power in places such as Venezuela, Bolivia, and Peru. Outspoken leaders, such as Venezuela's Hugo Chavez, vehemently oppose U.S. economic influence in the region. Venezuela, Bolivia, and Cuba signed the **Bolivarian Alternative for the Americas** (known by its Spanish acronym—**ALBA**) agreement, a political and economic pact based in part on the exclusion of the United States. Bolivia nationalized its oil and gas industries in May 2006, a move that hurt investors such as Brazil's government-owned oil and gas company, Petrobras. Brazil's territorial expanse and economic size created an internal perception in recent decades that Brazil was the regional leader in South America's political and economic affairs. Venezuela's oil reserves and current political influence, coupled with moves such as the Bolivian oil and gas nationalization, present formidable challenges to Brazil's desire for regional leadership.

Geopolitics and the Global Economy

Import Substitution   The global economic depression of the 1930s and World War II caused many decision makers in Latin American countries to strive to be more internally self-sufficient. The countries of the region established the goal of becoming less dependent on selling unprocessed or unrefined raw materials in exchange for high-priced manufactured goods from industrial countries. Under import substitution, Latin American countries attempted to use their raw materials in their own internal production of various manufactures for domestic markets. Governments established high tariffs, quotas, and bureaucratic barriers on goods arriving from countries outside of the region. The various country governments owned many of the new industries, and others, established earlier by foreign interests, were nationalized, or taken over by the government of the Latin American country in which they operated. The process proved to be costly for many Latin American countries as they incurred large debts to fund the construction of industry and purchase of manufacturing equipment. By the 1970s, this policy contributed to the rapid growth of one or two major urban centers in each country. Because of their larger home markets, the most populous countries (Argentina, Brazil, Colombia, Mexico) produced the most under this system. Economic growth continued to be uneven.

Oil Crisis and Debt   In the 1970s, geopolitical events in the Middle East contributed to rapidly rising oil prices and diminished supply to world markets. Oil producers initially incurred increased revenues from the higher oil prices. They invested their revenues in European and U.S. banks. These **petrodollars** were urged on Latin American countries in the form of loans. Such loans were used to pay for oil imports in the non-producing countries and major infrastructure projects in oil producing countries such as Brazil, Mexico, and Venezuela—and to improve some government officials' overseas bank accounts.

High oil prices contributed to a global recession in the more materially wealthy countries, weakened markets for products from Latin America, and resulted in much higher interest rates on the loans. The combination of debts and falling export income caused many Latin American countries to default on debt payments by the mid-1980s, resulting in the 1980s distinction as the **"Lost Decade."** The Brazilian government had to devote its large overseas trade balance, generated by import-substitution industries, plus sales of mineral and farm products, to servicing its extensive international debts. Brazil's debt servicing reduced its ability to invest in domestic production, thereby dramatically diminishing economic progress. The high interest rates caused foreign investment and aid to dry up in the late 1980s, and debt liabilities forced countries to emerge from reliance on their internal markets.

Economic Geography

The GNI PPP (Figure 9.14) and consumer goods ownership data (Figure 9.15) for Latin American countries reflect a wide range from economic giants such as Brazil to materially impoverished countries such as Haiti. The current global connections of many economies in the region are partially a legacy of past economic institutions and international relations and partially a response to 1990s changes in governmental approaches to participation in the world economy. Elected governments were in power everywhere (except Cuba), and emphasis shifted from inward-looking policies to the need for cooperation. The policies that focused on government-run industry and the protection of domestic products through tariffs, quotas, and red tape, or **protectionism,** involved high levels of government intervention and highlighted the lack

of capital available for internal investments. These policies gave way to those of structural adjustment, urged by the World Bank and the International Monetary Fund, which encouraged less government spending, more foreign investment, and more export industries. The new policies also opened markets to foreign products and privatized government corporations. Structural adjustment policies were partly forced on Latin American countries as a means of reducing debt burdens incurred during the 1970s and 1980s. As with other simplistic approaches to development, this economic restructuring superimposed on the Latin American regional culture was subject to potential disasters, as Mexico found in 1994–1995 when its economy opened too rapidly, sucking in imports and capital investments and creating a huge trade imbalance.

NAFTA remains the largest and most globally competitive trade bloc in the Americas. Regional trade blocs, such as the **Caribbean Community Common Market (CARICOM)** and the **Common Market of the South (MERCOSUR),** are unable to compete with much more globally significant organizations such as NAFTA, the EU, and ASEAN. Attempts by the United States, and several Latin American countries to create what would be the world's largest free trade bloc, a **Free Trade Area of the Americas (FTAA),** by 2005 were unsuccessful with a lack of ratification of FTAA agreements. The newly formed ALBA trade bloc presents a further challenge to MERCOSUR and other regional associations. Membership in ALBA requires a rejection of formal trade relations with the United States.

Mexico

Mexico is by far the most economically developed of the countries of Middle America. In 2005, Mexico had a total GNI PPP that accounted for over 82 percent of Middle America's total GNI PPP. Mexican GNI PPP per capita is one of the highest in Latin America and near the top of the World Bank's upper-middle income group. Ownership of consumer goods was well above that in the countries of Central America apart from Costa Rica. Mexico's HDI and GDI are much higher in rank than Brazil's, but not as high as Argentina's.

Farming employs one-fourth of Mexico's labor force. Manufacturing, combined with jobs in service and government account for the majority of the remainder of Mexico's labor. Tourism is a major and growing source of income and employment for Mexico, with 21.0 million visitors in 2005 (up from 17 million in 1990). Mexico is a regional leader in developing a tourism industry and is one of the first Latin American countries to have a separate ministry of tourism with dedicated funds within its federal governmental structure. Mexico City, Pacific coastal resorts such as Cabo San Lucas, Mazatlán, and Acapulco, well-preserved ruins of Mayan city-states, and the Caribbean coastal resorts of the Yucatán Peninsula are major tourist attractions for Mexicans, Latin Americans, and visitors from North America, Europe, and Asia.

While diversified urban-industrial economies develop in Mexico's center and north, the east coasts facing the Gulf of Mexico

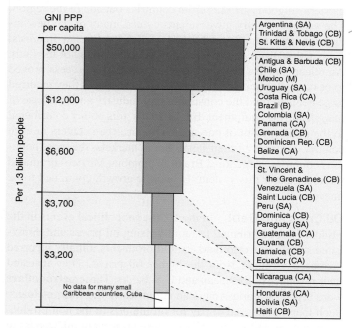

FIGURE 9.14 **Latin America: country average incomes compared.** The countries are listed in order of the GNI PPP per capita for 2005. M=Mexico; CA=Central America; CB=Caribbean; NA=Northern Andes; B=Brazil; SA=Southern South America. Many small Caribbean Basin countries are not included because of unavailable data. *Source: Data (for 2005) from* World Development Indicators, *World Bank; and Population Reference Bureau.*

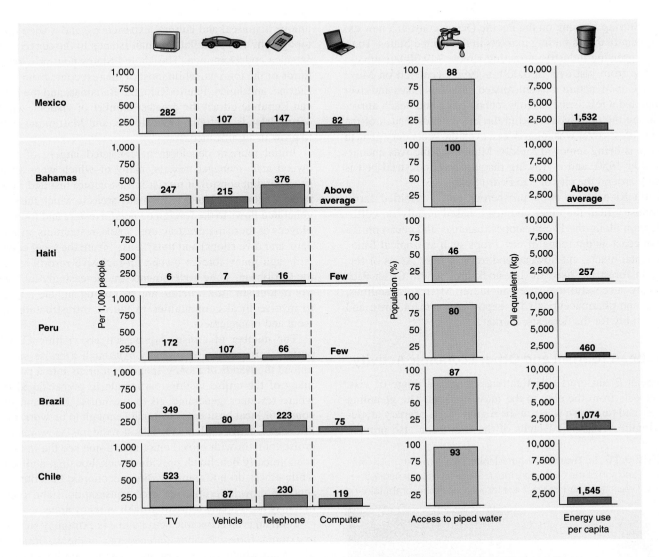

FIGURE 9.15 **Latin America: ownership of consumer goods, access to piped water, and energy usage.** Contrast conditions in the materially wealthy, moderately wealthy, and poor countries. *Source: Data (for 2002) from* World Development Indicators, *World Bank, 2004.*

supply much of Mexico's wealth from large oil and natural gas fields. Oil production had a major impact on the economy of the coastal area between Tampico and Campeche, and recent finds extended this oil-producing area inland. In the late 1990s and early 2000s, oil formed less than 40 percent of exports as more oil was used by growing industries within Mexico, and the products of those industries made up a higher proportion of exports.

Central America

In contrast to Mexico's diversifying economy, the Central American countries have a much narrower economic base, resulting in slower economic and social infrastructure development. Guatemala's capital, Guatemala City, has manufacturing industries, but recent influxes of people swamped the job markets and half of the

working-age population is unemployed. The lowland Petén area in northern Guatemala is forested and less populated than parts of the country to the south. Although dramatic archaeological relics of the Maya civilization attract tourists (1.3 million in 2005) and scholars from all global regions, concerns over kidnappings and crime challenge tourist developer's abilities to attract larger numbers of foreign visitors. Although tourism brings some revenue to the Petén, much of the forest resources are being cleared, which could ultimately hurt the tourism industry.

Honduras is one of the most materially poor countries in the Western Hemisphere. Barely 25 percent of its land can be used for farming. The ranching economy established in colonial times persists in parts of the west, and large banana and pineapple plantations dominate the Ulua River valley and the northern coastal plain. Coffee became a primary export in the

1960s. Shrimp farming on the Pacific Ocean coast is a new export specialty that is finding markets in the United States. Tourism is a growing industry, attracting almost 749,000 visitors in 2005 (up from just over 500,000 in 1995), centered on Maya ruins in Copán, nature tourism-related mountain hikes and river rafting, and a relatively thriving scuba diving and beach attraction along the north coast and in the Bay Islands. Much of the former forest cover was logged, and even the secondary growth of pines is being removed rapidly. Manufacturing was encouraged after 1950, and a growing maquiladora industrial belt is forming around the city of San Pedro Sula.

Costa Rica has a more prosperous and diversified farming industry than the other countries, based mainly on coffee production along the Pacific slopes, bananas and cacao on the eastern coast, and a range of new crops such as tropical fruits, ornamental plants, cut flowers, and tropical nuts. Areas of former rain forest are used for livestock production. Costa Rica has many successful industries, including Microsoft facilities, textiles and pharmaceuticals for export and cement, tires, and car assembly for the domestic market.

Caribbean Tourism and Other Economic Activity

Caribbean Basin tourist destinations cater to a range of visitor interests from the elite to the mass markets. The gleaming beaches and turquoise water of the region are its primary attraction (Figure 9.16). The islands of the Caribbean also provide

FIGURE 9.16   St. Thomas, Virgin Islands. While tourists enjoy the sea, sand, and sun of the Caribbean, much of the revenue generated from their vacations may not reach the local economy or community.

unique historical and cultural experiences, and a wide range in topography from the flat Cayman Islands to the rugged hills of Jamaica and Dominica. The United States now provides two-thirds of the tourists, while most of the rest come from Canada, Europe, and Japan. Puerto Rico, the Bahamas, and the Dominican Republic attract the greatest number of visitors, while the Virgin Islands, Antigua, Guadeloupe, and Martinique continue to grow in significance.

Initial tourism development consisted largely of foreign-owned and -managed resorts, many of which were all-inclusive. Foreign control of tourist infrastructure resulted in severe rates of **economic leakage** for many areas in which the revenue generated flows to the foreign owners/investors and foreign employees of tourism infrastructure (hotels, restaurants, gift shops, bars, and dive shops) and thus "leaks" from the local economy. Although many foreign-owned all-inclusive resorts remain in the Caribbean, some governments are increasingly taking control of tourism infrastructure and marketing and are attempting to involve local communities in tourist infrastructure investment and management.

The number of cruise ship passengers visiting Caribbean ports continues to rise annually. Although a cruise ship may unload thousands of money-spending tourists into a port town, many of the cruise visitors participate in prepackaged events where revenues generated go to the cruise lines and only a couple of local businesses fortunate enough to be working with the ship's activity planners. Island residents view resort and cruise tourism with mixed emotions. Some see the industry as economically beneficial, providing jobs, housing, and schools, while others do not reap any direct economic benefits and resent the crowds, crime, resource consumption, and pollution that may accompany tourism growth in many areas.

Tourism is an economic activity that is extremely susceptible to external forces. Economic recessions in the wealthier countries from which visitors originate, destination-based crime or terrorism, or media reports of natural disasters such as volcanic eruptions or hurricane damage all may produce immediate reactions in the form of dramatic declines in tourist visits. Although tourism brings in foreign currency, much of it is used to buy the foods and equipment expected by the tourists.

Agriculture is another major industry in the Caribbean Basin. Commercial cultivation of sugarcane (and corresponding rum industries) and bananas are among the region's dominant crops. Cattle ranching and coffee cultivation persist in upland areas of islands with topographic relief.

Mining is prominent in the Dominican Republic and Jamaica, as is manufacturing in Puerto Rico. A booming oil based economy exists in Trinidad and Tobago. Natural gas and oil production, together with the refining of oil from Southwest Asia, is a major sector of Trinidad's industrial base. In the mid-1990s oil and natural gas output rose by 10 percent. Manufacturing industries include petrochemicals and steel, and employ 15 percent of the labor force, while government and other service jobs provide 50 percent of employment. Tourism is not yet a rapidly developing industry.

Andean South America

All five of the countries of the Northern Andes subregion suffer from **export-led underdevelopment.** In the late 1800s and early 1900s, the countries of this subregion relied on a series of export-based booms as specific mining and agricultural products came into demand in Europe and North America. Local people and foreigners invested in these products, while the government taxed the exports as the basis of its own income. Any hard currency earned went to pay for imports, on which tariffs were not charged.

Venezuela and Colombia have the highest incomes in this subregion (see Table 9.1). Venezuela's income is directly related to global demand for its oil and mineral exports. Debts accrued during the expansion of oil production in the late 1970s restricted development in the 1980s, when oil prices declined. Colombia's income strength in the subregion relates to its diverse mix of economic products, including mineral extraction, agriculture, manufacturing, and the illegal drug trade. Dramatic economic growth in Colombia is restricted by political instability and civil strife. Ecuador suffers from a small home market. Peru's economy experienced a series of booms and busts as it developed different export products. Bolivia's mining output of tin and silver is less in demand on world markets than it used to be. The country lacks an ocean outlet since a war over phosphate deposits in the 1800s resulted in Chile taking over its route to the sea at Arica. Bolivia has South America's second-largest natural gas reserves (Venezuela has the continent's largest natural gas reserves). Natural gas is a highly contentious and politically charged natural resource in Bolivia, where elections are often determined by the candidates' positions regarding the natural gas industry.

Peru dramatically opened its economy to external investment in the 1990s, creating a boom in mining exploration. In 1992, Peru opened Latin America's largest gold mine at Yanacocha near Cajamarca. It began production at other mines by the late 1990s and continued to explore opportunities. Peru's rapid economic growth in the mid-1990s was later slowed by the combination of expanding demand for imports, increasing indebtedness, and continuing poverty among a large sector of the population. The country experienced a recession pattern similar to that which occurred in Mexico. Peru is among the world's largest producers of cocaine. Revenue generated from cocaine may cushion some of Peru's economic challenges, but distribution of wealth related to the drug trade is extremely uneven.

Exports of silver and tin dominated the Bolivian economy for decades. A 1980s fall in world tin prices hurt Bolivia's limited economy. The Bolivian economy today depends on other metallic ores, such as gold, which comprise up to 50 percent of Bolivia's exports. New development is occurring in the eastern interior plains where nutrient-rich soils and improved transportation make it possible to move more soybeans from field to consumer each year. The crops are trucked across the Andes to Peruvian ports for export. Oil from the eastern Andean slopes makes Bolivia self-sufficient in energy. Bolivia is the world's second-largest producer of cocaine. United States pressure on the Bolivian government to crack down on production in the Chaparé Valley east of Cochabamba has met little success due to internal political disagreement and an economically challenged population. Bolivia's 2006 nationalization of its oil and gas industries intensified preexisting investor fears, which may further challenge Bolivia's ability to attract foreign capital.

Ecuador is a small country with limited upland and interior economic potential. Economic growth occurs primarily on the coast around Guayaquil, where the expansion of fishing for tuna and white fish, together with shrimp farming, is adding to an increasingly diversified base of light manufacturing, commercial farming for sugarcane, bananas, coffee, cacao, rice, and tourism. The construction of an oil pipeline in the 1970s from the Andes to the port of Esmeraldas increased overall oil production. Petroleum products make up nearly half of Ecuador's exports. New laws opened mining prospects for foreign corporations from Canada, South Africa, France, and Belgium that are testing the viability of gold, silver, lead, zinc, and copper deposits.

Global attention remains focused on Ecuador's Galápagos Islands, where unique animal and plant species attract scientists and tourists from all parts of the world (Figure 9.17). The government of Ecuador is caught between the desire to fully exploit the tourism potential of the islands and the need to preserve a world-class natural heritage site, as well as to maintain the conditions that make Galápagos an attractive tourist destination. Visitor quotas, established by the government to protect the ecological balance of the islands, have been increased several times and are often exceeded. The economic potential of the islands could easily be destroyed if too many visitors degrade the health of the animals and natural resources that make the Galápagos such a desirable destination. An even greater challenge to the Galápagos environment may be taking shape as Ecuadorian migrants move from the mainland coast to the islands in search of tourism jobs or to illegally fish in the marine reserve's waters. Uncontrolled population growth and the fishing of protected resources may pose a more destructive and rapid threat to the fragile balance of the Galápagos ecosystem.

FIGURE 9.17 Galápagos. The volcanic landscape of the Galápagos Islands.

The region around metropolitan Caracas is the focus of Venezuela's manufacturing and service industries. Southern Venezuela is a source of mineral wealth from iron and bauxite mines. The mines provide a basis for a manufacturing zone in Ciudad Guyana on the Orinoco River. The manufacturing zone is powered by hydroelectricity that is generated in the Guiana Highland valleys. This region has potential for future economic development but remains isolated by poor transportation facilities and is affected by Venezuela's shortage of investment capital.

Colombia has a more diversified economy than the other Northern Andean countries. Medellín, and the capital, Bogotá, became Colombia's primary urban industrial centers for textiles, steel, agricultural equipment, and domestic goods. In the lower valleys and near Colombia's northern coast, large plantations raise tropical cash crops, including cotton, sugarcane, bananas, cacao, and rice for export. Colombia is the world's second-largest producer of coffee, after Brazil. Coffee, Colombia's largest legal export, thrives in the optimum conditions of well-drained mountain slopes, fertile soil, and sufficient rainfall. The country began to grow new varieties of corn and rice in the mid-1990s, which were bred for higher yields in the poor soils of the eastern savannas. These crops, coupled with livestock and locally grown cattle feed, are meeting with success in this sparsely settled area. Columbia is the world's largest producer of Cocaine (see "The Northern Andes and the International Drug Trade," p. 300). Colombia is also a mining country with considerable iron, coal, oil, natural gas, gold, and emerald resources. Overseas trade, however, is regularly hampered by continuous internal political strife.

Brazil

The Brazilian government continues to encourage agricultural production, but agriculture's proportion of GNI fell as that of manufactured goods rose. Coffee and sugarcane remain important exports, but coffee, once the mainstay of the economy, now makes up less than 10 percent of agricultural exports. New developments, such as the growing of oranges and other citrus in the south and tropical fruits and nuts farther north, added to the diversity of commercial farming. The efficient commercial production of soybeans rapidly positioned Brazil as the world's top soybean producer (Figure 9.18). The largest manufacturing sector in Brazil is the production of automobiles and trucks, concentrated in the southern São Paulo suburbs. In the 1990s, the lowering of import tariffs led to further competition, falling prices, and expansion of Brazilian car sales and production, especially of small cars.

The Brazilian government owns many large corporations that employ thousands of citizens. Electrobras is a nationalized corporation that produces and distributes electricity. Small hydroelectricity projects on the plateaus in the east gave way to huge projects in the interior. The world's second-largest hydroelectricity project (China's Three Gorges Dam is the largest) is at the Itaipu Dam on the Paraná River at the border with Paraguay (Figure 9.19). The dam, which opened in 1983 under

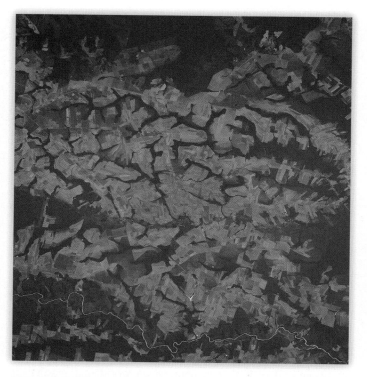

FIGURE 9.18 **Brazil: western Mato Grosso.** A space shuttle photo of new cattle ranches and soybean farms in a 60 km² (40 mi²) area. The plateau surface has been cleared and the steep, floodable intervening valleys left in forest. Lands along the Rio Sangue (bottom) are flooded up to 10 m (40 ft.) deep for three summer months.

FIGURE 9.19 **Brazil and Paraguay: hydroelectricity.** The spillway of the Itaipu hydroelectricity station on the Paraná River between Brazil and Paraguay. Eighteen massive turbines make this Brazil's largest hydroelectric investment. Although the electricity is shared between the two countries, Paraguay sells nearly all of its quota to Brazil, where there is greater demand.

the joint ownership of the governments of Brazil and Paraguay, produces more than one-fifth of Brazil's power and greater than three-fourths of Paraguay's.

Economic Activity in Southern South America

Agricultural products such as apples, soft fruits, grapes, and wine, together with fish, lumber, and new minerals, are growing components of the Chilean economy. The agricultural products and lumber come from central Chile, where a large proportion of the population lives. Copper mining is important in the Atacama Desert of northern Chile, with the world's two largest mines at Chuquicamata and Escondido along with a series of new large mines being prospected and developed in the late 1990s. After the highest-grade copper deposits were mined out by the early 1900s, Chile used U.S. technology to mine lower-grade ores. Its companies now lead the world in refining technology.

Argentina has a larger manufacturing base than Chile. The upturn in world economic activity and economic restructuring in the 1990s resulted in more investment, mainly from foreign sources, in Argentine manufacturing. Argentina became the fastest-growing economy in Latin America as it established what appeared to be a stable financial system. Most of the Argentine factories are based in and around the Buenos Aires metropolitan area. The farm products of the pampas are still important to Argentine trade, but oil and gas resources fuel new industries, mainly on the coast. Argentine service industries developed as part of this growth. In 2005, 3.9 million foreign tourists visited centers on the coast and in the mountains. Some older settlements such as the northern cities of Tucumán, Córdoba, and Mendoza, where the economy is still based on irrigation agriculture, do not grow as rapidly because of the dominance of Buenos Aires in attracting manufacturing investment. Some inland centers, such as Jujuy near the Bolivian border, remain poor and overwhelmed by Bolivian migrants. In the late 1990s, the Mexican crisis and fears of open competition with Brazil led to a slowing of growth in Argentina, which contributed to Argentina's economic crisis of 2001 and subsequent currency devaluation in 2002. Although the country's economy rebounded and grew strongly from 2002, a dramatic increase in federal spending and higher domestic prices for goods due to higher foreign demand of Argentine products, combined to produce inflationary increases in Argentina each year since 2002.

Geographic Diversity

Mexico

Mexico, which today dominates Middle America in terms of aerial extent, population, and economic size, gained its independence from Spain in 1821 through a series of rebellions (Figure 9.20). Political instability resulted in numerous power struggles and dozens of government turnovers in the decades after the end of Spanish rule. Stability came in 1871 with the harshly repressive dictatorship of Porfirio Díaz at the cost of many social freedoms. Oppression during this period, referred to as the Porfirioto, planted the seeds for revolution that would shape Mexico's political culture for most of the 1900s. Revolutionaries who overthrew the Porfirian government in 1911 created a new land tenure system with the purpose of easing rural poverty. The plots of land created in 1917 became known as *ejidos,* which were state-owned rural cooperatives developed to provide landless peasants with a degree of control over the land they farmed. The government granted the right to use land to individual *ejido* farmers, or *ejiditarios,* as well as the theoretical right to pass land use to their offspring. However, ownership of the land remained with the government. The *ejido* system resulted in the fragmentation of landholdings as the government took control away from some landowners and gave it to *ejiditarios.* In many cases, the government placed already impoverished farmers on marginal land, exacerbating their situation. Although rights to individual property ownership were defined more clearly in the 1990s, many rural, materially poor farmers continued to struggle while large-scale agribusiness took control of more land.

Another long-lived institution arising from the Mexican revolutionary period (1910–1920) is the Institutional Revolutionary Party (*Partido Revolucionario Institucional,* PRI). The PRI grew out of the revolutionary movement into the most powerful and lasting political party in the history of Mexico, dominating Mexican politics for more than 70 years. The party controlled the government, nationalized industries, and allegedly fixed many political votes to remain in power.

Serious debt problems in the 1980s contributed to a reversal of extreme protectionist government policies concerning Mexican industries, leading to policies permitting foreign competitors into Mexican markets. The Mexican government significantly reduced tariffs and trade restrictions, cut inflation, attempted to reduce deficit spending, privatized telecommunications, banking, and agriculture, and reduced a wide range of central government controls. Mexico signed the North American Free Trade Agreement (NAFTA) with Canada and the United States in 1992, and after passage in the respective legislatures, the pact was formally implemented in 1994.

In July 2000, a historic election overturned the PRI's dominance of Mexican politics with the election of Vicente Fox, who represented the National Action Party (*Partido de Acción Nacional,* PAN). Fox and the PAN inherited a vastly corrupt system supported in part by graft and money from illegal drug smuggling to the United States. Mexico's voters went to the polls in July 2006 to elect a replacement for President Fox as he neared the end of his second and final term (ending December 1, 2006). The election results were extremely close (within 0.6 percent) and forced a recount between the PAN, which claimed victory, and the leftist Democratic Revolution Party (*Partido de la Revolucíon Democrática,* PRD). On December 1, 2006, PAN candidate Felipe Calderon assumed the presidency of Mexico.

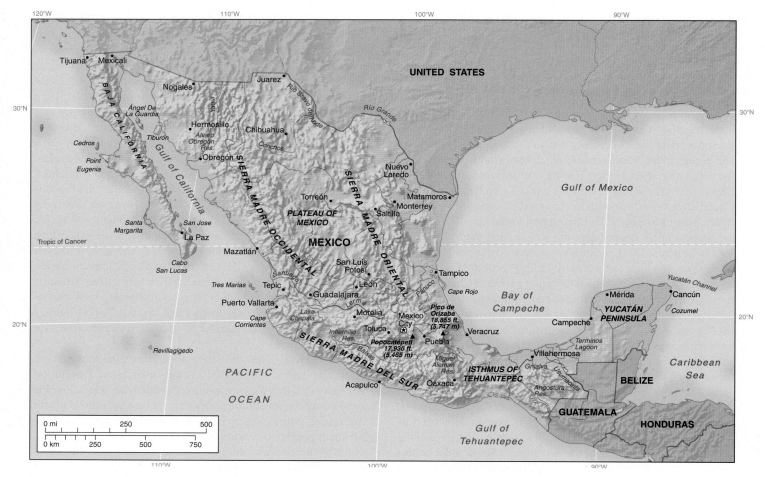

FIGURE 9.20 Mexico: a diversity of regions. Northward, the climate is drier, but economic opportunities increase close to the U.S. border. Southward, the land narrows and localities become increasingly remote from the influence of the Central Valley of Mexico.

Southern Mexico is less developed than northern areas of the country and more reminiscent of the countries of Central America. Most of the region is remote and populated by Native American subsistence farmers who grow corn on the hillsides and some wheat on the valley floors. Overgrazing by sheep is common. Poor transport facilities slow development in this region. Contrasting with this poverty, the government-planned tourist resorts such as Acapulco bring tourists to the coast, and new highway links connect the resorts to airports.

Indigenous identity is strongest in the Mexican states of Oaxaca, Yucatán, and Chiapas. Rapid tourism development in parts of the Yucatán Peninsula is dramatically altering the human geography for some while furthering a feeling of isolation for others (Figure 9.21). Over a million Native Americans in Chiapas do not speak Spanish. Local controversy surrounds government-sponsored oil and gas operations, power generation, and the construction of a road connecting Chiapas to Guatemala. Many members of indigenous communities assert a lack of local involvement in decisions made concerning land and resources in their state. Indigenous communities argue any financial benefit bypasses the residents of Chiapas while it

reaches more influential Mexican citizens in the Central Valley. Chiapas also houses thousands of refugees who fled earlier civil disturbances in Guatemala and El Salvador. The combination of dispossessed refugees and local disillusionment with the Mexican treatment of these states contributed to the Zapatista rebellion and regional unrest.

The 1994 challenge mounted by the Zapatista National Liberation Army in Chiapas drew attention to the social costs of economic reform. The Zapatistas made the point that the Native Americans, who comprise one-third of the Chiapas population and were already very materially poor, were further disadvantaged by the economic restructuring and NAFTA. The Native Americans in Chiapas are only a part of the 13.5 million Mexicans who live in "extreme poverty" and a further 23.6 million who are "poor." In the mid-2000s, the Chiapas situation remained unresolved. Reforms promised in 1996 bought time for the government but were not implemented. In response, Zapatistas set up their own autonomous municipalities in defiance of government orders. The Mexican government maintains a large military presence in the region but does little to stimulate dialogue.

FIGURE 9.21 Mexico: Cancún resort development. Tourism growth supports the diffusion of resort hotel development along the Caribbean coast of the Yucatán Peninsula, while economic conditions worsen in nearby interior villages. This area suffered great damage in the 2005 hurricane season.

Maquila

Towns along the U.S. border, from Tijuana in the west to Matamoros in the east, experienced very rapid growth. The Mexican government enacted the Border Industrial Program (BIP) in the mid-1960s to stimulate growth in northern Mexico and to relieve growth and population pressures from Mexico City and the surrounding Central Valley of Mexico. The BIP, more commonly referred to as the *maquiladora* program, made it possible for foreign-owned factories (*maquila*) situated in Mexico to import components for assembly without customs duties. Such factories utilized significantly cheaper Mexican labor and took advantage of Mexico's less stringent labor and environmental regulations, to assemble goods that could be exported, duty free, across the Mexican border and back into the producing country (most often the United States). U.S. corporations involved in textiles, apparel, electronics, and wood products built their factories in these towns, which grew rapidly through the 1990s. Asian and European corporations joined U.S. facilities in producing under *maquila* laws in the northern zone.

The concentration of economic activity along the border brought increased air and water pollution. Medical research continues to indicate a strong correlation between the industrial expansion of the maquiladora region and increased rates of cancer and other diseases among the people on both sides of the Mexico-U.S. border. Both Mexico and the United States are implementing programs to reduce industrial pollution in the region.

Central America

Central America is a land bridge, or **isthmus,** that narrows in width from the Mexican border in the northwest to the Colombian boundary in the southeast (Figure 9.22). The Central American isthmus provides a narrow physical boundary separating marine environments of the Caribbean Sea/Atlantic basin to the east and the Pacific Ocean to the west. Seven independent countries occupy the land of Central America. Belize was a British colony from the late 1800s until 1981 and has a relatively high percentage of people who trace their ancestry to the African continent. Its mainland location, increasing use of the Spanish language, and Roman Catholic populace link Belize to Central America rather than to the Caribbean Basin.

Regional independence from Spain came with a brief union between contemporary Mexico, El Salvador, Guatemala, Honduras, Nicaragua, and Costa Rica as the Mexican Empire from 1821 to 1823. In 1823, the incipient Central American countries separated from Mexico as the United Provinces of Central America, which lasted until 1838 when they each became independent. Panama existed as a province of Colombia until a U.S.-brokered canal deal created a separate country in 1903. Belize remained within the British sphere until 1981. Since 1838, unity has not figured prominently in the interactions among the countries of the subregion.

The four largest countries—Guatemala, El Salvador, Honduras, and Nicaragua—had 2006 populations ranging from Nicaragua's nearly 6 million to Guatemala's more than 13 million (see

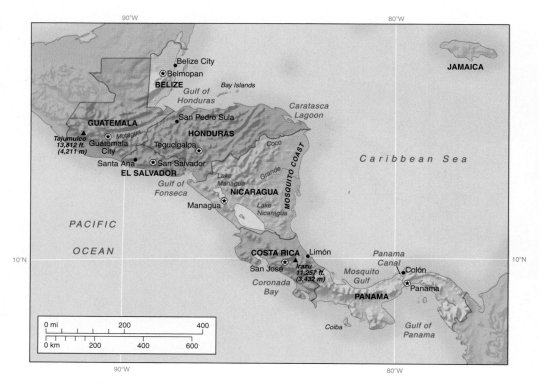

FIGURE 9.22   Central America: major geographic features, countries, and cities.

Table 9.1). Coffee plays an important role in the cash crop export-oriented economies of each. Civil unrest and natural disasters have wreaked havoc on the social fabric, physical infrastructure, and economies of all four countries. Deadly conflict, stemming in part from dramatic inequalities in land ownership in Guatemala, El Salvador, and Nicaragua, focused attention on disparity issues but did little to bring about an egalitarian solution.

Countries

Although El Salvador experiences frequent earthquakes, land reform conflict in the 1980s proved far more costly to the country than any natural disaster. Civil war erupted in when landowners of vast coffee estates challenged changes imposed by a reformist government. Thousands of people migrated in an attempt to escape the bloodshed. Such large numbers of war refugees moved to the capital, San Salvador, that the economy and infrastructure of the city virtually collapsed. Thousands of Salvadorans fled the country during the civil war and took up residence in the United States. These foreign-based communities send regular payments to relatives remaining in El Salvador. Such remittances are currently El Salvador's largest source of foreign revenue.

Nicaragua's population is concentrated around the lakes of Nicaragua and Managua in the structural depression that is subject to earthquakes. Other parts of the country have poor transportation links to this region. Decades of oppression under the Somoza family dictatorship in Nicaragua instigated revolutionaries to overthrow Somoza control. The Marxist-oriented Sandinista

revolutionaries took control in 1979 and inflicted another form of oppression and corruption on the people of Nicaragua for the next 11 years. The United States imposed a trade embargo on the Sandinista government, which added to the country's economic disruption and already impoverished social conditions. The United States also backed armies (Contras), partially comprised of former Somoza military personnel, who were fighting against the Sandinistas. The Sandinista party was eventually voted out of office in Nicaragua in 1990. The U.S.-imposed trade embargo devastated the economy, giving it the lowest GNI in the subregion in the late 1990s and undercutting Honduras as the poorest country in Central America.

Civil unrest in El Salvador, Guatemala, and Nicaragua eased considerably in the 1990s. The three countries began the slow process of repairing the physical and social damage inflicted by years of conflict. It remains difficult for the governments of these three countries to attract external investment because of their recent violent histories, and tourists are slow to return and spend their money.

Belize, Costa Rica, and Panama have the smallest populations of the Central American countries. Belize is by far the smallest country in the region, having only 300,000 citizens. The country became independent in 1981 after a long period of British colonial rule. Although years of British tenure established an English language base and some cultural elements different from those of neighboring countries, the Hispanic influence of the region exerts a strong force on Belize and is causing the development of a more regionally uniform human geography. Many Guatemalans assert a historic right to the lands and waters of Belize.

Costa Rica, with 4.3 million people in 2006, is the only country in the region that has had a long-term democratic government. This may be related to the legacy of its initial colonial settlement, when there were few Native Americans to be dominated and the people of European origin took up medium-sized farms. As a stable democracy, Costa Rica attracted manufacturing and free trade zones financed by U.S. and Taiwanese corporations. Costa Rica has little civil strife. It does not have an army, but the militarized police and relatively good quality of life help to maintain internal peace. Costa Rica is one of the top tourist destinations in Latin America. Costa Rica's beaches, volcanoes, and culture help the country to attract more than a million tourists per year. An extensive system of national parks incorporates a variety of ecological habitats and microclimate zones (through altitudinal zonation) and protects a vast array of flora and fauna. In the 1990s and early 2000s, Costa Rica increasingly became home for U.S. retirees seeking to take advantage of its climate, stability, and low cost of living compared to the United States. In the early 2000s, there were more than one-half million U.S. retirees living in the country.

Panama became a separate country when the United States facilitated its independence from Colombia in 1903, in preparation for the construction of the Panama Canal. The newly independent country had an immediate economic and military dependence on the United States, creating a relationship in which Panama functioned more like a U.S. colony than as an independent country. The construction of the canal brought migrants from the Caribbean islands, Asia, and Southern Europe. The canal created a dramatic dichotomy in the human landscape for Panama, which exists to varying degrees today. The corridor along the canal, known as the Canal Zone, is relatively wealthy and developed in contrast to areas of Panama to the east and west. Panama City, the country's capital, retains a cosmopolitan population and an international role in trade and finance. The United States virtually ran the country until 1979 and then invaded in 1989 to protect strategic and economic interests. Control of the Panama Canal reverted entirely to the Panamanian government in 1999, although the United States reserved the right in its treaty with the government to protect the canal should the United States perceive it to be threatened. The United States attempts to maintain close ties to Panama in an effort to combat the northward progression of narcotics trafficking through the region.

In 1991, Taiwanese investors and Panamanian officials established a free trade zone in Panama's Caribbean port city of Colón (the Colón Free Trade Zone), in which some 8,000 people are employed in bulk retail and value-added manufacture of consumer goods. The free trade zone serves regional customers, such as buyers for stores in Colombia and Venezuela, as well as clients from other world regions such as Europe. The free trade zone has enjoyed considerable growth since its inception, and is currently undergoing a major expansion. To the east of the Canal Zone, the Central American Highway tapers off in the impenetrable Darién, an area of extremely dense vegetation between the canal and the Colombian border that prevents completion of effective north-south highway links. Many believe completion of the highway would dramatically increase the flow of illegal drugs northward from the Northern Andes region of South America.

The Panamanian government is currently undertaking the expansion of the Panama Canal in order to facilitate the traffic of ships, referred to as **post-Panamax,** that are too large to pass through the existing locks (Figure 9.23a and b). The expansion will create a new channel for part of the transit, and two new sets of locks (current transit requires passage through three sets of locks). Environmentalists are concerned that expansion of the canal and increases in ship traffic will deplete fresh water from the Chagres River watershed (Panama's primary freshwater source), while the Panama Canal Authority asserts the expansion design will conserve and recycle a significant amount of the fresh water needed to fill the locks for each ship's transit. The expansion project will create thousands of temporary jobs in Panama and likely result in increased immigration of workers from Central American and Caribbean countries. Expansion-related immigration will create new human geographies in Panama.

FIGURE 9.23 **Panamax vessels transit the locks of the Panama Canal.** The ships pictured transiting the Pedro Miguel (a) and Miraflores (b) locks of the Panama Canal are known as Panamax vessels, which indicates they are the maximum size of vessel that will fit through the canal's locks. The world's largest retailers, such as The Home Depot, utilize vessels that are currently too big to transit the Panama Canal (known as post-Panamax vessels). The government of Panama won the approval of the country's citizens in the fall of 2006 and is moving forward with a canal expansion project that will enable post-Panamax vessels to transit the canal.

(a)

(b)

The Caribbean Basin and Environs

The contemporary human and physical geography of the Caribbean Basin reflects dramatic transformations imposed on the region by European countries during the colonial era. The loss of indigenous people to disease, the forced migration of Africans to the basin, and the establishment of extensive sugar cultivation on the islands laid the foundation for the social, environmental, and economic conditions present in the Caribbean today. The Caribbean Basin consists of a few large islands, several small islands, and three political units on the South American mainland (Figure 9.24). In 2006, more than 35 million people lived in the Caribbean Basin. The political geography of the Caribbean is one of the most diverse of any subregion. Caribbean countries vary in size and population from Cuba, with more than 11 million people in 2006, to numerous tiny islands with few inhabitants (see Table 9.1). There is a wide range in the basin's economic diversity from poverty-stricken Haiti to the more affluent Bahamas. The majority of the Caribbean's residents live on one of the four largest islands, known as the Greater Antilles: Cuba, Hispaniola (consisting of two countries: Haiti and the Dominican Republic), Puerto Rico, and Jamaica. The smaller Caribbean islands are commonly referred to as the Lesser Antilles. The Lesser Antilles chain is an arc of small islands around the eastern edge of the Caribbean Sea, with the Leeward Islands in the north (Virgin Islands to Guadeloupe) and the Windward Islands in the south (Dominica to Grenada). Many of the small island countries of the Caribbean have only a few thousand residents.

North of the Caribbean are the Bahamas, which include some 700 islands situated to the east of Florida. Another group of islands, including Trinidad and Tobago, lies along the northern shore of South America. Also included in this subregion are the three Guianas—Guyana, Suriname, and French Guiana. Although situated on the South American mainland, their histories, people, and present economies are more connected to

FIGURE 9.24   The Caribbean Basin. Numerous island countries and some European and U.S. territories.

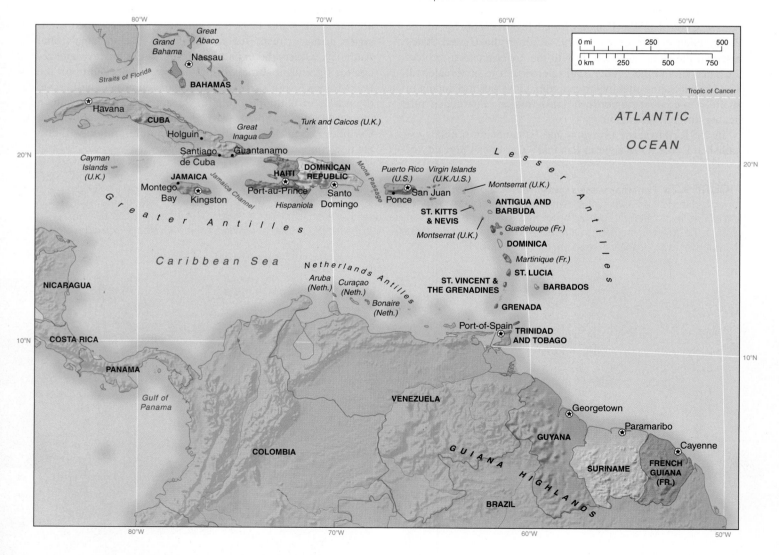

those of the Caribbean Basin than to those of other mainland South American countries.

National Identity

The people of the Caribbean have a very strong sense of belonging to the individual island or country of their birth, as opposed to strong feelings of regional unity or identity. Although the majority of the Caribbean residents are of African ancestral heritage, most people identify with their fellow citizens and perceive readily apparent differences between their nationality and those of neighboring countries. A Jamaican would identify with other Jamaicans, and see their national people as different from those of Barbados, rather than perceive any unity from ancestral heritage. National pride is often expressed through rivalries with neighboring island countries. For example, sporting events such as cricket and soccer among the people of Jamaica, Barbados, Trinidad, and Guyana are very competitive, prestigious, and emotional affairs.

Colonial Farming Heritage

The Caribbean colonies of Spain, France, Britain, and the Netherlands assumed a huge significance in the global economy when the Caribbean Basin took over the leading sugarcane production role from Mediterranean countries in the early 1600s. From 1500 to the 1800s, some 10 million African slaves were shipped to the Americas, of which half went to the Caribbean Basin, 39 percent to Brazil, and under 5 percent to British American colonies (including the United States). African peoples brought their own social customs, religious beliefs, handicraft skills, and various forms of artistic expression. At the height of sugar cultivation prosperity in the 1700s, the average plantation unit was 200 acres, used 200 slaves, and produced 200 tons of sugar each year.

Island and Country Distinctions

Puerto Rico is not an independent country nor is it a state within the U.S. political system. An official relationship between Puerto Rico and the United States began in the early 1900s and became more formally structured in 1952, with the establishment of commonwealth status. Puerto Ricans are considered to be U.S. citizens, yet they do not have true representation in Washington, D.C. Puerto Rico's representatives in the U.S. Congress do not have the right to vote on the passage of bills. Puerto Ricans have open access to migration into the United States, and many have taken advantage of this since the 1950s. U.S. statehood remains an issue, with distinct political views within the commonwealth regarding Puerto Rico's status. The smallest group would like Puerto Rico to become an independent country, free of any U.S. authority, another group supports statehood, and the largest group currently is content to maintain commonwealth status. Although the Puerto Rican government declared Spanish to be the official language in 1991, English

as a second language is mandatory in the Puerto Rican public school system.

Puerto Rico has a much higher per capita GNI than other Caribbean islands. Although its sugar industry expanded, its economy remained narrow until after World War II, when farming shifted to dairying and manufacturing increased rapidly. Machinery, metal products, chemicals, pharmaceuticals, oil refining, rubber, plastics, and garments provide a diverse product mix and have grown under special U.S. tax laws for the island. Tourism became a leading economic sector for Puerto Rico with 3.7 million visitors in 2005.

Haiti, occupying the smaller western part of the island of Hispaniola, is the poorest country in the Americas. Life expectancy is low, infant mortality high, and 50 percent of its adults are illiterate. Over three-fourths of the people are crowded onto poor-quality lands, the result of an imbalanced division of land resource holdings. Independence in 1804 was followed by the political instability that continues today. The sugar plantations deteriorated, and economic stagnation and decline set in. The United States occupied the country from 1915–1934 and again in 1994 to end military takeovers of the government. In between these dates, corrupt and repressive governments produced little economic development, leaving most people dependent on their subsistence plots of land. Some commercial farming for coffee and cacao produced export income, but a 1980 hurricane destroyed many trees and recovery was slow. In the 1970s, tax incentives attracted U.S. corporations to set up factories in Haiti. Exports of clothes, electronics, and sports equipment became more valuable than farm products. Haiti's mineral resources, however, remain undeveloped. Tourism is not developing in Haiti as it is in other Caribbean countries due to the extreme material poverty, political instability, and a high incidence of HIV/AIDS. Following the military coup against a democratically elected government in 1992, an international embargo on Haitian products caused the economy to collapse, including the loss of vital aid, default on its public debt, and closure of the new assembly industries. Exports and imports halved. A democratic government was reinstated in 1994 after U.S. intervention. Economic growth in Haiti in the early 2000s was further challenged by drug-traffic-related corruption.

The former French colonies of Martinique, Guadeloupe, and French Guiana are political subdivisions of the country of France, known as overseas departments. Their residents are French citizens and members of the European Union. The former Dutch colonies of Aruba, Bonaire, and Curaçao off the northern coast of Venezuela are low-lying and arid, obtaining their water supplies from desalination plants. They remain administratively linked to the Netherlands today. Although Aruba has had a separate status since 1986, it is moving back toward closer political ties with the Dutch government. The people are exceptionally cosmopolitan, including descendants of Africans, Indians, Dutch, Portuguese, Danes, and Jews. Residents are able to speak an old trading language, Papiamento, as well as Dutch (the language of their government), English (the language of tourist visitors), and Spanish (the language of influential neighbor Venezuela). Oil refineries linked to the Dutch development of Venezuelan oil, along with tourism and offshore banking, are among the main sources of income.

Guyana (former British Guiana), Suriname (former Dutch Guiana), and French Guiana all have small populations on relatively large areas of land. Dutch drainage engineers in the early 1800s made the coastal plains of the Guianas habitable. Indentured laborers were shipped in from South Asia to work on the vast sugar plantations near the coast. In both Guyana and Suriname, living standards declined after independence as the result of civil strife. Guyana and Suriname export bauxite from inland mines, while French Guiana exports timber to the EU. The reopening of gold mines by Canadian companies in Guyana caused exports to quadruple in the 1990s.

United States' corporate ownership of the sugar industry dominated Cuban economic geography in the early 1900s. When Fidel Castro led a Communist takeover in 1959, ties with the United States and U.S. companies were severed and dependency shifted to the former Soviet bloc countries. Large state farms and cooperative farms replaced confiscated private plantations. Although some farms diversified to produce citrus fruit for Eastern Europe, overall productivity remained relatively low. Manufacturing included import-substitution units for locally needed goods such as textiles, wood products, and chemicals. Cuba exported sugar, tobacco (cigars), and some strategic minerals to the Soviet bloc in exchange for oil, wheat, fertilizer, and equipment. During the 1980s, 85 percent of Cuba's trade was with the Soviet bloc, but this close link was broken with the 1991 dissolution of the Soviet Union. Cuba suffered as other Communist countries from the breakup of the Soviet bloc. From 1989 to 1994, Cuban GNI fell by 34 percent and its sugar crop fell from 8.4 million tons in 1990 to 3.4 million in 1995. Most farm workers found growing fruit and vegetables for the black market to be more profitable. The United States maintained its trade embargo through the early 2000s. Tourism, however, began to grow as a source of foreign currency, with 2.3 million visitors in 2005. Canadian and other non-U.S. mining companies took over Soviet-instigated mining projects, primarily for nickel, cobalt, and gold. Cuba formed close economic and political

ties in the early 2000s with oil and natural gas endowed countries such Venezuela and Bolivia, a move which provided some short-term economic relief from losses related to Soviet dissolution. After nearly five decades of rule, Fidel Castro suffered serious health challenges in the summer of 2006, and he turned over the reigns of government to his brother.

Northern Andes

The Andes Mountains are a dominant feature in all five countries of the Northern Andes subregion consisting of Bolivia, Colombia, Ecuador, Peru, and Venezuela (Figure 9.25). The world's second-highest mountain range creates a multitude of local environments at different heights throughout the subregion. The geologically active Andes are in the process of uplift, resulting in frequent earthquakes, landslides, and occasional

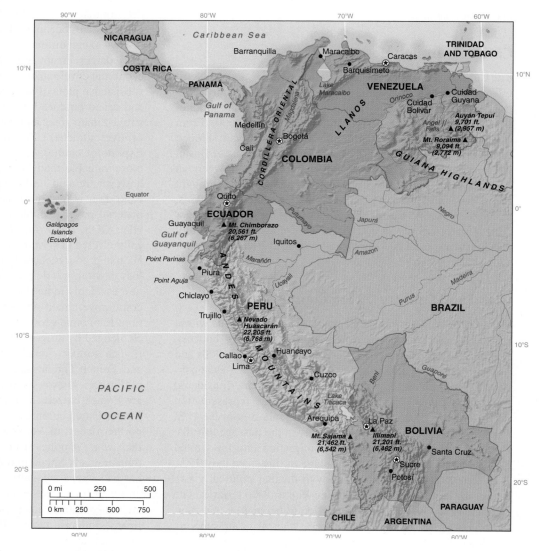

FIGURE 9.25 Northern Andes: main geographic features. Note the position of the Andes Mountain ranges in each country and how they isolate interior lowlands from the main centers of population and trade.

volcanic eruptions. The dramatic topography of the Andes Mountains isolates large interior sections of the Orinoco and Amazon River basins from global connections.

Mountain-dwelling culture groups exhibit centuries-old social customs in agricultural towns tucked into the valleys of the region's steep terrain. Cash from the illegal drug trade creates a unique elite with dramatically different lifestyles from those in rural mountain villages. Political stability is occasionally challenged by powerful criminal elements and intermittent border disputes, such as the disagreement between Ecuador and Peru over Amazonian territories.

Countries

Although rugged terrain and high relief are common features in each of the Northern Andes countries, topographic and environmental variety is present throughout. The high altitude of the Andes in the western part of the subregion contrasts with and isolates the lower lands to the east. Eastern areas are largely covered by tropical rain forest vegetation and experience high rainfall throughout most of the year. The interior reaches of Colombia, Venezuela, and Bolivia experience more seasonal rains and a tropical grassland vegetative cover. The most dramatic contrast of the subregion exists on the Pacific coast of Peru, where air circulation patterns, cold currents, and the rain shadow of the rugged Andes combine to produce one of the most arid climates in the world.

The informal and illegal economies of the countries of the Northern Andes, based on a globally significant role in the production of coca and cocaine, grew rapidly in the 1970s. Increasing demand from the wealthy U.S. market fostered dramatic growth and continues to support coca cultivation and cocaine production. Government efforts thus far have not been successful in eradicating coca production. The number of routes from Colombian processing centers through the Caribbean Basin to the United States is rapidly increasing. Coca-growing areas are diffusing from the eastern slopes of the Andes into Brazil.

Drug-related instability in Colombia worsened during the 1990s. Allegations of a political power base that is increasingly controlled by drug money are rampant. Clashes between some government officials who openly oppose the drug cartels and guerrilla groups that support narcotic

cultivation regions often result in kidnappings and executions on all sides. The United States infuses large amounts of money into fighting the Colombian drug war, yet marginal government relations between the two countries and corruption within the Colombian political system dilute the value of any external assistance efforts. Cease-fire efforts between the Revolutionary Armed Forces of Colombia, a formidable guerrilla group whose base is in the heart of a prime drug cultivation region, and the Colombian government crumbled in 2002, leaving many in the region and around the world wondering if full-scale combat would follow.

Brazil

Brazil is the largest country in both area and population in all of Latin America (see Table 9.1). Brazil has three times the area of Argentina, the second-largest South American country. In 2006, Brazil's population exceeded 186.8 million. Politically, Brazil is divided into states that are part of a federal government system (Figure 9.26). The people of Brazil have a strong

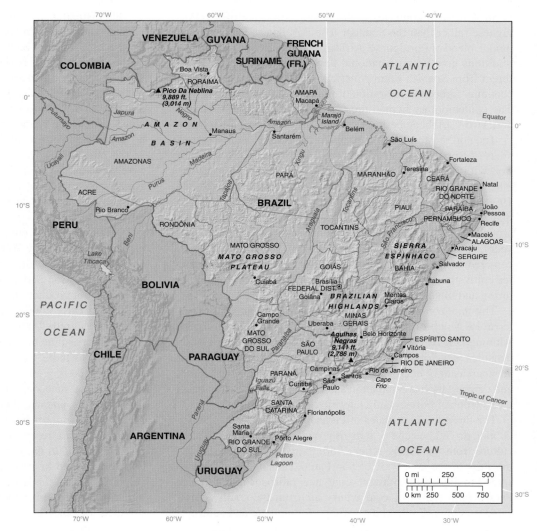

FIGURE 9.26 Brazil: major geographic features. The states, rivers, and major cities.

sense of national pride and a passion for cultural expression. Although the Portuguese language and Roman Catholic church provide some national homogeneity, the diverse contemporary human geography and cultural expression of Latin America's largest country are tangible extensions of the historical mixture of European, African, and indigenous influences. The physical expanse of Brazilian territory incorporates a vast array of topography, climate, and vegetative cover, and holds a diverse and abundant supply of natural resources. Brazil presents an immense economic market, has the greatest economic diversity of any Latin American country, and was among the world's ten largest economies in 2004. Despite decades of rapid economic growth between 1940 and 1980, and the size of its economy, Brazil faced increasing problems in the 1980s and 1990s that resulted in high inflation, declining economic growth, and growing numbers of materially poor people.

The establishment of sugar plantations along the northeast coast of Brazil during the colonial era began the course of European-induced human and physical geographic change experienced in so many locations in Latin America. What most distinguishes the colonial history of Brazil from that of other Latin American countries is the lasting influence of the Portuguese rather than the Spanish. The Portuguese language and unique blend of Portuguese, African, Caribbean, and indigenous traditions give Brazil a national flavor that is distinctively different from Spanish Latin America. The Portuguese plantation economy forced the migration of millions of Africans to Brazil. The contemporary human mosaic includes the largest African and Afro-Caribbean heritage of any Latin American country.

The northeastern coastlands and plateau were the first settled areas of Brazil. Sugarcane plantations prospered until competition with the Caribbean Basin began in the later 1600s. Periodic droughts devastate the interior of this region (as in the late 1990s), causing many to emigrate to other parts of Brazil. The main economic development in Brazil occurred in the southeast around Rio de Janeiro, the colonial and national capital for 200 years, and São Paulo. The early industrial focus of the region centered on mining and commercial agriculture (based on coffee cultivation). Today, this region is a major center of manufacturing and financial service activities for Brazil.

The Amazon River network drains 60 percent of Brazil's territory. Much of this area is covered by tropical rain forest. Although the Amazon River basin has relatively low population densities, development efforts led by the Brazilian government are increasingly encroaching on tropical forests of the region, and deforestation is rampant in some parts of the Amazon basin (see "Tropical Forests and Deforestation", p. 296).

The three southern states of Paraná, Santa Catarina, and Rio Grande do Sul became centers of growing population and agriculture from the 1930s. After 1950, the coffee crop exhausted the soils of the area west of São Paulo, causing cultivation to spread to the west-southwest along new railroads into northern Paraná until bursts of cold wintry air from the south demarcated its limits. Cattle and a variety of temperate and subtropical agricultural products, including oranges, are produced in these states, settled largely by immigrants who were encouraged to move from Germany, Italy, and Japan. Hydroelectricity generated on the Paraná River and its tributaries powers manufacturing in the area. Inland of São Paulo, straddling the drainage divide between the Amazon and Paraná-Paraguay River systems, is an area of rapid farming expansion known as the *Cerrado*. This vast stretch of savanna to the east and southeast of the Amazon basin became one of the world's main soybean producers, contributing more than 60 percent of Brazil's 60-million-ton soy harvest in 2006.

Modern Mining

The national iron ore company is developing the iron ore mining Carajás Project in the eastern Amazon River basin. A railroad built in 1985 takes the ore to a port near São Luis on the northern coast for export to Japan, the United States, and Europe, while the Tucurui Dam generates hydroelectricity for mining and industrial needs. The controversial project required the clearance of 3 million hectares (7.4 million acres) of rain forest, and much of that area is now used for ranching and small farms.

Other Brazilian mining developments include the production of manganese (used in hardening steel), tin, and bauxite. One-third of the world's bauxite resources occur east of Manaus in the Amazon River basin. In the upper reaches of the basin, small deposits of gold attract thousands of independent miners, but the use of mercury to separate the gold pollutes the rivers, and this activity forms a major intrusion in the lives of Amazon tribes such as the Yanomami.

Brazil was 90 percent dependent on foreign energy sources in the mid-1970s. The oil consumers' crisis of the 1970s severely challenged Brazil's economy and directly contributed to Brazil's incurrence of foreign debt that would last for several decades. An immediate response was the development of an ethanol fuel program (from sugarcane) in 1975. Brazil dramatically transformed its energy needs, gaining complete fuel energy independence by 2006 and becoming an energy exporter. The government steadily increased the acreage devoted to sugarcane cultivation and continued to invest in sugar mills and other ethanol production technologies and facilities during the 1990s and early 2000s. Seventy percent of new auto sales in Brazil in the mid-2000s were flex fuel cars that could function on regular gasoline or a variety of ethanol blend fuels. Exports of ethanol to India, South Korea, the United States, and other foreign markets grew steadily into the late 2000s. Petrobras, the Brazilian state oil company, was originally established to import, refine, and distribute oil products but now finds itself an oil producer with offshore wells along the eastern and northern coasts and major reserves in the western Amazon.

Connections to the global economy thrive in the **free trade zone** established in 1966 in the Amazon River city of Manaus (Figure 9.27). In this zone, foreign companies may import materials for assembly without tariffs and export the assembled products to other countries. Manaus is now Brazil's fastest-growing city and the second in manufactured goods value after

FIGURE 9.27 Brazil: Amazon River. The city of Manaus is the largest on the Amazon River. It has an opera house that was built in the early 1900s to cater to the 2,000 or so rubber barons who lived there and controlled the valuable rubber trade. Now refurbished, the opera house is in the center of a city of over 1 million people, with high tech industries in huge tariff-free areas and improved communications by river, road, and air.

gion. The Hispanic cultural imprint is a strong common element of the human geography of this subregion. The majority of the population of each country speaks Spanish as their first language and adheres to Roman Catholicism.

Southern South America has the highest percentage of European descendants of the subregions on the South American continent. Sparse indigenous populations and climatic environments more reminiscent of Europe led to the establishment of pervasive European-based populations in Argentina, Chile, and Uruguay. Argentina's population dominates the four countries of the subregion, with more than half of Southern South America's people living here in 2006.

The diverse physical and human geographic spatial patterns of Southern South America relate in part to the subregion's extensive latitudinal coverage from north to south, a relatively narrow width, the rugged Andes Mountains and geologic instability

São Paulo. The local human and physical landscapes are dramatically different from outlying forest areas. There are more than 6,000 factories in the huge industrial parks of Manaus, many of which are the production facilities of foreign-owned companies, including Honda, Sharp, Kodak, Olivetti, Toshiba, Sony, and 3M. Manaus was transformed by the number of visitors from elsewhere in Brazil coming to purchase goods that they were not allowed to import into Brazil. The "electronics bazaar" occupies a maze of streets in the city center. People come to buy foreign-made computers and electronic goods, products that Brazilian companies, protected by high tariffs, make poorly and sell at high prices.

Southern South America

Argentina, Chile, Paraguay, and Uruguay form the southernmost part of South America and are sometimes called the "Southern Cone" because of their combined shape on a map (Figure 9.28). Physically, this long and narrow subregion extends far south into the midlatitudes. Average temperatures are cool and rainfall amounts increase progressing from the north to the south in the subre-

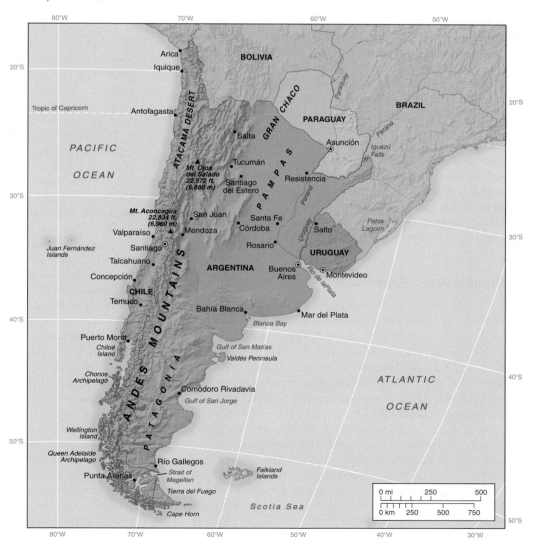

FIGURE 9.28 Southern South America: major geographic features. The countries, Andes Mountains, rivers, and major cities.

of the west, and the influences of the cold Pacific and warm Atlantic Ocean currents. All the physical properties of the region support a wide range of temperature and precipitation regimes, resulting in varied settlement and land use patterns. The physical geography of Southern South America coupled with the historic goals of Spain created a vibrant contemporary human mosaic in the subregion's countries. The Andes Mountains proved an important dividing factor between types of settlement.

In the 1990s, the Argentine economy was very robust and the citizens of the country were enjoying a relatively prosperous standard of living. A series of both local and global actions combined to devastate the economy in Argentina by the end of 2001 and bring the middle classes into the streets of Buenos Aires in protest. The combination of government policies, global trade relations, and government corruption necessitated devaluation of the Argentine currency. The Argentine economy now may be in an incipient stage of recovery, but any positive change is likely to be slow.

Paraguay remains a materially poor country. The larger portion of the country to the west of the Rio Paraguay, the mainly semiarid Gran Chaco, remains lightly settled apart from military camps and lumber operations. The quebracho ("ax-breaker") tree is its commercial timber product. The tree is very hard and in demand for railroad ties and its tannin extract.

The eastern part of Paraguay is more developed, with forest products being diversified by commercial agriculture. Livestock products and some industrial crops are exported. Paraguay gets its power from hydroelectricity and earns foreign exchange from its share of the Itaipu project on the border with Brazil by selling most of its power to Brazil. Transportation connections with Brazil have improved, and that is now the main direction of trade. Paraguayan border towns such as Ciudad del Este sell cheap consumer goods to Brazilians and are involved in smuggling between Brazil and Argentina.

Buenos Aires, Argentina

The rapid growth of Buenos Aires followed patterns that are common to other cities in Southern South America. After slow growth in its early history, Buenos Aires expanded in the late 1800s as commercial farming took hold in its pampas hinterland. The built environment developed a trading and industrial waterfront and a central thoroughfare at right angles along the route inland. Rapid growth occurred in both the economy and the immigrant population around 1900. This resulted in a middle class of skilled workers and office workers moving out to new suburbs, while the poor and most affluent remained in the inner city. Amenities came slowly to the suburbs. During the later 1900s, increasing rates of population growth produced squatter settlements and rising inner-city population densities as apartment blocks replaced mansions. Population growth stagnated in the coastal industrial areas. From the 1960s, much of the commercial and industrial activity moved out of central Buenos Aries to the city edges. Attempts were made to divert the overcrowding in the Buenos Aires metropolitan area by placing new projects and development in other centers farther inland from Buenos Aires, but they have not been successful.

Contemporary Geographic Issues

Tropical Forests and Deforestation

Deforestation, or the permanent clearing of forest vegetation, is a centuries-old land modification practice. Societies in all world regions and in both temperate and tropical latitudes used forest resources for fuel, shelter, and transportation. Dramatic increases in the permanent clearing of tropical forests, and especially tropical rain forests, in recent decades is causing significant alarm among a growing global body of scientists, medical researchers, government officials, and environmentalists. Global communities concerned with the potential ecological and human health impact from tropical forest clearing call on governments practicing or permitting tropical deforestation to cease. Locally, where deforestation is taking place, governments, business communities, and farmers assert their sovereign rights to resource use.

Tropical rain forests exist at latitudes where high temperatures and high levels of humidity year-round produce a fairly consistent precipitation and vegetation pattern each season. The three most significant locations of tropical rain forest in the world are in Southeast Asia (see Chapter 5), Central Africa (see Chapter 8), and the largest, the Amazon River basin in South America (Figure 9.29 a and b).

Tropical rain forest is the dominant vegetation and climate regime in the Amazon River basin of Brazil and adjacent parts of its neighboring countries, including eastern Colombia, Ecuador, Peru, Bolivia, southern Venezuela, and the Guianas. The tropical rain forest of the Amazon basin is the largest in the world. Although other locations in Latin America contain similar vegetation and climatic conditions, including the Pacific coast of Colombia, along the eastern coasts of the Central American countries, and down the northeast coast of Brazil, the Amazon dominates in size and global impact.

Tropical forests contain the highest plant and animal species diversity per unit of land of any ecosystem or biome in the world. Seventy percent of the world's known plant and animal species reside in tropical forests, with hundreds of distinct species existing in relatively concentrated areas. Contemporary global health care depends on the species diversity of tropical rain forests. Existing treatments and promising cures for various forms of cancer come from tropical forest species. Numerous human diseases and aging conditions may be treated, cured, or slowed through medicines derived from the tropical rain forest plants. Medical research communities assert the need to explore the unknown potential hidden in the diversity of tropical rain forest vegetation. International pharmaceutical research, development, and sales generate hundreds of millions in revenue from products related to tropical rain forest species. The potential benefit of hundreds to thousands of species present in the tropics has yet to be identified. Permanent clearing could eliminate countless medicinal cures and treatments yet undiscovered.

Trees absorb carbon dioxide from the atmosphere and return oxygen to the atmosphere. Tree respiration takes place in the troposphere, the lowest level of the Earth's atmosphere,

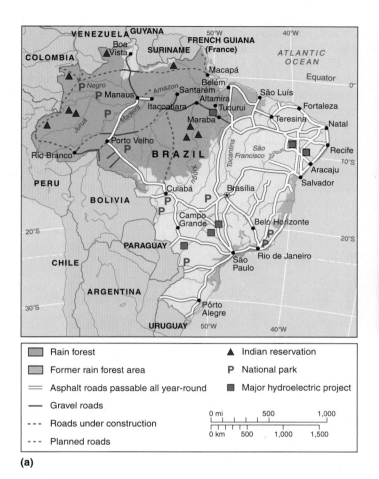

(a)

Rain forest

Former rain forest area

Asphalt roads passable all year-round

Gravel roads

Roads under construction

Planned roads

▲ Indian reservation

P National park

■ Major hydroelectric project

0 mi   500   1,000

0 km   500   1,000   1,500

(b)

FIGURE 9.29 **Brazil: Amazon rain forest.** (a) Map of the Amazon River basin in Brazil and the extent of tropical rain forest. (b) Aerial view of the tropical rain forest in Rondonia state, where the forest is cut into as farms are established along the roads.

where humans live and breathe. Removing huge tracts of tropical forest eliminates a primary source of oxygen and a mechanism for removing carbon dioxide at the same time. Burning forests, or drowning them under dammed waters, both common practices in Latin America, contributes to the release of carbon dioxide and methane, both considered to be culprits in global warming.

Tropical forests provide a number of local environmental stabilizers, holding moisture and releasing it into the local and downwind environment, and regulating stream flows and preventing downstream flooding. Tropical rain forests provide a source of water when global cycles bring drier periods to a region. Through the process of **evapotranspiration,** tropical rain forests can provide humidity and rainfall to the region they inhabit as well as far downwind. Tropical soils are not nutrient rich like many temperate soil regions. The vegetation in the tropical forests stores the life-sustaining nutrients within its green cover, so that clearing the forest removes the local nutrient base and diminishes the potential for future species diversity. The vegetation keeps the local soil intact. In its absence, soil will wash through the watershed and out through the drainage system. Trees on poorer soils maintain their existence by circulating the nutrients without letting them enter the soil: when leaves fall, insects and fungi soon break them down so that roots near and above the surface can capture the chemicals again.

Government debt, high population densities in the east, millions of materially impoverished and jobless people, farmers with no land to farm, and a wealth of unexploited natural resources prompted the Brazilian government to promote the development of the Amazon basin. The Brazilian government is hoping huge mining projects located in forested areas will bring in foreign capital and ease Brazil's massive foreign debt. The government claims sovereign rights to exploit its resources however it deems appropriate. Brazilian political leaders assert the need to develop resource potential fully in order to bring down an indebtedness structure that is so severe they are barely able to make the interest payments. Brazilian officials point to countries such as the United States, where unencumbered resource exploitation led to material wealth and prosperity. Many local and international environmental groups are calling for international reduction of Brazilian debt to help the country shift its development efforts away from the Amazon.

Critics argue that the development of large mining operations in the Amazon basin is leading to an influx of people to provide services. As supporting service and other industries grow, more people will move to the region. As roads are cut into the forest, farmers and ranchers settle along both sides. Poor farming practices coupled with the nature of the soils and vegetation often lead to rapid nutrient depletion, and the farms are either sold to large-scale enterprises or abandoned for new land. As new roads are built, new settlers move in, existing farmers move to more productive land, and the rain forest vegetation slowly disappears.

The government claims such projects produce jobs that employ people who would otherwise have no work or income. People working the mines and providing services to the developing regions are grateful for the opportunities presented. Some farmers claiming land in the region assert their pride in landownership and their ability to be self-employed. In places where better soils coincide with a good local knowledge and where farmers are not too indebted, there has been some success in growing commercial crops of beans and vegetables. In most parts, the cut areas gave way to cattle ranching, but few

cattle can be supported per hectare and the carrying capacity soon declines to uneconomic levels.

The future of the tropical rain forest in Latin America will be determined by the ability of local and global governments to understand the ecosystem better and cooperatively look for egalitarian uses that may serve the needs of countries like Brazil while sustaining the global ecological and human health value of the rain forest system. Table 9.3 provides a matrix with some of the positions taken by those urging tropical rain forest conservation and those who believe resource exploitation in the rain forest is in the best interest of the local and national communities.

TABLE 9.3   DEBATE: TROPICAL DEFORESTATION

Conserve Tropical Rain Forest	Use Tropical Rain Forest Resources
Tropical rain forest (TRF) resources provide a "sink" for carbon dioxide. Burning TRF vegetation adds carbon dioxide to the Earth's atmosphere. TRF areas are a source of oxygen in the lowest level of the Earth's atmosphere, where humans live and breathe.	There is incomplete carbon dioxide data for the Earth's atmosphere. Large portions of the Earth's surface are unreported. Ocean exchanges with the lowest levels of the Earth's atmosphere are more significant than TRF exchanges.
There is tremendous bio diversity in the plant life present in TRF ecosystems.	There is no conclusive evidence that TRF clearing will permanently change the biodiversity of the Earth as a whole.
Many medical treatments are derived from TRF products, and many disease cures come from TRF products, including current treatments and potential cures for cancer patients. Destruction may eliminate many undiscovered cures and treatments. TRF-derived pharmaceuticals earn billions internationally each year.	Medical treatment come from many sources. Many treatments and cures may be synthetically generated in laboratories and do not require the use of naturally growing species from TRF.
Governments permit the rapid clearing of TRF resources and sell them internationally, claiming rights to destroy domestic resources that impact the entire Earth. Yet, the same governments may be corrupt and waste other resources and spend their cash foolishly.	Debt-ridden and impoverished countries need to and have the right to use their natural resources for their own best interest. The wealthier countries of the world obtained high material living standards by depleting much of their own and others' resources as they grew. Now those countries want to hold back countries that have TRF resources wealth.
Indigenous tribes and local people are displaced by TRF clearing. In some cases, bloody conflicts ensue while government officials turn a blind eye.	Growing countries need to push their frontiers and develop their resources. "Productive" members of society have a right to use land in a manner that will benefit them and their country.
TRF resources provide an increasing tourism revenue potential. During the 1990s, travel to natural areas in the tropics was one of the fastest-growing components of the global travel industry. Although some governments claim to balance resource clearing for export sales and development goal with conservation for tourism growth, few have shown a true commitment to achieving such a balance.	Governments have the right to determine how they will earn revenue from their resources. Governments of TRF resource-wealthy countries assert their ability to balance resource depletion and extraction with conservation and replenishment.
TRFs provide a natural habitat for species found only in this biome. Removing the TRF would eliminate habitat and cause permanent loss to global species diversity. Loss of species could alter the ecological balance of the Earth.	A good source of income in a debt-challenged country with a large materially impoverished segment in its population is far more important than the conservation of a bird or a tree.

Urban Pressures in Mexico and Brazil

Mexico City is by far the largest urban agglomeration in Middle America. Its site was a major population center before the arrival of the Spaniards and even before it became the Aztec capital. It continued to be the focus of road and rail networks within New Spain and independent Mexico. Mexico City is the political capital and media center of the country, and has three-fourths of Mexico's manufacturing industry and nearly all of its commercial and financial establishments.

During the 1950s, 1960s, and 1970s, up to 1 million people per year migrated from rural areas in Mexico to metropolitan Mexico City. Housing supplied by government efforts, combined with that from private developers, could not nearly provide adequate accommodation for almost one-third of the rapidly growing city's population. The city attracted rural migrants with the expectation of greater economic opportunities, better education, more diverse recreation and cultural choices, and more substantial health care services. Squatter settlements exploded in many parts of the Central Valley of Mexico in and around the city. The massive shantytown city of Nezahulacóyotl, now with more than 1.5 million people living on flood-prone lands, is the valley's most notorious squatter settlement resulting from Mexico's migration decades. Squatter neighborhoods often lack amenities such as electricity, water, sewage, and even paved streets. In some cases, once such shantytowns were established, the authorities began to pave the streets, put in utilities, and provide access to schooling and health care, but amenities are often slow to arrive for many, and millions live well below poverty standards.

The overall result of such rapid growth is a combination of overcrowding, congestion, and air and water pollution. The physical geography of the Central Valley coupled with the intensely crowded living conditions make the urban center one of the most polluted, in terms of air quality, in the world. Although residents of metropolitan Mexico City have greater access to health care than rural Mexicans, infant mortality rates and other social health indicators are among the worst in the country due in large part to poor air and water quality. Depletion of underground water resources during the past several decades presents another problem for the city. Many areas within the urban system are subsiding, some more than 5 meters. Engineers struggle to stabilize historic structures that have been slowly sinking for decades. As job markets also failed to cope with the population explosion of the city, the informal sector of the economy grew from around 4 percent to 26 percent of people of working age from 1980 to the 1990s. Government efforts from the 1980s through the early 2000s to attract jobs and migrants to other regions in Mexico helped to slightly alleviate population pressure in Mexico City, but overcrowding and unemployment remain formidable challenges in the Federal District.

In Brazil's two largest cities, São Paulo and Rio de Janeiro, shantytowns, known as *favelas,* house millions of materially poor people who cannot be accommodated by the formal patterns of housing construction, infrastructure, and service provision (Figure 9.30 a and b). It is estimated that over 7 million people in São Paulo and up to 6 million people in Rio de Janeiro

FIGURE 9.30 Brazil: Rio de Janeiro. The dramatic physical geography of Rio de Janeiro contributes to the pride Brazilians feel for the city. (a) Contrasting with the tourist view and the expensive hotels and condominiums along Copacabana beach, the Rocinha favela crowds thousands of materially poor Brazilians into marginal structures on a steep hillside. (b) The harbor from the hilltop statue of Christ.

(a)

(b)

live in favelas. Favelas vary in their character. In and around the city centers, they occupy gaps in the built environment, in which families erect their own minimal accommodations. It is rare for these favelas to have water, electricity, waste disposal, or anything resembling a road. Old apartment blocks in the city centers, which were abandoned and taken over by impoverished peoples, resemble favelas in many ways, with several families occupying each room. On the outskirts of the cities, different types of favelas are pushing the built-up area outward at a rapid pace. These favelas spring up almost overnight and grow by thousands of people within months. Once established, favelas may be provided with some basic roads and may soon acquire utilities, shops, and schools. Crime is rampant in many favela areas. Gang violence and kidnappings severely challenge police authority and the justice system in Brazil's large cities. Strong allegations were made in the late 1990s and early 2000s that police officers were either bribed or frightened away from patrolling the streets of some favela areas where drug dealers and gang leaders took control of the streets.

The favelas are also the primary site where thousands of Brazilian street children base their nomadic and often shelterless existence. Economic, political, and social globalization forces have neither eliminated nor meaningfully diminished the conditions that combine to place hundreds of thousands of Brazilian youth in jeopardy. Some argue that multinational corporations, international banking and development institutions, and intergovernmental trade relations have increased Brazilian poverty rates rather than reduced them. Brazilian government debt and resource-export economic goals figure prominently in the social environment that produces so many children of the streets. The socioenvironmental conditions produced from such domestic and foreign relationships include extreme disparities in wealth, high poverty rates, relatively high death rates for women in childbirth, high rates of teenage pregnancy, high illiteracy rates, and high unemployment. Street children either literally live on the streets with no shelter or work the streets at a young age, trying to earn money for their survival or that of their families. The street children often end up in prostitution or an illegal narcotics trade, and many contract life-threatening diseases. The most challenging situation is the reported periodic murder of groups of street children (often believed to be the work of vigilantes), whose bodies are abandoned in ravines or concealed areas. Although local and global human rights groups are persistent in their attempts to attract significant attention to the plight of the street children, the problem remains very serious.

The Northern Andes and the International Drug Trade

The countries of the Northern Andes comprise one of the world's primary drug-producing regions (Figure 9.31). The initial system in the subregion centered on cultivation areas in Peru and Bolivia, with processing or production centers based in Colombia. The expansion of Bolivian cultivation coverage slowed in the early 1980s, and the Peruvian coverage declined in output following a

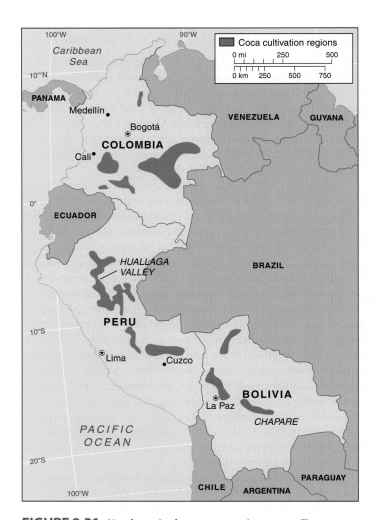

FIGURE 9.31 Northern Andes: coca-growing areas. The dominance of Peru and Bolivia in the 1980s gave way to rapid expansion of cultivation in Colombia in the 1990s and early 2000s.

fungus infestation of the crop. Colombian cultivation and production increased and includes coca, opium poppies, and marijuana, and is now greater than that in Bolivia and Peru.

The human and physical geography of Colombia enable drug cultivation, processing, and shipment to thrive. Colombia has three rugged mountain ranges that run roughly in a north-south axis through the country. Inland cities developed in relative isolation from one another, contributing to the establishment of more local identities within Colombia at the expense of a strong unified national identity. A history of political turmoil, including intense violence between left-wing and right-wing ideologues, further fragmented national cohesion. Climate, soils, altitude, and shade conditions are well suited to coca cultivation, and a lack of developed transportation infrastructure combined with dense vegetation covering vast rural and remote lands within Colombia's borders all combine to facilitate a thriving cocaine industry.

The global demand for cocaine from consumers in the United States, Europe, and Asia has a devastating impact on the local politics, agriculture, and the social fabric in the Northern Andes. An excess of production over market demand in the early 1980s caused the price of cocaine to fall by 75 percent. The huge drop in price dramatically opened the international market to those with less disposable income, resulting in a subsequent rapid increase in demand. The United States led efforts to prevent drug production in these countries, but its supply-based (rather than market-based) strategies were diluted by limited government cooperation, corruption, coercion, and challenges in crop eradication. United States investment in combating narcotics trafficking from the region makes Colombia the third-largest recipient of U.S. financial aid (after Israel and Egypt) in the world. Although Bolivian farmers planted other commercially valuable crops on former coca-growing land as a result of U.S. aid, the acreage under coca remained the same. Peruvian government attempts to halt cocaine production are limited in part by fears of the resurgence of guerrilla groups among disaffected former coca growers. Subsistence and small-scale farmers are attracted to working on large coca farms where average wages are far in excess of what they would otherwise earn. Other farmers are forced by threats of violence to become part of the drug labor force, and in many cases, their farms are forcibly taken over by the drug cartels.

In the early 2000s, the Colombian government took a tougher stance on narcotics production and appeared to be making limited progress in stemming the violence among and between the various groups related to the cocaine industry. Programs of eradication, however, have to compete with the high prices farmers get for their drug crops and with the guerrilla-backed disorder that such farmers may support, including the strategic occupation of isolated oil-pumping stations or airports. Chemical defoliation of coca-growing did not decrease coca production. A doubling of the output of coca occurred simultaneous to aerial spraying efforts. Efforts to stem the flow of cocaine from Colombia and other Northern Andes countries are continually challenged with the relatively high income from coca-cultivation that lifts farmers out of poverty and the long-established custom of coca-leaf chewing among Native Americans living at high altitudes in the Andes.

GEOGRAPHY AT WORK

Floodplain Erosion in Mexico

Dr. Paul Hudson (Figure 9.32) is a physical geographer who specializes in fluvial geomorphology, the study of how flowing water interacts with the Earth's surface, particularly as it relates to erosion, flooding, and floodplains. Paul's expertise is large lowland coastal plain river systems, and most of his research occurs along the Gulf Coastal Plain. His current research focus is primarily in eastern Mexico on the Rio Panuco system, which drains portions of Mexico's Altiplano, Sierra Madre Oriental, and Gulf Coastal Plain. Paul's ongoing research projects include examining controls on floodplain evolution, river erosion, and flood processes. Mexico's Gulf Coastal Plain is a hot and humid region with extensive swamps and large river systems. The Panuco Valley, like most of the world's large river systems, is an agricultural landscape, primarily for sugarcane, citrus, and cattle ranching. The location of land use types within the valley, however, is influenced by the floodplain geomorphology. Older floodplain deposits represent the effects of landforms and river movement on flooding and soils, which limit specific agricultural activities. The Panuco Valley also represents the heartland for the Huastec culture region, the northern extent of prehistoric complex culture in Mesoamerica. Unlike the Maya and Aztec, very little is known about the Huastec. However, because prehistoric Huastec resided primarily within the floodplain, archaeological materials are found within flood deposits. Thus, the region's rivers become very important for understanding prehistoric human-environment interaction, as well as the management and preservation of this cultural resource.

To understand the effects of water and erosion in the lower portions of the basin it is important to understand land use in the basin headwaters, the eastern Sierra Madre Oriental. A project Paul conducted with one of his graduate students examined the impact of slash and burn (swidden) agriculture on soil erosion. This research is also important because the economy of small communities, many of which are *ejidos*, are dependent on farming. Soil erosion results in a change in the land suitable for farming, and thus directly influences the sustainability of their economic system, in addition to ecological recovery.

FIGURE 9.32 **Dr Paul Hudson** (right) of the University of Texas at Austin Department of Geography and the Environment, carrying out field research in the Panuco Valley, Mexico.

These research projects involve state-of-the-art digital geographic technologies, such as GIS, remote sensing, and GPS. However, while these technologies are useful, the research activities are driven by field evidence. Field research activities frequently involve topographic surveying and sampling floodplain deposits, which are then analyzed in Paul's geomorphology laboratory. Other samples may be shipped to laboratories for radiocarbon dating, which provide an age control on flood deposits.

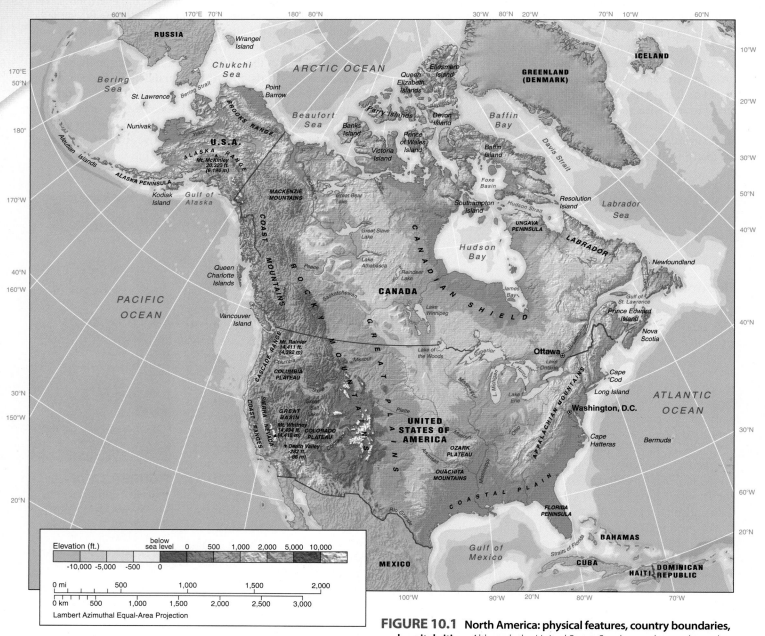

FIGURE 10.1 **North America: physical features, country boundaries, and capital cities.** Although the United States Southwest shares a boundary with Mexico, the majority of the region is surrounded by ocean.

In the pre-dawn hours of the morning, five to six days per week, millions of people who were born in Latin America rise from their beds and make their way to the nearest bus stop as they begin their commute to jobs they hold in metropolitan Los Angeles, New York City, Washington, D.C., Houston, Chicago, and dozens of other large U.S. cities. In the United States they are labeled "Hispanics"—a broad demographic classification that transcends notions of race and nationality, yet is weighed down by stereotypes. Some Latin Americans are legal citizens of the United States, some are on work visas, and others have illegally entered the country and bypassed U.S. immigration and naturalization laws and red tape.

Hispanic Americans are the fastest growing ethnic minority in the United States, comprising more than 14 percent of the U.S. population in 2006. They are the largest ethnic group in many of the country's largest cities. While thousands of Hispanic peoples in the United States have assimilated into professional positions in the U.S. economy, working as physicians, lawyers, accountants, government officials, and teachers, the majority of the community serve in lower-skilled positions such as bussing tables in restaurants, maid service in hotels or suburban homes, sanitation collection, landscaping, and construction. Although their market influence and political clout are increasing in parts of the United States, controversy surrounds their employment, housing, health care, schooling, and citizenship.

Chapter Themes

Defining the Region

Distinctive Physical Geography: mountains and plains; glaciation, major rivers and the Great Lakes; tropical to polar climates; natural vegetation and soils; natural hazards, regional environmental issues.

Distinctive Human Geography: population patterns; contributing history; political geography; economic geography.

Geographic Diversity
- The United States.
- Canada.

Contemporary Geographic Issues
- Immigration.
- North American Free Trade Agreement (NAFTA).
- The challenge of Québec.

Geography at Work: Using Geography to Aid Public Policy

Defining the Region

The North America region is comprised of two countries, Canada and the United States (Figure 10.1). The United States and Canada are among the world's largest countries in terms of area, have relatively low population densities, a wealth of natural resources, and high material living standards. Canada's huge land area is sparsely inhabited with a 2006 population of 32.6 million (Table 10.1). The United States is the world's third-largest country in terms of population, yet its 2006 figure of 299.1 million is substantially smaller than the Earth's two population giants, China and India, each with more than 1 billion residents. More than three-quarters of North America's residents live in urban areas (Figure 10.2).

The very high living standards in Canada and the United States attract migrants from all world regions who come to North America seeking prosperity and stability. The majority of North America's immigrants congregate in the region's vast urban areas where opportunities are greatest. Internal migration patterns further add to the growth of the region's cities and, coupled with international immigration, create increasingly plural yet often spatially segregated metropolitan centers. Illegal immigration into North America and the corresponding urban congregation of illegal immigrants is creating a growing controversy surrounding job competition and perceived increases in gang violence, among other problems.

Although Canada and the United States are closely linked culturally and economically, Canadians have a strong sense

TABLE 10.1	NORTH AMERICA: Data by country, area, population, urbanization, income (Gross National Income Purchasing Power Parity), ethnic groups						
		Population (millions)		**%Urban**	**GNI PPP 2005**		
Country	Land Area (km²)	mid 2006 Total	2025 est. Total	2006	Total (US$ billions)	Per Capita (US$)	Ethnic Groups (%)
NORTH AMERICA							
Canada	9,976,140	32.6	37.6	79.4	1,049.8	32,220	British origin 35%, French origin 25%, other European 20%, First Americans 3%
United States of America	9,809,460	299.1	349.4	79.0	12,547.7	41,950	White 83%, African American 12%, Asian 3%, Native American 1%

Source: World Population Data Sheet 2006, Population Reference Bureau. Microsoft Encarta 2005.

(a)

(b)

FIGURE 10.2 **North America: urban and rural landscapes.** (a) The urban setting of Toronto, Canada. (b) The contrasting open wilderness of Zion National Park, Utah, U.S.A.

of national pride, which has thus far prevented their Canadian identity from being overwhelmed by their politically and economically powerful neighbor.

Mainland North America is like a geographic fortress, relatively isolated and naturally protected from much of the rest of the world. Canada and the United States are bound to the east by the Atlantic Ocean, separating them from Europe and Africa by thousands of miles. Canada's northern border is the Arctic, while Canada itself serves as the entire northern border of the conterminous United States. The western border of both countries is the vast Pacific Ocean Basin, which also puts thousands of miles between the region and East and Southeast Asia. The U.S. southern border is divided between Mexico and the Gulf of Mexico.

Distinctive Physical Geography

Mountains and Plains

North America is part of the North American plate (see Figure 1.5). New plate material forms along the divergent margin at the Mid-Atlantic Ridge, forcing the plate to move westward and clash with the Pacific plate at the convergent and transform plate margins along the western coast of North America. The younger and more geologically active west coast of North America has higher mountains and is affected by frequent earthquakes and periodic volcanic activity, while the less-disturbed eastern coasts are low-lying or hilly with older mountains inland and infrequent earthquake activity. The **Canadian Shield** rocks, the oldest in North America, cover roughly half of Canada's surface area and contain major deposits of mineral ores.

The western third of Canada and the United States is primarily comprised of rugged mountains consisting of several different elements. The highest peak in the region is Mount McKinley in the Alaskan Range (6,194 m, 20,231 ft.). The Rocky Mountain ranges extend from Alaska and northwestern Canada southward to New Mexico. To their west in the United States are extensive high plateaus from the Colorado Plateau in the south to the lava layers of the Columbia Plateau just south of the Canadian border. West of the plateaus are the Sierra Nevada mountains of California and the Cascade ranges (Northern California through Oregon and Washington to the U.S.-Canada border).

Lowlands with little relief dominate southern Ontario, the Prairie Provinces, and the central United States. Layers of sedimentary rock cover the shield rocks, forming plateaus and escarpments such as that over which the Niagara Falls plunge. In the United States Midwest and in Ontario, deposits from the melting ice sheets cover most of these rocks. Similar deposits occur over part of the Canadian Prairies, but in areas not covered by the ice sheets, distinctive features such as groups of small mounds and depressions were formed by exposure of the land to an atmosphere of extreme cold.

East of the Rockies, the Mississippi drainage basin (the world's third-largest watershed) drains an area of 3,225,000 km² (1,245,000 mi²). The northern lowlands within the Mississippi watershed were covered by rock fragments and particles dropped by the ice sheets blanketing the area during the Pleistocene Ice Age, while other areas within the watershed have deposits of wind blown silt, or **loess,** which accumulated as far south as Louisiana and Mississippi.

East of the Mississippi lowlands, the Appalachian Mountains form a continuous chain of rolling hills and mountains extending from northern Georgia into the Adirondacks of New York, the Green and White Mountains of New England, and the Atlantic

Provinces of Canada. Few Appalachian mountain ridges exceed 2,000 m (6,500 ft.). The rocks were deformed and uplifted by ancient plate movements and subsequently eroded. In the northeastern United States and eastern Canada, the glaciers scraped away much of the surface rock and soil, carried it southward, and deposited it along the coast in the low ridges (moraines) that form much of Long Island and Cape Cod. West of the Mississippi, the Ozarks and Ouachitas form similar upland areas.

Glaciation, Major Rivers and the Great Lakes

During the Pleistocene Ice Age, the last 2 million years of geologic time, advancing ice sheets covered much of Canada and the northern United States. The greatest effects occurred in the last half million years, but most of the ice melted beginning about 15,000 years ago, leaving glaciers only in the highest and coldest areas. Distinctive ice-eroded landscapes are relic features in high mountains and in northern Canada, while the melting ice sheets left depositional features mainly in the Midwest.

The Great Lakes formed as glacial ice gouged depressions between piles of marginal moraines built from rock fragments dropped at the margins of melting ice masses. As the ice sheet retreated northward at the end of the last glacial phase, meltwater accumulated in the depressions, which became the Great Lakes. Today, the Great Lakes function as inland seas with coastlines, ports, and recreation areas. The St. Lawrence Seaway opened in 1959 as a cooperative project between the Canadian and United States governments. Ocean-bound ships, from as far inland as Chicago and Duluth, continue to utilize the seaway.

North America's surface **hydrology** (surface water drainage) has a major impact on both the physical and human geographies of the region. The major rivers and the Great Lakes provide sources of fresh water and transportation routes. The Mississippi River was the basis of early interior transportation and continues to play a role in the transport of bulk materials. Its largest tributary, the Ohio River, was a transportation route at the heart of manufacturing developments in the United States during the late 1800s. The Colorado River in the arid Southwest is harnessed for power and irrigation water (Figure 10.3). The Columbia River is used to irrigate farmland and to generate hydroelectricity in the northwestern United States. In the 1950s, the Great Lakes–St. Lawrence Seaway brought easier trade and associated industries deep into the interior of Canada and the United States, although this function is much reduced owing to the increased size of ocean ships (Figure 10.4). In northwest Canada, the Mackenzie River is a major summer transportation route.

The largest area of lowland and fertile farmland occurs in the combined drainage basin of the Mississippi-Missouri-Ohio rivers and the Great Lakes, covering nearly one-third of Canada and the United States. Although this is a continental interior region, the rivers provide relatively easy outlets to world ocean trade routes to the Gulf of Mexico in the south and along the St. Lawrence and Hudson rivers to the Atlantic Ocean in the east.

FIGURE 10.3 North America: water resources. The generation of hydroelectric power at the Hoover Dam on the Arizona-Nevada border is a significant source of energy in the region. The dam controls the flow of the Colorado River and creates Lake Mead, a source of fresh water for metropolitan Las Vegas, and a national recreation area.

FIGURE 10.4 North America: water-based transport. The St. Lawrence River, which connects the Great Lakes to the Atlantic Ocean, is an important waterway for recreation and commerce in Canada. This container ship is docked alongside a wharf in Montréal.

The Mississippi-Missouri River system was controlled for flood protection, improved water transportation, and hydroelectricity generation by building a series of dams and levees from the early 1900s. The flood protection system prevents most floods from spilling onto surrounding land; however, extreme rain events prove to be too much for the levees and dams to handle (Figure 10.5). The levees prevent the deposition of sediment on the floodplain in southern Louisiana resulting in

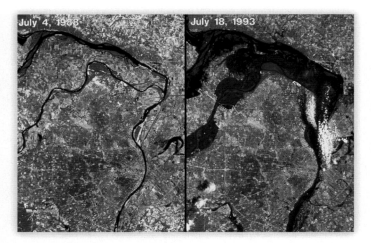

FIGURE 10.5 **United States: flood hazard.** The 1993 floods on the Mississippi River near St Louis. The satellite photo on the left was taken before the floods. In the satellite photo on the right the dark area shows the extent of floodwaters. The reddish areas are highways and buildings.

surface subsidence in metropolitan New Orleans and the river's delta region and in the intrusion of salt water into the region's coastal wetlands.

Tropical to Polar Climates

The majority of North America's climatic environments are temperate midlatitude in type (see foldout map inside back cover). The western coast is dominated by oceanic temperate climate that gives way eastward to continental temperate climates. The west coast climates of Canada and the United States are restricted in their inland impact by the north-south mountain ranges.

Although the southwestern United States is situated north of the Tropic of Cancer, warm and arid conditions dominate most of the year. Southernmost Florida has tropical, humid conditions, and the Gulf coast from Texas to Florida and the Atlantic coast as far north as the Virginia tidewater have subtropical conditions that bring warm spring and fall temperatures, and hot, humid summers. In northern Canada and Alaska, the long and harsh Arctic winter gives way to a brief, mild summerlike climate from late May to early August.

The central plains and lowlands enable bursts of Arctic air to move southward in winter, bringing extreme cold especially to the Midwest and northeastern United States but also extending in shorter bursts to the Gulf coast. Warm, dry southwestern air or humid air from the Caribbean and Gulf of Mexico moves northward in spring and summer. The frequent confrontation of cold Arctic air and humid subtropical air in the central lowlands leads to the formation of active cold fronts with thunderstorms that may produce violent tornadoes mainly in spring and summer. Tornado activity moves northward from Texas in February to the Great Lakes in June. The United States experiences more tornadoes each year than any other country.

Natural Vegetation and Soils

Figure 10.6 depicts the spatial distribution of major vegetation types in North America. The hot deserts of the Southwest support mainly drought-resistant varieties such as cactus and low shrubs. The continental temperate Great Plains and prairies of the western Mississippi River basin and south-central Canada historically contained prairie grasslands that are now largely plowed to make use of the underlying black earth soils. Eastward, the more humid conditions, from tropical southern Florida to the more temperate Northeast, supported **deciduous** (broadleaf) forest, which gave rise to brown earth soils of moderate to good fertility. North of the deciduous forests and the prairies, a wide band of **coniferous** (needle-leaf) forest has poor podzolic soils and gives way to the tundra along the Arctic Ocean shores.

Today, the eastern mountains have thin soils on steep slopes. Before extensive clear-cutting by loggers in the late 1800s and early 1900s, the eastern mountains supported a profusion of trees that contained a much greater variety of species than those in Western Europe. The best farming soils formed under forest and grassland in the interior plains, where the old glacial and windblown materials combining with the moderate amounts of precipitation produced rich brown (forest) or black (grassland) soils. In the southeastern United States, the frequent presence of sandy soils with low nutrient content caused some areas to be dominated by pine trees. Along the western coast of the United States, north of San Francisco, huge firs and cedars, growing over 100 m (300 ft.) tall, formed a massive timber resource.

Natural Hazards

A greater range of natural hazards than in any other country in the world marks the natural environments of the United States. The conterminous United States is affected by hurricanes, severe thunderstorms with accompanying tornadoes, lightning, hail, earthquakes, volcanoes, floods, blizzards, ice storms, and wildfires. Alaska and Hawaii are affected by various natural hazards including extreme winter weather and earthquakes in Alaska and nearly continuous volcanic activity in Hawaii. All the natural hazards experienced within the borders of the 50 United States have a major impact on the human activities of U.S. residents.

Each hazard has a geographic area of greatest impact.

- Hurricanes threaten the mid- and southern Atlantic coast and Gulf coast states in late summer and fall, when water temperatures are warmest.
- Tornadoes are most common in the Plains states, the Mississippi Valley, and Florida, where the necessary climate ingredients meet to support their formation. Tornadoes affect Texas in February and move northward to the Great Lakes area by June.
- River floods occur most commonly in the Mississippi River valley, following the spring melting of snow on the surrounding hills or heavy summer rains. Flash flooding

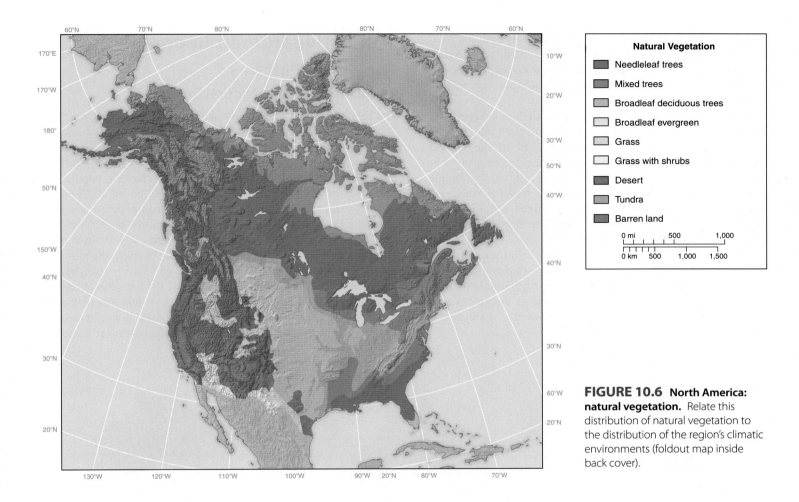

Natural Vegetation

- Needleleaf trees
- Mixed trees
- Broadleaf deciduous trees
- Broadleaf evergreen
- Grass
- Grass with shrubs
- Desert
- Tundra
- Barren land

FIGURE 10.6 **North America: natural vegetation.** Relate this distribution of natural vegetation to the distribution of the region's climatic environments (foldout map inside back cover).

occurs in all regions of the United States and is particularly dramatic in arid regions in the West after sudden rains fill dry streambeds.

- Lightning is a year-round event across the southern part of the United States and is especially prevalent in Florida.
- Severe winter storms affect the upper Plains states, the Midwest, the western mountains and the northeast coast of the United States. Ice storms pelt many areas east of the Rocky Mountains.
- Earthquakes and volcanoes occur mainly along the west coast from Alaska southward through California. The meeting of the North American plate and the East Pacific rise causes horizontal displacements along the San Andreas Fault, which produces frequent earthquake activity. Volcanic activity is nearly continuous in Hawaii, and it also occurs in the Cascades where the Juan de Fuca minor plate plunges beneath part of North American plate.

Canada is less troubled by most of the natural hazards plaguing the United States, primarily due to its position in the high latitudes of the Northern Hemisphere. The waters off the Canadian coasts are much too cold to support hurricanes. The extreme temperature and moisture contrasts needed to support severe thunderstorms and accompanying tornadoes, hail, and

lightning are also less common in Canada due to its high latitudinal position. Canada does experience earthquake activity and the threat of tsunami ("tidal" waves) along the west coast, and very cold temperatures and high snowfall totals are common across the country during the winter.

The Canadian provinces of Quebec, Ontario, and New Brunswick were devastated by an ice storm in January 1998 when 7–11 centimeters (3–4 inches) of ice accumulated over a six-day period. The heavy ice brought down trees, power lines, utility poles, and transmission towers. Millions of Canadians went without power and electric heat for days (some had no power for a month). The 1998 ice storm was Canada's costliest natural disaster.

Regional Environmental Issues

Rapid economic development in North America created many ongoing environmental challenges. Agricultural expansion and population growth in the arid southwestern United States continues to deplete regional freshwater sources and creates intense competition among the states in that part of the country. Cities like Washington, D.C., Baton Rouge, and Los Angeles, became plagued by smog and high concentrations of ground level ozone, generated from the exhaust gases of large numbers of vehicles and thermal power plants. Acid rain, derived from power plant emissions particularly along the Ohio River valley, affected trees and caused rivers and lakes downwind in the northeastern United States and eastern Canada to become more acidic.

Canada, with a much smaller population and a larger land area, has fewer environmental problems than the United States. Industrial pollution largely created in the United States travels through air currents and reaches the ground in parts of Canada in the form of acid rain. Industrial effluents from both Canada and the United States foul the waters of the Great Lakes. Although Canada reduced polluting outputs from smelters on its territory near the British Columbia border and along the northern shore of Lake Superior, more significant reduction is needed from both countries to ensure lasting health of the region's freshwater systems. Canadian attempts to diminish industrial effects on the natural environment through federal legislation resulted in the imposition of heavy fines on pollution-creating industries, such as the nickel smelters at Sudbury in Ontario.

Distinctive Human Geography

Population Patterns

Population Change: Natural Growth

The relatively rapid annual population increase in the United States is unusual among the world's materially wealthy and technologically advanced countries, due in part to the young immigrant communities within the U.S. who have larger families and higher birth rates than other groups. The U.S. population of 299.1 million could rise to 349.4 million by 2025. In the 115 years from 1891 to 2006, the Canadian population increased by more than 27 million, from 4.8 million to 32.6 million. Both countries received immigrants in the early 1900s and after World War II. In the post–World War II era, population growth in Canada was rapid, doubling from 1941 to 1971 before slowing to a one-fourth increase from 1971 to 1991. Canada's total fertility rate of 1.5 in 2006, is less than that of the United States (2.0), but its population growth rate is similar due to immigration (Figure 10.7a and b). Canada has lower death

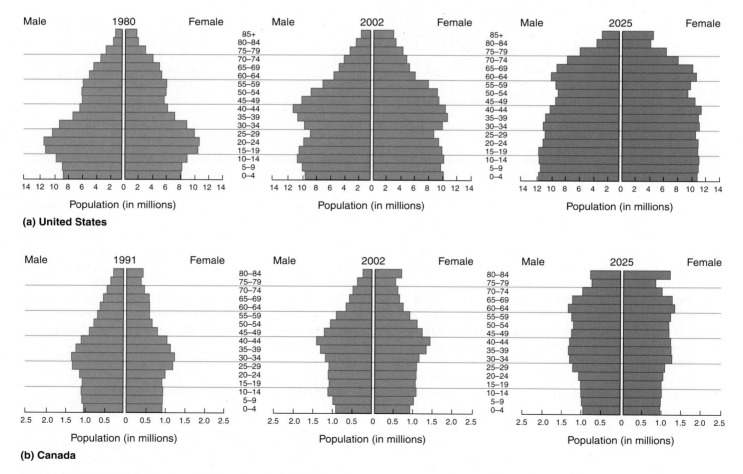

FIGURE 10.7 United States and Canada: age-sex diagrams. (a) The 1980/1991 and 2002 graphs show how the baby boom of the 1950s and 1960s produced a progressing bulge in the corresponding age groups. The 2025 graph shows how the baby boom will lead to a significant growth in numbers of older Americans. (b) The 2002 graphs detail maturing populations with the largest numbers of people in the 35- to 44-years of age groupings. *Sources: U.S. Census Bureau; International Data Bank.*

rates than the United States, and its continued immigration rates match those of the United States.

Population Distribution: Increasing Urban Density

The United States of America is a highly urbanized country with 79 percent of its 2006 population living in urban areas and just over half in metropolitan centers of over 1 million people (Table 10.2). More than 90 percent of the U.S. population lives within a two-hour drive of a large city of over 300,000 people. More than 32 million people live along the eastern seaboard in an urban corridor from Washington, D.C., to Boston, Massachusetts, known as the Megalopolis, and another 23 million people live in the metropolitan areas of the western Manufacturing Belt and Midwest. In addition to the Megalopolis, urban clusters exist around the Great Lakes from Chicago to Detroit, in Florida and westward along the Gulf coast to southeastern Texas, and along the Pacific coast in the west (Figure 10.8). Large cities are the essential feature of the country's contemporary geography. Urban population clusters are more numerous in the eastern contiguous United States. Vast areas of the western United States, away from the Pacific coast, contain very low population densities. Hawaiian population is clustered near Honolulu and resort towns, and Alaska, which is the largest U.S. state in area, has the lowest population density in the country.

Canada has approximately one-tenth the population of the United States on a larger land area, which gives Canada a much lower average population density than that of the United States. It is necessary to consider the distribution of people in Canada in order to better understand Canada's population geography.

The vast majority of Canadians are concentrated in a belt across the southern part of the country nearly parallel to the border with the United States. Thus, portions of this populated belt across southern Canada have high population densities, while most of the remainder of the country is virtually empty. Canada is experiencing a trend toward metropolitan expansion and also had 79.4 percent of its population classified as urban in 2006. The two largest metropolitan areas, Toronto and Montréal, rival many U.S. metropolitan areas in size. Together with Vancouver on the west coast and Ottawa, the capital, these cities contain one-third of Canada's population. Winnipeg, Edmonton, and Calgary are other major Canadian cities. Canadians are less spatially divided by ethnic segregation than Americans, but congregation is important among newer immigrant groups in big cities such as Toronto and Vancouver.

Patterns of Migration in the United States

The continuation of relatively high rates of population increase for the United States is primarily a result of immigration, which has been vital to the country's economic growth since the days of European exploration. Immigration currently accounts for one-third of U.S. population growth (see "Immigration," p. 332). Following the early domination by people of British

TABLE 10.2 — **Population of Major Urban Centers in North America (in millions)**

City, Country	2003 Population	2015* Projection
NORTH AMERICA		
New York, USA	18.3	19.7
Los Angeles, USA	12.0	12.9
Chicago, USA	8.6	9.4
Philadelphia, USA	5.3	5.7
Miami, USA	5.2	6.0
Toronto, Canada	4.9	5.8
Boston, USA	4.2	4.8
Houston, USA	4.1	4.9
Washington, DC, USA	4.1	4.6
Atlanta, USA	4.0	5.3
Detroit, USA	4.0	4.2
Montréal, Canada	3.5	3.7
San Francisco-Oakland, USA	3.3	3.6
Phoenix, USA	3.2	4.0
San Diego, USA	2.8	3.1
Minneapolis-St. Paul, USA	2.5	2.8
Tampa-St. Petersburg, USA	2.2	2.5
Baltimore, USA	2.1	2.4
Denver, USA	2.1	2.6
St. Louis, USA	2.1	2.3
Vancouver, Canada	2.1	2.4
Cleveland, USA	1.8	2.0
Pittsburgh, USA	1.8	1.9
Portland, USA	1.7	2.1
Cincinnati, USA	1.6	1.7
Las Vegas, USA	1.6	2.2
Riverside-San Bernadino, USA	1.6	1.9
San Jose, USA	1.6	1.8
Sacramento, USA	1.5	1.8
Kansas City, USA	1.4	1.5
San Antonio, USA	1.4	1.6
Virginia Beach, USA	1.4	1.6
Milwaukee, USA	1.3	1.4
Columbus, OH, USA	1.2	1.4
Orlando, USA	1.2	1.5
Providence, USA	1.2	1.4
Ottawa, Canada	1.1	1.2
Austin, USA	1.0	1.4
Calgary, Canada	1.0	1.3
Memphis, USA	1.0	1.2
New Orleans, USA	1.0	1.1

*estimated

Source: United Nations Urban Agglomerations 2003, with estimates for 2015 (2003).

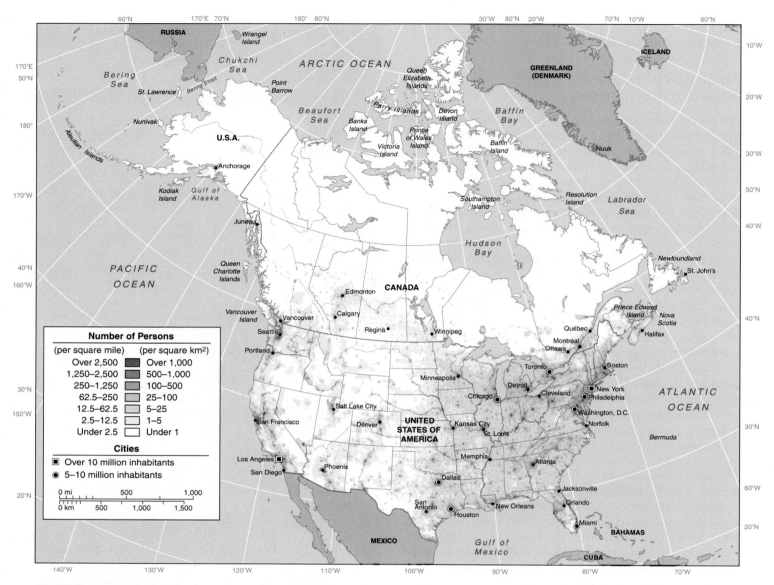

FIGURE 10.8 North America: population distribution. Compare the distribution of population in the eastern and western halves of the United States and note the proximity of most Canadians to the U.S. border.

origin, Irish and German people immigrated in greater numbers during the middle of the 1800s. In the later 1800s, the highest numbers of immigrants came from southern Europe, particularly Italy, and at the turn of the century, large numbers came from the Slavic countries of Eastern Europe, the Balkans, and Russia.

Peoples from the African continent were forced to migrate as slaves to North America in the 1700s. During the 1800s, the African American population continued to grow through natural increase after the end of the Atlantic slave trade.

From the mid-1900s, Latin American peoples from Middle and South America, and Asian peoples from South, Southeast, and East Asia became major sources of immigrants. The Chinese, who came to the United States as laborers in the late 1800s,

and the Japanese, who moved in as farmers in the early 1900s, were later joined by other Asian groups from 1960, particularly from Vietnam, Korea, and India. Most recently, influxes of Africans and the 1990s immigration of Russians following the demise of the Soviet Union further diversified the cosmopolitan American population.

The dynamics of population geography in the United States involve internal migrations as well as international immigration (Figure 10.9a). From the early to mid-1900s, large numbers of African Americans moved from the rural and urban South to northern cities. Those from the Atlantic coastal plain mostly moved to Washington, D.C., Baltimore, Philadelphia, New York, and Boston; those from the Mississippi Delta moved primarily to Cleveland, Cincinnati, Indianapolis, and Chicago.

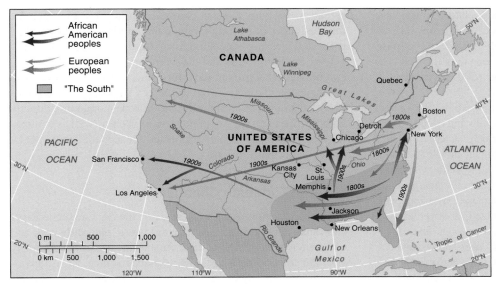

(a)

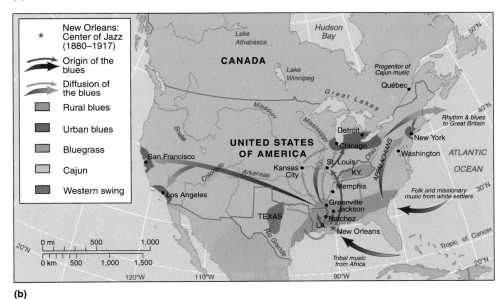

(b)

FIGURE 10.9 **United States: internal migration.** (a) The major movements of African and European Americans in the 1800s and 1900s. (b) Compare the developments in blues music and their diffusion to the movement of African Americans.

the 1980s, but recessions in that decade slowed southern economic growth and reduced such migrations.

Canadian Patterns of Ethnic Integration

Government policy protecting multiculturalism and well-known high living standards combine to make Canada an extremely attractive destination for peoples emigrating from their homelands. The Canadian federal government instituted policies outlawing ethnic discrimination and protecting the rights of immigrants. For several decades, Canada has been one of the leading immigrant-receiving countries of the world. Contemporary immigration trends for Canada exhibit significant increases in the numbers of people migrating from Asia and the Americas (including the Caribbean) and an overall percentage decrease in the European immigrant population.

The large urban centers of Toronto, Vancouver, and Montréal are experiencing dynamic cultural processes that increase the diversity of each metropolitan area. Canada's largest city, Toronto, attracts more immigrants from all reaches of planet Earth than any other Canadian city. Asian migration is most notable along the west coast of the country. The city of Vancouver and its hinterlands in the province of British Columbia are the hosts to two types of immigrant from Asia, those who settle year-round and those who use the city and environs as a second home or home for their families while they "commute" to work in Hong Kong.

Smaller numbers of African Americans moved to the West Coast. In 1900, nearly 90 percent of African Americans lived in the South, but this proportion declined to just over half by the 1990s. The movements also stimulated the dissemination of cultural features of African American society, such as the varied types of blues music (Figure 10.9b).

Most of the movements of African Americans out of the South were completed by 1970. After that, smaller **countermigration** or return movements, often on retirement, balanced or exceeded the northward flow. Movements of European Americans from the Northeast to both the West Coast and the South continued into

Contributing History

Indigenous groups of people, who are referred to today as **Native Americans** in the United States and **First Nations** in Canada, inhabited North America for centuries before 1500. The first Americans probably migrated from Siberia to present-day Alaska over 20,000 years ago. By 1500, they lived in hierarchically structured ethnic groups, commonly called tribes. The combination of natural resources, environmental conditions, and a sedentary settlement structure enabled groups such as the Mississippians in present-day southwestern Illinois, to flourish for centuries.

After the arrival of the Europeans, many Native Americans perished from the introduction of diseases to which they had no immunity. The survivors were increasingly pushed to the most marginal lands in the region. This process began in the east and southwest in the 1600s and lasted into the 1800s in the central and northwestern parts of North America.

In the three centuries after Spain's initial exploration of the Americas, the region became a series of colonies and occupied territories governed by the French, Spanish, British, Dutch, Russians, and Swedes. The British became the dominant power in North America by the mid-1700s. Almost immediately, settlers south of the St. Lawrence River valley fought to become the independent United States of America by signing the Declaration of Independence in 1776, and forcing Britain to recognize U.S. sovereignty.

Canada arose mostly out of the colonial territories first established along the St. Lawrence and the Great Lakes and remained a British colony far longer than the United States. It achieved a degree of independence through the British North America Act in 1867, while maintaining legal ties to Britain until 1982, when it gained full control of its constitution. Today, Canada is a fully independent country that enjoys membership in the Commonwealth of Nations.

U.S. Development

Three primary economic and sociostructural patterns emerged during the British Colonial development of the eastern portion of the incipient United States. After independence, areas along the Atlantic coast formed springboards for thousands who moved westward within the United States as new lands were acquired. The New England community subculture diffused into areas around the Great Lakes. The Pennsylvania system of individual family farms became the basis of farming in the southern Midwest. The plantation system that was first established in the Virginia tidewater, along with the accompanying use of slaves, was extended south and then southwestward along the Gulf lowlands to Texas. The North-South tensions that culminated in the Civil War of the 1860s arose from the spatial diffusion of these early differences.

The United States tripled its area by 1850, buying land from France and Spain and acquiring the western third by negotiation and military conquest (Figure 10.10). To ensure the rapid occupation of these new areas, land was surveyed and sold by the U.S. government at lower and lower prices. The **Homestead Act** of 1862 provided families with very inexpensive and at times even free farmland. Settlers

spread to the vast interior plains. Families able to overcome the challenges of starting farms in the interior, such as building roads and clearing vegetation from acquired lands, took advantage of the fertile soils and the warm, moist summer climate, a combination that proved ideal for growing corn and wheat and raising cattle and pigs. From the early 1800s, the United States exported a range of farm produce to Europe, competing directly with European farmers in bulk grain markets as well as continuing to export crops such as cotton and tobacco that could not be grown in northern Europe.

Gradual Canadian Changes

The settled region along both sides of the St. Lawrence River and estuary (part of modern Canada) was a materially poor and often environmentally hostile remnant left to British authority after the United States declared its independence. British loyalists who voluntarily emigrated or were forced out of the incipient United States joined small numbers of Native Americans and French-speaking European settlers in Canada. Those loyal to the British Crown mostly settled on the east coast or farther inland around the northern and western shores of Lake Ontario, due in part to a concentration of French settlers along the lower St. Lawrence who resented British rule. Westward of the Great Lakes there was little settlement in an area still then largely administered by the Hudson's Bay Company, which was created by a Royal British Charter in 1670 to administer the develop-

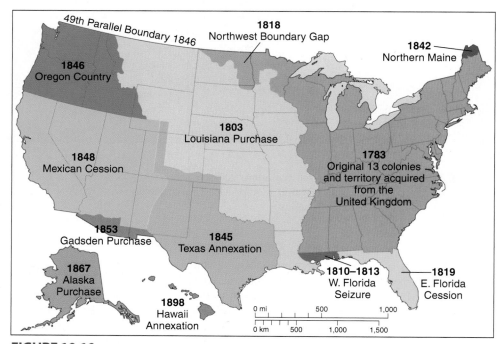

FIGURE 10.10 United States: rapid expansion of territory. Land first occupied by Native Americans, and later assumed by European colonial empires, were gradually incorporated into the expanding United States. The eastern lands were acquired from the United Kingdom at independence in 1783. The Louisiana Purchase brought back lands that France had just garnered from Spain. Texas claimed independence after ousting the Mexicans in the early 1830s, but then joined the United States in 1845. The western lands of the conterminous United States included the southwestern lands acquired through the 1848 Mexican Cession after a rapid U.S. military victory. Alaska and Hawaii came later.

ment of nearly one-third of present-day Canada. The interior remained the realm of indigenous peoples and isolated groups of French settlers attempting to escape the watchful eye of British rule. When the Hudson's Bay Company fur trappers began to compete with Americans for land along the lower Columbia River near the coast of the Pacific, negotiations in 1846 set the boundary between the two countries west of Lake Superior at the 49th parallel.

Growth in U.S. Manufacturing

The first Industrial Revolution began in Western Europe (see Chapter 2), coinciding with the United States' independence from Great Britain. Independence and British hostility motivated Americans to establish their own manufacturing industries, providing goods that were at first not obtainable from Britain. Textiles, metal goods, and leather goods were among the first industries to develop, often powered by water mills in New England, New Jersey, and eastern Pennsylvania. By 1850 Pennsylvania's coal replaced charcoal in iron making and subsequent steel production in Pittsburgh that set off further industrial developments. The northeastern United States soon emerged as a dominant region of the country, housing the bulk of the population with growing material wealth and related political power based on manufacturing industries.

The primary symbol of the early steel-based Industrial Revolution—the railroad—spread across the country to the Pacific coast. The Manufacturing Belt of the northeastern United States, which initially developed based on the regional resources of coal, iron, and the agricultural produce of the Midwest, later drew in mineral resources from other parts of the country. Railroads linked the mines of the western mountains to the eastern economic and political core. The mining of gold, silver, and then copper and zinc in the new western lands provided capital for further development and widened the scope of the late-1800s metal industries.

In the 1900s availability of such natural resources, utilization of rivers for irrigation, navigation, and hydroelectricity, and discoveries of oil and natural gas drew more people and manufacturing corporations southward and westward from the country's core. Although most Americans viewed natural resources in terms of the potential for economic gain, influential groups from the late 1800s persuaded the federal government to set aside large areas of attractive wilderness or places of historic significance as national parks, often before much settlement had occurred.

The United States also sought to develop conditions in which its population could flourish by making the most of their freedoms under the democratic Constitution. It established compulsory education much earlier than European countries. Its emphasis on technology transfer began in the late 1800s with land grants for establishing engineering and farming universities in many states.

These natural and human resources, along with innovative manufacturing structures and processes, quickly brought the United States prominence in world economic activity by the late 1800s. The expanding internal market of the United States provided the demands that stimulated many of the developments. Americans achieved **economies of scale** by building larger factories for increased output. **Horizontal integration** occurred when financiers bought up several producers of the same product, giving the new owners a large share of the total production of goods and enabling them to set market prices. **Vertical integration** took this process further by uniting the producers of product inputs with the product manufacture in a single corporate structure. As an example of these processes, Andrew Carnegie built larger steel mills to gain economies of scale; then he bought out other steelmakers in a time of economic recession in the 1870s to integrate the steel industry horizontally. In the next phase of his vertical integration scheme, he purchased both coal and iron mines that provided the raw materials and the heavy engineering corporations that used the steel. By 1900, Carnegie's United States Steel Corporation dominated the industry. Henry Ford took the concept of the factory **production line** to a new level. He applied it to the assembly of a wide range of components in the production of automobiles. His company produced thousands of a limited number of models for a growing market. His methods, called **Fordism,** were widely adopted by other industries and in other countries. The mechanization of production was another feature that made the United States the world leader in output of manufactured goods by the mid-1900s.

Canada Emerges

Not until the late 1800s did Canada begin the process of becoming a major industrial country. British rule that suppressed internal development, as well as a lack of capital and expertise, hindered the creation and expansion of Canada's global connections. After the British North America Act of 1867, Canada became a dominion within the British Empire with increased responsibility for its own affairs. At that stage, Canadian leaders began to integrate their vast country by building transcontinental railroads to encourage the settlement of the prairie grasslands and the Pacific coast. Manufacturing industries, including wood processing and steelmaking, were established behind tariff walls to protect them from American competition in particular.

Political Geography
Indigenous Peoples and Federal Governments

The stability and wealth of North America creates the perception for many that there are no regional geopolitical issues. Despite very strong economies and the presence of stable governments in Canada and the United States, many segments of the respective populations feel disenfranchised, lacking in a political voice, unable to compete for jobs, or marginalized by mainstream society.

Native American peoples are arguably the most spatially segregated community in the United States. Many Native

Americans feel abandoned by the federal government. Indigenous Americans are primarily concentrated on federally sequestered reservation lands that are often arid and unsuitable for agricultural productivity. Native American peoples within the United States are among the country's most materially impoverished communities. Complicating the reservation lifestyle was an administration structure imposed by and monitored directly from the federal government rather than state or local levels. Some 50 percent of Native Americans are unemployed and 90 percent are on welfare.

Attempts were made in the later 1900s to reduce the poverty imposed on Native Americans who had often been forced to live on marginal and largely unwanted lands. Native American communities with natural resources attempted to capitalize on the economic value of desired commodities such as fresh water and timber, while others filed lawsuits seeking compensation for being placed on resource-deficient land. Lucrative deals with governments and water projects, however, did little to change the economic structure of the Native American reservations.

New enterprises brought limited satisfaction. Some Native Americans developed tourist facilities on their reservations, but others chose to forgo such opportunities—often to avoid copying the commercial practices of European Americans. Casino gambling is now a significant source of income for several Native American groups. Among the casino-related challenges are managing the influx of capital, lack of high-wage jobs, combating gambling addiction, avoiding an increasing dependence on a single employment and revenue stream, and numerous social ills that often accompany a gaming-based culture. While casino-related development on reservations remains controversial, such enterprises may encourage some Native American groups to engage with the wider U.S. economy. To capitalize on their water, minerals, and tourist opportunities, Native Americans need to agree on how to administer and develop the lands that are held in trust for them. Native American peoples began to increase their proportion of the U.S. population in the 1970s for the first time since census records began in 1790.

In 1973, the Canadian government opened the possibility of negotiating indigenous land claims with organizations representing native peoples. Until that date, little had been done in much of Canada to implement treaties negotiated with native peoples in the 1800s. Today there are several areas throughout Canada where indigenous communities have local or regional governmental control. The largest area, Nunavut ("Land of the People" in Inuit language), became a new territory with its own elected government in 1999, although it will remain subject to federal control. Other agreements were reached in northern Québec and with the Inuvialuit people in the northwest Arctic. Further discussions are underway, although many of the smaller claims may take several years to resolve. Some are complex because of overlapping land claims and because bargaining involves the often-opposing interests of native groups, the federal government, and provincial governments.

North America and Geopolitics

The United States wields a great degree of political influence throughout all world regions primarily due to the size of its economy and its military prowess. The United States and Canada participate in a number of intraregional and interregional diplomatic, economic, and military alliances. Both the United States and Canada, along with Japan, Germany, the United Kingdom, France, Italy, and Russia, are part of the **Group of Eight** (G8), which is an economic discussion forum of the world's eight most materially wealthy countries. The United States wields significant influence in the United Nations (UN), the North Atlantic Treaty Organization (NATO), the World Bank, the International Monetary Fund (IMF), the Organization of American States (OAS), and the World Trade Organization (WTO), among many others. The UN, World Bank, IMF, and OAS are all headquartered in the United States.

The United States enjoys a distinctive role as the host country of the UN. The international headquarters of the UN is in New York City. The **UN Security Council** is the most powerful branch of the organization. The purpose of the Security Council is the "maintenance of international peace and security." The United States (along with China, France, the United Kingdom, and Russia) is one of five countries granted a permanent seat on the Security Council. The permanent members of the Security Council have the power to veto council decisions. The decisions of the Security Council are binding for the entire UN General Assembly.

Economic Geography

The United States of America is the world's most developed country in terms of economic prosperity, size, and influence. The United States maintains the largest total GNI of any country in the world, although Switzerland, Japan, and some Scandinavian countries exceed its GNI per capita (Figure 10.11a and b). The United States had the world's largest aggregate economy in 2005, over twice the size of China's, the second largest (see Figure 4.9). Canada, with a much smaller population, was eighth (when adjusted for purchasing power parity, Canada's aggregate GNI PPP is not among the world's top 10).

The international economic influence of the United States began to significantly expand after World War II during the economic recoveries in Europe and East Asia. Investment growth and influence diffused into many developing countries during the Cold War. After 1950, the United States established a huge lead in the initial development and use of computers. Although Western Europe and especially Japan now challenge this economic and technological superiority, no other single country is close to rivaling the total GNI PPP of the United States or the size of its internal market for products.

In 1988, the United States and Canada established the United States and Canada Free Trade Agreement, which led to the North American Free Trade Agreement (NAFTA) in 1994

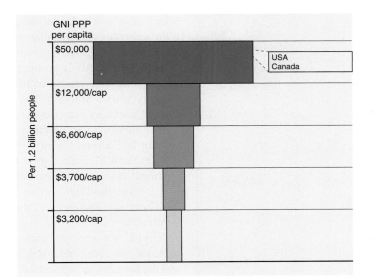

(a)

FIGURE 10.11 North America: country incomes and well-being compared. (a) GNI PPP figures for each country. (b) A comparison of the ownership of consumer goods and access to piped water and energy. The figures for the United States and Canada are among the world's highest. *Sources: (a) Data (for 2005) from* World Development Indicators, World Bank; *and Population Reference Bureau; (b) Data (for 2002) from* World Deevelopment Indicators, World Bank, 2004.

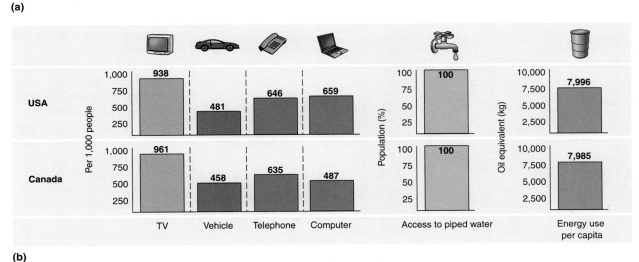

(b)

by adding Mexico to the arrangement (see "NAFTA," p. 334). The agreement between the United States and Canada continues to thrive under NAFTA, despite a series of disputes over individual items.

The Changing U.S. Economy

Manufacturing developed in the early 1800s in southern New England, New Jersey, and eastern Pennsylvania. It was based mainly on water mill power and produced metal goods, textiles, and leather goods. Numerous small mill factories required new transportation facilities to take their products to widespread markets, which contributed to the expansion of rail service in the United States. Railroads reached west to Chicago (Figure 10.12a) from several East Coast cities by 1860, and their construction formed the basis of iron industry expansion.

Major federal investments in transport infrastructure stimulated much of the spread of economic growth. Beginning in the 1860s, the transcontinental railroads were financed by federal and state governments granting lands along the routes to the railroads, which sold them and encouraged homesteading.

In 1860, the value of U.S. manufactured goods exceeded the value of commercial farm products for the first time. Agriculture became industrialized with the increasing use of mechanization and chemicals; markets for crops and livestock products were linked to the growing railroad network (Figure 10.12b).

Until 1950, manufacturing was the primary engine fueling the expansion of the U.S. economy. New consumer goods and transportation vehicles, including cars, trucks, and airplanes developed. These industries were less tied to sources of raw materials than the 1800s industries, and many were market-based or assembly industries where the best locations were central to a range of component producers or large markets.

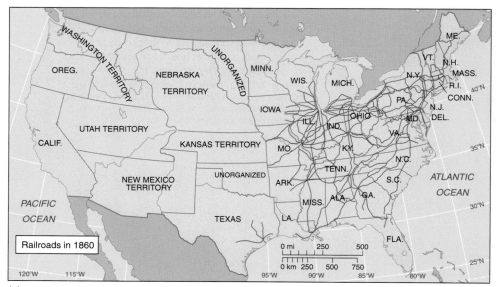

(a)

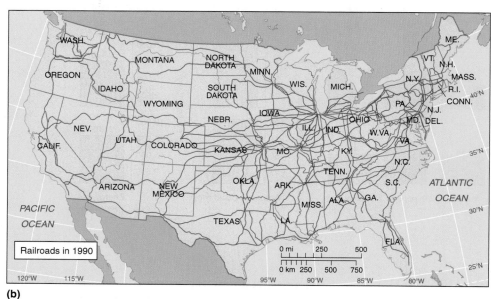

(b)

FIGURE 10.12 **United States: railroad networks.** (a) 1860; (b) 1990. The network was almost complete by 1900. The extension across the continent between the 1860s and 1900 supported settlement in the U.S. West. The building of transcontinental routes was held back until after the Civil War (1861-1865). In the later 1900s, many lines in the more densely populated eastern United States were closed as truck and air transportation were supported by federal investment in interstate highways and airports. *Source: Data from Guinness et al., 1985.*

States is a world leader in applying high technology to manufacturing and service industries. An increasing range of high tech goods is produced in the newer industrial areas, including the globally known Silicon Valley of California; metropolitan Boston (Figure 10.13); metropolitan Washington, D.C. (primarily Fairfax County, Virginia); the Research Triangle region in North Carolina; metropolitan Austin, Texas; the Denver-Boulder region in eastern Colorado; and areas of the Pacific Northwest.

Private non-good-producing industries, such as retail and wholesale trade and the services, accounted for approximately 70 percent of U.S. economic activity in 2005. More than half of U.S. economic activity in 2005 was generated through services such as health care, media and communications, banking, investments, securities, commodities exchange, accounting, and other financial services, publishing, computer-related services, arts, entertainment, and recreation industries among others. Among numerous service industries experiencing growth in the early 2000s were investment advice and wireless telecommunications services. By 2006, the employment structures and revenues generated in most major U.S. urban centers were dominated by service industries. Tourism and leisure-based services also became a leading part of many local economies. In 2005, 50 million tourists from other countries visited the United States, while 43 million went from the United States to other countries, and many millions of Americans were tourists in their own country. Such industries locate economic activity away from the workshop and factory. The United States, with its economic and technological lead, has so far been able to take greater advantage of this new economic base than any other country and is investing to maintain this lead.

Large U.S.-based corporations increasingly replaced higher paid service jobs in the United States with relatively less expensive employees working in offices in India in the early 2000s (see Chapter 6). Millions of educated, English-speaking workers in India are hired by big U.S. companies to fill "back-office" positions such as customer service call centers, accounting and tax services, computer software design, and technology and medical research. White-collar workers in technology-related positions in

While products diversified and became more technologically sophisticated, production and management techniques developed. American-based multinational corporations took their products to the world and opened factories in many countries in Europe and other continents.

In the later 1900s, U.S. manufacturing continued to be important, although it employed fewer than 20 percent of the work force compared to nearly 40 percent in 1950. Products were even more diverse, with some heavy industry and producer units moving to countries with lower labor costs. The United

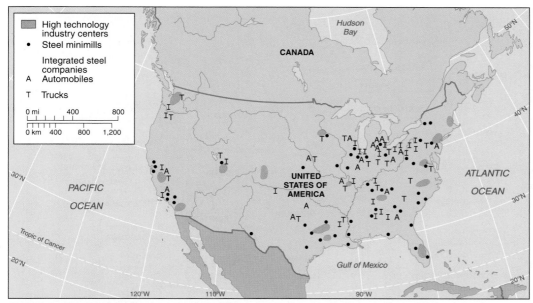

FIGURE 10.13 **United States: distribution of some major manufacturing industries, mid 1990s.**
The continuing concentrations of steelmaking and vehicle assembly in the Manufacturing Belt contrasts with the widespread nature of high tech industry centers. Steel production in minimills became more widely distributed from 1970.

Industrialization in Canada was a much later phenomenon than in the United States. Beginning before World War II and developing rapidly during the 1940s, the production of aluminum, vehicles, and consumer goods became important as Canada developed import-substitution industries to supply its own markets and avoid total economic domination by the United States. Hamilton at the western end of Lake Ontario became a steelmaking center. Montréal, Toronto, and Vancouver became centers of financial services and a wide range of commercial enterprises. The signing of NAFTA augmented trade between the two countries, resulting in increased U.S. investment in Canadian industries.

the United States are increasingly facing greater competition for fewer U.S.-based positions. Critics of this trend say corporations are exporting U.S. economic growth to India, while advocates argue that it makes U.S. companies stronger, and with an aging U.S. population, it is a way of supplying young, skilled workers to meet the needs of the U.S. economy without encouraging further immigration.

Canadian Economic Development

Canadian economic development generally followed that of the United States but often with a lag of several years until the later 1900s. In the 1990s, Canada combined a continuing emphasis on its natural resource base with being an affluent, high tech society. It had one of the highest proportions of trade per capita in the world. Canada's GNI PPP per capita rivaled that of many European countries in 2006 but was still not quite equal to that of the United States.

Until the mid-1900s, Canada's economy depended mainly on primary products such as grain, timber, and minerals. Canada remains a major world producer of newsprint, wood pulp, and timber, and is one of the world's leading exporters of minerals, wheat, and barley. Canada produces 30 percent of the world's newsprint, which, with wood pulp, is manufactured mainly along the lower St. Lawrence River valley. Most timber output comes from the west coast forests. The minerals that place Canada in the forefront of world mining countries include coal, oil, and natural gas from Alberta, iron ore from Labrador and Québec, uranium from Ontario and Saskatchewan, nickel from Ontario and Manitoba, and zinc from several places. Agriculture remains significant in the Prairie Provinces, southern Ontario, and the specialized fruit-growing districts of British Columbia, where wine production is gaining international attention.

Canada lagged behind its neighbor through the early 1900s, partly because of restrictions resulting from its colonial ties to Britain and partly because of the smaller size of its home market for goods and protection of its own industries. It gradually became more closely enmeshed with the U.S. economy and dependent on it for financing and markets. In the post-World War II era, rich stocks of metal ores in Canada's north along with coal, oil, and natural gas in Alberta and hydroelectric power generation in northern Québec ignited new geographic directions of development and provided the basis for growing Canadian affluence. Canada's centers of manufacturing production around Toronto and Vancouver, and to a lesser extent around Montréal and in Alberta, now rival individual centers in the United States in size and diversity of output. Canadians enjoy material living standards almost equal to those in the United States, and many who live in Canada strongly support the notion that they enjoy an even better quality of life. Some Americans prefer the Canadian way of life and live across the border from Detroit or Buffalo. The 2000 United Nations human development and gender-related development (HDI and GDI) indices, based on 1998 statistics, both placed Canada in first position above European countries, the United States, and Japan.

Geographic Diversity
The United States

The United States of America dominates the midlatitudes of continental North America with Canada to the north and Mexico to the south (Figure 10.14). Alaska and Hawaii are separated from the conterminous 48 states by Canada and the Pacific Ocean respectively.

FIGURE 10.14 The United States of America: the 50 states and major cities.

Metropolitan areas of the United States house the majority of the country's transportation and communications connections, the larger manufacturing facilities, corporate headquarters, and financial and business services. Some 85 percent of U.S. real estate value occurs in the 2.5 percent of the country's land that is occupied by urban areas.

The United States market economy, which is the world's largest, significantly influences economic activity across the globe. The United States remains a global leader in high technology and is the world's entrepreneurial leader. The United States was the world's leader in raising venture capital in 2004 with more than $17 billion raised in that year.

Problems of Affluence

The country's material wealth does not reach all within its borders. Those with advanced educations or double income families continue to garner greater material wealth, while those who are located in impoverished regions or lack access to educational opportunities struggle to acquire the basic necessities for nutrition, shelter, and health care.

The gap between the materially wealthy and the materially poor widened from the 1970s to the early 2000s in the United States. Some of the increasing inequality may be explained by the following factors. The first is lightly regulated labor markets that react to world conditions by depressing wages for unskilled jobs to compete with poorer countries, and increasing salaries for skilled workers as demand from service industries and professions rises. A second factor is polarized household incomes as the greater number of two-income homes contrast with a growing number of single parent households, which make up more than one-third of the poorest 20 percent of the population. Finally, lower paying service jobs have replaced relatively higher paying manufacturing jobs, while technological advances have increased the need for more highly skilled and specialized workers. Increasing numbers of lesser skilled immigrants are filling low-paying service jobs and have become a growing component of materially poor residents of the region.

The materially poor areas of many inner cities have lower-quality buildings and services such as education and health because of the inability of small jurisdictions with predominantly poor populations to support such services. The contrasts produced by **uneven development** are illustrated by the example of Boston, where some of the best public secondary education in the country is found in the suburbs while some of the worst is in Boston's inner city—both ends of the spectrum in the same metropolitan area.

The income and material wealth gaps are often linked to perceived ethnic and racial differences. High income-earning people have many choices with respect to where they live. Many choose to congregate with those of similar socioeconomic standing in trendy downtown areas or established suburban neighborhoods. They may live close to shopping, ethnic restaurants, cinemas, places of worship, or recreation facilities. Less-affluent groups in U.S. society have fewer opportunities for choosing where they live and become segregated into communities that more-affluent groups avoid. Such segregated groups often occupy inner-city areas that contain high proportions of African American, Latin American, or recent immigrant communities, most of whom have poor access to high-quality education and job opportunities.

Other problems of affluence result from the environmental impacts of intensive use and extraction of resources. In the early 2000s, the United States, with just under 4.6 percent of the world's total population, consumed 40 percent of the world's oil production. Such disproportionate consumption often places the United States at the forefront of international protests.

In the beginning of the 1970s, the United States passed stringent environmental legislation to improve its air and water quality. Following the rise of the environmental movement in the United States and related legislative changes, air and water quality standards improved in the 1980s and 1990s. Such regulations, however, sometimes led to the relocation of the most polluting industries to more materially impoverished countries.

Urban Landscapes

Urbanization within the United States produces distinctive landscapes through the types of building, the differentiation of land uses, and the distribution of groups of people. Human variation in the landscape includes the materially wealthy and the materially poor, African Americans, Hispanic Americans, European Americans, Asian Americans, and many others. American cities grew from colonial ports and inland market centers to 1800s industrial and 1900s commercial metropolitan centers.

Industrial and Commercial Cities   In the later 1800s and early 1900s, many cities in the eastern and central United States expanded into large conurbations of housing, factories, shops, and offices with the rapid growth of manufacturing industry and the railroad network. The larger factories and mills required many workers, who were accommodated in surrounding housing units.

By 1920, American cities often evolved into a **concentric pattern of urban zones** around the central business district (CBD), with poorer housing in inner suburbs and more-affluent zones beyond (Figure 10.15a). In cities built across a local relief of hills and valleys, the development of railroads, roads, and commercial activities was concentrated along the valleys, giving the city geography a pattern of wedge-shaped urban sec-

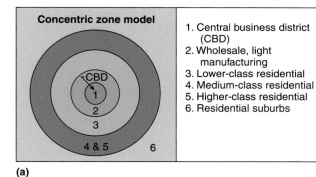

(a)

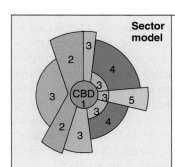

(b)

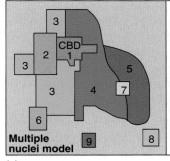

(c)

FIGURE 10.15   United States: urban landscapes. (a) The concentric zone model, based on a 1922 study of Chicago. (b) The sector model, based on 1930s census information, emphasizing the impact of railroads and highways on the location of land uses. (c) The multiple nuclei model of 1945 depicts a large central business district and smaller peripheral business district. *Source: Data from Harris et al., 1945.*

tors of different land uses (Figure 10.15b). In some cities, the distinction between land uses was clear, but the distribution was not in such regular patterns, establishing multiple nuclei (Figure 10.15c) in which industrial and commercial activities occurred in several specialized and separated zones, instead of around a single CBD.

An explosion of house and road building opened up huge suburban areas of single-family homes for those with rising incomes after World War II as the United States dominated the world economy. New federal roads that linked major cities, followed by the interstate highway system established in 1956, made greater mobility possible and encouraged suburban housing and commercial development.

These trends gave rise to a new form of city that developed from the monocentric (single center) before 1940 to the multicentric pattern of today. The CBD became more specialized around financial businesses as many of its shopping and industrial functions moved to the suburbs. Competition for space elevated land prices, which precipitated the construction of tall office buildings primarily in each city's commercial center. New highways, often elevated or in tunnels, circled the CBD.

Beyond this sector of intense commercial activity, the older pre-1940 suburbs, crossed by railroad yards and older highways with commercial strip developments, were often taken over by expanding African American, Hispanic, or other ethnic groups. Such groups either did not have the resources to move to newer suburban homes, faced discrimination by financial institutions or home sellers when they tried to do so, or preferred to stay in an area that was familiar to them. The neighborhood concentrations of African Americans in northern cities did not appear until influxes of African Americans from the U.S. South combined with the suburb-bound movements of the European American population of northern cities in the 1950s.

The changing face of U.S. cities since 1950 resulted in many movements of people within the cities. The growth of the suburbs involved the movement of wealthier families out of the older inner-city areas to new suburban single-family homes. This trend was labeled "white flight," since it resulted in low-income African Americans and other minority groups being left in the older areas that lost many services and jobs to the growing suburbs. Property values in the older areas declined, sometimes leading to abandonment and dereliction.

Postindustrial Cities
From the late 1970s, expanding suburban growth created several smaller suburban cities around the primary urban center. Such suburban developments contain significant commercial space and service industries. The term **"edge city"** was coined for new exurban developments relying primarily on car and truck transport and secondarily on air travel. Edge cities include large developments of shopping malls, offices, warehouses, and in some cases factories, located on the edge of major metropolitan areas (Figure 10.16). Since the 1970s, there were some movements of people from the suburbs farther out from the city into surrounding rural areas, and other movements back to the city center in a process known as gentrification.

Gentrification
Gentrification is the movement of higher-income residents and business owners into materially poor areas of inner cities, leading to the physical and economic improvement of property and potentially dramatic changes in the quality of life in the districts affected.

The availability of inexpensive property coupled with the proximity to in-town jobs attracted many suburbanites to move back into inner-city areas in the 1970s. The proximity of inner-city neighborhoods to a wealth of urban amenities, including ac-

FIGURE 10.16   An edge city: Bethesda, Maryland. New blocks of offices and apartments mingle with parks, highways, and shopping centers, typical of new urban developments on the edges of major cities.

cess to leisure-time facilities and theaters, symphony halls, opera houses, museums, restaurants, improved parks and waterfront areas, and major sports facilities fueled dramatic urban renewal and inner-city growth for many metropolitan centers through the early 2000s. The primary drawbacks consisted of higher crime rates and less funding for public schools. Crime rates have decreased in many inner-city areas resulting from changes in police and city government attitudes as well as changes brought about by the gentrification process itself. Schools have improved in some areas, while in others, it remains common for younger affluent families, who begin living in the inner city, to move out to suburban locations when their children are of school age, or to send their children to private schools.

The movement of high-income people into city centers often causes the displacement of more materially impoverished occupants of the urban zone that is undergoing the gentrification process. As money moves into previously poor neighborhoods, old buildings are renovated or replaced with new structures. The cost of commercial and residential property in the gentrified areas escalates dramatically. Housing opportunities for the materially poor, who can no longer afford to live in the gentrified neighborhoods, are often not part of the process. Proponents argue that the process dramatically increases the economic and social well-being and standards of the gentrified neighborhoods and the CBD as a whole.

The extent of gentrification is great in terms of the number of cities affected by it, but often modest in relation to the total built environment of each inner city—several blocks rather than larger tracts. In Manhattan, New York, for instance, an area of the Upper West Side between Central Park and the Hudson River and near the Lincoln Center and museums experienced gentrification. Similar areas have been identified in Baltimore, Boston, Philadelphia, Pittsburgh, Washington, D.C., Atlanta, and San Francisco.

Regions of the United States

The large areal extent of the United States makes it a country of many internal geographic regions that developed from the interaction of people with the natural environments and resources. Regions rarely have perfect boundaries of cultural change, as exhibited in the overlapping set of regions in the U.S. Northeast (Figure 10.17) which are a reflection in part of the initial geographic concentration of European settlements and the subsequent industrial and urban development emanating from early settlements.

New England   Airplane engine-related industries, high tech computer design and manufacturing, and Cold War defense-related industries dominated New England's economy during WWII and for several subsequent decades. The region's large number of higher education institutions combined research innovations with the local availability of entrepreneurial capital to create a post-WWII boom of economic development. This is sometimes referred to as the "Massachusetts Miracle" because of the numbers of new jobs created. New England's financial and professional services continue to grow today and its natural

environment of rocky and sandy coasts, wooded hills, interior mountains, and well-preserved early American history, appeals to many Americans as a place to vacation or as a high-quality living environment.

Megalopolis   The **Megalopolis** region stretches from the greater Washington, D.C., area (by some definitions, as far south as Richmond, Virginia) at the southwest end of its long urban arm to Boston at the northeastern end. The urban chain of the Megalopolis includes the huge metropolitan areas of Baltimore, Philadelphia, and New York, as well as many smaller urban centers, positioned between metropolitan Washington, D.C., and Boston. Approximately 50 million people—nearly one-sixth of the total population of the United States—live in this region. The Megalopolis overlaps the Middle Atlantic and New England regions. The major cities of this region grew out of the country's initial port, commercial, and transportation hubs.

Despite the migration of some corporation headquarters out of the Megalopolis cities to other parts of the United States, this region still dominates the American economy, federal politics, and the modern media industry. Other groups of cities within the United States have been identified as new versions of the megalopolis

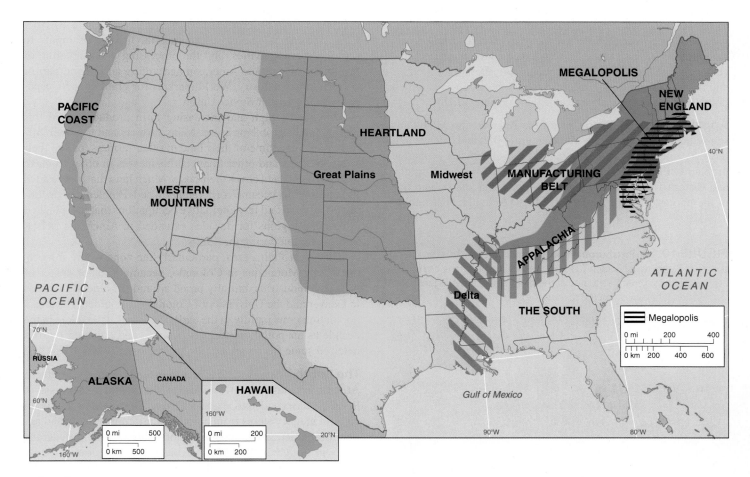

FIGURE 10.17 United States: the regions. The regions reflect differences in natural environments and patterns of human occupation. The overlap of regions in the northeast resulted from early concentrations of people and changing economic and social emphases.

phenomenon, but none matches this region in its numbers of people or its economic, political, or social leadership.

Manufacturing Belt   Between the 1800s and mid-1900s, an area stretching from Boston on the East Coast through New York and Philadelphia westward to Pittsburgh, Cleveland, Detroit, and Chicago became so dominated by manufacturing industries that it was termed the "Manufacturing Belt." This region produced over two-thirds of U.S. manufactured products until challenged by the West Coast and southern states during and after World War II.

Although the old Manufacturing Belt remains dominant in manufacturing output within the United States, service industries continue to take on increasing significance in this region. Chicago became a major center for financial services arising out of its futures trading in agricultural products. Pittsburgh, once considered the industrial furnace of the United States, underwent a dramatic transformation during the 1970s and 1980s, and in 2006 is one of the best examples of urban renewal in the United States. Pittsburgh played a prominent role in the growth of the United States through the development of an intense concentration of iron and steel production and related heavy industry. In addition to being known for its significant industrial output, the city became infamous for its pollution. During the city's renaissance of the 1970s and 1980s, many manufacturing jobs were replaced with service positions. Selected factories closed to make way for Fortune 500 companies. The city emerged as an international center for medical research and health care. Pollution-filled air and the glow from the blast furnaces have been replaced with a dramatically modern and vertical skyline (Figure 10.18).

Appalachia   The hilly region of the Appalachian Mountains spans the middle part of the Manufacturing Belt and extends into the U.S. South. Each part of Appalachia has distinctive economic problems. Eastern Kentucky, West Virginia, western Virginia, and eastern Tennessee form central Appalachia and remain one of the poorest parts of the United States. Although coal mining

FIGURE 10.18 Pittsburgh, Pennsylvania. The modern vertical skyline of Pittsburgh's "Golden Triangle," the local name for the city's central business district, is situated at the point where the Allegheny and Monongahela rivers merge to form the Ohio River.

continues to be somewhat important, it is subject to significant price fluctuations that produce inconsistent regional conditions. A measure of economic diversity was introduced into southern Appalachia with the production of cheap electrical power through the **Tennessee Valley Authority** from the late 1930s. Major aluminum companies and federal facilities, such as the Oak Ridge atomic laboratories and the Huntsville rocket center, were joined by many other manufacturing concerns. Cities such as Asheville, Knoxville, Chattanooga, and Huntsville expanded, developing a wide range of service industries. Contrasts of material wealth, infrastructure, health care, and educational opportunities remain between the urban centers and the surrounding, more materially impoverished rural mountain communities.

U.S. Heartland: Midwest and Great Plains   The lowland area between the Appalachians and Rockies that is north of the Ohio River and the Ozark Mountains is an essentially agricultural region, including eight of the top 10 U.S. farming states.

The term "Midwest" is mostly applied to the eight states of the Corn Belt and Great Lakes area. The eastern part overlaps with the Manufacturing Belt, so that cities such as Chicago, Detroit, Cleveland, and Cincinnati stand amidst the world's most productive farming region. The Great Plains include North and South Dakota, Nebraska, Kansas, Oklahoma, and parts of Montana, Wyoming, Colorado, and northern Texas (Figure 10.19).

Agribusiness—the close commercial linking of inputs to farming, farm activities, and the processing and marketing of farm products—made American agriculture an integral part of a much larger industry. World markets for grain opened, especially in the former Soviet Union.

Despite the high and increasing farm productivity of this region, major problems arose. Many farmers had to obtain off-farm jobs, rent some of their land to other farmers, or sell their lands to banks and other financial businesses in order to make loan payments and cover their costs. Much land that had resided in family farms became corporately owned. Problems of declining population and few alternative occupations that had plagued other rural regions for decades extended to America's richest farming region during the 1980s.

The Great Plains area, particularly the zone where it meets the Rocky Mountains in Colorado, became a major center of high tech industries, initially based around the federal facilities in Denver and the University of Colorado at Boulder. The area's population grew rapidly in the early 1990s but slowed late in the decade as air pollution, water shortages, and crowded schools deterred some from moving into the region.

The South   The American South extends from the southern Atlantic coastal plain westward to the southern Mississippi River and beyond into eastern Texas. The region's common cultural identity stems from colonial migration patterns and agricultural practices, as well as membership of the Confederacy states in the Civil War. Most of the South was dominated by the plantation economy, established in colonial times. All of the southern states suffered after the Civil War, experiencing decades of poverty at a time when the northern states made huge strides in industrial expansion and garnered significant material

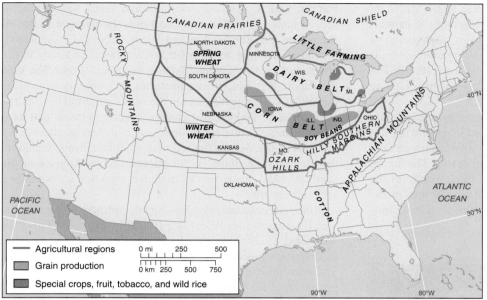

FIGURE 10.19 **United States Heartland: agricultural regions.** The Corn Belt is the most productive and versatile sector, where corn and soybeans ("grain production" areas on map) are grown for direct sale or feeding livestock (cattle in the west, swine in the east). In the cooler north dairying is important, and in the drier west, wheat and other small grains are dominant, while cattle feedlots dominate toward the south. High-value crops are grown on high-value land close to cities. Special conditions favor fruit growing on the eastern shores of the Great Lakes. Poor soils, short growing seasons, and mountains mark the boundaries of the world's most productive farming region.

wealth. Black slaves were "freed" by the outcome of the Civil War, but soon became debt slaves of the agricultural lands on which they had been working.

The African American descendants of the former slaves did not gain political and social rights for more than 100 years until the civil rights movement of the 1960s. This development coincided with the building of the interstate highway system that opened up much of the South to new economic opportunities and the expansion of cities such as Miami, the urbanizing area from Charlotte to Atlanta, New Orleans, Houston, Dallas–Fort Worth, Memphis, and Louisville. Service industries, as well as manufacturing, moved into the area, and its changing fortunes were highlighted by local boosters who took up the term "Sun Belt" to refer to the U.S. South with its mild winters and sunnier climate—in contrast to the "Frost Belt" of the U.S. Northeast. Although this term was also applied farther west and much of the economic growth associated with the Sun Belt was geographically patchy—more a set of "sun spots" than an evenly prosperous belt—the changes in outlook in the region were real and lasting.

Rural areas remain the poorest parts of the South, with the Mississippi Delta lowlands forming one of the most economically challenged areas in the United States. The Delta lowlands extend up the Mississippi River valley from Louisiana and Mississippi through Arkansas and into southeastern Missouri. The Delta was a major area of cotton production until the mid-1900s. Today, the region remains materially poor with relatively less infrastructure and fewer educational opportunities.

Most Southerners now work in manufacturing and service industries. In the environs of Greenville, South Carolina, production of Michelin tires (from France), BMW cars (from Germany), Lucas automotive electronics (from Britain), and Hitachi electronics (from Japan) provides jobs for many residents. Such investments were based on the perception of this region as having skilled labor, often without labor unions, good local support for industry, and pleasant environmental conditions for management. In Texas, manufacturing paralleled the accumulation of oil wealth and services development around Houston, Dallas–Fort Worth, and Austin partly related to the National Aeronautics and Space Administration center in Houston and other high tech research industries.

The service industries in the South include tourism, with Florida leading the subregion in visitor numbers and tourism revenue, health care, education, and financial services. Some service industries pay higher wages than semiskilled jobs in manufacturing.

Several strong hurricanes made landfall in southern states along the U.S. Gulf coast in 2004 and 2005. The costliest storm in United States history, Hurricane Katrina, struck southeast Louisiana and southern Mississippi on August 29, 2005. Katrina left more than 1,000 people dead and produced tens of billions of dollars in damages. Gulf coast states from Texas to Florida will need several years to recover from the destruction.

Western Mountains The western one-third of the United States is a mountainous region. The highest ranges are the Rockies on the east and the Sierra Nevada of California and the Cascades on the west, separated by broad plateaus (see Figure 10.1) and areas of basin and range. The mountains block much of the humid westerly winds flowing into the region from the Pacific Ocean resulting in vast arid rain shadow areas throughout the interior. Settlement focuses on irrigated farming and towns that grew up as markets, mining centers, or nodes on the transcontinental railroads. A large proportion of the land between such centers is still sparsely settled. Much of the land in this region is owned by the federal government, including more than 70 percent of the state of Nevada.

Most of the water available in the region falls on the mountains as snow and runs off in swollen streams during the spring season. Many federally funded irrigation projects manage this water supply, with the largest group of projects along the Colorado River in the south and the Columbia River in the north. Irrigated farming and cheap electricity created a series of productive oases that formed the basis for the growth of cities such as

Las Vegas, Nevada, Phoenix, Arizona, and Salt Lake City, Utah. Rapidly growing metropolitan Las Vegas continues to capitalize on its gaming and entertainment industries, sunny and dry climate, and proximity to national parks and government military bases, which serve as large employers (Figure 10.20). The Las Vegas metropolitan area enjoys a healthy business climate and a relatively low cost of living.

In the southern part of the region, Hispanic people make up a large proportion of the population. Defined in part by the Rio Grande, the border with Mexico is largely a mountainous desert where few people live outside the border towns. The border cities are twinned (Figure 10.21) and hold three-fourths of the Americans and Mexicans living within 80 km (50 mi.) of the border—a population that rose from 9 million to over 15 million people since 1980.

FIGURE 10.20   Las Vegas: New York, New York hotel and casino.
New York, New York is one of several themed resort and casino complexes that were built in Las Vegas in the 1980s and 1990s—an era when the local gaming and tourism developers attempted to "reinvent" the city to attract more visitors. In the early 2000s, developers began to move away from the themed properties and back to more traditional resort-style developments.

Pacific Coast   The American Pacific coast has become a second national core, close in economic importance to that of the region between Boston and Washington, D.C. Settlement increased after the arrival of transcontinental railroads in the 1870s and 1880s. National defense needs for World War II in the Pacific then placed manufacturing and military centers on the West Coast. Puget Sound in the north and San Diego in the south became major naval centers. Seattle grew in part as a manufacturing center for military and commercial aircraft. By the early 2000s Seattle housed many software firms, including the corporate headquarters of Microsoft. Los Angeles in the south also had major aircraft manufacturers and an expanding service industry.

Both San Francisco and Los Angeles attracted financial service businesses after World War II, and by the mid-1990s high tech jobs, a more diversified range of service industries, and increased overseas exports created thriving economies in each. The metropolitan Los Angeles economy in the early 2000s was greater than the economy of South Korea, and the city now contains a thriving motion-picture industry, a growing major port, and more computer software jobs than Silicon Valley, the best-known high tech center situated just south of San Francisco. California has the world's eighth largest economy and is America's most productive farming state.

Important to economic growth in the southern half of California was the supply of water to its largely arid area (Figure 10.22). Federal and state government funds supported huge water storage and distribution projects following the early appropriation of water by the cities of Los Angeles (Owens Valley) and San Francisco (Hetch-Hetchy).

Years of drought stretch limited water supplies. It is likely that farmers, who use 85 percent of the water to produce less than one-tenth of the state's economic output, will have to pay more in the expectation that they will use it more carefully. At present, farmers pay only half the cost of delivering water to them, while city dwellers pay 20 times as much as the farmers pay.

Arizona, California, and Nevada, all compete for fresh water from the Colorado River basin. The states have all signed an

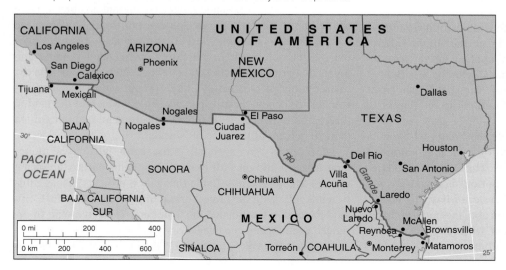

FIGURE 10.21   The United States-Mexico borderlands. The contacts between the two countries led to a series of twinned towns with related economies. U.S. manufacturers situate factories in Mexico to take advantage of cheap labor costs, as well as less regulation and enforcement of worker safety standards and polluting effluent.

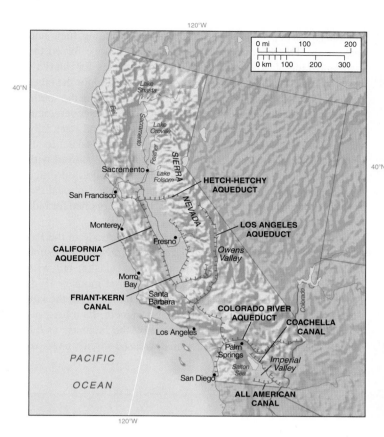

FIGURE 10.22 **Pacific Coast: water flows in California.** Water flows from the high precipitation areas of the north, where reservoirs store the water, toward the arid south. Water supplies for San Francisco (Hetch-Hetchy) and Los Angeles (Owens Valley) in the early 1900s were later enhanced by federal and state water projects in the Central Valley. The federal Imperial Valley project and Colorado River Aqueduct in the far south took water from the Colorado River to irrigate dry lands and provide water for Los Angeles and San Diego.

(a)

(b)

FIGURE 10.23 **Growth and development in the desert.** Rapid growth and development in metropolitan Las Vegas expands across the desert valley floor. A new golf course community (a) and several new subdivisions (b) required large amounts of water from very limited resources. Nevada, Arizona, and California all compete for fresh water from the Colorado River.

agreement, which limits each to an allocated percentage of fresh water from the river. Arizona is in the process of creating its own distribution system, the Central Arizona Project. Las Vegas is one of the fastest growing metropolitan areas in the United States. The water needs for metro Las Vegas increase significantly each year as both the population and the number of tourists visiting the region continue to grow (Figure 10.23). Increasing water demand in the Las Vegas Valley encroaches on the supply of water downstream in Southern California. As Southern California's water needs increase, officials are looking to alternative sources in the water-scarce region.

The West Coast developed a new importance through an increasing volume of business and personal movements with Asia and other Pacific countries. Seattle and especially Los Angeles/Long Beach are major world ports. The West Coast attracts most of the growing Asian investment in the United States. Five of California's top 11 banks are Japanese owned; 30 of the smaller ones are backed by Chinese money. Asian capital, particularly Japanese, backs car design centers, media indus-

try corporations, and leisure industries. Half of the 1,400 Taiwanese companies in the United States are based in California, and research-marketing-manufacturing links across the Pacific Ocean are increasing.

California's proximity to Mexico, and the state's trade links with the Latin America region continue to attract Hispanic peoples from south of the U.S. border. Southern California has the highest proportion of Hispanic Americans in the country, with

Los Angeles having doubled its middle-class Hispanic population since 1980 to more than half a million; its Hispanic businesses have doubled since 1993.

Alaska and Hawaii

In 1959, Alaska and Hawaii (Figure 10.24) were added to the United States. They contribute to the extreme variety of environments and resources within the United States but have their own issues.

Alaska is a huge area of northern land that was bought from Russia in 1867. High mountains, ice fields, and glaciers mark its southern coast. Inland, the Yukon lowlands have long cold winters, and to the north, the Brooks Range and North Slope leading down to the Arctic Ocean are even colder. Any potential for commercial farming is severely limited by climatic conditions. The Yukon lowlands widen westward to the Bering Sea, but there is little mining or manufactured output to trade in that direction. The dramatic mountain scenery and abundant wildlife widespread in the state do support a growing tourist industry.

Alaska's Pacific location and proximity to Russia make it a strategic site for U.S. military bases, and the U.S. Government owns a majority percentage of Alaska's land area. Alaska also contains vast deposits of natural resources including fuels and metal ores. The discovery of oil on the North Slope forced the state and federal governments to develop policies for Alaska's native peoples, who live mainly in coastal areas, and for the other settlers, most of whom live in and around Anchorage in the south. Residents of the state receive payments from the state government from taxes on oil production. The oil discoveries, and especially the oil spill from the Exxon *Valdez* tanker, made Alaska a focus of environmental concern. Although oil exploration slowed with the prevention of drilling in the Arctic National Wildlife Refuge on the north coast east of Prudhoe Bay, parts of the National Petroleum Reserve to the west of Prudhoe Bay were opened for bids in 1998, with pumping expected to begin in 2010.

Hawaii was a long-time hinge-point of U.S. trade in the Pacific, a naval port, and a source of agricultural products such as pineapples and sugar. Tiny compared to Alaska, but with more people, it has a major international tourist industry (Figure 10.25). Hawaii experienced economic challenges in the late 1990s and early 2000s with high unemployment. Its tourist industry suffered from cutbacks on travel by the Japanese as the yen lost value. Sugarcane production decreased slightly as Asian

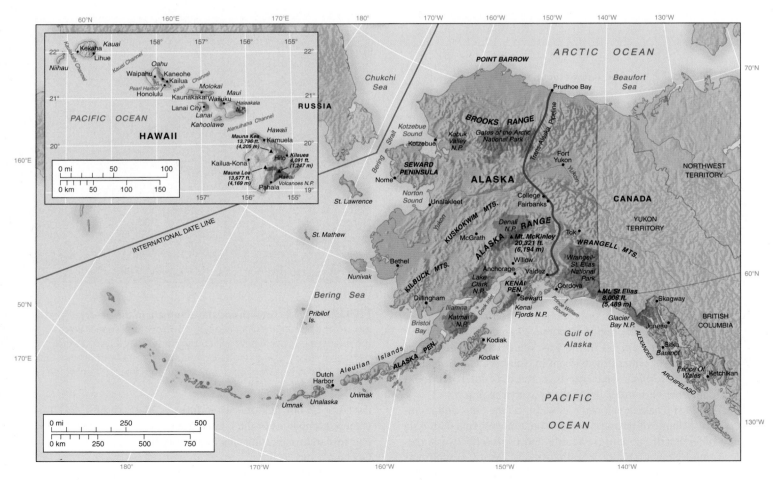

FIGURE 10.24 Alaska and Hawaii: the last two states to join the United States of America. They contrast in size, climates, and resources. The much smaller Hawaiian Islands have twice the population of Alaska.

FIGURE 10.25 Hawaiian Luau. Tourists visiting the Polynesian Cultural Center on the island of Oahu, watch as center staff prepare a roasted pig during a Luau ceremony.

competition and changing U.S. consumption habits reduced the market. Hawaii also suffers from a poor business climate due to many regulations, poor schools, a high union presence, and its geographic isolation from globally connected business systems. Coastal areas of the state attract wealthy retirees from the mainland United States and other parts of the world, creating dramatic disparities in real estate values and material wealth.

Canada

Three thousand miles of distance from Victoria, British Columbia, in Canada's southwest to Halifax, Nova Scotia, on Canada's southeast coast, coupled with the narrow width of the concentrated region of population along the country's southern border results in a series of regionally individualistic groups of people (Figure 10.26). Geographic logic might suggest the likelihood of Canadians interacting more easily with cities and systems across the border in the United States. Although Canadians do interact with urban systems across their border, the people function internally in domestic subregions and on a country level through their federal government. The distinctive subregional groups within Canada's borders include the people of the Atlantic Provinces, the French-speaking peoples of Québec, the multicultural population of Ontario and especially of Toronto, the scattered farming populations and industrial cities of the Prairie Provinces, the people of Vancouver and the British Columbia coast who look increasingly toward Asia, and the indigenous

Canadians, or First Nations, of the northlands. The Canadian provinces were created a century after the U.S. states, are generally much larger, and have greater internal political control than U.S. states do with respect to the federal government.

In the 1990s, a growing minority of independence-seeking people in Québec attempted to secede from Canada, potentially fracturing the Canadian state into at least two or more individual countries. In addition, the closeness of the major centers of population and commercial development to the United States and the extensive coverage of entertainment and news media from across the border in the United States have powerful influences on modern Canadian life that could blur the clarity of a cohesive Canadian national identity. The unique population distribution of Canadians, combined with living in the shadow of the globally mighty United States, actually appears to provide strong incentive for, rather than detriment to, the formation of a national identity (the Québec issue aside). Perceptions of people living outside the region, as well as many residents of the United States, that Canada is nothing more than a "satellite" or "subsidiary" of the United States may provide momentum for Canadians to show the world they are separate and different from the United States.

The United States continues to be seen as both a friend and an opponent. Canadians enjoy many benefits of affluence and national security because of their close integration with the U.S. economy and defenses. Canada, along with the United States, is a member of NATO. Canada has a history of military alliance and cooperation with the United States. Both countries worked together to maintain a line of radar warning stations facing the Soviet Union across the Arctic Ocean during the Cold War. Canada and the United States participate in open free trade and (with Mexico) are each other's most important trade partners. The two countries first entered into formal open trade relations with the United States–Canada Free Trade Agreement of 1987–1988, which eliminated most tariffs and trade barriers. Both countries then joined with Mexico to sign and formally participate in the NAFTA in 1993–1994.

Political and economic ties are not the only areas in which Canadians cooperate with the United States. Much of Canada's television programming is U.S. produced. The two countries are integrated into many of the same professional sports leagues, including the National Hockey League, the National Basketball Association, and the professional baseball leagues. Hockey, basketball, and baseball teams representing the larger cities in Canada participate equally with the U.S. professional teams representing U.S. metropolitan areas.

Although Canada benefits from its geographic proximity to and societal commonalities with the United States, situations resulting from this proximity elicit criticism from the Canadian population toward the United States. Many Canadians object strongly to being the recipients of acid rain from pollution generated by heavy industry in the U.S. Ohio River valley. In the late 1990s, problems arose on Canada's west coast over salmon fishing yields in Alaska, British Columbia, and Washington. Canadians enjoy very low crime rates and have willingly accepted

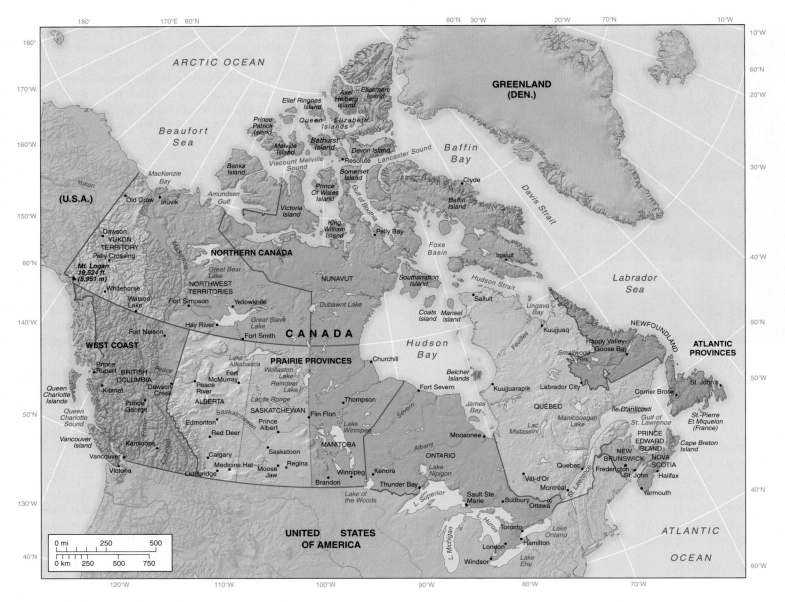

FIGURE 10.26 Canada: regions, provinces, territories, and cities. The Canadian provinces have greater powers than the individual states of the United States. Their geographic distribution in a line north of the U.S. border makes communications difficult among different parts of the country. The red lines indicate the division of regions within Canada.

more stringent gun-related legislation from their government. There is a perception in Canada that the United States is exceptionally crime ridden, and many Canadians joke that everyone in the United States is walking around with a gun.

Although living next to an economic giant generates some resentful and even fearful attitudes among Canadians, a few Canadian activities present challenges for the U.S. government. One area of Canadian activity that is of great concern to the United States is the smuggling of high-quality marijuana into the United States from its production area around Vancouver. In the 1990s, Canadian companies took over the development of Cuban mining from the former Soviet groups, despite a U.S.

embargo against Cuba. Nonetheless, the positive economic aspects of the juxtaposition of the two countries, the shared media and trade, the political cooperation, and the generally friendly attitudes the residents of the two countries have toward one another far outweigh any negative attitudes or friction emanating from either country.

Canadian City Landscapes

Of Canadian cities, only Québec has a historic heart with buildings older than the 1800s. Most cities from Toronto westward were built almost entirely in the 1900s. Other major differences between Canadian and American cities include the level of planning involved and the relationships of government units within metropolitan centers.

Toronto is the largest Canadian city and the center of Canadian financial services and manufacturing industries, as well as the provincial capital of Ontario, Canada's wealthiest province (Figure 10.27). When Toronto began to extend suburbs into surrounding jurisdictions in the 1950s, the province of Ontario required that these jurisdictions and the city plan cooperatively to make possible the amalgamation of services and regional road construction (Figure 10.28). Further growth in metropolitan Toronto was encouraged around hubs outside downtown, including North York, the area adjacent to the international airport, and a new center with coordinated and concentrated development. A major project was also undertaken to redevelop the waterfront.

Toronto was able to cope with increasing immigrant groups moving into older inner suburbs: Italian, Greek, Portuguese, and Chinese districts form distinctive ethnic enclaves. The city capitalizes on the increasing diversity of its residents. Restaurants, stores, art galleries, and festivals reflecting the vast international heritage of the citizenry continue to multiply throughout the city. Locals and tourists alike crowd ethnic business establishments, helping the diversifying economy of the metropolitan region. Toronto retains a busy downtown and surrounding older suburbs, and it has the lowest homicide rate

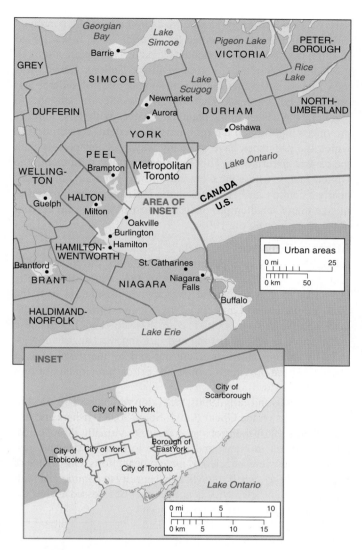

FIGURE 10.28 **Spatial extent of metropolitan Toronto.** The constituent jurisdictions and surrounding counties in southern Ontario. Toronto's development and urban planning are carefully controlled by the province of Ontario.

of large North American cities, uniformly good schools, good public transportation systems, and controlled urban sprawl. Many cities in the United States envy Toronto's development situation, since they have problems of multiple jurisdictions and limited overriding planning controls that affect only a few functions. Some U.S. citizens, however, find Toronto too ordered and even monotonous.

Regions of Canada

Canada has a federal government that links the 10 provinces, the Northwest Territories, the Yukon, and Nunavut—the country's main political regions. This arrangement was decided by Britain in the 1867 Act of Confederation, uniting the different

FIGURE 10.27 **Central Toronto.** The Skydome, CN Tower, and downtown commercial buildings seen from Lake Ontario. The lake frontage made Toronto a major port with the opening of the St. Lawrence Seaway in 1959. By the late 1990s many dock areas along the waterfront underwent extensive redevelopment to provide public recreation and tourism opportunities.

parts of what became modern Canada in the face of a perceived military threat from the United States after its Civil War ended in 1865. In 1982, Canada ended its legal ties to Britain, although it remained within the Commonwealth of Nations. Canada was then free to determine the constitutional roles of federal and provincial governments. That was not easy. Several attempts to reconcile the wishes of the people of Québec with those of other provinces highlighted the problems of a federal constitution in which the provinces have, in many ways, greater powers than the central government. This situation contrasts with that in the United States, where the Constitution was generated internally nearly a century earlier. States in the U.S. are generally smaller in size than Canadian provinces and lost powers to the federal government during the 1900s.

Atlantic Provinces Canada's political tensions are enhanced by its regional economic and cultural differences. The east coast was settled first and forms the small hilly Atlantic Provinces of Newfoundland (which has jurisdiction over the almost uninhabited Labrador), Nova Scotia, Prince Edward Island, and New Brunswick. The small-scale economy of these areas, based at first on fishing and farming, was augmented locally by mining and manufacturing and by the naval base at Halifax. These provinces remain the primary recipients of federal regional aid.

In the 1980s and early 1990s, the region was hit badly by declining fish stocks on the Grand Banks. Some 30,000 fishers and fish plant workers lost their jobs as cod stocks virtually disappeared. Unemployment in Newfoundland rose to over 20 percent. Few new industries were attracted by government efforts. One source of hope could be to switch some fishing capacity to commercial sealing, although environmental concerns must be assessed against the economic plight. Other prospects, such as pumping oil from the Hibernia field 315 km (200 mi.) offshore in the main iceberg lanes, would be very costly and environmentally risky. The development of nickel and cobalt mining at Voisey Bay, Labrador, which began construction in 2002, is expected to be in production by 2007 and could bring wealth to the owners and possibly to a relatively small population of local workers. It is significant on a world scale, however, that the cobalt from Voisey Bay replaces the falling output from the previously dominant mines in the Democratic Republic of Congo.

Although the Atlantic Provinces contain some groups of French speakers, particularly in New Brunswick, sympathy for them is eclipsed by the potential secession of Québec from Canada (see "The Challenge of Québec," p. 336). Nova Scotia, Prince Edward Island, and,

to a lesser degree, New Brunswick are developing tourist industries in an attempt to replace lost components of the economy. Nova Scotia has a well-structured and geographically comprehensive tourist industry covering all parts of the province. New jobs in the region may be found in retail, hotel, and restaurant services.

Québec The province of Québec was settled shortly after the first Atlantic coast settlements by French people who developed a distinctive type of long-lot land settlement that maximized land ownership along either side of the St. Lawrence River. Québec became part of British Canada following General James Wolfe's defeat of the French army under Montcalm at Québec City in 1759. The French settlers and their descendant Québecois resented living under their conquerors. The economics of their culture were barely above subsistence, but traders bringing furs from far inland supported a merchant class during the 1800s. The combination of French language and Roman Catholic religion forged a strong loyalty that shifted to political activism in the 1900s as a reaction against Anglo-Canadian control. The erosion of French speech in the rest of Canada as new immigrants settled the prairies fostered a resolution that this would not happen in Québec.

As a result of such political activities, French was accepted as an equal national language, and Québec gained other concessions from the remainder of Canada despite not acknowledging English as an alternative language within its province. Québec looks beyond Canada to other French-speaking countries and takes a leading role in developing a new global French technical language rather than simply accepting English words.

The province of Québec covers a large area, extending northward to include most of the peninsula east of Hudson Bay (Figure 10.29). The majority of the population lives along

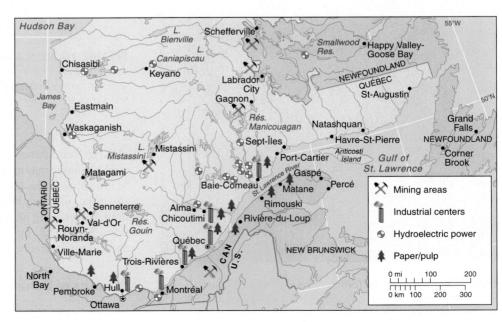

FIGURE 10.29 **Québec Province, Canada. Major cities, mining areas, industrial centers, and hydroelectricity sites.** Nearly all economic activity crowds into the St. Lawrence River lowlands.

the St. Lawrence River estuary. The two largest cities, Québec and Montréal, have a range of manufacturing and service industries that place them among the world's great cities (Figure 10.30). A series of industrial towns use local hydroelectricity to power timber industries, pulp and paper mills, and aluminum refineries along one of the world's main shipping lanes. With global market production of wood pulp extending to tropical areas, the prices—and hence, the well-being of Québec's major industry—fluctuated since the late 1980s. Expansion of global demand, in Asia, Europe, and North America, does not always balance the production output.

North of the St. Lawrence River estuary, the land is bleak, eroded by former ice sheets and covered by coniferous forest, lakes, and tundra. The rocks contain large mineral resources that are gradually being exploited as world markets and transport facilities make this possible. One of the major potential resources is hydroelectricity, and Québec is investing heavily in new facilities to export power to the United States. Québec produces 25 percent of Canada's manufacturing output and has a strong economy in its own right.

Ontario   The province of Ontario contrasts with Québec. Although it was settled later than the provinces to the east, Ontario became the center of British rule after 1776 and possessed the best farming land. Southern Ontario between Lakes Huron, Ontario, and Erie has a relatively mild climate, considering its interior location on the North American continent. The regional climate is modified by a maritime influence from the very large lakes and from being situated farther south than most other parts of Canada. Soils and climate suitable for pasture, grain crops, and tobacco made this the most attractive part of Canada for immigrant farmers in the 1800s. The city of Toronto arose at that stage, but its main development occurred from the middle of the 1900s, when the opening of the St. Lawrence Seaway turned it into an ocean port, the French-language policies of Québec drove English-speaking businesspeople westward from its former rival, Montréal, and its role as capital of Ontario expanded.

FIGURE 10.30 Québec City. The provincial capital, Québec City is situated at the confluence of the St. Lawrence and St. Charles rivers. The waterfront area is the Lower Town. The Upper Town is built on a bluff with the imposing Chateau Frontenac hotel dominating the cityscape. Québec City is designated a World Heritage City by the United Nations.

Northern Ontario extends to the shores of Hudson Bay, but its ancient rocks were scraped bare of soil by ice sheets and are covered partially by coniferous forest. Local deposition of clays in meltwater lakes provided usable soils, but growing seasons are short. Most of the settlements in northern Ontario are mining settlements, including the huge complex around the nickel mines of Sudbury. Settlements in the areas west of Sudbury and north of Lakes Huron and Superior are few and far between and constitute a major gap in the linear east-west Canadian settlement pattern.

Prairie Provinces   The Prairie Provinces of Manitoba, Saskatchewan, and Alberta were settled following the building of the Canadian Pacific and Canadian National Railroads in the late 1800s. In the early 1900s, they became major wheat-growing areas. The crop was planted in the spring and ripened in the short growing season, which was as little as 90 days on the northern margin of the plowed area. Toward the west, there was insufficient moisture for wheat, and cattle ranching took over. The area is essentially a northward extension of the Great Plains in the United States. The ground rises westward in a series of steps, further reducing the growing season and precipitation. Many of the agricultural districts had out-migrations of people since 1950 as a result of the mechanization of farming and the closing of marginal farms.

Winnipeg, Regina, Edmonton, Saskatoon, and Calgary were railroad towns that became grain and meat markets, and three of them became provincial capitals. After 1950, the discovery of coal, oil, and natural gas in Alberta brought extractive and manufacturing industry wealth and more people to that province. The large distances east and west to Canadian ports raised the price of exports and led to some of the main energy markets being southward in the United States.

West Coast   British Columbia faces the Pacific Ocean and includes all the mountainous west of Canada apart from a range shared with southwestern Alberta. It is a province of geographically concentrated economic activity, including pockets of mining, fruit growing, and tourism in small lowland areas that are separated by high mountains. Most of the people of the province live in the Vancouver area, extending to Vancouver Island. Attempts to develop the port of Prince Rupert farther north were not so successful.

The people of British Columbia often distance themselves mentally as well as physically from the rest of Canada. Some of them act more British than the British themselves, with elaborate rituals of afternoon tea common in Victoria, while others emphasize the Pacific connections of Vancouver, which has an increasing Asian population and growing economic ties to Asia. Asian migrants brought money to metropolitan Vancouver, which facilitated the growth of many metropolitan industries. The growing Asian community further diversifies the culture of the city. In some Vancouver neighborhoods, however, tension developed between some Canadians of British descent and Asian migrants. Many Asians living in Vancouver garnered wealth in business endeavors in Hong Kong, and they tore down traditional dwellings in parts of Vancouver and erected large Asian-style homes in their place, which frustrated long-time residents who preferred the traditional architectural styles of the residential areas of the city.

Northern Canada   The northern parts of all the provinces from Saskatchewan to British Columbia end in the barren rock, trees, and tundra that characterize northern Canada. The Northwest and Yukon Territories long formed the largest part of Canada—a federally controlled zone where few people live permanently apart from American Indian and Inuit (Eskimo) groups of Native Americans. In 1999, Nunavut became a new territory.

Mining settlements produce a range of metallic ores, and there are few communications in this wilderness area apart from those linking mines to their markets. In the late 1990s, a new diamond mine near Yellowknife, Northwest Territories, was expected to produce up to 3 percent of the world's diamonds. The diamond mine needs to address local community concerns about caribou herd migration paths and threats to hunting and fishing if it is to establish a lasting positive impact. On a global scale, the mine, together with others outside South Africa, could reduce the South African De Beers company's domination of global diamond markets. Native Americans attracted to the mining settlements are often unable to relate to the contrasting lifestyles and in the past have created a social problem that the Canadian government did not tackle well until the 1980s. The future of this region is not bright as mining activities and federal subsidy incomes decline. A possible alternative income is from tourism.

Contemporary Geographic Issues

Immigration

The contemporary immigration picture for the United States is extremely dynamic. Immigrant communities from all world regions are occupying space in the country's metropolitan centers, such as the growing Ethiopian communities in Los Angeles and Washington, D.C., and in smaller cities and towns throughout as exhibited by the dramatic growth in the Hispanic communities of western Michigan. Although distinctive ethnic spatial patterns are evident as detailed in Figure 10.31, rapid growth in some communities creates new ethnic maps almost daily. While many highly educated people migrate to the United States to fill employment demands in high tech and computer-related industries, medical research, and numerous fields within the physical sciences, more significant numbers come with no material wealth or education in search of a life different from the environment in their country of origin.

Over half of the African Americans in the United States live in the South in both rural and urban areas. The majority of the re-

maining African American population is clustered in northern and western metropolitan centers. The African American percentage of the total U.S. population is diminishing as the percentage of Asian Americans and especially Hispanic Americans increases.

Asian immigrants are currently concentrated along the West Coast of the United States, especially in the larger urban centers. There are also significant concentrations of Asian Americans in parts of the upper Midwest and in large cities in the eastern United States. The climate and ecosystems of the Gulf Coast region attracted and now house a significant population of people who have emigrated from Vietnam. Vietnamese Americans living along the Gulf coast engage in agricultural and fishing practices similar to those of Vietnam.

The United States Census Bureau 2005 estimates reported 198.4 million people (67 percent) in the United States were of European origin, 36.3 million (12.2 percent) were African Americans, 2.2 million (0.7 percent) were Native Americans, and just over 12.4 million (4.1 percent) were classified as Asian. Some 42 million (14.4 percent) were Hispanics, an ethnic designation based mainly on a Latin American heritage and the Spanish language.

Hispanic peoples in the United States were the country's largest and fastest growing minority group in 2006. Hispanic Americans are located mainly in the states bordering Latin America, including Florida, Texas, New Mexico, Arizona, and California; most large cities of the northeastern part of the United States, and in the expansive urban reaches of metropolitan Chicago.

People of European ancestry in the United States have the country's lowest birth rates. Without immigrant communities from Latin America and other world regions, the United States might experience a population decline as in some European countries. Young immigrant communities with relatively higher birth rates maintain a replacement population for jobs, economic growth, and the infusion of money into the tax base for infrastructure expenditures.

Immigration into the United States became one of the country's most controversial issues in 2006. While it is important to understand that the majority percentage of people of Latin American heritage in the United States pay taxes, legally reside within the U.S. borders, and directly contribute to the U.S. economy, the number of illegal immigrants was on the rise. Increasing numbers of illegal immigrants, or "undocumented aliens" (estimated by the PEW Hispanic Center to be as high as 12 million in 2006) mainly of Hispanic origin, live and work in the United States.

Millions of legal and illegal Hispanic immigrants in the United States are laborers in commercial agriculture, construction, landscaping, housekeeping, and fast-food service positions. Although their wages are relatively low by U.S. standards, they may earn more per week working in the United States then they do for a month in their county of origin. Many send a significant percentage of their wages, or remittances, to relatives living in Mexico, El Salvador, Guatemala, Honduras, Mexico, and Nicaragua—where such receipts may be the most significant foreign revenue earned (as is the case for El Salvador).

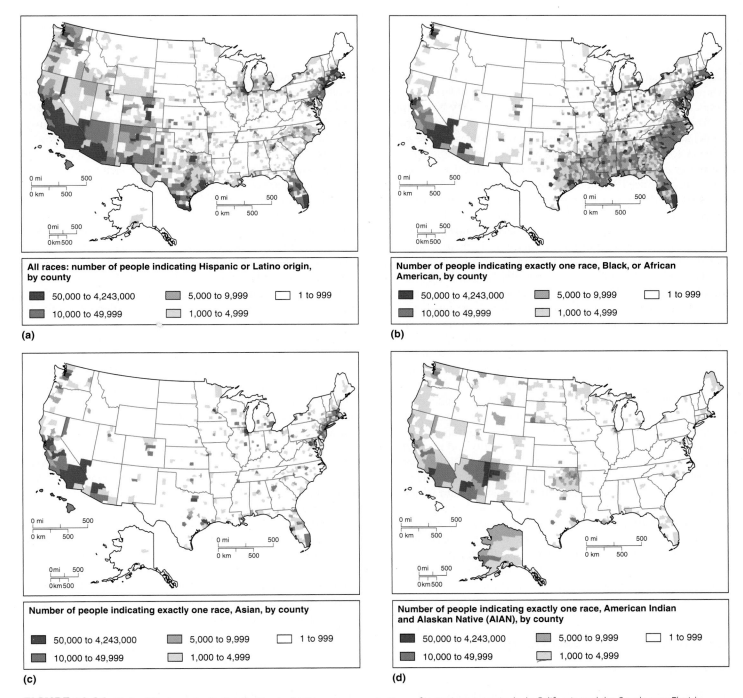

(a)

All races: number of people indicating Hispanic or Latino origin, by county

- 50,000 to 4,243,000
- 10,000 to 49,999
- 5,000 to 9,999
- 1,000 to 4,999
- 1 to 999

(b)

Number of people indicating exactly one race, Black, or African American, by county

- 50,000 to 4,243,000
- 10,000 to 49,999
- 5,000 to 9,999
- 1,000 to 4,999
- 1 to 999

(c)

Number of people indicating exactly one race, Asian, by county

- 50,000 to 4,243,000
- 10,000 to 49,999
- 5,000 to 9,999
- 1,000 to 4,999
- 1 to 999

(d)

Number of people indicating exactly one race, American Indian and Alaskan Native (AIAN), by county

- 50,000 to 4,243,000
- 10,000 to 49,999
- 5,000 to 9,999
- 1,000 to 4,999
- 1 to 999

FIGURE 10.31 **United States: ethnic distributions.** (a) Primary concentrations of Latin Americans include California and the Southwest, Florida, Chicago, and the cities of Megalopolis. (b) Primary African American concentrations include California, the South from east Texas through North Carolina, Florida, and major cities of the Great Lakes and Megalopolis. (c) Primary concentrations of Asian Americans include the Pacific Coast, Chicago, and the cities of Megalopolis. (d) Primary concentrations of Native American and Alaskan Native peoples include Alaska, California, Arizona, New Mexico, and Oklahoma. *Source: Data from U.S. Census Bureau, Census 2000 Redistricting Data (PL 94-171) Summary File. Cartography: Population Division U.S. Census Bureau, American Factfinder at factfinder.census.gov provides census data and mapping tools.*

Increasing public dialogue concerning immigration and undocumented workers led to a series of significant events during the spring of 2006. Millions of Hispanic Americans rallied and staged "sit outs" from work in U.S. cities to assert their importance to the U.S. economy and to voice their concerns over existing and proposed immigration laws. President George W. Bush announced plans in May of 2006 to increase, by several thousands, the number of National Guard troops stationed along the U.S.-Mexico border to oversee the heightening of border security. The president also announced plans to fortify structural barriers along the border in urban and other more populated corridors.

Another controversial trend surrounding Hispanic immigrants is the day laborer phenomenon. Day laborers are workers lacking formal contractual relationships with employers and who operate in the informal economy. Homebuilders, for example, might pick up groups of workers (who may be waiting in front of a convenience store in the hopes of encountering work) and take them to a building site where they will be paid to work for that day. Day laborers do not enter into any type of formal or contractual position, pay taxes, or receive any benefits. This practice became very common in many U.S. cities in the early 2000s. Controversy is generated by those asserting the laborers should be paying taxes, by neighborhood residents and business owners who are concerned by large groups of laborers who "hang out" in front of their business or in their neighborhoods while waiting for prospective employers to pick them up. Fairfax County, Virginia, in suburban Washington, D.C., used county funds to build a day laborer center where the workers could wait to be picked up for employment. Some residents of the county did not approve of their tax dollars being used to support what they see as an illegal practice, while other county residents felt it was the perfect solution. Nearby Montgomery County, Maryland, was debating the need to build such county-supported centers during the spring of 2006.

Controversy among voters, business owners, and government officials focuses on the responsibility of the government in terms of resolving issues related to illegal immigration. Members of the domestic Latin American community and many employers would like to see the U.S. Government grant amnesty to illegal Hispanic peoples within the U.S. borders. Others propose temporary "guest worker" arrangements where undocumented immigrants, in conjunction with secured employment, could obtain work permits allowing their residence for a stated period (with an end date). Those against the presence of illegal immigrants within the United States would like the government to track them down, deport them, and dramatically strengthen border security in order to prevent their return.

North American Free Trade Agreement (NAFTA)

The formal implementation of the North American Free Trade Agreement (NAFTA) in 1994 created one of the world's largest trading blocs. The United States, Canada, and Mexico moved toward an extensive elimination of trade barriers to increase the economic activity between them and strengthen their economic and political positions at the global scale. The three countries tied by NAFTA are now each other's largest trading partners.

NAFTA has many proponents and critics in each of the three countries (Table 10.3). Similar living standards in the United States and Canada contrast sharply with poverty conditions common to millions in Mexico. Proponents of NAFTA claim it creates jobs, strengthens and expands business and industry, diversifies the economies of each of the three country partners, enhances foreign revenues, and fortifies each country's ability to compete in a

global trade arena. Those opposing NAFTA assert the agreement results in the exportation of jobs and revenues from Canada and the United States to Mexico, the perpetuation of harsh labor conditions for Mexican workers, the degradation of the natural environment in all countries, and the granting of nearly supreme powers to big business and industry at the expense of workers, consumers, and those living in the shadow of production facilities.

Political leaders from the three NAFTA partners (Mexican President Carlos Salinas de Gortari, 1988–1994; U.S. President George H. W. Bush, 1988–1992, and, later, U.S. President Bill Clinton, 1993–2001; and Canadian President Brian Mulroney, 1984–1993) signed the 1992 trade pact. The respective legislative bodies each then passed the agreement, which was formally implemented in 1994. The agreement effectively removed most tariffs immediately and established a timeline by which all tariffs would be eliminated within 15 years of implementation.

The United States led efforts to push NAFTA forward in hopes of creating a trade bloc to compete with the European Union. The majority of NAFTA proponents, who pushed through an accelerated legislative process in the United States, were mostly representatives of large corporations. Labor unions, human rights groups, and environmentalists were strongly united against the rapid development of a trade agreement between the three countries and hoped to stop the fast-track legislation by arguing that they were largely excluded from the negotiations. Environmental critics of NAFTA claim that many of the key negotiators who rapidly pushed NAFTA through the U.S. Congress were among the United States' biggest polluters.

Proponents of NAFTA assert that the trade agreement forces the equal treatment of corporations and industry throughout the region. Business leaders claim that the package promotes fair and equal business opportunities for all companies in the three countries. Corporate leaders also assert that equal treatment reduces the risks of governmental interference in free market flows. Mexico has a long history of protectionism, or imposing restrictions, taxes, and quotas on goods and services produced outside of Mexico, and nationalization of industry (government takeover and control of business and industry, thus removing private ownership). Business proponents claim that protectionist policies are curtailed under NAFTA. Business leaders firmly believe that capitalism thrives best in NAFTA-created conditions in the region, and this promotes and fortifies economic and political stability for all three countries.

Strong opposition to NAFTA emanates from labor unions in the United States and advocates of human and worker's rights in Mexico. Union leaders argue that the agreement exports thousands of jobs from the United States to Mexico. Working conditions in production facilities in Mexico are dramatically different from those in the United States. Mexican wages are, on average, one-eighth to one-tenth of U.S. wages, and benefits and other labor costs are nonexistent or significantly lower in Mexico. Laws restricting or regulating environmental pollution from production facilities, and those concerning labor safety,

TABLE 10.3   DEBATE: THE EFFECTS OF NAFTA

Supporters of NAFTA	Critics of NAFTA
Opened new markets for the three countries.	Primarily opened new markets for Mexico and Canada.
Competition of lower-priced goods produced in Mexico forces down the price of goods produced in the United States and Canada; thus, U.S. and Canadian consumers "win" with lower-priced goods.	Low labor costs in Mexico and lower-priced goods in the United States and Canada cause U.S. and Canadian companies to move their production operations to Mexico; thus, thousands of jobs are lost ("exported") to Mexico, hurting employment in the United States and Canada.
Strengthens the global economic weight of the three countries and makes them better able to compete with the EU and other trade blocs and countries of the world.	Significant economic disparity exists between the affluent United States and Canada and the relatively materially impoverished Mexico, placing Mexico at a disadvantage and creating more of a service role for Mexico to Canada and especially the United States, rather than truly making it an equal trade partner and stronger international economic player.
Promotes democracy and political stability in Mexico and strengthens the Mexican economy, thus ensuring greater stability for North America.	Perceptions by some Mexicans of heightened economic disparity in their country due to NAFTA result in political instability such as the Zapatista uprising.
Creates thousands of jobs in Mexico. Cities and towns in northern Mexico, where the majority of NAFTA-related production (*maquila*) takes place, enjoy much higher living standards and higher rates of employment than most other parts of the country.	Cultural distinctions are blurred in all three countries. U.S. culture may overpower parts of Canada and especially northern Mexico. Fast food is replacing traditional food; U.S. holiday celebrations are replacing traditional celebrations. Areas in the U.S. Southwest are developing a watered-down culture that is a mixture of U.S. and Mexican elements. The increased use of the Spanish language in parts of the United States, especially areas in the Southwest, increases tension with some English-speaking residents.
Strengthens Mexican environmental conditions through environmental side agreements negotiated along with the primary trade agreement, resulting in a healthier environment for Mexico and especially the border region; reverses environmental damage on the U.S. side of the border.	Side agreements negotiated with NAFTA fall far short of strengthening environmental laws in Mexico due to lax enforcement. The wording of NAFTA facilitates environmental abuse by companies in all three countries as NAFTA protects the companies' rights to free trade over the rights of people living in areas polluted by factories and other production facilities.
Forces the equal treatment of corporations in the three countries.	Corporations are too powerful under NAFTA.

are fewer and less enforced in Mexico. United States-based companies are able to manufacture products in Mexico with significantly lower labor costs and with far fewer environmental or labor restrictions. United States labor leaders say these factors are far too attractive for U.S. companies to forgo, and they claim numerous companies are shifting production to Mexico and therefore replacing U.S. workers with Mexican ones.

The Mexican Border Industrial Program, first implemented in 1965, slowly opened the door for foreign firms to establish production facilities in Mexico. The idea of the program was to stimulate growth in northern Mexico (and thus relieve some pressure on the then rapidly growing metropolitan area of Mexico City), create jobs, and infuse capital into a region that had operated under protectionist and nationalist economic policies for decades. The program created a manufacturing zone across northern Mexico known as the *maquila* zone (the factories are referred to as *maquiladoras*). The idea was that foreign firms, such as U.S. companies, could import raw materials or parts, pay Mexican workers low wages to refine or assemble them into finished goods, take the goods back across the border (to the United States for a U.S. firm) relatively tax free, then sell them to consumers. Although *maquila* enterprise is now permitted virtually anywhere in Mexico, the greatest concentration of *maquiladoras* remains in the border region in the north.

Environmental conditions in the maquila zone in northern Mexico are among some of the worst in the world. Multidisciplinary research suggests an increasing array of health problems and ecological damage due to the concentration of factories in a country with loose environmental and worker safety laws. The health and environmental issues present in northern Mexico also exist across the border in the southwestern part of the United States. Cancer rates along the Texas-Mexico border are some of the world's highest.

The controversy surrounding NAFTA is far from over. Further trade barriers will be reduced or eliminated in the next few years in accordance with the progression plan of the agreement. Activist groups both for and against NAFTA continue to lobby the respective governments, to promote their views on the future of the agreement. There are proponents and critics in each of the three countries involved. Depending on the source, convincing statistics touting its value and success, or its detriment to society, the economy, and the health of the region's citizens all make compelling arguments.

The Challenge of Québec

The greatest ethnicity-related challenge facing the Canadian federal government and the unity of the country is the devolutionary pressure from French-speaking Canadians who desire greater autonomy. Approximately 25 percent of Canadians are French-speaking descendants of the French settlers who came to the area during the earliest stages of European development in the region. Most French Canadians are distributed along the St. Lawrence River, primarily in the province of Québec. Most **Francophones,** or French-speaking Canadians, have an extremely strong sense of being a separate national group within the larger Canadian political framework. They see themselves as having a different ethnic history, different religion, and different culture than the English-speaking Canadian majority.

Although the majority of the French descendants adhere to Catholicism, in contrast to the primarily Protestant English Canadians, Québec's Francophones see their French language as the strongest symbol they have for discerning their uniqueness from the English-speaking Canadian majority. The perception of national identity loss is enhanced by the choices new immigrants make in metropolitan Montréal, Québec's largest city and the largest French-speaking city outside of France. Many immigrants settling in Montréal, who spoke neither English nor French as their first language, are choosing to learn English rather than French and are choosing to conduct business in English and to send their children to private English-language schools.

The Québec separatist movement gained momentum during the 1980s and became especially strong during the 1990s. Decades of perceived oppression from the federal government as well as the perception of a threat to their cultural identity and language have unified members of the **Parti Québécois,** a political party formed with the purpose of achieving Québec's independence from the federal government. The Parti Québécois seeks the establishment of the independent country of Québec. A 1995 referendum resulted in an almost 50 percent vote by the residents of Québec province in favor of separation.

Several groups inside the provincial borders of Québec do not share in the enthusiasm for separation. English-speaking and other business owners fearing revenue losses are adamantly opposed to independence. The more recent diverse immigrant population residing in the metropolitan Montréal area also wishes to remain part of Canada. One of the most spatially organized internal challenges to the Francophones' movement for independence comes from an indigenous Canadian people known as the **Cree,** who claim the northern third to northern half of Québec as their ancestral land. The Cree do not wish to be part of an independent Québec and threaten to secede from any newly formed Québec state and remain with Canada. The northern half of Québec is the site of significant natural resource potential. Most notably, the generation of hydroelectric power is so successful in northern Québec that the province is able to export energy to the United States and generate profitable revenue streams.

The Québec issue raises questions of whether the varied regions, separated by great overland distances, can continue to provide a complementary unity or will embark on a course that will tear the country apart. Other provinces in Canada have grown weary of the federal government's attempts to appease the provincial government of Québec. The provincial governments of the Prairie Provinces have successfully lobbied the federal government for greater control over their resources and decisions using the perceived "special treatment" given to Québec as a bargaining tool. Some speculate that Canada's Atlantic Provinces, which would be physically separated from the rest of Canada should Québec secede, might also seek independence and try to economically align itself better with the New England states of the United States.

The issue may demographically resolve itself. Birth rates within the French-speaking Québécois population are among Canada's lowest—the population is experiencing a natural decline. Immigrant communities' decisions to learn English and ignore the provincial government's offered financial incentives to learn the French language is increasingly significant as immigrant birth rates are significantly higher than those of the French Québécois.

GEOGRAPHY AT WORK

Using Geography to Aid Public Policy

A growing opportunity for geography students in the coming decades is to find employment within the federal government using geography skills to inform public policy decisions. As policymakers are faced with more complex decisions, and are called up by stakeholders to use the best available science and information, an increasing number of federal agencies recognize the need for geographic skills. Geographers have desirable skills in creating and updating geographic databases, selecting and evaluating data for potential use from a variety of specified sources, producing maps and charts, and documenting sources and procedures. Geographers can also produce graphical and tabular data demonstrating relationships between policy and outcomes, and can prepare visual representations to show policymakers how best to carry out activities in support of their agency's mission.

For example, while working at the White House, Dr. Brian Hannegan (Figure 10.32) headed an interagency group established to improve the permitting process associated with a new oil and gas pipeline testing and repair program enacted by Congress in 2002. Geographic Information Systems (GIS) displaying pipeline maps overlaid on maps of ecologically sensitive areas and ranges of threatened and endangered species, provides important information at a glance to pipeline operators on what permits and information they will need to gather, and which federal and state agencies will be involved. The database also supports visual display of the Best Management Practices intended for that area, and allows operators to communicate seamlessly with the permitting agencies to indicate when a test shows the need for a repair, as well as when a repair has been completed.

A GIS database of this type also provides a basis of information for several other policy areas that rely on complete and verifiable environmental data. For example, by highlighting locations where ecologically sensitive areas overlap with the range of a threatened or endangered species, GIS enables a Fish and Wildlife Service biologist to quickly prioritize land conservation areas or management practices in the highlighted

FIGURE 10.32 Geography and Public Policy. Dr. Brian Hannegan utilizes Geographic Information Systems technology and geographic skills to better inform the public policymaking process.

areas that provide the greatest protection with the least economic impact. When this type of data is available to the public, stakeholders may easily identify priority areas for cooperative conservation efforts and are able to work with state and federal officials to create appropriate policy recommendations to protect those areas.

In addition to environmental policy benefits, GIS offers tools to display any kind of data in a format useful to policymakers, whether it be drug-use incidence for the Office of National Drug Control Policy or fundamental census data, which the Department of Commerce and a number of other federal agencies use to set priorities and make funding allocations. As the federal government looks more and more to integrate scientific data into its decision-making process, the need for trained geographers to help communicate this data to policymakers will only increase.

The key terms are highlighted in bold in each chapter. The definition given here is referred to the (chapter; page number) of each term's first occurrence.

absolute location (1; 3). Location of a place on Earth's surface as defined by latitude and longitude or by distance in kilometers (miles) from another place.

acid deposition (2; 42). Dry or wet deposition of acidic material from the atmosphere, often resulting from sulfur and nitrate gases and particles emitted into the air from coal combustion in power plants.

African Union (AU) (8; 260). The organization that combines all African countries since 2001. Formerly the Organization of African Unity (OAU).

agglomeration economies (2; 52). The total economies achieved by a production unit because of a large number of related economic activities in the same area.

agribusiness (2; 54). The large-scale commercialization of agriculture that places farming within the broader context of inputs of seeds, fertilizer, machinery, and so on, and of outputs of processing, marketing, and distribution.

alluvial fan (6; 167). A fan- or cone-shaped river deposit, often formed where a stream issues from a mountain gorge into an open plain.

alluvial soils (4; 110). Soil materials deposited in valley floors by annual river floods.

alluvium (6; 167). Deposits of rivers in their flood plains, often composed of mud, silt, sand, or granel.

altiplano (9; 268). High plateau in the Andes Mountains of Peru and Bolivia.

altitudinal zonation (9; 271). Zones of climate, vegetation, and crops, which change with height above sea level.

animism (8; 232). Traditional religious beliefs based on the worship of natural phenomena and the belief in spirits separable from bodies.

apartheid (8; 235). The separation of people of different ethnic groups, affecting housing, education, and jobs. It was government policy in South Africa until 1994.

Arab League (7; 206). Organization created in 1945 to encourage the united action of Arab countries for their mutual benefit.

Asia-Pacific Economic Cooperation (APEC) (5; 145). Asian and Pacific countries dedicated to trade liberalization.

Association of South East Asian Nations (ASEAN) (5; 145). Established in 1967 as a defensive alliance among Indonesia, Malaysia, the Philippines, Singapore, and Thailand in response to the advance of communism in Southeast Asia. Now increasingly a trading group that admitted Vietnam, a Communist country, as a member in 1995 and Myanmar in 1997.

atmosphere-ocean environment (1; 6). The combination of atmosphere and oceans, in which movements that cause weather and climate are controlled by solar energy.

atoll (5; 137). A coral island with a central lagoon.

badlands topography (6; 167). Closely spaced networks of deep gullies often carved by occasional streams in soft sediments unprotected by vegetation cover. They destroy the usefulness of the land.

barrier reef (5; 137). A coral reef structure surrounding an island and separated from it by a lagoon. The Great Barrier Reef lies off Queensland, Australia.

Benelux (2; 63). An acronym for **Be**lgium, the **Ne**therlands, and **Lux**embourg. The term was coined in recognition of the close working relationship that these countries have with one another.

biome (1; 12). A world-scale ecosystem type, such as tropical rain forest or savanna grassland.

birth rate (1; 19). The number of live births per 1,000 of the population in a year.

black earth soils (chernozems) (3; 76) Highly fertile soil type in which organic matter accumulates near the surface, commonly beneath temperate grassland communities.

Black Triangle (2; 42). A heavily polluted industrial area straddling the Polish, Czech, and German borders.

Bolivarian Alternative for the Americas (ALBA) (9; 279). A political and economic pact signed by Venezuela, Bolivia, and Cuba in the early 2000s with the focus of excluding the United States.

British East India Company (6; 172). A British company that traded with and conquered much of the Indian subcontinent. After an 1857 mutiny in India, the British government took over political control.

British Indian Empire (6; 172). Established after 1857 on the Indian subcontinent, including Ceylon and later extended to Burma. Lasted until independence and partition in 1947.

brown earth soils (3; 76). Fertile soil type in which plant matter replenishes nutrients in the upper layers, commonly forming beneath temperate deciduous forest.

Buddhism (6; 170). A religion that began in South Asia but became the major religion of East Asia. Followers of Buddhism have a greater social openness than do followers of Hinduism and accommodate other philosophies and religions such as Confucianism and Shinto.

Canadian Shield (10; 304). The ancient rocks of northern Canada, the oldest in North America, which contain many mineral deposits.

capital city (1; 21). The city in which central government functions are concentrated; sometimes the largest city but often one that is specifically designed for the purpose.

Caribbean Community Common Market (CARICOM) (9; 280). Formed in 1973 by 13 former British colonies to provide special entry to U.S. markets.

cartel (7; 207). An organization that coordinates the interests of producers (such as OPEC).

caste (6; 170). The basis of social class divisions in South Asia.

caste order (6; 170). A social class system associated with Hinduism that is based on the supremacy of Aryan peoples. The Aryan castes include priests (Brahmans), warriors, and other Aryan people; non-Aryan castes include cultivators, craftspeople, and untouchables.

central planning (3; 89). The Soviet Union practice in which the government decided how many goods and services were needed by society, and gave instructions for their production almost without cost considerations.

centrally planned economic system (1: 26). A system in which planning and decisions regarding a country's economy are made by central government. In the Cold War this system competed with Western capitalism.

chaebol (4; 126). Conglomerates of family-owned companies in South Korea, characterized by diversified production and services.

Christianity (2; 43). A religion that developed out of Judaism, based on the belief that God came to Earth in the form of Jesus of Nazareth. Main religion of the Western world.

class (1; 17). A stratification of society that is based on economic, religious, or social criteria.

climate (1; 6). The long-term atmospheric conditions of a place.

collectivization (4; 121). The transformation of rural life in Communist countries such as China and the Soviet Union, in which individual farmers were grouped in cooperatives that took ownership of their land and labor.

colonialism (2; 50). The system by which one country extends its political control to another territory to improve local conditions and/or economically exploit the human beings and natural resources of the subordinate territory.

command economy (3; 89). The Soviet practice whereby the government ran the economy, owned all industries, and set quotas, favoring heavy industry over the production of consumer goods.

Common Market of the South (MERCOSUR) (9; 280). Trading group established in 1991 among Argentina, Brazil, Paraguay, and Uruguay.

Commonwealth of Independent States (CIS) (3; 74). A political and economic organization created in 1991 by 11 republics of the former Soviet Union. Members included Russia, Belarus, Ukraine, Moldova, Armenia, Azerbaijan, Kazakhstan, Kyrgyzstan, Tajikistan, Turkmenistan, and Uzbekistan. Georgia became the twelfth member in 1993. The CIS coordinates relations between member countries, including issues involving economics, foreign policy, and defense matters.

Communauté Financière Africaine (CFA) (8; 243). Economic links between France and its former colonies in Africa. From independence until 1994, it involved links to the French franc, but this arrangement ended with the devaluation of the CFA franc.

commune (4; 121). The organization that controls rural life in some Communist countries, including agriculture, industry, trade, education, local militia, and family life.

communism (2; 50). A system in which the workers govern and collectively own the means of economic production. When spelled with a capital "C," Communism refers not to the system as it was originally envisioned but rather to the totalitarian systems adopted in countries such as the Soviet Union and China, where small elite groups rule or ruled under the guise of communism.

concentration (2; 54). Agricultural production carried out on fewer and larger farms and limited to smaller areas of higher productivity.

concentric pattern of urban zones (10; 319). A pattern of urban geography with a central business district, surrounded by a hierarchy of residential zones.

Confucius (Kong Fuzi) (4; 119). A Chinese administrator who established a system of efficient and humane political and social institutions that became the basis of procedures and ways of life in much of East Asia.

coniferous forest (10; 306). Forest of needle-leaf trees, commonly occurring in areas with long winters and on poor sandy soils.

continental temperate climate (2; 40). The climates of continental interiors in midlatitudes. They are often drier and have greater extremes of summer and winter temperatures than the climates of coastal areas in these latitudes.

continentality (3; 76). Especially cold winters and hot summers resulting from locations on landmasses that are far from the moderating effects of large water bodies such as oceans.

convergent plate margin (1; 6). In plate tectonics, where two plates move toward each other, causing earthquakes, volcanic activity, and mountain-building as they meet.

core country (1; 30). A materially wealthy country that plays a major part in controlling world economic processes.

countermigration (10; 311). Movement of people returning to regions they once left, as in the return of African Americans from northern U.S. cities to the South.

country (1; 21). A self-governing political unit having sovereignty within its borders and recognized by other countries.

Cree (10; 336). A "First Nation" of people, who claim the northern part of Québec province in Canada.

creole (8; 234). A language that is used commonly to overcome problems of communication among peoples with different languages; it may be composed of elements of other languages.

crony capitalism (5; 149). The economic system in which entrepreneurs gain advantages through close links to government officials and ministers.

cultural geography (1; 14). The study of spatial variations in cultural features such as material traits, social structures, languages, or belief systems.

Cultural Revolution (4; 121). The attempt by Mao Zedong between 1966 and 1976 to change the basis of Chinese society. The disruption held back the country's economic development.

cultural rights (1; 33). The rights to protect one's cultural traditions.

culture fault line (1; 18). A line or zone of tension between contrasting cultural regions across which conflict may occur.

culture hearth (1; 18). A small region of the world that acted as a catalyst for developments in technology, religion, or language, and as a base for their diffusion.

Daoism (4: 119). The teachings of Chinese philosopher Laozi, who disliked the organized and hierarchical social system of Confucius and advocated a return to local, village-based communities with little outside interference.

death rate (1; 19). The number of deaths per 1,000 of the population in a year.

deciduous (10; 306). Plants with broad leaves that mostly lose their foliage in cold or dry seasons.

decolonization (2; 69). The process by which mainly European countries enabled their colonies to become independent countries.

deindustrialization (2; 52). A rapid fall in manufacturing employment and the abandonment of factories in a once-important industrial region.

democratic centralism (2; 50). The practice of sole governance by the Communist Party, the political party of the working class, because it is believed that only the

Communist Party is the true representative of the people.

demography (1; 19). The study of human populations in terms of numbers, density, growth or decline, and migrations from place to place.

deposition (1; 11). The dropping of particles of rocks carried by rivers, wind, or glaciers when they stop flowing or blowing, or melt, respectively.

desalination plant (7; 197). A mechanism that extracts fresh water from seawater by evaporation and condensation.

desert (8; 231). An area that does not support vegetation, often associated with an arid climate.

desert biome (1; 12). Major biome-type characterized by a discontinuous plant cover or none.

desertification (8; 231). Processes that destroy the productive capacity of an area of land.

development (1; 29). The process by which human societies improve their quality of life, including economic, cultural, political, and environmental aspects.

devolution (2; 66). The process by which local peoples desire less rule from their national governments and seek greater authority in governing themselves.

diaspora (7; 200). The scattering of a people, such as the Jews and Chinese, to other countries.

direction (1; 4). The position of one place relative to another, measured in degrees from due north or by the points of the compass.

distance (1; 4). The space between places, measured in kilometers (miles), travel time, or travel cost.

divergent plate margin (1; 6). In plate tectonics, a zone between two plates that are moving apart, often marked by an ocean ridge.

diversified economy (7; 216). An economy in which manufactures are more important than primary products and where there is a variety of manufactured products and a growing service sector.

earth surface environments (1; 6). Where atmosphere-ocean environments interact with solid earth environments to produce relief features such as river or glacial valleys and coastal cliffs and beaches.

East African Community (EAC) (8; 246). A grouping of Kenya, Tanzania, and Uganda, launched in 1967 and relaunched in 2000.

Economic Community of West African States (ECOWAS) (8; 243). Organization of Western African countries, founded in 1975 and involved in peacekeeping, a con-

sultative parliament (2002), free trade and movement of labor, and banking.

economic geography (1; 23). The study of the spatial aspects of material wealth and poverty, the use of resources, and the production of goods.

economic leakage (9; 282). Foreign ownership and control of industry, such as tourism in many parts of Latin America, where the revenue generated by the industry flows to the foreign owners and foreign employees and largely bypasses local communities and thus "leaks" from the local economy.

economies of scale (10; 313). Increased productivity gained by building larger factories and gaining access to larger markets.

ecosystem (1; 11). The total environment of a community of plants and animals, including heat, light, and nutrient supplies.

ecosystem environments (1; 6). The environments of living organisms, where weather, mineral nutrients, and solar energy combine.

edge city (10; 320). A post-industrial city, usually in the suburbs of a major city, containing significant commercial space and service industries.

El Niño (5; 137). The phenomenon in which the warm equatorial waters push back the cold Peruvian current off the western coast of tropical South America, producing local fish kills and having wider world climatic effects.

El Niño Southern Oscillation (ENSO) (9; 270). The full name of the El Niño phenomenon.

emerging countries (1; 32). In the early 2000s, a group of developing countries with economic growth that is beginning to challenge wealthier countries.

encomienda system (9; 278). The Spanish colonial system employed in Latin America in which lands were allotted to Spaniards who were responsible for exploiting their wealth and had jurisdiction over native peoples.

entrepôt (4; 130). A port that collects goods from several countries to trade with the wider world and distributes imports to its immediate area or hinterland. Examples include Hong Kong and Singapore.

erosion (1; 11). The wearing away of rocks at Earth's surface by running water, moving ice, the wind, and the sea to form valleys, cliffs, and other landforms.

ersatz capitalism (5; 149). The economic system that implies an inferior substitute for locally based economic development when the capital, skills, and management are all imported to produce goods for export.

estuary (2; 39). A wide river mouth that experiences changes in tidal water level and quality.

ethnic cleansing (2; 68). The process by which a dominant group of people in a country causes another ethnic group to leave a region, often using threats or military force (*see* **genocide**).

ethnic group (1; 16). A group of people with common racial, national, religious, linguistic, or cultural origins.

ethnic religion (1; 14). A religion that is linked to a particular ethnic group, such as Judaism and Hinduism.

European Union (EU) (2; 63). Name adopted by the European Community in 1993, suggesting both an expansion to other European countries following the end of the Cold War and the possibility of a future closer political federation.

evapotranspiration (9; 297). The combination of evaporation from water and transpiration through plant leaves that produces humidity in the atmosphere.

export-led underdevelopment (9; 283). The process by which countries export their raw materials while their governments charge taxes on exports to pay for their imports. There is no stimulus to develop manufacturing and so diversify the economy to bring more wealth into it.

extensification (2; 55). In agriculture, the production of fewer livestock or crops from the same area.

favela (9; 299). A shantytown in a Brazilian city.

federal government (1; 21). A division of central government functions among states or provinces.

First Nations (10; 311). Indigenous or native Canadians.

First World (1; 30). The Western countries, led by the United States.

five-year plan (3; 89). A comprehensive economic planning scope in the former Soviet Union. These plans were followed from 1928 until 1991.

fjord (2; 39). A formerly glaciated valley that was flooded with ocean water after the sea level rose in the postglacial age.

flow (1; 3). The movement of materials, as in water moving along a river. It is also used to describe movements of people, goods, ideas, and finances.

Fordism (10; 313). The application of production lines in the assembly of a wide range of components in various manufacturing industries based on the production system established by Henry Ford for the automobile industry.

forest biome (1; 12). A major biome-type characterized by closely spaced trees.

formal economy (1; 31). The economic sector in which workers have recognized or licensed jobs, receive agreed-upon wages, and pay taxes.

Francophone (10; 336). In North America, a French-speaking Canadian.

free market enterprise (capitalist) system (1; 25). The economic system that is based on competition and pricing of goods determined by the market. It is the basis of capitalism, but market "freedom" is often reduced by government and actions of major producers.

Free Trade Area of the Americas (FTAA) (9; 28). An attempt by the United States and several Latin American countries to create the world's largest trading group.

free trade zone (9; 294). An area within a country where components can be imported without tariffs for assembly with a view to exporting the finished goods.

friction of distance (1; 4). The relative difficulty of moving from one place to another, which increases with kilometers (miles), cost, or travel time.

fringing reef (5; 137). A coral reef along a coast without a lagoon.

gender (1; 17). The cultural implications of being male or female, with particular reference to the inequalities suffered by females in human societies.

genocide (2; 50). The systematic extermination of an ethnic group, nation, racial, or religious group.

gentrification (2; 47). The movement of higher-income groups to occupy and improve residences in older and poorer parts of cities.

geographic inertia (2; 52). Once capital investments are made in factories and infrastructure that give a region agglomeration economies, production will continue there for a period of years after other areas emerge with lower production costs.

geographic information system (GIS) (1; 3). The computer-based combination of maps, data, and often satellite images that is a foundation for geographic analysis.

geography (1; 2). The study of spatial patterns in the human and physical world: where and how the human and natural features of Earth's surface are distributed, are related to each other, and change over time.

glasnost (3; 88). The Soviet Union policy of the late 1980s designed to create greater openness and exchange of information.

global city-region (1; 28). A region that is dominated by one or more cities with major involvements in the global economy.

global positioning system (GPS) (1; 3). A system with access to satellites that provides a fix of a place on Earth's surface.

global warming (1; 13). The process by which average temperatures in Earth's atmosphere rise over a period of several decades or centuries, leading to the melting of ice masses and a rising sea level.

globalization (1; 5). The growing interconnections of the world's peoples and the integration of economies, technologies, and some aspects of cultures.

Gondwanaland (5; 136). The former huge continent, from which the southern continents and Indian peninsula broke away.

governance (1; 22). The coordination and regulation of human activities at different levels of geographic scale, often outside the powers of sovereign countries.

grassland biome (1; 12). A major biome-type characterized by grasses, with few, scattered trees.

Great Artesian Basin (5; 138). The area in eastern Australia where rain falling on the Great Dividing Range flows through the rocks to provide water under the lowlands that naturally comes to the surface under pressure.

Great Leap Forward (4; 121). The attempt by the Chinese Communist government in the late 1950s to increase the pace of industrialization. It failed by ignoring food production at a time when bad weather brought poor harvests and famine.

Green Revolution (5; 149). The result of introducing high-yielding strains of wheat and rice. The outputs of commercial farms in South and East Asia increased by several times, but the costs of seeds, fertilizers, and pesticides were too high for smaller farmers.

greenhouse effect (1; 8). The natural process of heating Earth's atmosphere. Solar rays of short wavelength reach Earth's surface and are absorbed and reradiated as long wavelength (heat). This radiation is partially absorbed by and heats the lower atmosphere, which contains water vapor and carbon gases. When humans add to the carbon gases, they enhance this process, raising temperatures above "natural" levels.

gross domestic product (GDP) (1; 24). The total value of goods and services produced within a country in a year. Often expressed as GDP per capita, when the total GDP is divided by the country's population.

gross national income (GNI) (1; 24). The total value of goods and services produced within a country in a year, together with income from labor and capital working abroad, minus deductions for payments to those living abroad.

Group of Eight (G8) (10; 314). An economic discussion forum consisting of the world's eight most materially wealthy countries; it includes the United States of America, Canada, Japan, Germany, the United Kingdom, France, Italy, and Russia.

growth pole (1; 30). An urban center designated for scarce investment resources in order to influence the surrounding area.

guest worker (2; 70). A foreigner who has permission to reside in a country to work but is not a citizen of that country. From the German word *Gastarbeiter*.

gulag (3; 101). Short for the Russian name *Glavnoe upravlenie ispravitel no-trudovykh lagerei* (Main Directorate for Corrective Labor Camps), a collection of prison camps in the former Soviet Union where criminals and those who opposed the Communist government were sent to perform hard labor as punishment.

Gulf Cooperation Council (7; 216). Formed in 1981 under the leadership of Saudi Arabia to focus on political problems raised by the Iran-Iraq War.

Han Chinese (4; 111). The largest group (94 percent) of people in China. An ethnic grouping based on the administrative culture spread by the Chinese empire in the AD 200s and 300s.

heartland (3; 94). The area of a country that contains a large percentage of the country's population, economic activity, and political influence.

Hinduism (6; 170). A religion of South Asia, observed mainly in India, that includes the worship of many gods related to varied historic experiences and is associated with the caste system.

hinterland (3; 94). The areas of a country that lie outside the heartland. The hinterland usually has a relatively small percentage of a country's population, economic activity, and political influence compared to the heartland, though it may be well-endowed in natural resources.

HIV/AIDS (8; 252). Disease affecting the immune system that is particularly prevalent in Southern Africa. It is regarded as a long-term pandemic that can be alleviated, but not cured, by drug treatments.

Homestead Act (10; 312). The U.S. act of 1862 that provided land cheaply or freely to families settling the U.S. West.

horizontal integration (10; 313). The combining of producers of the same product in a single corporation to create economies of scale and the control of prices.

household responsibility system (4; 121). The replacement for communes in rural China after 1976, returning ownership and decision-making to individuals and groups that could sell surpluses in open markets.

human development (1; 30). A broader view of development that focuses on people rather than economic change, which is regarded as a means to the end of enabling people to enlarge their capabilities so that they can enjoy the richness of being human.

Human Development Index (HDI) (1; 24). A measure of human development based on income, life expectancy, adult literacy, and infant mortality.

human geography (1; 5). The study of geographic aspects of human activities, often with a population, political, economic, cultural, or social focus.

Human Poverty Index (HPI) (1; 25). A measure of human poverty, linked to HDI, which indicates levels of personal deprivation.

human rights (1; 33). The rights that should be part of normal human experience, including justice, a decent standard of living, personal security, and freedom of thought and speech.

hydrology (10; 305). The surface water drainage of a region, including lakes and rivers.

imperialism (2; 50). The practice of extending the rule of an empire over foreign lands.

import substitution (1; 31). A policy in which countries develop manufacturing industries to fulfill internal market demands, often protected by high tariffs to exclude foreign competition.

indigenous people (1; 21). The first inhabitants of an area or those present when the area is taken over by another group.

Industrial Revolution (2; 50). The period of the late 1700s and early 1800s when increasingly complicated machines and chemical processes, fueled by inanimate power sources such as water and coal, replaced traditional ways of making goods by hand with simple tools. The mass production of goods resulted, as did the need for raw materials. The Industrial Revolution began in England and then spread to other areas of Europe and the world.

infant mortality (1; 19). The number of deaths per 1,000 live births in the first year of life.

informal economy (1; 31). The economic sector in which workers act outside the formal sector.

intensification (2; 54). In agriculture, the increased output of crops or livestock per area unit of land.

irredentism (2; 49). The desire to gain control over lost territories or territories perceived to belong rightfully to a group; associated with nationalism.

Islam (7; 200). A religion of Northern Africa and Southwestern Asia, and parts of South and East Asia, based on the teachings of Muhammad as recorded in the Qu'ran. *Islam* means "submission to the will of God."

isthmus (9; 287). Naturally occurring land bridge.

Jainism (6; 170). A religion, mainly in India, that involves a nonviolent code and has laws against harming animals (thus forbidding farming).

Judaism (7; 200). The religion of the Jewish people who worship Yahweh as the creator and lawgiver.

kibbutz (pl. kibbutzim) (7; 219). A communal farming village in Israel, the social and spiritual basis of the new Israeli nation after 1948, but now the home of fewer than 5 percent of the Israeli population.

Kyoto Protocol (1; 13). Statement adopted by the United Nations Convention on Climate Change in 1997, setting targets for reducing primary greenhouse gases.

language (1; 14). The means of communication among people by speaking, writing, and signing.

laterite (8; 231). Soils that form in seasonal tropical conditions. The cementation of clay and iron minerals form an impenetrable layer that is used as building blocks.

latitude (1; 3). The distance of a place north or south of the equator, measured in degrees.

lingua franca (5; 150). A language that aids communication among people of varied languages, often at first used for trade, but later for varied purposes.

localization (1; 5). The geographic differentiation of places among and within countries.

location (1; 3). A place's location is defined by its position on Earth's surface in terms of latitude and longitude (*see* **absolute location**) or by its level of interaction with other places (*see* **relative location**).

loess (2; 304). Fine-grained and fertile soils developed from windblown deposits.

longitude (1; 4). The distance of a place east or west of 0° longitude, measured in degrees.

Lost Decade (9; 280). In the 1980s, country economies in Latin America stalled because of high interest rates and debts.

map (1; 3). The representation of the features of Earth's surface on paper at varying scales.

maquiladora (9; 287). Mexican government program that encourages foreign-owned factories to be sited in Mexico by not charging import duties on raw materials for assembly.

market gardening (truck farming) (2; 54). The commercial production of high-cash-value, specialty fruit and vegetable crops such as table grapes, raisins, oranges, grapefruits, apples, and lettuce.

marsupial (5; 137). A mammal that raises its young in a pouch instead of a womb. Mainly found in Australia, these include the kangaroo and koala.

medina (7; 204). The crowded streets of the older sections of Arab towns in Northern Africa and Southwestern Asia.

megalopolis (4; 110). An expanded urbanized area that includes several metropolitan areas with over a million people and dominates the economy of surrounding areas. First identified in the northeastern United States, covering the area between Boston and Washington, D.C.

meridian of longitude (1; 4). An imaginary line joining places of the same longitude on Earth's surface.

microclimate (9; 269). The climate of a small area such as a valley, but may also include the climate of a single plant.

migration (1; 19). The long-term movement of people into (in-migration, immigration) or out of (out-migration, emigration) a place.

Ministry of Economy, Trade, and Industry (METI) (4; 116). The Japanese government body responsible for assistance and advice to industry, including the export marketing office.

modernization (1; 30). The theory of development in which poorer countries attempt to follow the stages through primary and secondary to tertiary sectors that the countries of Western Europe and North America followed in the AD 1800s and early 1900s.

monotheism (7; 200). A religion based on belief in a single god.

Monroe Doctrine (9; 279). An 1823 declaration asserting U.S. rights to control activity in the Americas over those of countries outside the region. In the Monroe Doctrine, the United States vowed to resist any intervention from outside countries in Latin American affairs.

monsoon climatic environment (5; 137). A tropical climatic environment in which there are wind shifts between summer and winter, bringing heavy rains from oceanic air in the summer and dry winds of interior continental air in winter.

Mughal (Mogul) dynasty (6; 171). Turkish invaders of India from Persia in the AD 1500s who conquered most of the region and left a heritage of magnificent buildings such as the Taj Mahal.

multinational corporation (1; 26). A corporation that makes goods and provides services in several countries but directs operations from headquarters in one country.

Muslims (7; 200). "Those who submit to Allah." Followers of Islam.

nation (1; 21). An "imagined community" in which a group of people believe that they share common cultural features, often linked to a specific area of land.

nationalism (1; 21). The desire of people to have their own self-governing country.

nation-state (2; 49). The linking of a separate and distinct people (nation) and a politically organized territory with its own sovereign government (state).

nation-state ideal (2; 49). The belief that each people (nation) must have its own country (state) in order to be free and govern itself as it desires.

Native Americans (10; 311). Indigenous people who inhabited the Americas before the European arrival, including Amerinds and Inuits (Eskimos).

natural hazard (1; 13). A natural event, such as a volcanic eruption, earthquake, tornado, hurricane, or flood, that interrupts human activities by causing extensive damage and deaths.

natural resource (1; 13). Materials present in the natural environment and recognized by humans as of practical worth (minerals, soils, water, building stones, timber).

neo-colonialism (9; 279). A relationship of economic, rather than political, dependence.

New Partnership for Africa's Development (NEPAD) (8; 261). A South African organization dedicated to African development through African-based efforts.

Nile Waters Agreement (7; 209). Agreement between Egypt and Sudan in 1959 to share Nile River waters, with 70 percent allocated to Egypt.

nongovernmental organization (1; 22). Groups of people who act outside government and major commercial agencies, mainly in advocacy roles such as delivering aid and lobbying for particular causes.

nonrenewable resource (1; 13). A natural resource that is used up once it is extracted; for example, coal or metallic minerals.

North American Free Trade Agreement (NAFTA) (9; 279). An economic agreement among Canada, Mexico, and the United States, signed in 1994.

North Atlantic Treaty Organization (NATO) (2; 50). A military alliance of non-Communist European countries and the United States, founded in 1949 to counter the military threat of the Soviet Union. In recent years, former Communist countries have joined the alliance, and Russia has formed a partnership with NATO.

northern coniferous forest (3; 76). A forest composed of coniferous trees (firs, pines, cedars) common in the northern parts of temperate continental interior climatic environments.

ocean biomes (1; 12). The oceans as ecosystems with energy pathways, nutrient cycles, and food chains.

oceanic temperate climate (2; 40). The climates of the western margins of midlatitude continents, in which cool, moist air from the ocean brings precipitation and moderates air temperatures.

Organization of Petroleum Exporting Countries (OPEC) (7; 207). Established in 1960 to further the interests of oil and gas producers throughout the world, often in materially poorer countries and often to resist the overriding power of multinational oil corporations.

Organization of the Islamic Conference (OIC) (7; 206). Established in 1970 by foreign ministers of Muslim countries throughout the world. It has 45 members but advances individual countries' interests rather than pursuing a common agenda.

orographic lifting (1; 8). Hilly areas, usually facing oceanic moisture sources, cause uplift of air and enhanced precipitation totals.

outsourcing (1; 25). The placing of component factories or services such as call centers in regions or countries with cheaper costs such as labor.

Palestine Liberation Organization (PLO) (7; 206). An organization that promotes the reestablishment of a country of Palestine. It has a secular, left-wing political basis.

Pan-Arab country (7; 206). The idea of Arab countries joining to form a single country. Egypt and Syria were joined for a short while in the 1960s, but few attempts have been made to achieve this end since.

pandemic (8; 252). A disease occurring over a wide area and affecting a high proportion of the population.

parallel of latitude (1; 3). An imaginary circle joining places of the same latitude on Earth's surface.

Parti Québécois (10; 336). A political party formed with the purpose of achieving Québec's separation from the Canadian people and political independence from the Canadian federal government.

perestroika (3; 88). The Soviet Union policy of the late 1980s designed to reconstruct the political and economic structure of the country so that it could compete in the capitalist world economic system.

peripheral country (1; 30). A materially poor country that is dependent on the world economy and materially wealthy countries.

permafrost (3; 77). Permanently frozen ground extending several hundred meters below the surface in Siberia and northern Canada. In summer, water in the surface active layer, 50–100 cm deep, melts.

petrodollars (9; 279). Dollars invested by Middle East oil producers during high oil prices in the 1970s.

physical geography (1; 4). The study of geographic aspects of natural environments.

place (1; 2). A point or area on Earth's surface having a geographic character defined by what it looks like, what people do there, and how they feel about it.

planned economy (2; 50). The Communist practice of the government, rather than the free market, deciding what goods and services need to be produced within a country.

podzol (3; 77). Soils of low fertility in which plant nutrients are removed by water passing through. Commonly develop beneath temperate coniferous forest and on sandy soils.

polar biome (1; 12). Major biome-type where plant growth is inhibited by extreme cold.

polar climate (1; 10). Climate typical of the polar regions: extremely cold all year.

political geography (1; 21). The study of how governments and political movements influence the human and physical geography of the world and its resources.

political rights (1; 33). The rights to vote and participate in one's own government.

population density (1; 19). The numbers of people per given area.

population distribution (1; 19). The spread of people in a region, incorporating areas of high, medium, and low density.

population doubling time (1; 20). The number of years taken to double the population of a country.

post-Panamax (9; 289). The government of Panama's expansion of the Panama Canal in the early 2000s to enable larger ships to pass through the canal.

primary sector (1; 25). The sector of an economy that produces output from natural sources, including mining, forestry, fishing, and farming.

primate city (1; 19). A city that contains a large proportion of the urban population of a country, often several times the population of the second city.

producer goods (2; 51). Industrial goods used by other industries to make consumer goods.

producer services (2; 54). Service industries that are involved in the output of goods and services, including market research, advertising, accountancy, legal, banking, and insurance industries.

production line (10; 313). The system of manufacturing in which components are made and assembled into the final product in a sequence of factory-based processes.

productive capacity (2; 51). The total amount of goods a country's industries can produce during a given period.

productivity (2; 54). The measure of the amount of product generated or work completed per hour of labor.

protectionism (9; 280). Governmental policies that focus on government-controlled industry and the protection of domestic products through tariffs, quotas, and red tape.

purchasing power parity (PPP) (1; 24). The measure of GNI or GDP that is based on internal country costs of living rather than external exchange rates related to the U.S. dollar.

quaternary sector (1; 25). The sector of an economy that specializes in producer services, including financial services and information services.

Qu'ran (7; 200). The holy book of Islam.

race (1; 16). A biologic stock of people with similar physical characteristics, or a group of people united by a community of interests.

rain shadow (9; 269). Low rainfall in an area to the lee of a mountain range, where winds warm and get drier as they descend after flowing across the mountains.

regional geography (1; 5). The study of different regions at Earth's surface in their country and global contexts.

relative location (1; 4). The direction and distance of a place relative to others, often affected by factors that slow or increase contacts among people.

relief (1; 10). The physical height and slope of the land, as in hills, mountains, and valleys.

religion (1; 14). An organized system of practices that seeks to explain our purpose on Earth and may include a set of values and/or worship of a divine being.

remittance (9; 279). Funds sent home by workers in foreign countries.

renewable resource (1; 13). A resource that is replaced by natural processes at a rate that is faster than its usage.

responsible growth (1; 32). A term originating in the 2002 World Summit on Sustainable Development. The summit urged development policies that linked economic growth, environmental sustainability, and social equity.

rift valley (8; 229). A deep valley caused by the rocks of Earth's crust arching and cracking to let down a section of crust to form the valley floor.

rural area (1; 19). Land outside urbanized areas, often having an economic emphasis on farming, mining, and/or forestry. Such areas may dominate population distribution in materially poorer countries.

Russification (3; 79). Policies directed at making non-Russians into Russians by encouraging or forcing non-Russians to adopt Russian cultural characteristics such as the Russian language.

Sahel (8; 231). The zone immediately to the south of the Sahara Desert in Africa that suffers droughts as the arid area expands by natural or human-induced actions.

salinization (7; 197). The process by which soils become unproductive because of an accumulation of alkaline salts near the surface. Often associated with poorly managed irrigation systems in arid areas.

savanna (8; 231). A tropical ecosystem that forms in areas of seasonal rainfall; dominated by grasses and home to large herbivores and carnivores. Many such areas show signs of burning by humans to restrict tree growth.

scale (1; 2). The relationship of horizontal ground distance to map distance, quoted as a fraction (1/10,000) or as a ratio (1:10,000), in which one unit on the map represents 10,000 units on the ground.

sea breeze (5; 136). As air rises over heated land, cooler air from the ocean replaces it at ground level.

Second World (1; 30). The Communist countries, led by the Soviet Union, until 1991. The term is now redundant.

secondary sector (1; 25). The sector of an economy that changes the raw materials from the primary sector into useful products, thus increasing their value, as in chewing gum or parts for airplanes.

sedimentation (9; 272). The deposition of rock debris, including in offshore areas, where too much fine debris may kill coral reefs.

separatism (2; 66). The desire by an ethnic group for independence, as in the case of the Basque group on the French-Spanish border.

shantytown (8; 255). An unplanned residential sector of urban areas in poorer countries. Housing is often built of any materials that come to hand and does not have electricity, water, or waste disposal.

shatter belt (6; 187). A zone between distinctive cultures and political groups that experiences conflict and a slowing of development.

Shia Muslims (7; 200). Also known as Shiites. Muslims who are partisans of the imam Ali (not acceptable to Sunni Muslims) and look to his return. They make up 90 percent of the Iranian population and 60 percent of Iraqis.

Shinto (4; 115). The traditional religion of Japan, built on ancient myths and customs that promote the national interests and identity.

Sikhism (6; 170). A Hindu-related religion with a strict code of conduct. Its temple kitchens provide food for all.

social rights (1; 33). The rights to have a job and earn a living with basic material standards.

soil (1; 12). Weathered rock material that develops by the actions of water, animals, and plants into a basis for plant growth.

solid earth environment (1; 6). The solid planet Earth, incorporating core, mantle, and crust, in which movements of solid rock are caused by interior heat energy.

Southern African Development Community (SADC) (8; 249). Established by countries in Southern Africa that were opposed to South Africa's apartheid policy in order to organize alternative trade outlets. Now the community encourages trade among the constituent countries, including South Africa.

spatial view (1; 3). A geographic view that focuses on differences among places.

specialization (2; 54). The concentration on fewer commercial products within a farming region.

state (2; 49). A country, or division of a country within a federal government.

state socialism (2; 50). The Communist Party actively running the political, social, and economic activities of the people.

steppe grasslands (3; 76). Temperate grasslands typical of the transition between forest and arid areas of temperate continental interior climatic environments.

structural adjustment (1; 31). The theory of development in which a country becomes more involved in the world economic system by producing export goods, reducing tariffs, encouraging privatization, and developing good government and a balanced budget.

subsistence agriculture (8; 242). The growing of crops to satisfy local needs.

subtropical rainy climate (4; 108). A climate found on east coasts, characterized by hot, rainy summers and cool winters. Such areas experience tropical cyclones (hurricanes, typhoons).

subtropical winter rain (Mediterranean) climate (2; 40). A climate that is characterized by summer drought and winter rains.

Sunni Muslims (7; 200). Also known as Sunnites. Traditional, conservative followers of Islam, forming the majority in most Muslim countries. Includes strict sects such as the Wahhabi.

supranationalism (2; 66). The idea that differing nations can cooperate so closely for their shared mutual benefit that they can share the same government, economy (including currency), social policies, and even military.

sustainable human development (1; 30). A level of development in which resources are exploited at a rate that is sustainable for future generations.

systematic (thematic) geography (1; 5). The study of spatial relationships within a specific theme of physical or human geography.

tectonic plate (1; 6). A large block of Earth's crust and underlying rocks approximately 100 km thick and up to several thousand kilometers across. Earth's interior heat causes plates to move apart and crash together, forming major relief features including ocean basins, mountain systems, and continental areas.

temperate climate (1; 10). The climates of mid-latitudes, in which there are summer-winter temperature contrasts without the extremes of lengthy very hot or cold periods.

temperature inversion (9; 272). Naturally occurring periods during the winter months when cold dense air remains "trapped" at the surface under warmer air for several days or even weeks. The cold air is usually trapped by mountains or some related physical barrier. Temperature inversions may produce stagnant air, which may complicate pollution problems in urban areas.

Tennessee Valley Authority (10; 322). Established by the U.S. government in 1933 to stimulate economic growth in southern Appalachia through the damming of rivers to improve transportation, flood control, and electricity costs.

tertiary sector (1; 25). The sector of an economy concerned with the distribution of goods and services, including trade, professions, and government employment.

Thar Desert (6; 168). Arid region extending from Afghanistan through Pakistan into westernmost India.

Third World (1; 30). The poorer countries, mostly in the Southern Hemisphere, not part of the First or Second Worlds.

total fertility rate (1; 19). The number of births per woman in her childbearing years.

transmigration (5; 140). The name given to the process through which Indonesia attempted to ease overpopulation on Java by moving large numbers of people to the less inhabited islands.

Treaty of Tordesillas (5; 140). The 1494 demarcation line (approximately 46°W longitude) dividing land rights in the Americas between Spain and Portugal. The line was created under the authority of the pope. Spain gained control of lands to the west of the line, and Portugal gained control of lands to the east.

Tropic of Capricorn (5; 137). The Southern Hemisphere tropic, a line of latitude at 23.5°S.

tropical climate (1; 8). Climatic environment typical of the tropical zone, having high temperatures all year.

tropical rain forest (8; 163). A forest ecosystem characterized by a great variety of species that are dependent on high temperatures and rainfall.

tsunami (5; 12). A huge wave generated by an earthquake that changes the ocean floor shape suddenly. The wave travels fast across the ocean and causes destruction on entering shallow coastal waters and confined valleys.

tundra (1; 12). Ecosystem type occurring in cold polar environments, consisting of mosses, grasses, and low shrubs.

typhoon (5; 137). A tropical storm of hurricane type experienced in Southeast and East Asia.

UN Security Council (10; 314). One of the principal divisions of the United Nations. The purpose of the Security Council is the maintenance of international peace and security. The United States, China, France, the United Kingdom, and Russia are the five countries granted permanent seats on the Security Council. The permanent members of the Security Council have the power to veto council decisions. The decisions of

the Security Council are binding for the entire UN General Assembly.

uneven development (10; 318). The increase in the gap between poor and wealthy regions in a country and the shifting locations of economic growth and decline over time, seen by Marxists as an outcome of capitalism.

unitary government (1; 21). Government of a country administered from a single center.

universalizing religion (1; 14). A religion that seeks to be global in its application, such as Islam and Christianity.

urban area (1; 19). An area with high densities of people, buildings, transportation linkages, and human activities of a high economic, political, and cultural order. Urban areas dominate population distribution in materially wealthy countries.

vertical integration (10; 313). The combining of producers of raw materials, manufacturers that process the materials, and those that assemble the products in a single corporation to achieve economies of scale.

Virgin Lands Campaign (3; 91). A Soviet agricultural campaign begun in the 1950s. It promoted farming in lands where it had never taken place before, primarily in lands that were very marginal because the soil was poor or not enough water or heat was present to grow crops. Much of the land was in the semidesert and desert areas of southern Siberia and Central Asia, especially in the Kazakh Republic.

weathering (1; 11). The action of atmospheric forces (through water circulation and temperature changes) on rocks at Earth's surface that breaks the rocks into fragments, particles, and dissolved chemicals.

world region (1; 34). This text recognizes nine world regions that each include a number of countries linked by cultural, political, economic, and environmental conditions.

CREDITS

PHOTO CREDITS

CHAPTER 1
Figure 1.6: © Brand X/Punchstock; 1.9a: NASA; 1.9b: NOAA; 1.10: © Corbis RF; 1.11: © Michael Bradshaw; 1.12: © Jill Wilson; 1.14b: © Paul A. Souders/Corbis; 1.14c: © Tibor Bognar/Corbis; 1.17: © Stone/Getty Images; 1.26: Courtesy Alexander Murphy.

CHAPTER 2
Figure 2.2a: © Mike Camille; 2.2b: © Alasdair Drysdale; 2.3a-b: © Photo Klopfenstein-Adelboden; 2.4: © Michael Bradshaw; 2.5a: © Jerzy Jemiolo; 2.5b: © George W. White; 2.10a: © Michael Bradshaw; 2.10b: © Jerzy Jemiolo; 2.15: © Ray Juno/Corbis; 2.16: © Alexander B. Murphy; 2.17a: © Michael Bradshaw; 2.17b: © Emily A. White; 2.19: © Yann Arthus-Bertrand/Corbis; 2.20b: © Heike Alberts; 2.21, 2.22: © Loren W. Linholm; 2.23a: © Alasdair Drysdale; 2.28: © David C. Johnson; 2.30a1: © John Cogill/AP/Wide World Photos; 2.30a2: © David Crausby/Alamy Images; 2.30b: © Franck Prevel/AP Photo; 2.34: © Pascal Le Segretain/ Getty Images News; 2.35: Courtesy of Anne Hunderi.

CHAPTER 3
Figure 3.2a-c, 3.6, 3.7a-b, 3.12, 3.13: © Ronald Wixman; 3.14: © M. Blinnikov; 3.16, 3.22: © Ronald Wixman; 3.26a-b: © Musa Sadulayev/AP/Wide World Photos; 3.31: Courtesy Robert Hutton.

CHAPTER 4
Figure 4.2, 4.3: © Michael Bradshaw; 4.10b: © Alasdair Drysdale; 4.13, 4.14: © Michael Bradshaw; 4.15: © Toyota Motor Corp/Handout/Reuters/Corbis; 4.16: © Vander Zwalm Dan/Corbis; 4.17: © Alasdair Drysdale; 4.18: © Vol. 111/Corbis; 4.19: © Jose Fuste Raga/Corbis; 4.20: © Alasdair Drysdale; 4.22: © John Ruwitch/Corbis; 4.23: Courtesy Erle Ellis.

CHAPTER 5
Figure 5.2a: © Ian Coles; 5.2b: © Penny Tweedie/Corbis; 5.2c: © Vol. 8/Corbis; 5.3: © Vol. 8/Corbis; 5.4: © Patrick Ward/Corbis; 5.8: © Vol. 8/Corbis; 5.14a: © Vol. 111/Corbis; 5.14b: © David Zurick; 5.17: © Jack Fields/Corbis; 5.18: © Getty Images; 5.20: © Vol. 112/Corbis; 5.21: © Mission Aviation Fellowship; 5.22c: © Vol. 28/Cobis; 5.22d-e: © Ian Coles; 5.26a-b: © Vol. 111/Corbis; 5.27: © John Russell/AFP/Getty Images.

CHAPTER 6
Figure 6.3: © Alasdair Drysdale; 6.5: © Jagdish Agarwal/The Image Works; 6.6: © Amit Bhargava/Corbis; 6.7b: © Munir Khan/Harappa.com; 6.8a: © Bill Westermeyer; 6.11: © David Zurick; 6.12: © B. P. Wolff/Magnum Photos, Inc.; 6.13: © Phillipe Lissac/Godong/Corbis; 6.14: © Parvinder Sethi; 6.16: © James P. Blair/National Geographic Image Collection; 6.17: © Alasdair Drysdale; 6.20: © David H. Wells/Corbis; 6.21: © Sherwin Crasto/Reuters/Corbis; 6.22: © Reuters/Corbis; 6.25: © Kapoor Baldev/Sygma/Corbis; 6.26: Courtesy Dr. Mohammed Ali.

CHAPTER 7
Figure 7.2a: © Joseph P. Dymond; 7.2b: © Heike Alberts; 7.2c: © Alasdair Drysdale; 7.3: NASA; 7.2d: © Alasdair Drysdale; 7.6: © Heike Alberts; 7.8: © Joseph P. Dymond; 7.11: © Alasdair Drysdale; 7.13: © Lee Frost/Robert Harding World Imagery/Getty Images; 7.14: © Vol. 145/Corbis; 7.18: © AFP/Getty Images; 7.19b: © Ed Kashi; 7.20: © Kess van der Berg/Photo Researchers, Inc.; 7.23: © Heike Alberts; 7.26: Photo courtesy Space Imaging Middle East; 7.27: Photo © Nabeel Turner/Stone/Getty Images; 7.28: © Paul Souders/Corbis; 7.30: NASA; 7.32a: © Mike Camille; 7.32b: © Michael Bradshaw; 7.36: Courtesy of Mark Corson.

CHAPTER 8
Figure 8.2: © Michael Bradshaw; 8.5a: © Torleif Svensson/The Stock Market/Corbis; 8.5b: © Vol. 35/Corbis; 8.5c: © Vol. 145/Corbis; 8.13: NASA; 8.15a: © Bill Westermeyer; 8.15b: © James Marshall/Corbis; 8.18: © Mission Aviation Fellowship; 8.19d: Image courtesy of Earth Sciences and Image Analysis Laboratory, NASA Johnson Space Center; 8.20: © Eric Natha/Alamy; 8.22: © Gideon Mendel/Corbis Images; 8.24: © AP/Wide World Photos; 8.26: © George Esiri/Reuters/Corbis; 8.29b: Courtesy Elio Spinello.

CHAPTER 9
Figure 9.2: © Owen Franken/Corbis; 9.3: © Bruce Dale/National Geographic Image Collection; 9.5: © James P. Blair/National Geographic Image Collection; 9.7: Courtesy of Space Imaging, Thornton, CO, USA; 9.9: © Bill Gentile/Corbis; 9.10: © John Maier, Jr./The Image Works; 9.16: © Neil Rabinowitz/Corbis; 9.17: © David Zurick; 9.18: NASA; 9.19: © Michael Bradshaw; 9.21: © RF/Corbis; 9.23a-b: © Joseph P. Dymond; 9.27, 9.29b: © Michael Bradshaw; 9.30a: © RF/Corbis; 9.30b: © Stephanie Maze/Corbis; 9.32: Courtesy Dr. Paul F. Hudson.

CHAPTER 10
Figure 10.2a-b, 10.3, 10.4: © Joseph P. Dymond; 10.5a-b: Space Imaging, Thornton, CO, USA; 10.16: © Jon Feingersh/Stock Boston; 10.18, 10.20, 10.23a-b, 10.25: © Joseph P. Dymond; 10.27: © Wolfgang Kaehler/Corbis; 10.30: © Joseph P. Dymond; 10.32: Courtesy of Bryan Hannegan.

Bradshaw, Essentials of World Regional Geography

Europe's role, 50–51, 52
impact on Africa south of Sahara, 256–58
impact on Latin America, 279–80
overview, 26–28
role in human development, 31–32
Global positioning systems, 3, 226
Global warming
current consensus about, 13
impact on Antarctica, 156
potential impact on Europe, 41
rain forest role in limiting, 297, 298
threat to South Asia, 169
Golan Heights, 220, 221
Golden Temple at Amritsar, 171
Gold mining
in Africa south of Sahara, 235, 251
in Latin America, 283, 292, 294
Gondwanaland, 136
Gorbachev, Mikhail
agricultural reforms, 92
economic reforms, 89, 90
political reforms, 88
Governance, 22
Government corruption. *See* Corruption
(government)
Government of National Unity, 249
Governments, basic features, 21
GPS. *See* Global positioning systems
Grameen Bank, 31, 174
Grand Banks decline, 330
Grassland biomes
in Africa south of Sahara, 231
basic features, 12
in Latin America, 271
in Russia and Neighboring Countries, 76, 77
Great Artesian Basin, 138
Great Britain. *See also* Europe
African colonization, 234–35
Australian colony, 161
devolution in, 66–67
Latin American colonies, 273, 279
North American colonization, 312
role in Northern Africa and Southwest Asia,
206, 224–25
rule of India, 166, 172–73
Southeast Asian colonization, 140
Great Dividing Range, 153
Greater Antilles, 290
Great Lakes, 305
Great Leap Forward, 121, 131
Great Limpopo Transfrontier Park, 257
Great Plains, 322
Great Trek, 235
Great Zimbabwe, 228, 232
Greenhouse effect, 8, 13. *See also* Global warming
Green Revolution
impact on farm wealth, 174
in India, 177
irrigation with, 168
in Southeast Asia, Australia, and Oceania, 149
Gross domestic product, 24, 116
Gross national incomes
in Africa south of Sahara, 260
countries compared, 23

defined, 24
in East Asia, 113
in Europe, 67
in Latin America, 280
in North America, 314, 315
in Northern Africa and Southwest Asia, 207
in Russia and Neighboring Countries, 90–91
in South Asia, 173
in Southeast Asia, Australia, and Oceania,
145, 146
Group of Eight
European representation, 57
North American representation, 314
Russian participation, 93, 100
Growth poles, 30
Guadeloupe, 291
Guano, 137
Guatemala, 273, 281
Guayaquil, 276, 283
Guest workers, 70
Guianas, 268–69, 290, 292
Gulags, 101
Gulf Cooperation Council, 216
Gulf War (1991), 197, 206, 225
Guyana, 292

H

Habyarimana, Juvenal, 242
Haciendas, 278
Haiti, 291
Halla, 126
Han Chinese, 111
Hannegan, Brian, 337
Hashemite clan, 225
Hassan Mosque, 195
Hawaii, 326–27
Hazards, natural. *See* Natural hazards
Health care. *See also* Diseases
HIV/AIDS treatment, 252, 253, 254
rain forest importance, 296, 298
in Russia and Neighboring Countries, 83
Heartland (Russia), 94
Heartland (U.S.), 322, 323
Heart of Borneo, 151
Heavily Indebted Poor Countries initiative, 260
Hebrew language, 199
Hezbollah, 222
Higher education in Oman, 194
High-tech industries, 186–87, 217. *See also*
Technology
Himalayan Mountains, 108, 165–66
Hindi language, 172
Hinduism
as major world religion, 14, 15, 16
in South Asia, 170
Hindutva, 187
Hinterland, 94–95
Hispanic immigration, 303, 325–26, 332–34
Hitler, Adolf, 88
HIV/AIDS
in Africa south of Sahara, 251, 252–54
in South Asia, 183–84
in Southeast Asia, 144

Holocaust, 43, 69
Homestead Act, 312
Honduras, 273, 281–82
Honecker, Erich, 58
Hong Kong, 113, 118, 130–31
Hordaland County council, 71
Horizontal integration, 313
Houphouët-Boigny, Félix, 246
Household responsibility system, 121–22
Huang He river, 108
Huastec culture, 301
Hudson, Paul, 301
Hudson's Bay Company, 312–13
Hu Jintao, 256
Hukou, 113
Human activities, environmental impacts of, 12–13
Human development, 29–33
Human Development Index, 24–25
Human geography, 5
Human Poverty Index, 25
Human rights
in Africa south of Sahara, 258–59
in East Asia, 111, 127–29, 131–33
elements of, 33–34
in Europe, 68–69
in Northern Africa and Southwest Asia, 222–24
in Russia and Neighboring Countries, 101–2
in Southeast Asia, Australia, and Oceania, 140
Hunderi, Anne, 71
Hurricane Katrina, 10, 323
Hurricanes
formation of, 10
in Latin America, 269
in North America, 306, 323
Hussein, Saddam, 206, 211, 224, 225–26
Hutton, Robert, 105
Hutu people, 228, 242
Hyderabad, 187
Hydroelectric power
in Latin America, 284–85, 294
in North America, 305, 331
in South Asia, 190–91
Hydrology, 305
Hyundai, 126

I

Ibadan, 254
Ice ages, 10–11, 305
Ice storms, 307
IKEA, 86
Illegal immigration, 303, 332–34
Illiteracy, 17, 223
Immigration
to Australia and New Zealand, 143, 153, 161
to Europe, 69–70
to Israel, 202, 219
to North America, 303, 308–10, 311, 329,
332–34
to Southwest Asia, 204
Immigration and Asylum Act (UK), 69
Imperialism. *See also* Colonialism
of China, 120
Europe's role, 50

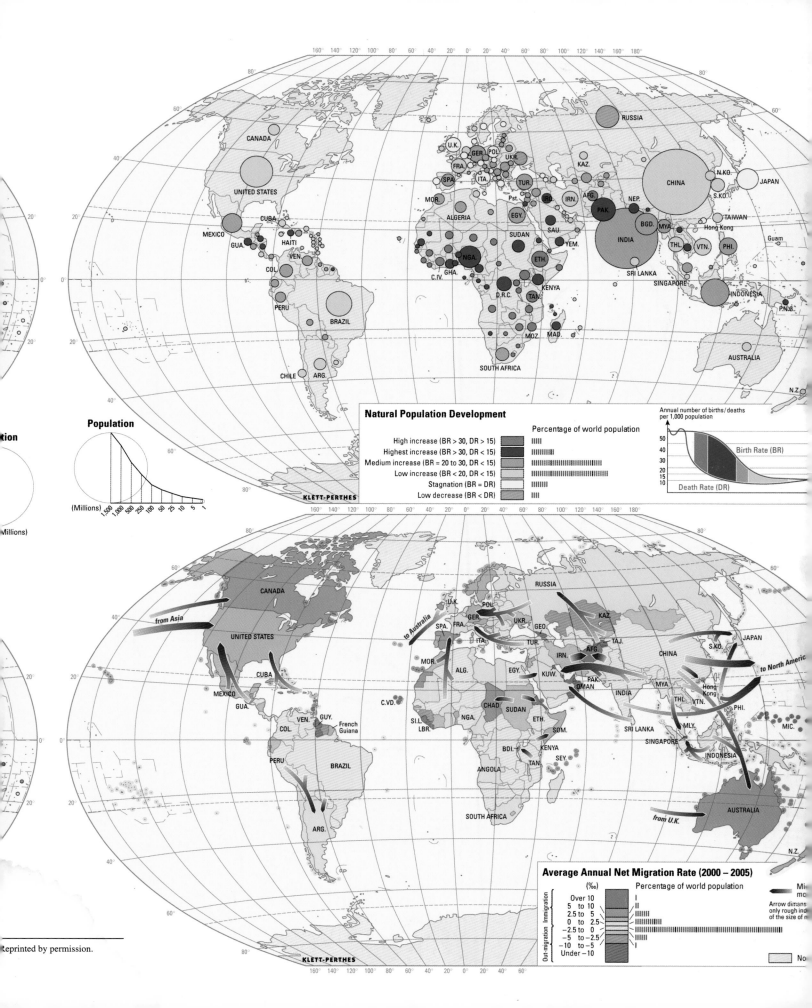

Natural Population Development

	Percentage of world population
High increase (BR > 30, DR > 15)	‖‖‖
Highest increase (BR > 30, DR < 15)	‖‖‖‖‖‖‖‖‖‖‖‖‖‖‖‖
Medium increase (BR = 20 to 30, DR < 15)	‖‖‖‖‖‖‖‖‖‖‖‖‖‖‖‖‖‖
Low increase (BR < 20, DR < 15)	‖‖‖‖‖‖‖‖‖‖‖
Stagnation (BR = DR)	‖‖‖‖‖
Low decrease (BR < DR)	‖‖‖

Population

(Millions) 1,500 1,000 500 250 100 50 25 10 5 1

Annual number of births / deaths per 1,000 population

50 40 30 20 15 10

Birth Rate (BR)

Death Rate (DR)

KLETT-PERTHES

Average Annual Net Migration Rate (2000 – 2005)

(‰)   Percentage of world population

Immigration
- Over 10
- 5 to 10
- 2.5 to 5
- 0 to 2.5

Out-migration
- -2.5 to 0
- -5 to -2.5
- -10 to -5
- Under -10

Arrow dimens
only rough ind
of the size of n

No

from Asia

to Australia

to North Americ

from U.K.

Reprinted by permission.

KLETT-PERTHES

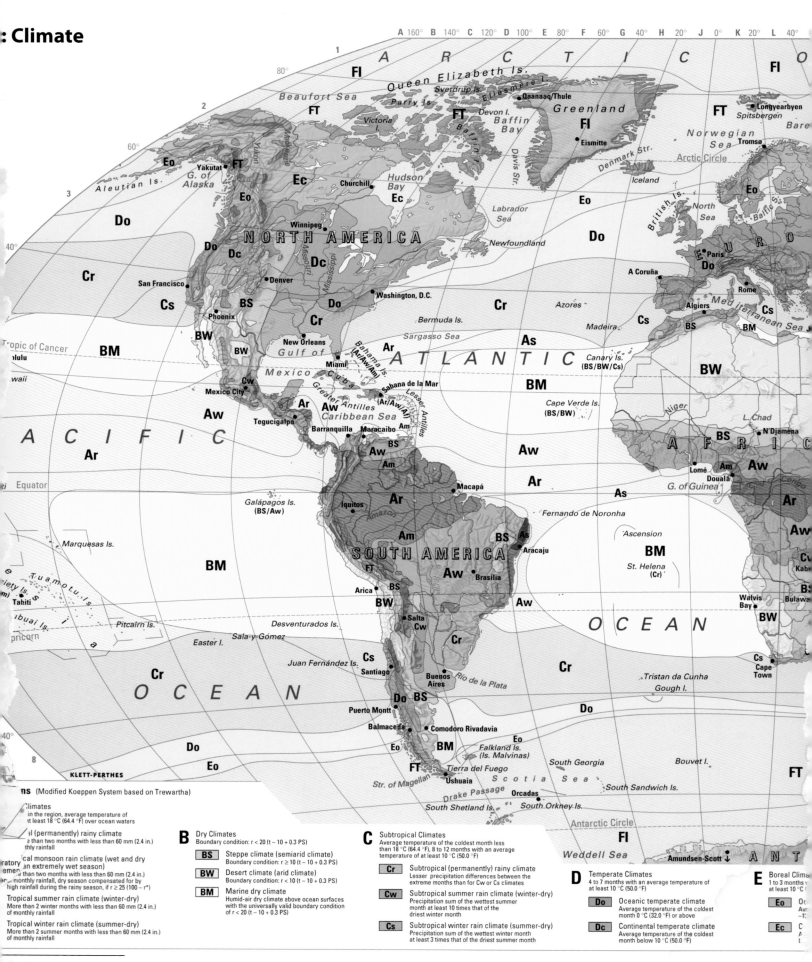

: Climate

ns (Modified Koeppen System based on Trewartha)

...limates
...in the region, average temperature of
... at least 18 °C (64.4 °F) over ocean waters

...al (permanently) rainy climate
...e than two months with less than 60 mm (2.4 in.)
...thly rainfall

...ratory
...en extremely wet season)
...d than two months with less than 60 mm (2.4 in.)
...r monthly rainfall, dry season compensated for by
...high rainfall during the rainy season, if r ≥ 25 (100 − r*)

Tropical summer rain climate (winter-dry)
More than 2 winter months with less than 60 mm (2.4 in.)
of monthly rainfall

Tropical winter rain climate (summer-dry)
More than 2 summer months with less than 60 mm (2.4 in.)
of monthly rainfall

B Dry Climates
Boundary condition: r < 20 (t − 10 + 0.3 PS)

BS Steppe climate (semiarid climate)
Boundary condition: r ≥ 10 (t − 10 + 0.3 PS)

BW Desert climate (arid climate)
Boundary condition: r < 10 (t − 10 + 0.3 PS)

BM Marine dry climate
Humid-air dry climate above ocean surfaces
with the universally valid boundary condition
of r < 20 (t − 10 + 0.3 PS)

C Subtropical Climates
Average temperature of the coldest month less
than 18 °C (64.4 °F) with an average
temperature of at least 10 °C (50.0 °F)

Cr Subtropical (permanently) rainy climate
Lesser precipitation differences between the
extreme months than for Cw or Cs climates

Cw Subtropical summer rain climate (winter-dry)
Precipitation sum of the wettest summer
month is at least 10 times that of the
driest winter month

Cs Subtropical winter rain climate (summer-dry)
Precipitation sum of the wettest winter month
at least 3 times that of the driest summer month

D Temperate Climates
4 to 7 months with an average temperature of
at least 10 °C

Do Oceanic temperate climate
Average temperature of the coldest
month 0 °C (32.0 °F) or above

Dc Continental temperate climate
Average temperature of the coldest
month below 10 °C (50.0 °F)

E Boreal Clima
1 to 3 months
at least 10 °C

Eo

Ec

KLETT-PERTHES

...s. Reprinted by permission.

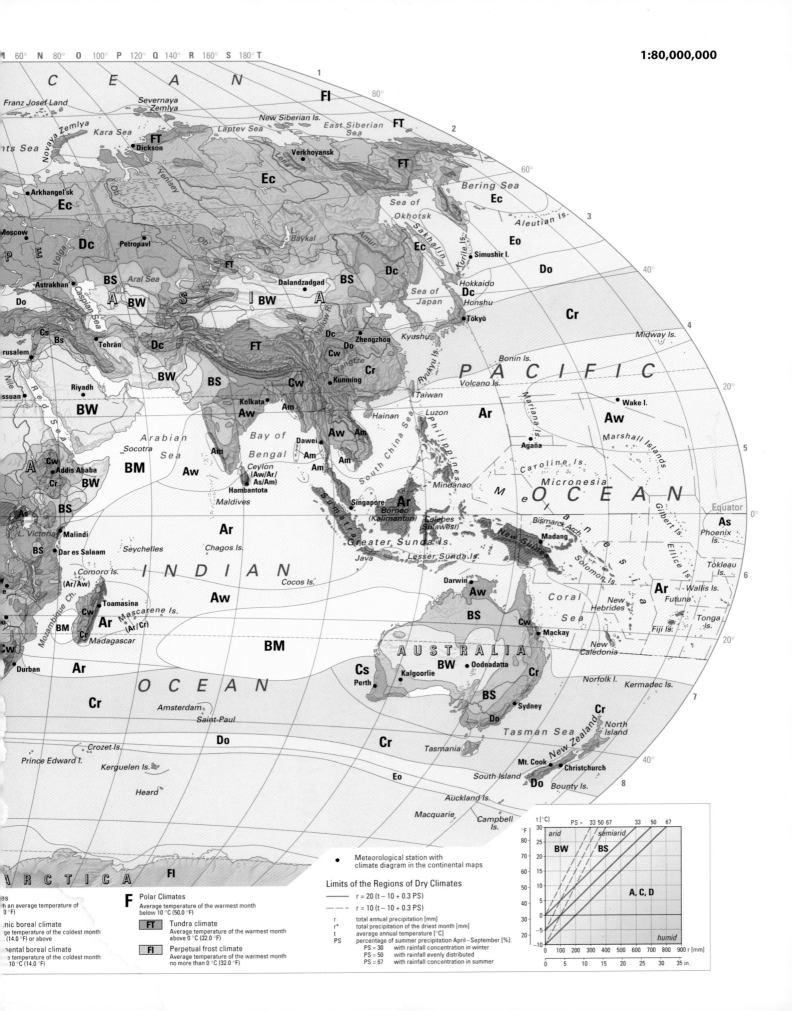

Urbanization and Migration

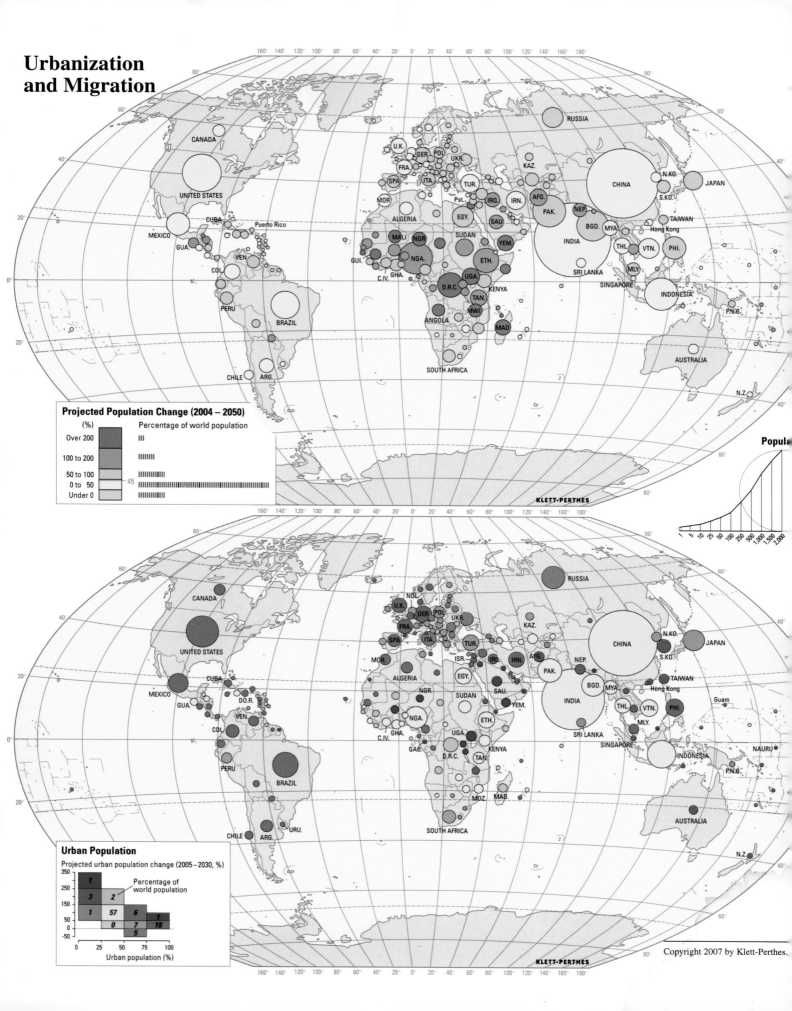

Projected Population Change (2004 – 2050)

(%)		Percentage of world population
Over 200		ǀǀǀ
100 to 200		ǀǀǀǀǀǀǀǀ
50 to 100		ǀǀǀǀǀǀǀǀǀǀǀ
0 to 50	45	ǀǀǀǀǀǀǀǀǀǀǀǀǀǀǀǀǀǀǀǀǀǀǀǀǀǀǀǀǀǀǀǀ
Under 0		ǀǀǀǀǀǀǀǀǀ

Popula

1 5 10 25 50 100 250 500 1,000 1,500 2,000

KLETT-PERTHES

Urban Population

Projected urban population change (2005–2030, %)

Percentage of world population

1			
3	2		
1	57	6	
0		7	1
		5	18

350
250
150
50
-50

0   25   50   75   100
Urban population (%)

KLETT-PERTHES

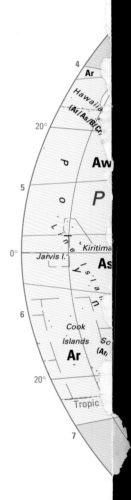

4
Ar
Hawaiia
(Ar/As/B/Cr
20°
P
Aw
o
P
L
i
n
5
e
Kiritima
I
0°
Jarvis I.
s
As
l
a
n
6
Cook
Islands
Sc
(Ar
Ar
20°
Tropic
7

Clima
A

Aw

As

A Guide to Map Reading

Prepared by Peter Konovnitzine, Chaffey College

Maps can be compared to computers—they contain a lot of information that needs to be viewed, interpreted, and decoded. Every map should have several essential bits of information.

The Parts of a Map

A **name** or **title** prevents confusion by clearly stating what it is you are looking at.

A date tells the map viewer how current the information is. Without a **date** on a map, you need to look for other clues, such as political boundaries and countries. For example, if a map shows a country called the Soviet Union, you know that the map was printed before 1991.

A **legend** or **key** box—usually placed at the bottom of a map—decodes all the colors and symbols used on the map. For example, most students who see the color green on a map associate it with vegetation, such as tress, grass, or forests. However, on many maps that show elevation, the color green indicates low elevation. The color blue is almost always used to indicate water—oceans, seas, lakes, streams, or rivers. In addition, the legend box contains vital information useful in decoding other map markings.

Direction is usually indicated by placing a compass rosette on the map showing where the major cardinal points are: north, south, east, and west. However, today most cartographers (mapmakers) omit this symbol, assuming that the map viewer knows that the top of the map is north, the bottom is south, the left-hand is west, and the right-hand is east.

Location refers to the geographic grid that is usually overlaid on every map. The geographic grid is the "netting" that consists of latitude and longitude lines. Parallels of latitude lines run east to west across maps, while meridians of longitude lines go north to south. At the intersection points of these lines you will find geographic coordinates. An example would be the geographic coordinates for the city of Los Angeles, California: 34 degrees north (of the equator) and 118 degrees west (of the prime meridian). This geographic coordinate is unique to Los Angeles. No other place on Earth has this geographic coordinate.

Scale helps in understanding the relationship between map distances and actual Earth distances. You will usually find scale in the legend or key box. There are three types of scales, and most maps made today have all three types:

> The linear or bar scale is a horizontal line drawn with markings placed at specific intervals indicating distances. The spacing between the markings indicates actual Earth distances. For example, if you take a ruler and measure the distance between two spacing ticks on the bar line, this distance will help you understand the actual distance between the places on the Earth.

> Verbal scale is simply a sentence that states the relationship of distances on the map to actual distances on the Earth. For example, it may say, "One inch equals 100 miles." This means that 1 inch on the map would actually equal 100 miles on the Earth's surface.

> A representative fraction (RF) scale is the most useful, since it does not require prior knowledge of any particular distance measuring system. Let's say that you are not familiar with the metric system, and the map you are looking at has both the linear and verbal scales in metric notation. It may state "1cm = 100 km." If you are not familiar with centimeters or kilometers, you will not be able to relate the scale used on the map to actual distances on the Earth's surface. The representative fraction scale has two advantages over the other two types of scale. First, it allows you to choose the distance measuring system that you are familiar with—either inches and miles or centimeters and kilometers (or any other system you want to use); and second, it always uses the same units both on the map and as it translates to actual Earth distances. For example, a common RF would be 1:62,500. To decode this, you would use one unit of your choosing—let's say inches—so that 1 inch on the map would be 62,500 inches on the Earth's surface. (By the way, this would come out to be about 1 inch to 1 mile.) Another way an RF is shown is as a fraction; 1/125,000. Notice again that the first number is always 1. This indicates that one unit on the map equals 125,000 identical units on the Earth's surface.

Additional Map Information for Geography Students

One of the key map features you can easily remember as you learn about geography is the relationship between major Earth grid lines and at the Earth's vegetative regions. Here is a quick summary of this unique relationship. (Note: There are some minor exceptions.)

Major Earth Grid lines

Name	Degree Value	Significance
Equator	0 degrees	Earth's main rainforest belt; also starting point for paralels (lines of latitude)
Prime Meridian	0 degrees	Starting point for meridians (lines of longitude) and Earth's **time zones**
Tropic of Cancer	23.5 degrees N	Major **northern desert belt** (except southeast Asia)
Tropic of Capricorn	23.5 degrees S	Major **southern desert belt**
Arctic Circle	66.5 degrees N	Tundra usually found north of this grid line; taiga forest south of this grid line
Antarctic Circle	66.5 degrees S	World's **storm belt**

About These Two Maps

The map on this page shows the climate regions of the world. The map on the reverse side provides information about the world's urbanization and migration.